Advanced Micro & Nanosystems
Volume 1

Enabling Technology for MEMS and Nanodevices

Related Wiley-VCH titles:

Baltes, H., Fedder, G.K., Korvink, J.G. (eds.)

Sensors Update Vol. 13
2004
ISBN 3-527-30745-1

Baltes, H., Fedder, G.K., Korvink, J.G. (eds.)

Sensors Update Vol. 12
2003
ISBN 3-527-30602-1

Hesse, J., Gardner, J.W., Göpel, W. (series eds.)

Sensors Applications:

Marek, J., Trah, H.-P., Suzuki, Y., Yokomori, I. (eds.)

Sensors for Automotive Technology
2003
ISBN 3-527-29553-4

Tschulena, G., Lahrmann, A. (eds.)

Sensors in Household Appliances
2003
ISBN 3-527-30362-6

Gassmann, O., Meixner, H. (eds.)

Sensors in Intelligent Buildings
2001
ISBN 3-527-29557-7

Tönshoff, H.K., Inasaki, I. (eds.)

Sensors in Manufacturing
2001
ISBN 3-527-29558-5

Ajayan, P., Schadler, L.S., Braun, P.V.

Nanocomposite Science and Technology
2003
ISBN 3-527-30359-6

Decher, G., Schlenoff, J.B. (eds.)

Multilayer Thin Films
Sequential Assembly of Nanocomposite Materials
2003
ISBN 3-527-30440-1

Gómez-Romero, P., Sanchez, C. (eds.)

Functional Hybrid Materials
2003
ISBN 3-527-30484-3

Komiyama, M., Takeuchi, T., Mukawa, T., Asanuma, H.

Molecular Imprinting
From Fundamentals to Applications
2003
ISBN 3-527-30569-6

Korvink, J.G., Greiner, A.

Semiconductors for Micro- and Nanotechnology
An Introduction for Engineers
2002
ISBN 3-527-30257-3

Pearce, T.C., Schiffman, S.S., Nagle, H.T., Gardner, J.W. (eds.)

Handbook of Machine Olfaction
Electronic Nose Technology
2003
ISBN 3-527-30358-8

10 0410296 X

Advanced Micro & Nanosystems
Volume 1

Enabling Technology for MEMS and Nanodevices

Edited by
H. Baltes, O. Brand, G. K. Fedder, C. Hierold, J. G. Korvink,
O. Tabata

WILEY-VCH Verlag GmbH & Co. KGaA

Editors

Prof. Dr. Henry Baltes
Physical Electronics Laboratory
ETH Zürich
Hönggerberg, HPT-H6
8093 Zürich
Switzerland
baltes@iqe.phys.ethz.ch

Prof. Dr. Oliver Brand
School of Electrical and Computer Engineering
Georgia Institute of Technology
Atlanta, GA 30332-0250
USA
oliver.brand@ece.gatech.edu

Prof. Dr. Gary K. Fedder
ECE Department &
The Robotics Institute
Carnegie Mellon University
Pittsburgh,
PA 15213-3890
USA
fedder@ece.cmu.edu

Prof. Dr. Christofer Hierold
Chair of Micro- and Nanosystems
ETH Zürich
ETH-Zentrum, CLA H9
Tannenstr. 3
8092 Zürich
Switzerland
christofer.hierold@micro.mavt.ethz.ch

Prof. Dr. Jan G. Korvink
IMTEK-Institut für Mikrosystemtechnik
Universität Freiburg
Georges-Köhler-Allee 103/03.033
79110 Freiburg
Germany
korvink@imtek.de

Prof. Dr. Osamu Tabata
Department of Mechanical Engineering
Faculty of Engineering
Kyoto University
Yoshida Honmachi,
Sakyo-ku
Kyoto 606-8501
Japan
tabata@mech.kyoto-u.ac.jp

Cover picture
Poly-Si lateral resonator coated with thin SiC film by C. R. Stoldt et al. (top left); IFX DNA sensor array biochip by Infineon Technologies (bottom right).

Library of Congress Card No.: applied for

British Library Cataloguing-in-Publication Data
A catalogue record for this book is available from the British Library.

Bibliographic information published by Die Deutsche Bibliothek
Die Deutsche Bibliothek lists this publication in the Deutsche Nationalbibliografie; detailed bibliographic data is available in the Internet at <http://dnb.ddb.de>

Printed in the Federal Republic of Germany
Printed on acid-free paper

Composition K+V Fotosatz GmbH, Beerfelden
Printing strauss GmbH, Mörlenbach
Bookbinding Litges & Dopf Buchbinderei GmbH, Heppenheim

ISBN 3-527-30746-X

Preface

We are proud to present you with this first volume of *Advanced Micro & Nanosystems*, which we call *AMN* for short. *AMN* addresses the needs of engineers and technologists who turn scientific ideas and dreams into product reality, and of entrepeneurs who keep these companies running. It also addresses graduate students who will evolve to either of these roles. There is much hype about Nano these days, just as there once was for Micro. Ultimately, however, the hard work of very creative individuals turns hype into realistic visions and finally into manufacturable devices. We have taken the word 'system' into the title for this reason. We emphasise the next step, the step beyond getting it right the first time, the deciding step that enables science to evolve to a viable technology. So much for our goal with *AMN*, and back to the present volume. We have grouped the eleven contributions into two sections.

The first section of seven articles is entitled *MEMS Technologies and Applications*, and starts off with Pasqualina Sarro's article on the third dimension of silicon, which addresses the fact that much of silicon processing is still planar, and of course MEMS technology offers ways around this limitation. This is followed by an article by Jim Knutti on MEMS commercialization. Jim has done it many times over, and his article is rare as it is difficult to teach this subject in a university atmosphere, and most companies see the process as a trade secret. Vladimir Petkov and Bernard Boser review capacitive interfaces for MEMS in the third chapter. Packaging is discussed by Victor Bright, Conrad Stoldt, David Monk, Mike Chapman and Arvind Salian, a very important task, especially if we remind ourselves of the role that packaging plays in the success and manufacturing cost of any integrated device, and the complexity of challenge that MEMS and NEMS present.

Mobile communication is steadily moving up the frequency scale, as well as the commercial importance scale, and in the next article Farrokh Ayazi discusses MEMS contributions to the area of high frequency filters and resonators, with their promise of reducing the chip count of a typical application whilst improving the component quality. MEMS is also making inroads into the mass storage area, with the promise of zero energy non-volatile devices, very high density data storage, and robustness with regard to external influences, so critical for security applications, as discussed in the article by Thomas Albrecht, Jong Uk Bu, Michel

Despont, Evangelos Eleftheriou, and Toshiki Hirano. The section closes with an article on scanning electrochemical probes by Christine Kranz, Angelika Kueng, and Boris Mizaikoff. These devices open our eyes not to photons or other particles, but to electrochemical gradients, opening up a new world of nanoscale information and imaging.

The second section of four articles is entitled *Nanodevice Technologies and Applications*. Two articles discuss the emerging topic of nanofluidics. The first, by Martin Geier, Andreas Greiner, David Kauzlaric and Jan G. Korvink, discusses the needs, and some solutions, for simulation tools with nanoscale resolution but multiscale in their approach. The second article, by Jan Lichtenberg and Henry Baltes, looks at nanofluidic devices and their uses. Next, Joseph Stetter and G. Jordan Maclay examine carbon nanotube (CNT) sensors. We can expect CNTs to make many inroads into the sensor field with its promise as a small electrical conductor and a chemical catalyst. The section ends with an article by Roland Thewes, Franz Hofmann, Alexander Frey, Meinrad Schienle, Christian Paulus, Petra Schindler-Bauer, Birgit Holzapfl, and Ralf Brederlow on CMOS-based DNA sensor arrays.

First and foremost we thank our authors for their hard work and timely contributions. The editors are grateful to the publishers, Wiley-VCH, for their support of the book series. In particular, the editors thank the publishing editor, Dr. Martin Ottmar, for his management of this enterprise and to Hans-Jochen Schmitt and Dr. Jörn Ritterbusch for their support.

We also look forward to welcoming you back, dear reader, to the next volume of *AMN*, which will take a timely look at recent advances in using CMOS technologies for MEMS. The articles are provided by an impressive team of experts in the field from around the world, so that we can expect this volume to condense a vast range of expertise in a handy reference format.

Henry Baltes, Oliver Brand, Gary K. Fedder, Christofer Hierold, Jan G. Korvink, Osamu Tabata

May 2004
Zurich, Atlanta, Pittsburgh, Freiburg and Kyoto

Contents

List of Contributors

Dr. Thomas R. Albrecht
Hitachi Global Storage Technologies
San Jose Research Center
650 Harry Road
San Jose, CA 95120
USA
thomas.albrecht@hgst.com

Dr. Henry V. Allen
Silicon Microstructures, Inc.
1701 McCarthy Blvd.
Milpitas, CA 95035
USA
info@si-micro.com

Prof. Dr. Farrokh Ayazi
School of Electrical
and Computer Engineering
Georgia Institute of Technology
Atlanta, GA 30332-0250
USA
farrokh.ayazi@ece.gatech.edu

Prof. Dr. Henry Baltes
Physical Electronics Laboratory
ETH Zürich
Hönggerberg, HPT-H6
8093 Zürich
Switzerland
baltes@iqe.phys.ethz.ch

Prof. Dr. Bernhard Boser
Department of Electrical Engineering
and Computer Science
Berkeley Sensor & Actuator Center
University of California at Berkeley
572 Cory Hall
Berkeley, CA 94720-1770
USA
boser@eecs.berkeley.edu

Prof. Dr. Oliver Brand
School of Electrical
and Computer Engineering
Georgia Institute of Technology
Atlanta, GA 30332-0250
USA
oliver.brand@ece.gatech.edu

Dr. Ralf Brederlow
Infineon Technologies
Corporate Research
Otto-Hahn-Ring 6
D 81730 Munich
Germany
ralf.brederlow@infineon.com

Prof. Dr. Victor M. Bright
Department
of Mechanical Engineering
University of Colorado
427 UCB
Boulder, CO 80309-0427
USA
victor.bright@colorado.edu

Dr. Jong Uk Bu
LG Electronics Institute of Technology
16 Woomyeon-dong, Seocho-gu
Seoul 137-724
Korea
jbu@lge.com

Dr. Mike Chapman
Motorola Sensor Products Division
M/D AZ01 Z207
5005 E. McDowell Road
Phoenix, AZ 85008
USA
mike.chapman@motorola.com

Dr. Michel Despont
IBM Zürich Research Laboratories
Säumerstraße 4
8804 Rüschlikon
Switzerland
dpt@zurich.ibm.com

Dr. Evangelos Eleftheriou
IBM Zürich Research Laboratories
Säumerstraße 4
8804 Rüschlikon
Switzerland
ele@zurich.ibm.com

Prof. Dr. Gary K. Fedder
ECE Department
& The Robotics Institute
Carnegie Mellon University
Pittsburgh, PA 15213-3890
USA
fedder@ece.cmu.edu

Alexander Frey
Infineon Technologies
Corporate Research
Otto-Hahn-Ring 6
D 81730 Munich
Germany
alexander.frey@infineon.com

Martin Geier
IMTEK-Institut
für Mikrosystemtechnik
Universität Freiburg
Georges-Köhler-Allee 103
D-79110 Freiburg
Germany
geier@imtek.de

Dr. Andreas Greiner
IMTEK-Institut
für Mikrosystemtechnik
Universität Freiburg
Georges-Köhler-Allee 103
D-79110 Freiburg
Germany
greiner@imtek.de

Prof. Dr. Christofer Hierold
Chair of Micro- and Nanosystems
ETH Zürich
ETH-Zentrum, CLA H9
Tannenstr. 3
8092 Zürich
Switzerland
christofer.hierold@micro.mavt.ethz.ch

Dr. Toshiki Hirano
Hitachi Global Storage Technologies
San Jose Research Center
650 Harry Road
San Jose, CA 95120
USA
toshiki.hirano@hgst.com

Dr. Franz Hofmann
Infineon Technologies
Corporate Research
Otto-Hahn-Ring 6
D 81730 Munich
Germany
franzhofmann@infineon.com

Birgit Holzapfl
Infineon Technologies
Corporate Research
Otto-Hahn-Ring 6
D 81730 Munich
Germany
birgit.holzapfl@infineon.com

David Kauzlaric
IMTEK-Institut
für Mikrosystemtechnik
Universität Freiburg
Georges-Köhler-Allee 103
D-79110 Freiburg
Germany
kauzlari@imtek.de

Dr. Jim Knutti
Silicon Microstructures, Inc.
1701 McCarthy Blvd.
Milpitas, CA 95035
USA
info@si-micro.com

Prof. Dr. Jan G. Korvink
IMTEK-Institut
für Mikrosystemtechnik
Universität Freiburg
Georges-Köhler-Allee 103/03.033
79110 Freiburg
Germany
korvink@imtek.de

Christine Kranz
Georgia Institute of Technology
School of Chemistry and Biochemistry
Applied Sensors Laboratory (ASL)
Boggs Building
770 State Street
Atlanta, GA 30332-0400
USA
christine.kranz@chemistry.gatech.edu

Angelika Kueng
Georgia Institute of Technology
School of Chemistry and Biochemistry
Applied Sensors Laboratory (ASL)
Boggs Building
770 State Street
Atlanta, GA 30332-0400
USA
angelika.kueng@chemistry.gatech.edu

Dr. Jan Lichtenberg
Physical Electronics Laboratory
ETH Zürich
Hönggerberg, HPT H4.2
8093 Zürich
Switzerland
lichtenb@iqe.phys.ethz.ch

Prof. Dr. G. Jordan Maclay
Quantum Fields LLC
20876 Wildflower Lane
Richland Center, WI 53581
USA
jordanmaclay@quantumfields.com

Prof. Dr. Boris Mizaikoff
Georgia Institute of Technology
School of Chemistry and Biochemistry
Environmental Science & Technology
Building
Office L 1240
311 First Drive
Atlanta, GA 30332
USA
boris.mizaikoff@chemistry.gatech.edu

Dr. David J. Monk
Motorola Sensor Products Division
M/D AZ01 Z207
5005 E. McDowell Road
Phoenix, AZ 85008
USA
dave.monk@motorola.com

Christian Paulus
Infineon Technologies
Corporate Research
Otto-Hahn-Ring 6
D 81730 Munich
Germany
christian.paulus@infineon.com

Vladimir P. Petkov
BSAC-EECS #1774
497 Cory Hall
University of California at Berkeley
Berkeley, CA 94720-1774
USA
vlpetkov@eecs.berkeley.edu

Dr. Arvind Salian
Motorola Sensor Products Division
M/D AZ01 Z207
5005 E. McDowell Road
Phoenix, AZ 85008
USA
arvind.salian@motorola.com

Prof. Dr. Pasqualina M. Sarro
Faculty of Electrical Engineering
Department of Microelectronics
P.O. Box 5053
2600 GB Delft
The Netherlands
sarro@dimes.tudelft.nl

Dr. Meinrad Schienle
Infineon Technologies
Corporate Research
Otto-Hahn-Ring 6
D 81730 Munich
Germany
meinrad.schienle@infineon.com

Dr. Petra Schindler-Bauer
Infineon Technologies
Corporate Research
Otto-Hahn-Ring 6
D 81730 Munich
Germany
petra.schindler-bauer@infineon.com

Prof. Dr. Joseph R. Stetter
Department of Biological, Chemical &
Physical Sciences
Life Sciences Building, Room 182
3101 South Dearborn St.
Chicago, IL 60616
USA
stetter@iit.edu

Prof. Dr. Conrad R. Stoldt
Department
of Mechanical Engineering
University of Colorado
427 UCB
Boulder, CO 80309-0427
USA
conrad.stoldt@colorado.edu

Prof. Dr. Osamu Tabata
Department
of Mechanical Engineering
Faculty of Engineering
Kyoto University
Yoshida Honmachi, Sakyo-ku
Kyoto 606-8501
Japan
tabata@mech.kyoto-u.ac.jp

Dr. Roland Thewes
Infineon Technologies
Corporate Research
Otto-Hahn-Ring 6
81730 Munich
Germany
roland.thewes@infineon.com

1
M³: the Third Dimension of Silicon

P. M. Sarro, Laboratory of Electronic Components, Technology and Materials, Delft University of Technology, DIMES, The Netherlands

Abstract
Microsystems technology is a fascinating and exciting field that has played and will continue to play a very important role in building a bridge between science and society. Physical properties and material characteristics are translated into structures and devices that can have a large positive impact on people's lives. Silicon micromachining has been largely responsible for the expansion of sensors and actuators into more complex systems and into areas not traditionally related to microelectronics, such as medicine, biology and transportation. The shift to 3D microstructures has not only added a physical *third dimension* to silicon planar technology, it has also added a *third dimension* in terms of functionality and applications. In this chapter, the basic issues and fundamental aspects of this field are briefly introduced. A few examples that illustrate *the power of the small world* are given and possible ways to pursue further miniaturization and/or increase in functionality are discussed.

Keywords
microsystems technology; MEMS; silicon micromachining; 3D microstructuring.

Advanced Micro and Nanosystems. Vol. 1
Edited by H. Baltes, O. Brand, G. K. Fedder, C. Hierold, J. Korvink, O. Tabata

ISBN: 3-527-30746-X

1.1 Introduction

The fascination for the small world has been a constant in the research in physics, biology and engineering. Many scientists share a great interest in miniaturization technologies and in studying the behavior of materials and structures in the micro- and nanometer range. The investigation of the small world of matter is crucial to understanding how things work and this knowledge can be used to create novel microstructures and devices, thus offering the necessary tools and components to realize applications of great societal importance.

Microsystems technology is a fascinating and exciting field that has played and will continue to play a very important role in building a bridge between science and society to translate physical properties and material characteristics into structures and devices that can have a large positive impact on people's lives. In this chapter, the basic issues and fundamental aspects of this field are briefly introduced and an attempt is made to indicate where we are now and where we are going and how the small world makes a big difference.

1.2 M^3: Microsystems Technology, MEMS and Micromachines

Microsystems technology (MST) generally refers to design, technology and fabrication efforts aimed at combining electronic functions with mechanical, optical, thermal and others and that employ miniaturization in order to achieve high complexity in a small space. Microsystems are thus intelligent microscale machines that combine sensors and actuators, mechanical structures and electronics to sense information from the environment and react to it. These tiny systems are or soon will be present in many industrial and consumer products and will have a huge impact on the way we live, play and work.

In addition to MST, there are a number of other terms and acronyms that are used to describe this field, referring either to technologies, design concepts or integration issues. The most frequently used or encountered terms come from the three major geographical areas involved in this field:

- Europe → Microsystems technology (MST);
- USA → Microelectromechanical systems (MEMS);
- Japan → Micromachines (MM).

Apparently in Europe the accent is placed on the miniaturization of the entire systems, in the USA on the mechanical components being brought into the microelectronic world and in Japan on the miniaturization of a machine. Maybe this reflects in some way the cultural backgrounds of the three regions.

Although the names are somewhat different, basically they all accentuate both the miniaturization and the multi-functionality and system character. Some groups consider the presence of a movable part in the system necessary to be able to talk about MEMS, but in most cases the multi-functional character and miniaturization are the essential ingredients or prerequisites.

In view of the truly global character of research and the many transnational co-operations in this area, a new way to address this field collectively is M^3:

$$\mathrm{MST} \cdot \mathrm{MEMS} \cdot \mathrm{MM} = M^3$$

The exponential factor 3 can also be seen in relation to the *third dimension*. In fact, it not only combines all definitions identifying this field, it also stresses and symbolizes the importance of the introduction of an active 'third dimension' to silicon, literally and figuratively, and the impact these truly three-dimensional (3D) microsystems have in applications ranging from health care to consumer products.

We could also see the M^3 as summarizing three key characteristics of this field: *M*ultidisciplinary, *M*iniaturization, *M*ankind needs:

$$M^3 = M \cdot M \cdot M$$

Each of these aspects will be addressed in the following sections. Let us now take a closer look at the general concept of microsystems (or MEMS or MM) technology.

1.2.1 Microsystem Technology

Microsystem technology (MST) has experienced about two decades of evolution, mainly driven by a few key applications. It is likely to drive the next phase of the information revolution, as microelectronics has driven the first phase. A multi-bil-

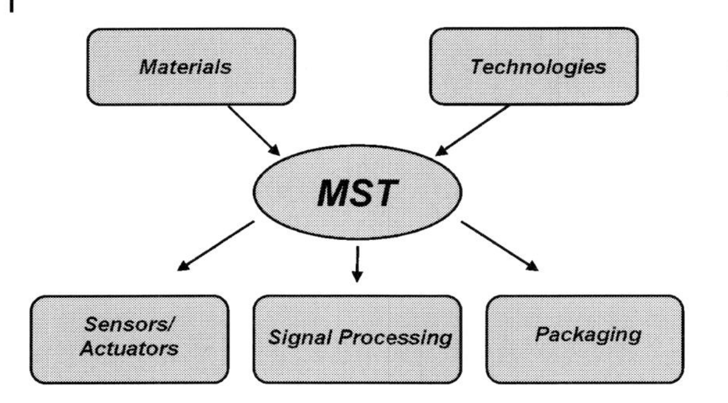

Fig. 1.1 Microsystems technology: the next phase in information evolution

lion dollar market by the middle of the next decade has been forecast and although there is some discrepancy in the figures presented by different bureaus and agencies, a common belief in a consistent growth is shared by all.

The general concept of MST, schematically depicted in Fig. 1.1, is to combine new materials with microprocessing technology (mostly well suited for low-cost mass production purposes) and micromachining technologies to form the three basic building blocks of every microsystem: sensing/actuation element, signal processing, package.

Advances in material science and processing are at the base of each MST product. Three main groups of materials are to be distinguished: materials for the package, materials for the actual device and the electronics and materials for the mechanical/electrical connection between these. Progress in semiconductor processing has evolved in a number of substrate materials predestined for use in microstructured devices, such as silicon, silicon-on-insulator, silicon carbide and gallium arsenide. Pricing and reliability considerations have led to the almost exclusive use of silicon-based micromachined devices. Packaging and assembly have focused on ceramics, printed circuit board (PCB) technology and multi-chip modules (MCMs) [1].

1.2.2
MST versus IC

The continuous advances in silicon-based integrated circuit (IC) technology, in terms of both processing and equipment, have definitely contributed to microsystems developments. At the same time the enormous growth in microsystems applications has stimulated the development of dedicated equipment and generated a larger knowledge of material and structure characteristics, especially in the mechanical area, which have been of great help to the IC world.

MST is often envisioned as being similar to semiconductor microelectronics and, although they possess many similarities, there are also some strong differences, as indicated in Tab. 1.1. Some of the most important ones are the lack of a 'unit cell' (no transistor-equivalent) and of a stable front-end technology (no CMOS equivalent). Moreover a multidimensional interaction space (not only elec-

Tab. 1.1 Important differences between MST and IC

	MST	*IC*
Unit cell	No unit cell	Transistor
Front-end technology	No single stable technology	CMOS
Interaction space	Multidimensional	Electrical
Basic disciplines	Multidisciplinary	Physics and engineering

trical connections) is present and it is a very multidisciplinary field as next to physics and engineering other disciplines such as chemistry, material science and mechanics play an important role.

Therefore, research is evolving toward a MST unit that is not a single unit cell, like the transistor, but small, specifically designed, components libraries that could be refined over time to become standard building blocks for each MST device domain.

1.3 M^3: Multidisciplinary, Miniaturization, Mankind Needs

Microsystems technology has a strong *multidisciplinary* character. Integration across several disciplines takes place. Next to physics and engineering, the basic disciplines of microelectronics, we find that chemistry and biology are becoming more and more a part of MST as new materials and phenomena play a major role in the development of new microsystems. Also, of course, as movable or flexible parts are often essential components of the system, the role of mechanics or rather micromechanics is much larger than it ever was in conventional microelectronics. Although the broad range of expertise and know-how that this field requires might make the path to problem solving and product development more difficult, it can also be seen as enrichment in the engineering world.

In fact, microelectronics is entering many industrial sectors that are becoming increasingly multidisciplinary environments, such as biotechnology, health care and telecommunication. Consequently, growing interest in an interdisciplinary educational program is observed as it will become extremely important to prepare a new generation of engineers capable of operating in such multi-disciplinary environments. Another positive aspect of the way in which research and development in this field is carried out is the development of important social skills such as dealing with people from different fields and speaking different languages (literally and figuratively), something that is more the norm than the exception. This learning process could be very useful in the global world in which we operate nowadays.

1.3.1 Miniaturization

Another key aspect of MST is *miniaturization*. Miniaturization is necessary

- to achieve increased functionality on a small scale;
- to utilize particular effects and phenomena that are of no specific relevance at the macroscale level;
- to increase performance in order to make new areas of application possible;
- to interface the nanoworld.

It is generally pursued by using silicon IC-based technologies, with proper modification or the addition of specifically developed modules. A key process is the 3D machining of semiconductor materials, leading to miniaturized structures constituting the sensing, actuating or other functional parts. The main processes are bulk micromachining (BMM) and surface micromachining (SMM).

1.3.1.1 Bulk micromachining

Bulk micromachining covers all techniques that remove significant amounts of the substrate (or bulk) material and the bulk is part of the micromachined structure [2]. Typical BMM structures are shown in Fig. 1.2. This microstructuring of the substrate is often done to form structures that can physically move, such as floating membranes or cantilever beams. Other types of structures that can be realized by bulk micromachining are wafer-through holes, often used for through wafers interconnects in chip stacks and very deep cavities or channels to form microwells or reservoirs for biochemical applications. The substrate (generally silicon) can be removed using a variety of methods and techniques. In addition to a number of processes using wet (or liquid) etchants, techniques using etchants in the vapor and plasma state (generally referred to as dry) are available.

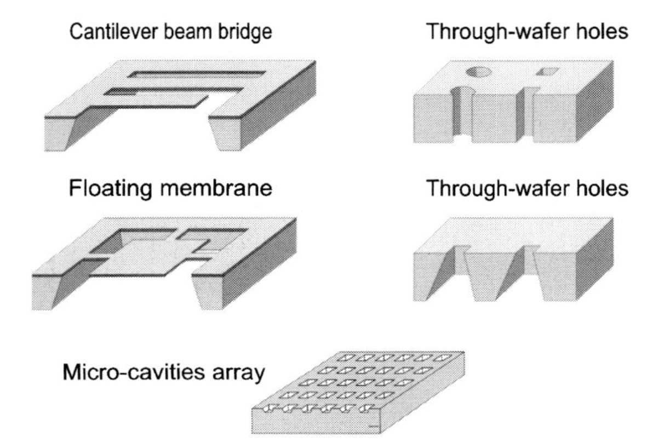

Fig. 1.2 Typical bulk micromachined structures

1.3.1.2 Surface Micromachining

Surface micromachining is a very different technology, which involves the deposition of thin films on the wafer surface and the selective removal of one or more of these layers to leave freestanding structures, such as membranes, cantilever beams, bridges and rotating structures [2]. Examples of typical SMM structures are shown in Fig. 1.3. In recent years, a number of new processes have been developed which use the upper few microns of the substrate (often an epitaxial layer) as a mechanical layer. The basic principle of surface micromachining shows two types of layers, the sacrificial layer and the mechanical layer. The sacrificial layer is removed during subsequent processing to leave the freestanding mechanical structure. The sacrificial layer is accessed from the side of the structure or through access holes.

Other processes or techniques relevant to microsystems fabrication are wafer-to-wafer bonding techniques, 3D lithography and some other high aspect ratio techniques such as LIGA and laser machining [3].

The choice of the 3D micromachining technology depends strongly on the application field and design of the MST product to be manufactured. The use of techniques compatible with standard semiconductor processing is often preferred as this permits batch fabrication, potential cost efficiency and system integration.

In Delft, the importance of integrating multi-elements on a single chip while focusing on the system aspects has been recognized for a long time [4]. Research has therefore focused on silicon (the major material for ICs) and silicon-related materials and the development of post-process modules following the approach shown schematically in Fig. 1.4. This approach allows on the one hand the addition of more functions on a chip, offering some flexibility and application-specific variations, and on the other hand preserves the compatibility with basic IC processes (Bipolar or CMOS), thus allowing the realization of complete systems (sensor + signal processing + actuator).

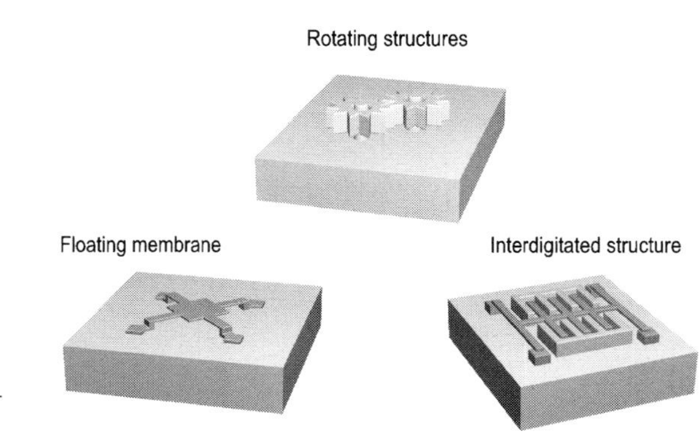

Fig. 1.3 Typical surface micromachined structures

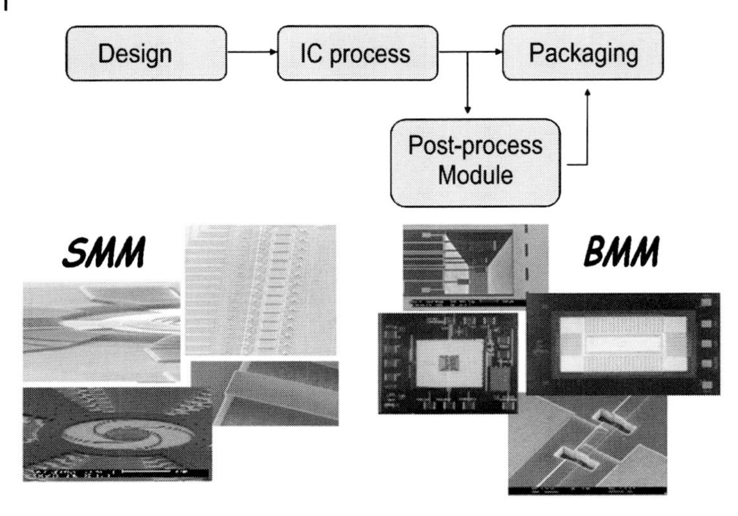

Fig. 1.4 IC-compatible post-process modules: the approach followed at DIMES and examples of structures realized with the developed BMM and SMM post-process modules

1.4 The Power of a Small World

Great progress has been made in the field of microsystems. Initially sensors or actuators built in bulk silicon with either simple dedicated processes or built-in conventional CMOS processes were introduced, more as an extension of the possibility of microelectronics than as a separate field. The real expansion into more complex systems and into areas not traditionally related to microelectronics, such as medicine, biology and optical telecommunication, came much later. Silicon micromachining has been crucial for this. In fact, the shift from planar, essentially 2D components, to vertical or 3D microstructures has added not only a physical *third dimension* to silicon planar technology, but also a *third dimension* in terms of functionality and applications (see Fig. 1.5).

Micromachining has also moved from silicon to other materials, such as glass, polymers and metals, thus creating an even larger pool of possible configurations. However, the level of maturity and the advantages related to the use of silicon are still predominant.

Let us examine a few examples that illustrate the 'power of the small world', i.e., the realization of applications, components and systems that would not be available or would not be as functional, as light or as small as they are now thanks to MST.

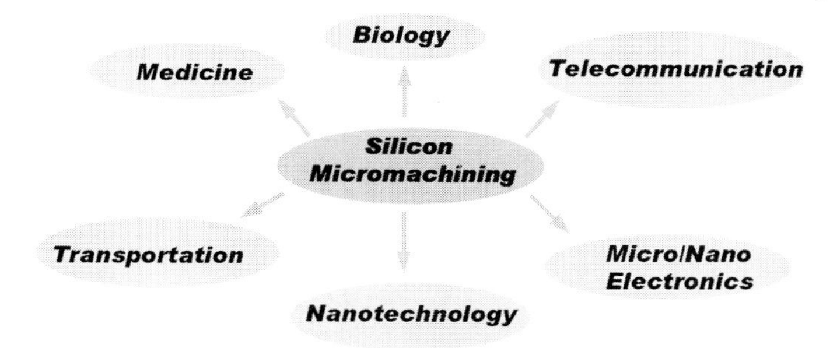

Fig. 1.5 Silicon micromachining: a crucial technology for an increase in functionality and application areas

1.4.1
MST in Transportation

An area where MST has made a major difference is transportation. Not only vehicles are equipped with an increasing amount of MST devices, creating smart cars, planes, boats, etc.; MST is also used in efforts to monitor highways, roads and bridges, leading to smart roads and safer skies. However, it has been the automotive industry, and thus the car, which has been and still is playing a key role in promoting MST technology and development.

It actually began with pressure sensors, with the introduction of the manifold air pressure sensor (MAP) for engine control in 1979, but one of the biggest commercial applications of MST is the accelerometer for air bags. In the case of a crash, the accelerometer sends the information to the inflation system. The bag is inflated by the large volume of gas created by an extremely rapid burning process. The bag then literally bursts from its storage site at up to 200 mph – faster than the blink of an eye. A second later, the gas quickly dissipates through tiny holes in the bag, thus deflating the bag so that driver and passengers can move.

At present, microsystems are most frequently used in safety applications, followed by engine and power train applications and on-board vehicle diagnostics. A major concern is reliability. Sensors working for safety features and leading to functions controlling at least part of the driving need to be of the highest quality. On the other hand, reliability in the automotive environment with temperatures up to 125 °C and aggressive media is an important cost issue. Packaging solutions and contactless sensors can help to reach the targets.

A revolution in this sector is starting to take place, implying a complete transition from the mechanically driven automobile system to a mechanically based but ICT-driven system. Microsystems are indispensable for fulfilling these ambitions. The current 7 Series BMW has a total of 70 MST devices installed in the car – a new milestone in the use of MST components in this application area [5].

1.4.2 MST in Medicine

MST developments have had and continue to have a remarkable impact on biology and medicine. Miniaturization on the one hand and system integration on the other are responsible for the presence or improvement of a number of applications that are all around us.

Biomedical sensors permit the reliable generation of essential physiological information required to provide therapeutic, diagnostic and monitoring care to patients. One example related to chronic cardiac disease is the pacemaker. This small apparatus, which has been implanted in more than two million people worldwide and is a true life saver, is based on a sensing and actuating unit, signal processing and a power source. In the newest generation of pacemaker-defibrillators, a MEMS accelerometer capable of tracking changes in motion, in this case the heart beat, sends the information to the processing unit and the proper amount of electric shock to restore the natural rhythm of the heart is delivered through the pacing lead [6]. The electronics of these microsystems, as for all implantable devices, must operate on low power. Novel powering alternatives to reduce further the size and weight of the apparatus and to prolong its lifetime are important research themes.

Microsurgery or minimally invasive surgery, an almost unknown notion a few years ago, is also an area where progress has been remarkable and tangible results are seen every day. Smart medical devices can be realized by embedding MST-based sensors at the most effective place on or within the device. Microsystems technology is the only method available that produces sensors small enough to be embedded in surgical devices without changing the basic function or form. MST-laden scalpels, needles and drills would give surgeons an unprecedented level of control or even the flexibility to perform entirely new procedures [7]. Tools such as the 'smart' scalpel schematically shown in Fig. 1.6, incorporate sensing and measuring devices to track and record information during surgery. The sensors, placed as close as possible to the edge of the cutting tool, for instance, can

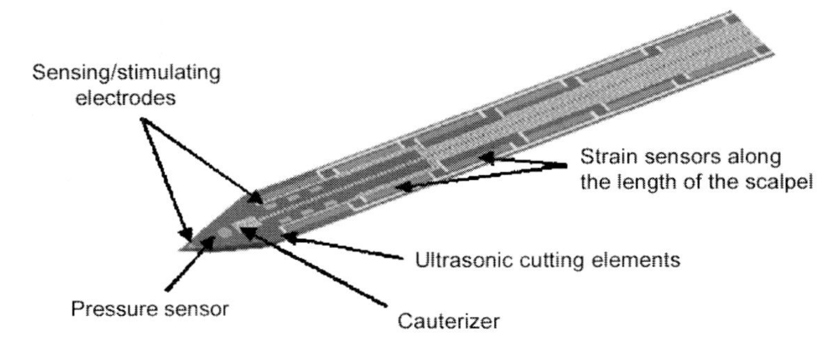

Fig. 1.6 The 'data knife' concept incorporates sensing and data-gathering capabilities on the edge of various surgical tools (adapted from [7])

tell how close a scalpel is to a blood vessel and shut the tool down if it approaches too close.

Using proven semiconductor manufacturing processes, microneedles to deliver small, precision dosages of drugs to localized areas are realized. A microneedle with a hollow core can easily fit into the bore of one of the smallest hypodermic injection needles commercially available today. Furthermore, the tip of the microfabricated microneedle is sharper and smoother than that of the steel needle. The smaller the diameter and the sharper the tip, the less pain and tissue damage occur during use. Initial tests have shown the microneedle to be nearly painless [7].

1.4.3 MST in Biology

Another area that is experiencing enormous development and where MST plays a crucial role is the miniaturization of (bio)chemical assays that are used for quality management in the biotechnology and food industry, medical diagnostics, drug development, environmental monitoring and high-throughput biochemical screening. In a successful multidisciplinary project [8], recently completed at our university, relevant progress has been achieved in the design and realization of an intelligent analytical system that measures many different molecular analytes simultaneously using specific molecular interactions on the surface of specially constructed microchips. Images of microchips containing arrays of microwells of different size and shape are shown in Fig. 1.7. Besides allowing a large number of different analyses simultaneously in a very short time, the system uses only extremely small amounts of reagents and samples.

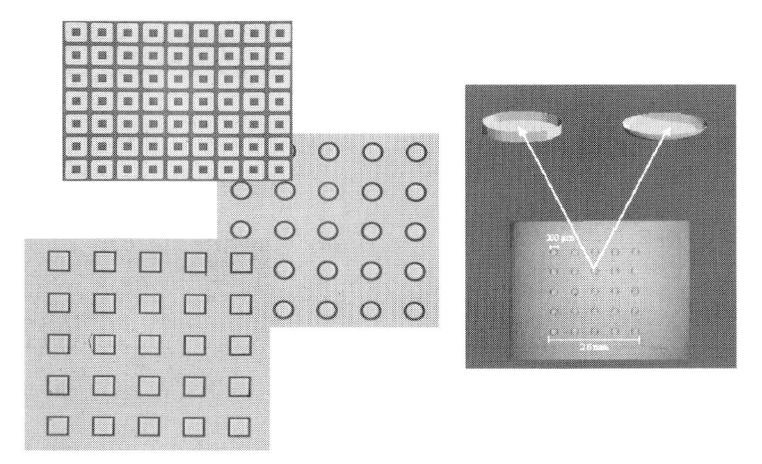

Fig. 1.7 Microchip containing picoliter reaction chambers developed in the DIOC-IMDS project [8]

1.4.4
MST in Telecommunication

Twenty-four hours a day, 365 days a year, data bits race through the optical fiber networks that span our globe. This vital link between people both for professional and private matters cannot afford any failure. Protection switches have a key role in guaranteeing network safety and in redirecting the data stream around a section that is not functioning. Switching in the optical domain is preferable because it avoids electro-optical conversion and is independent of data and encoding format. MST technology and specifically silicon micromachining have made it possible to realize fiber-optic switches that have many advantages over the conventional mechanical-relay type (no fatigue; miniaturization = faster switching time). Although the optical MEMS industry has not lived up to the high expectations with respect to implementation of photonic MEMS-based switching subsystems, the developed technology seems promising for other applications. Optical MEMS are, in fact, well suited for a variety of displays to be used in consumer electronics products, such as digital cameras and TVs, and also as medical instrumentation and car information systems [9].

However, by far the largest contribution that MST can offer is in wireless communication by providing more function and power with smaller parts. Radiofrequency (rf) systems for telecommunication are expected to be the next major application for MST [10]. Microsystems for rf applications, also known as rf MEMS, cover a large variety of devices, such as microswitches, tunable capacitors, micromachined inductors, micromachined antennas, microtransmission lines and mi-

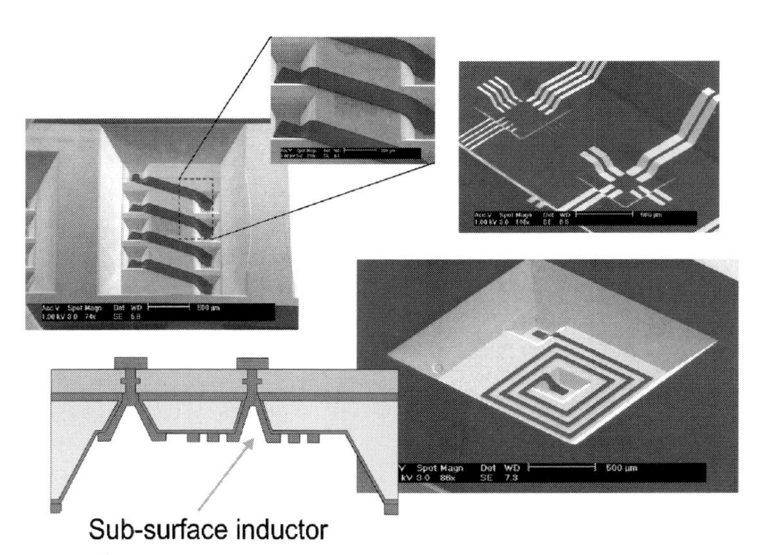

Fig. 1.8 Integrated rf components realized with the DIMES BMM post-process module

cromechanical resonators. Manufactured by conventional or novel 3D microstructuring techniques, they offer increased performance, such as lower power consumption, lower losses, higher linearity and higher Q factors.

The importance of silicon micromachining for rf applications has been recognized in DIMES. Novel post-process modules based on innovative techniques have been developed to address some of the limitations of planar silicon IC technology, making it possible to enhance the performance of integrated rf devices and systems [11]. Examples of integrated rf devices realized using BMM, such as sub-surface inductors, 3D solenoid inductors and through-chip transmission lines, are shown in Fig. 1.8.

1.5 Future Perspectives

MST has indeed had and continues to have a strong impact on almost every aspect of our lives. What are the challenges that lie ahead and which are the areas on which the microsystems community will focus? Let us look at some important aspects and mention some of the activities we are currently pursuing or intend to explore in our laboratory.

1.5.1 Autonomous Microsystems: More Intelligence in a Small Space

It is clear that the future generation of microsystems will have to satisfy challenging demands. These systems have to be capable of self-regulation and wireless communication, should be compact in size and should operate at low power, often in harsh environments. These systems will provide, among others, the future front-ends of information networks and links from microelectronics to the cellular world.

An integrated autonomous microsystem, schematically depicted in Fig. 1.9a, needs to contain several basic functional modules to interact with its environment. It should be able to *sense* the perturbation in an environment (hearing or sight), and also to actuate perturbation to the environment for response (motion). It also needs to *communicate* with other microsystems and with a central point to establish collective and coordinated functions (Fig. 1.9b). Many of the operations involve computing and control for complex information processing. Finally, the autonomous microsystems must contain a *power generation/conversion unit*. These functions of the autonomous microsystems can be potentially realized by integrating memory/microprocessors with MEMS in a power-efficient manner (system-on-a-chip).

The effective 3D microstructuring offered by MST together with the introduction of new materials into microelectronics will be essential for this future generation of integrated microsystems. In particular, attention should be paid to novel concepts and principles for power generation and/or conversion and for operation in harsh environments.

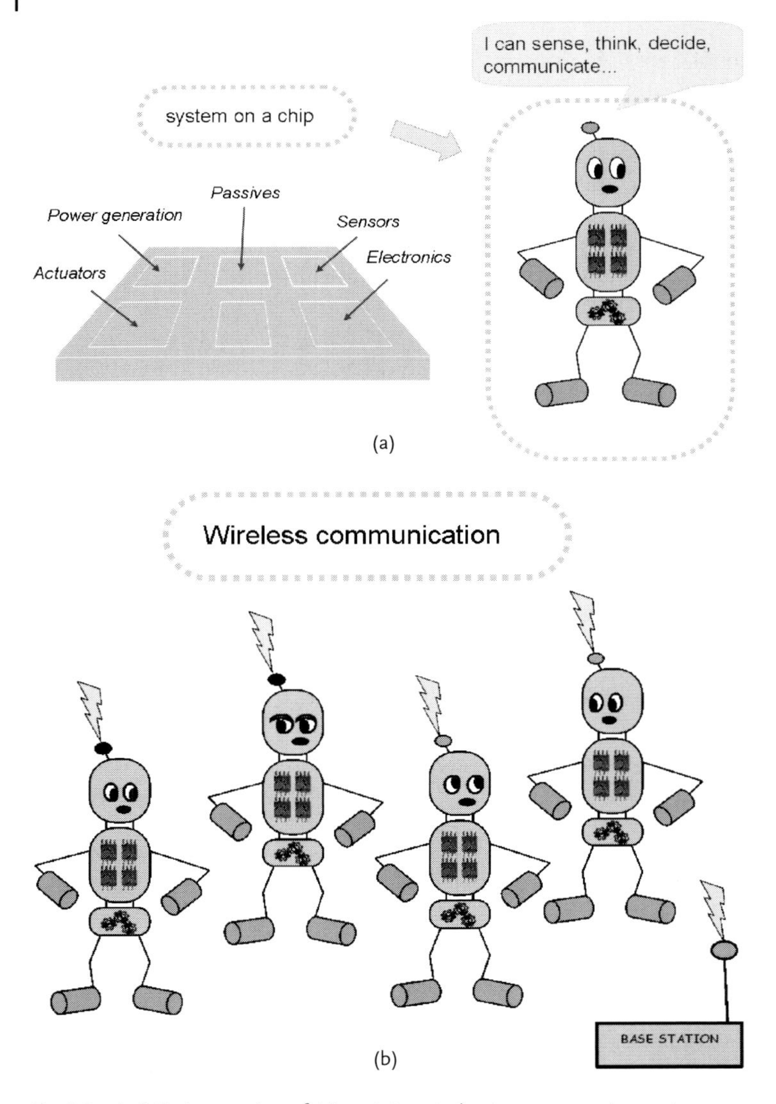

Fig. 1.9 Artist's impression of (a) an integrated autonomous microsystem and (b) wireless communication among autonomous microsystems (© P. M. Sarro, M. v. d. Zwan, 2003)

Applications can include any measurement requirements in remote locations or where access is difficult, from monitoring weather patterns to tracking human movement. After space applications, where reliability, power consumption and size are very critical, a large application area can be envisioned in implantable sensors for medical applications and home monitoring systems for the elderly. Through these systems, for example, patients could be sent home earlier and monitored via the Internet and elderly people could continue living in their homes, with direct connections to a central point. Besides improving the quality of life and extending independence, this will lead to considerable saving in health costs.

1.5.2 Top Down or Bottom Up: the Next Phase in Miniaturization

Although much progress has been made and exciting results have been achieved, there is still, as Richard Feynman wisely said about half a century ago, '*Plenty of room at the bottom*' [12]. After application-specific issues, a more generic discussion addresses the approach to follow in order to pursue further miniaturization: top down or bottom up?

The top down method focuses on downscaling to miniaturize devices and systems to the nanometer scale. The 'conventional' microsystems technology/MEMS belong to this approach. The bottom up approach relies on rather different technologies and materials to generate novel miniature systems. Atoms and molecules are integrated to form devices, shaping the system atom by atom.

These two approaches can also be combined to create a new converging technology, thus further increasing future perspectives. As illustrated in Fig. 1.10, we are at the point where both approaches can be used to create new systems, devices and even new materials.

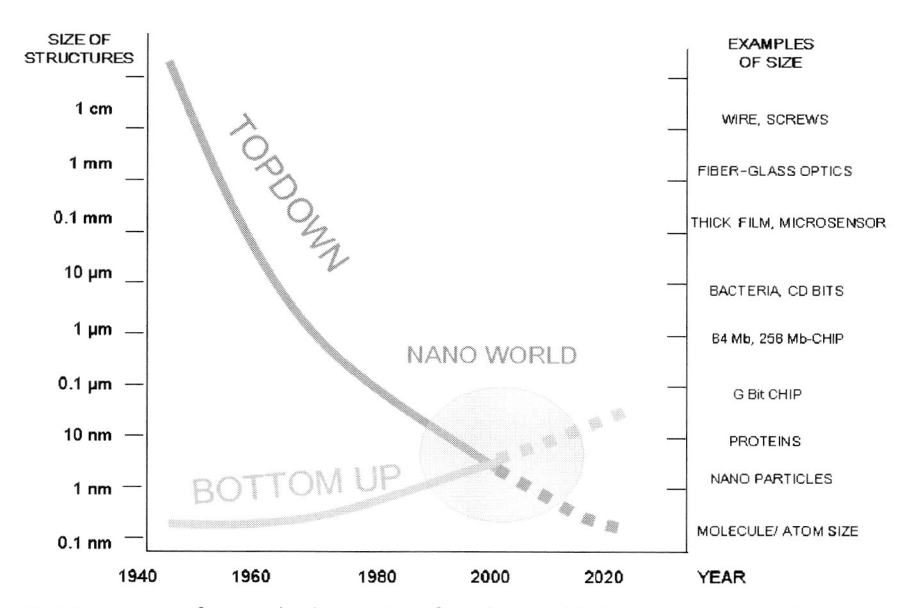

Fig. 1.10 History of nanotechnology: status of top down and bottom up approaches (adapted from [13])

1.5.2.1 **Top down**

Top down mostly means shrinking dimensions either to improve functionality or to create complex systems in a small space. In order to achieve this, novel micromachining technologies are required. These include various etching techniques (DRIE, laser ablation), resulting in well-controlled, accurate definition of the microstructures; specific lithographic processes (spray coating or electrodeposition of resist, front-to-back alignment (FTBA), direct write) to transfer fine patterns onto multi-level, truly 3D microstructures and coating or plating techniques (electroplating, electroless plating, vapor deposition) to deposit insulating, sensing or conducting layers on these large topography (highly non-planar) surfaces. The *M* from microsystems, MEMS or micromachine can slowly become an *N*, i.e. Nanosystems technology, nanoelectromechanical systems, nanomachines:

$$M^3 \rightarrow N^3$$

1.5.2.2 **Bottom Up**

In the *bottom up* approach, the focus is on other materials (polymers, carbide, diamond, organic materials), new 'building' techniques and the use of different effects. Inspired by biology, where the precise (self) organization of molecules permits the many functions carried out in living cells, scientists have started to explore means to build molecular machines, wires and other devices in a very precise manner, by stacking individual atoms or molecules. Research into methods for such assembly is still in its infancy, but its promise is great. Moreover, the potential use of DNA – with its ability for 'programmed self-assembly' – for biomolecular electronic devices (nanowires, molecular memories) in combination with advanced microsystems technology, various protein molecules and nanoparticles is also a challenging and promising future development.

Combination of both approaches is also possible. Our group participated in a European project that aims at developing an enabling technology for 3D computer structures, such as a fault-tolerant, 3D, retina-cortex computer (CORTEX) [14]. One objective of this project is to demonstrate the feasibility of very high-density, 3D molecular 'wires' (bottom up approach) between electrical contacts on separate, closely spaced, semiconductor chips or layers, as illustrated in Fig. 1.11. The high-density array of through-wafer interconnects with a very high aspect ratio is a good example of the top down approach, whereas the building of molecular wires for chip-to-chip connections well represents the bottom up method. Whether a top down or a bottom up approach is used (or a combination of both), major developments are envisioned for the coming decades.

1.5.3
The Link Between Nanoscience and the Macro World

Although remaining on the micron or submicron scale, MST will continue to be of significant importance for nanotechnology and nanoscience. In fact, microsystems,

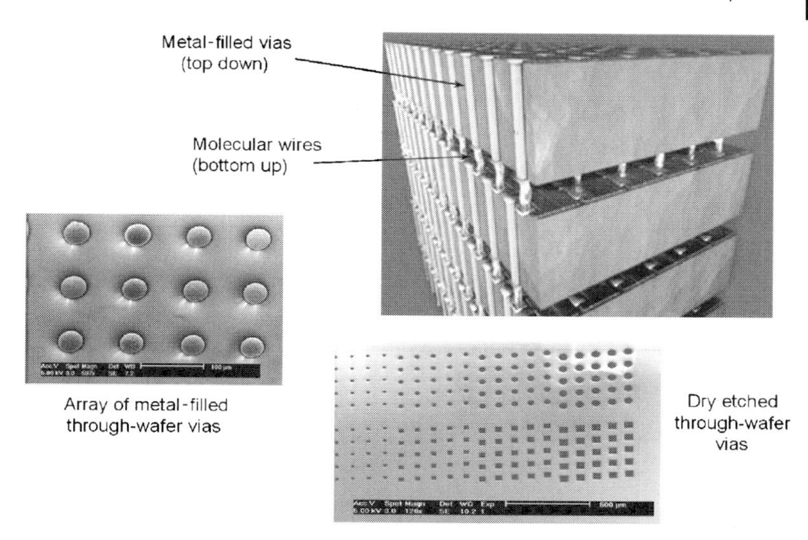

Fig. 1.11 The CORTEX project: a combination of top down and bottom up approaches

either sensors, actuators or mechanical microstructures with the necessary interfacing, are often the unavoidable bridge between the atom or molecular structure and the macro-world. Moreover, specific tools that are indispensable for building molecular systems or investigating phenomena at the atomic level heavily rely on MST.

The use of silicon-based technologies or derivatives can have an extreme impact on studying phenomena or realizing crucial tools for complex molecular systems. A good example is atom optics on a chip (atomic laser). At the University of Colorado, USA, a very active group involved in atom optics is investigating the use of Bose-Einstein condensates (BEC) to make practical devices [15]. An essential component of the system is the MOT (magneto-optic trap), the largest physical structure of most BEC and atom guiding systems. It is highly desirable to miniaturize this component. MST technology can play a crucial role here by offering small, truly 3D structures. The possibility of using the third dimension is fundamental to acquiring full control over the magnetic field applied to the atoms.

Another example is related to 3D atomic resolution microscopes, the scanning tunneling microscope (STM) and the atomic force microscope (AFM). Scanning probe microscopy is a key technology for nanotechnology research and fundamental in studying phenomena at the atomic/molecular level. Scanning probe microscopes also allow the manipulation of atoms and molecules for the creation of nanodevices. Very often the tips of these tools, see Fig. 1.12, a recognized symbol for nanotechnology, are realized with the help of silicon-based microsystem technologies [16].

New developments in MST will further contribute to building the necessary interface between nanodevices or nanostructures and microinstrumentation and will help in studies of phenomena at the nanoscale level. On the other hand, future develop-

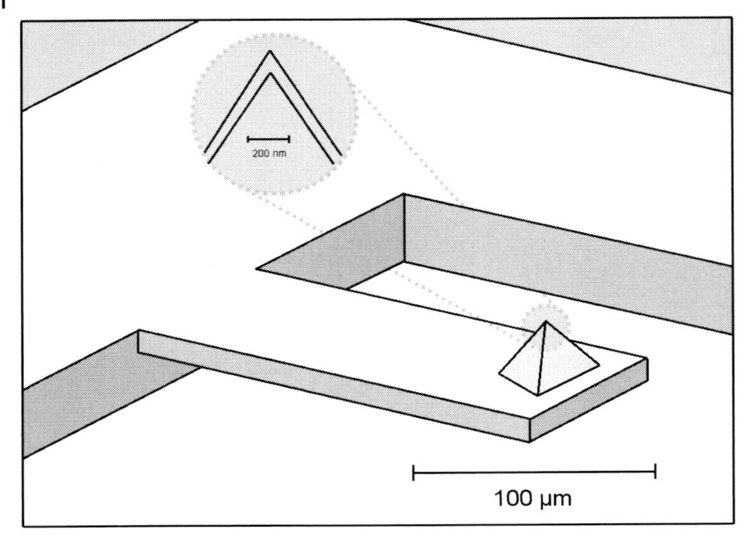

Fig. 1.12 The tip of a scanning probe microscope fabricated by silicon micromachining

ments in nanotechnology will permit the realization of highly sophisticated microsystems, such as miniaturized drug-delivery systems and miniaturized computers.

1.6 Conclusions

MST research activities continue to focus on innovative technological processes and material findings to create the necessary environment to address the many challenges in areas currently recognized to be of strategic importance and of high societal relevance. In particular, increasing the autonomous and wireless character of microsystems will be intensively pursued. Mostly a top down approach will be followed to realize truly 3D microstructures, essential to many microsystems. Research on silicon-related/compatible materials such as SiC, polymers and alternative metals to expand application areas further or improve system performance will also be carried out. The potential of the bottom up approach will be increasingly explored as this approach can offer more possibilities and proper solutions in new areas as well as stimulating new developments.

Whatever the approach followed or the application area targeted, research in microsystems technology will continue to provide further developments in many scientific disciplines and at the same time make a tangible contribution to society.

1.7 Acknowledgments

Such a multidisciplinary field and such a multidisciplinary environment require substantial support and cooperation. I thank all the undergraduate and graduate students and colleagues both at Delft University as in the many groups with whom I frequently cooperated for some of the material shown, for constructive discussions and for fruitful cooperation. Many thanks are due to the process engineers of the DIMES clean room for the realization of many of the devices and microstructures presented. The creative and skilful preparation of the illustrative material of this chapter by Michiel van den Zwan is greatly appreciated.

1.8 References

1 S. Krueger, F. Solzbacher, *mstnews* **2003**, *1*, 6–10.

2 P.J. French, P.M. Sarro, in: *Handbook of MEMS*, O. Paul, J. Korvinck (eds.); Norwich, NY: William Andrew Publishing, **2004**, Chapter 17.

3 M. Madou, *Fundamentals of Microfabrication*; Boca Raton, FL: CRC Press, **1997**.

4 P.M. Sarro, *Sensors and Actuators A* **1992**, *31*, 138–143.

5 *http://www.bmw.com.*

6 *http://www.medtronic.com/brady/patient/medtronic_pacing_systems.html.*

7 *http://www.smalltimes.com/print_doc.cfm?doc_id=5405 http://www.verimetra.com.*

8 *http://www.ph.tn.tudelft.nl/Projects/DIOC/Progress/DIOC.Progress.html.*

9 M. Bourne, A. el-Fatatry, P. Salomon, *mstnews* **2003**, *4*, 5–8.

10 J. Bouchard, H. Wicht, *mstnews* **2002**, *4*, 39–40.

11 N.P. Pham, *Silicon Micromachining for RF technology*, PhD Thesis, Delft University of Technology, **2003**.

12 R. Feynman, *Caltech's Eng. Sci.* February **1960** (*http://www.zyvex.com/nanotech/feynman. html*).

13 T. Iwai, *Proc. IEEE-MEMS 2003*, Kyoto, Japan, 19–23 January **2003**, pp. 1–4.

14 N.T. Nguyen, E. Boellaard, N.P. Pham, V.G. Kutchoukov, G. Craciun, P.M. Sarro, *J. Micromech. Microeng.* **2002**, *12*, 395–399; CORTEX web site: *http://ipga.phys.ucl.ac.uk/research/cortex.*

15 D.Z. Anderson, V.M. Bright, L. Czaia, S. Du, S. Frader-Thompson, B. McCarthy, M. Squires, *Proc. IEEE-MEMS 2003*, Kyoto, Japan, 19–23 January **2003**, pp. 210–214.

16 *www-samlab.unine.ch/Activities/Projects/Snom/snom.htm.*

2
Trends in MEMS Commercialization

J. W. Knutti, H. V. Allen, Silicon Microstructures, Inc., Milpitas, CA, USA

Abstract
This chapter looks at various aspects of MEMS commercialization starting with a breakdown into various commercial phases from the early university concepts that spawned industrial research activity through early self-contained start-ups and ending with the current phase of a MEMS infrastructure. Next, various market forces are described which dictated much of the commercialization activity. The commercial success of MEMS structures depends first and foremost upon a motivated marketplace where MEMS technology provides a clear competitive and enabling advantage. Funding and corporate structure have been another important aspect of commercialization. The wide breadth of applications, the initial parallels to the integrated circuits industry and the myriad of possible angles to approach commercializing the technology have spawned a broad range of successful and not-so-successful business models. This chapter summarizes the points of these sections with an outlook on important factors that will continue the commercialization of MEMS technology.

Keywords
MEMS; commercialization; infrastructure; investment.

Advanced Micro and Nanosystems. Vol. 1
Edited by H. Baltes, O. Brand, G. K. Fedder, C. Hierold, J. Korvink, O. Tabata

ISBN: 3-527-30746-X

2.1 Introduction

MEMS technology has gone through many phases of naming. In its early days, it was primarily referred to as silicon micromachining, making mechanical structures in silicon and a number of early acronyms. All referred to the basic concept that the silicon manufacturing technology that was being used at the time to great advantage in signal processing could be used to the same advantage to make mechanical structures. Further, these mechanical structures could act as the physical interface between the 'real world' and the ever-evolving complex signal processing of the time.

The concept of a mechanical technology with the same advantages and opportunities as the integrated circuit industry was a very compelling story. This, coupled with the very photogenic nature of the resulting devices, sparked a lot of imagination and garnered a lot of interest along the path to commercialization.

Unfortunately, commercial success is linked to market and operational success and financial viability. Throughout its evolution, MEMS technology has had its share of successes in a wide range of industries including automotive, medical, industrial and consumer markets. At the same time there have been many examples of over-hype of the technology and some spectacular disasters.

This chapter looks at various aspects of MEMS commercialization starting with a breakdown into various commercial phases from the early university concepts that spawned industrial research activity through early self-contained start-ups and ending with the current phase of a MEMS infrastructure.

Next, various market forces are described which dictated much of the commercialization activity. The commercial success of MEMS structures depends first and foremost upon a motivated market place where MEMS technology provides a clear competitive and enabling advantage. While there are many examples of such an advantage, the glamour of the technology also has led to many cases of a solution looking for a problem and examples where a MEMS technology could also be used, but with no particular advantage.

Funding and corporate structure have been another important aspect of commercialization. The wide breadth of applications, the initial parallels to the integrated circuits industry and the myriad of possible angles to approach commercializing the technology have spawned a broad range of successful and not-so-successful business models.

Like the much larger integrated circuits industry, the MEMS industry has many of the same infrastructure requirements for technology, design tools, manufacturing technology, equipment, quality and reliability. Often these are more demanding than those of the signal processing markets. At the same time, there is less of a revenue base.

Finally, this chapter summarizes the points of these sections with an outlook on important factors that will continue the commercialization of MEMS technology. While many market studies and analyses have been completed, history has shown that the MEMS industry has been driven by unidentified market requirements and new applications that quickly emerge outside of the realm of surveys of existing applications. These dominating applications have driven the phases of the MEMS industry to date and undoubtedly are the hidden potential for future commercialization.

2.2 Phases of MEMS Commercialization

2.2.1 Early Research Phase

While the concepts of patterning and etching thin films are inherent in the development of integrated circuit design, using a silicon wafer as a substrate for a wide variety of mechanical structures started to attract major attention in the early

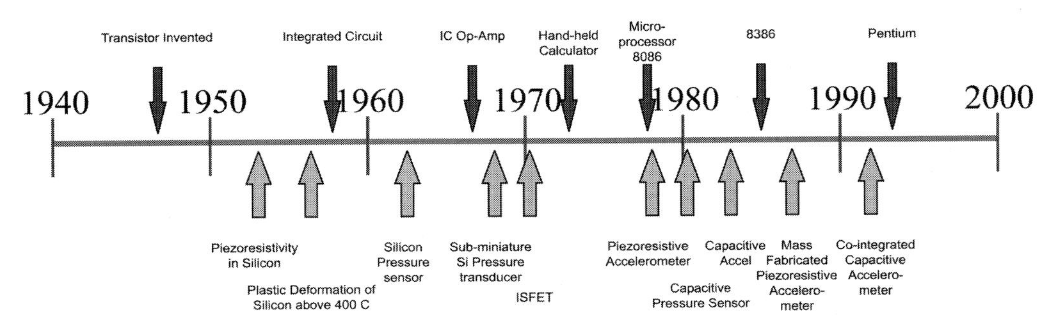

Fig. 2.1 Sensor evolution timeline [1–7]

1970s. One of the first applications of silicon as a mechanical structure used the piezoresistive properties of bulk silicon as a simple strain gauge. Very soon more complex structures were being reported and proposed, including thinned diaphragms with integrated piezoresistors for pressure sensing, suspended mass structures as acceleration sensors, more complex strain gauge structures for medical applications and silicon as a source of miniature integrated plumbing. Proposed applications included implantable drug infusion pumps and miniature gas chromatography systems (Fig. 2.1).

Much of the early development was funded by government agencies such as the National Institute of Health and NASA. As a result, much of the initial industrial interest focused on medical and aerospace applications. At the same time, the automotive industry was motivated by fuel efficiency and environmental considerations to add engine control electronics, which pushed requirements for low-cost pressure sensors (such as manifold absolute pressure (MAP) and fuel vapor control). Industrial pressure sensor manufacturers also started looking for less costly and higher performance alternatives to conventional hand-assembled sensor structures. As a result, the university research activities in MEMS had a diverse and strange assortment of industrial collaborators with an equally diverse set of design criteria and motivations. Small divisions sprouted within semiconductor companies, who quickly found that the ostensibly similar lithography and furnace steps did not fit within traditional IC manufacturing operations. Silicon wet etching caused even more compatibility issues (Fig. 2.2).

2.2.2
Bulk Silicon Pressure Sensor Phase

The next major phase of commercialization may be best referred to as the bulk silicon pressure sensor phase because of the dominance of etched diaphragm pressure sensors in commercial activity. Starting in the late 1970s and early 1980s, pressure sensors were the primary source of business and revenue.

(a)

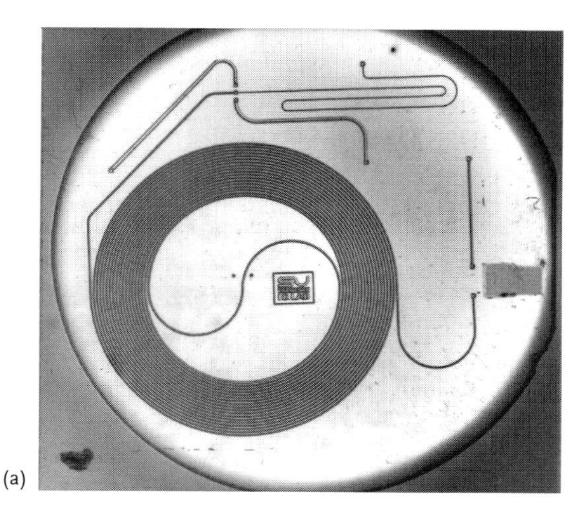

(b)

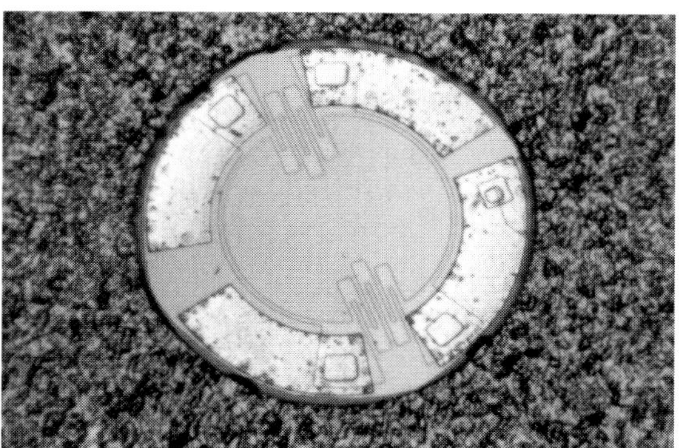

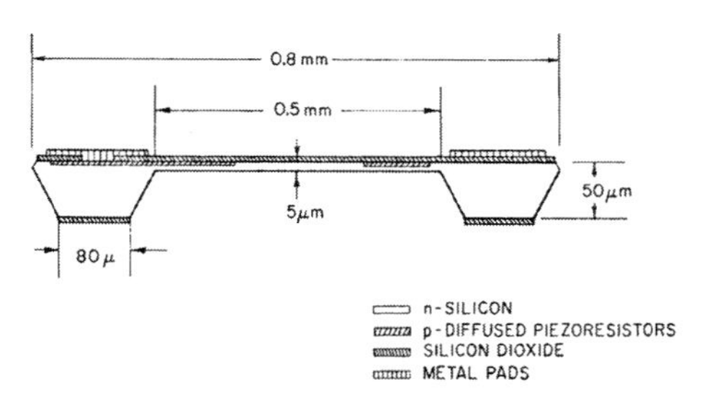

Fig. 2.2 (a) Gas chromatograph system by Steve Terry, Stanford University Micromachining Program. Professor Jim Angell, Micromachining Program Faculty Research Advisor. Photograph courtesy of EG&G IC Sensors, Inc. (b) Differential pressure sensor by Samaun, Stanford University Micromachining Program. Professor Jim Angell, Micromachining Program Faculty Research Advisor. Diagram courtesy of Stanford University. Photograph courtesy Stanford University

Three primary market segments drove this early phase. First, the automotive industry continued to push development of high-volume manufacturing. Second, the medical industry was driven by health and cost concerns from reusable to disposable (single-use) pressure sensors. Third, long-term stability and cost were motivating a shift from traditional hand-assembled strain gauge on steel industrial pressure sensors towards integrated silicon pressure sensors whose manufacturing costs were typically an order of magnitude lower.

The dominant technology was bulk micromachining (devices based on etching through or deeply into a silicon wafer). Manufacturers typically adapted standard semiconductor equipment. Lithography consisted of standard top-side tools, plus a series of self-designed tools for aligning features to the back side of the wafers. Wet etching (typically timed KOH) dominated the etch processes. Many different etch fixture recipes were developed and electrochemical etch stops (ECE) plus plasma etching emerged as a technique for etch control driven by the manufacturing needs of the bulk pressure sensor.

Proposed bulk silicon structures received a lot of press exposure (for example, the June 1984 cover story of *Popular Science* entitled 'The Coming Age of Micromachines,' by Jeanne McDermott, pp. 87–91). The devices were very photogenic and made it easy to imagine a myriad of opportunities. However, the industry was young and the system integration and applications infrastructure were limited. The 'killer applications' that would ramp up quickly to huge markets were also limited and the revenue base grew slowly.

While many other prototype parts were being proposed and developed, much of the commercial activity was driven by a handful of start-ups and university laboratories. These activities were funded by development contracts and alliances with end-user companies that recognized potential product advantages that could be used in the representative industries. In the early to mid-1980s, applications such as early tire-pressure monitoring systems were reported (tied to the initial 'run-flat tires' that had a limited range when deflated), and also several optical switches and micromirrors, early chemical analysis on a chip (electrophoretic separation, flow restrictors and microvalves) and a number of early inertial devices. Sensor bus systems and devices with integral signal processing were also reported, but no substantial market emerged.

The financial and commercial driver remained the pressure sensor, which represented the main revenue source for these early companies.

2.2.3 Accelerometer Phase

During the late 1970s, the first silicon accelerometers were developed (Fig. 2.3). This coincided with automotive safety requirements for airbags and suddenly the advantages of silicon structures to replace earlier purely mechanical accelerometer switches became a driving force in MEMS technology. The size of the potential accelerometer market allowed it to become the dominant factor in MEMS business directions and decisions during this period. This was the 'killer application' that drove the next stage in business, technology and market development.

In the early 1980s, several bulk accelerometer structures were funded and developed. These devices were heavily driven by the aerospace industry where there was a requirement for very small devices in flight applications. A silicon seismic mass was etched into the silicon using bulk silicon (whole wafer thickness) wet etches. The mass was usually suspended on thin silicon springs, also etched into the silicon. Piezoresistive, capacitive and resonant beam structures were developed. Volumes were small and targeted at high performance and cost was a secondary driver.

In the mid-1980s, the first low-cost mass-produced silicon accelerometers were introduced. These were bulk devices that were heavily based on aerospace-funded

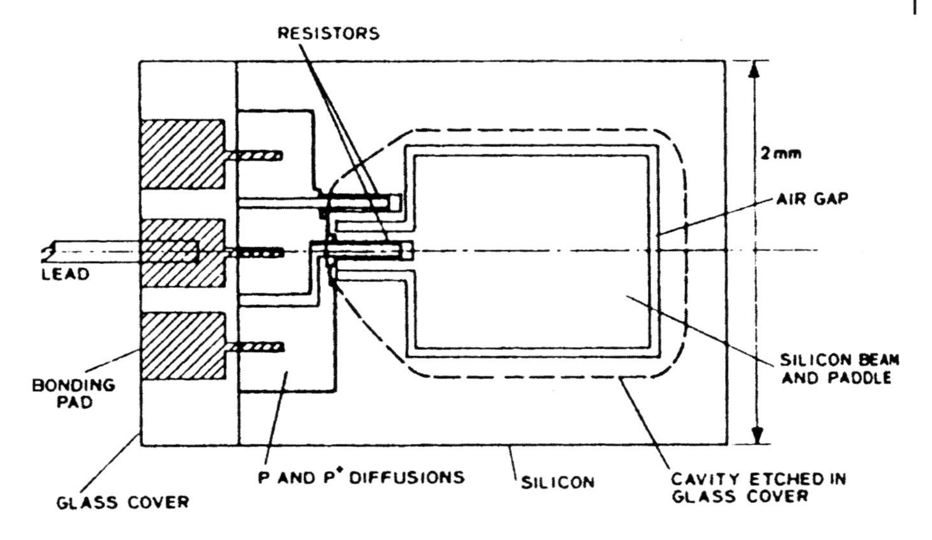

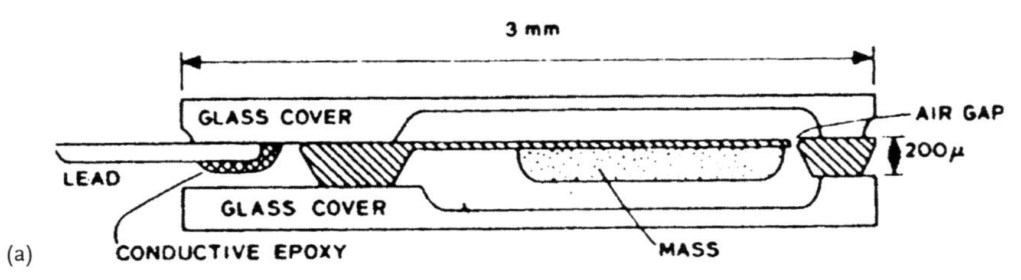

(a)

(b)

Fig. 2.3 (a) Roylance accelerometer – prototype for piezoresistive accelerometers of the 1980s. Accelerometer by Lynn Roylance, Stanford University Micromachining Program. Professor Jim Angell, Micromachining Program Faculty Research Advisor. Photograph courtesy of Stanford University. (b) Roylance accelerometer before cap. Micromachined accelerometer by Lynn Roylance, Stanford University Micromachining Program. Professor Jim Angell, Micromachining Program Faculty Research Advisor. Photograph courtesy Stanford University

developments. Initial versions were targeted at industrial applications to replace more expensive discrete component parts.

At the same time, the automotive industry was responding to mandatory air-bag requirements. Early air-bags were triggered by mechanical accelerometer switches. These were typically made of individually machined and assembled metal mass, spring and contact arrangements. Manufacturing control, cost reliability and accuracy were major limits of the mechanical parts. Silicon accelerometers promised much better performance at lower cost and automotive suppliers suddenly focused on silicon devices as the answer to air-bag needs.

The requirements for air-bag sensors combined with the capabilities of silicon immediately introduced many new concepts, including self-testing for safety, manufacturing test and mass production, and generated a whole new era in quality, reliability and commercial stability requirements for the silicon micromachining industry. New corporate level funding was also generated based on this new high-volume application. During this phase (where the term MEMS also started to emerge), the industry was driven to a new level of infrastructure and maturity to meet the requirements of the automotive safety system markets. During this phase, the automotive industry was experimenting with ride control. This potential market, along with the air-bags, provided the background for substantial automotive industry investment into MEMS. A few new large suppliers emerged.

At the same time, concepts for surface micromachining were emerging. The concept of applying thin films to the surface of a silicon substrate and then removing a sacrificial layer with an undercut etch was particularly attractive to the mass production of small low-cost air-bag sensors. Circuitry was required to process the air-bag signal and combining both on the same chip provided a number of advantages to air-bag suppliers.

As a result, much of the surface micromachining technology development was motivated by the market need for air-bag accelerometers during this phase of the MEMS industry evolution.

The financial and commercial drivers for this phase of MEMS development were centered around requirements for automotive air-bag sensor elements.

2.2.4 System Level Phase

One of the byproducts of the accelerometer phase was a motivation to develop both dedicated MEMS interface circuits (two-chip systems) and also complete systems on a single chip. This was driven by several factors, including the chip set and co-integrated signal and sensor combinations of the air-bag requirements. Many of the suppliers who had evolved technology to match the air-bag requirements realized that system level technology could be applied to other markets. At the same time, several circuit manufacturers observed that the air-bag markets had developed other applications where interface chips that could drive sensor and actuator structures could provide correction and outputs.

Simultaneously, several high-volume, low-cost consumer applications were evolving (such as hand-held tire gauges, sports modules and diving modules) and distributed industrial control.

During this phase in the 1990s, a number of CMOS interfaces – circuits for amplifying, driving, compensating and calibrating MEMS structures – were developed by the integrated circuit industry. Prior to this phase, MEMS manufacturers and users were virtually on their own in developing these circuits as needed to make a complete device. Traditionally this was another barrier to entry for prospective customers and users of MEMS devices.

Dedicated ASICS and general-purpose interfaces and a series of standard packaged MEMS structures opened the way for many industry-specific companies outside the MEMS industry to use standard MEMS components in their products. This infrastructure marked another phase in MEMS growth where the customer could exploit the technology without a need for captive experts.

The financial and commercial drivers for this phase focused on spreading the breadth of MEMS applications into a variety of applications of standard and semicustom parts that individually would not have supported the required technology and infrastructure development.

2.2.5 Optical Switching Phase

In the late 1990s, growth in the requirement for wide-band communications provided the economic driver for another major investment phase into the MEMS industry. Optical switching was identified as another potential 'killer application' of MEMS where potential market size could justify investing large sums.

This was the first phase where infrastructure segmentation became prevalent. A significant amount of funding during this phase went to companies that focused on specific aspects of the technology. MEMS equipment and materials supply, modeling and design software, foundries, fabless component developers and system integrators all emerged. This was in stark contrast to the very early phases, where such infrastructure did not exist and a MEMS supplier was forced to be vertically integrated and self-sufficient in each of these aspects.

While the optical switching market stumbled and a substantial part of the investments were lost, the MEMS landscape was again dramatically modified by the business forces of this phase. Much of this phase resulted in a large investment in technology that never reached the manufacturing phase. Much of the benefit has been adapted by MEMS manufacturers in other industries who have been able to exploit this influx of technology and infrastructure investment for other markets.

2.2.6
Nanostructure and MEMS Infrastructure Phase

The current phase of MEMS commercial development can best be characterized as the nanostructure and infrastructure phase. Nanostructure devices have captured the same imagination as the microstructure parts of many years ago. The current MEMS industry can provide a model to develop a nanostructure industry with a much shorter evolution. The market, economic and industry characteristics between the nanostructure and microstructure industries are very similar. Companies who support the MEMS infrastructure are ideally positioned to support the nanoindustry. During MEMS commercialization, many attempts were made to use the development of IC industry as a model. The significant size, technologies requirements, simultaneous market development requirements, diverse customers and sales channels and vertical integration made this a poor match. The MEMS industry was forced to develop on its own. The nanoindustry has a chance to accelerate its development by relying on MEMS for a fast path to commercialization.

During this phase, the market has grown to the level where segmentation is possible and many industry-specific companies are emerging to address specific markets. Companies now focus on specific markets such as communications, chemical and medical analysis, test and measurement, consumer products or automotive systems without being MEMS experts. They can also develop products and markets for specific components, microrelays, rf devices, acoustics or medical probes and exploit the capabilities of the MEMS industry.

The commercialization during this phase is market-driven by industry application, requiring considerably smaller investments to open new channels, and is backed by the reliability, quality and substantial investments of the MEMS infrastructure which supports many different markets (Fig. 2.4).

2.3
Marketing – MEMS Push Versus Market Pull

2.3.1
Disconnect Between Development and Commercialization

One characteristic of the MEMS industry is that it has been an intriguing technology. It has often been viewed as providing the link between the physical world and signal processing. However, a problem with this situation is that MEMS technology often has been ahead of the market demand. This is the classical situation of a solution looking for a problem and there are numerous examples where MEMS products failed as a result. In many cases, MEMS solutions replaced an alternative technology with no clear advantage to justify the development cost. In other cases, the primary requirements for success were not all present, resulting in a partial solution that was not commercially successful.

(a)

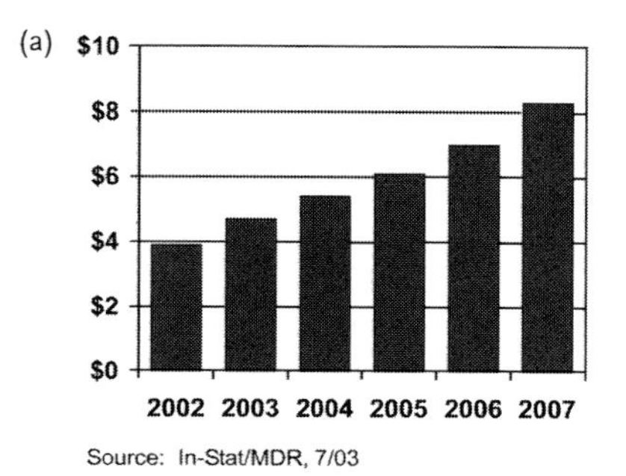

(b)

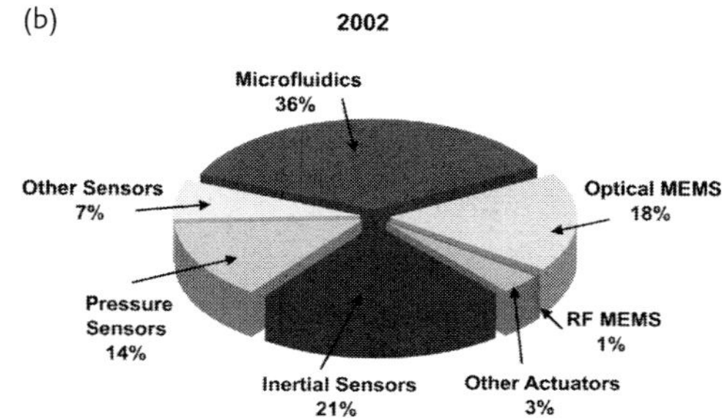

Fig. 2.4 (a) Worldwide revenue forecast for MEMS 2002–2007 (US$ billion) and (b) share of MEMS revenue by device, 2002 vs. 2007

2.3.2
Characteristics of a Successful Application

The characteristics of a successful application include the following:

1. A need that cannot be met with existing technology, such as cost, size performance and manufacturability – Examples abound where a silicon microstructure met a specific need. For example, for disposable medical pressure sensors, MEMS structures provided an order of magnitude reduction in cost. This cost made the application feasible. In the air-bag accelerometer industry, the silicon device was required to meet cost, performance and reliability. However, if the microstructure only provides a comparable solution, the added cost will typically not justify the development. A classical example is in medical flow restrictors and in disposable biosensors, where a molded plastic solution can be much more cost-effective.
2. Collateral system developed – Often the microstructure requires collateral circuitry and system functions to complete the marketable system. For example, air-bag sensors require an entire control system, tire pressure sensors require telemetry and optical devices require the remainder of the controls to be effective.
3. Successful package, calibration scheme, test and measurement – One characteristic of a MEMS sensor or actuator compared with traditional signal processing ICs is that the MEMS device typically must be mounted or located where

it is exposed to the environment. Package limitations and requirements on testing can be a primary obstacle to successful applications. For example, packaging stress on a pressure die can prevent accurate measurements and exposure to corrosive liquids can destroy the sensor.

4. Favorable market timing – Market exists and matches technology ramp up. In many cases the MEMS structure has preceded the market. For example, tire pressure sensors were developed in the early 1980s long before the market demand evolved. The optical switching phase is another classical example, where the market existed long before the MEMS structures were developed and the market was saturated with products manufactured with competing technologies long before the MEMS structures were available. An additional timing dilemma exists when the application is enabled by the MEMS part. In these cases, market development will not evolve until the device is developed and then a long market development time results in an extended period of no revenue for the MEMS company.
5. Advantageous market size – MEMS structures can enable a wide variety of applications. Although the MEMS structure is critical to the application, the wafer volume may not support an entire wafer facility. Historically, MEMS devices have been designed with a wide range of special processes. This lack of standardization has limited foundry effectiveness, often resulting in underutilized and inefficient captive wafer fabs. The vertical specialty supplier with a captive fab lacks the throughput for process control and base load to absorb factory costs.

A successful application requires all of these items to be met to provide a winning business model that can recover the development costs and offer a revenue stream to sustain the development period (Tab. 2.1).

In addition, the MEMS system integrator must also be positioned to be a specialist in the application. Examples of successful applications typically have a solid MEMS developer and supplier teamed with a system integrator (such as an automotive component supplier, HVAC specialist or medical device company).

2.3.3 Evolutionary Versus Revolutionary Products

The MEMS industry has examples of both evolutionary and revolutionary product designs. The pressure industry has been very evolutionary. The primary character-

Tab. 2.1 Requirements for a successful MEMS application

A need that cannot be met with existing technology	✓
Collateral system already developed: circuitry and system functions	✓
Planning for package, calibration scheme, test and measurement	✓
Existing market and match of technology ramp-up time	✓
Market size understanding to ensure wafer volume can support business	✓

istic for suppliers is a steady source of revenue to fund continued development of new products. At the opposite extreme, the digital light mirror of Texas Instruments is an example of a revolutionary approach where there was virtually no revenue for over 10 years while the technology, system aspects and market were developed.

2.3.4
Evolution of Pressure Sensor Markets and Milestones

As an example of an evolutionary product area, the MEMS pressure industry developed around non-silicon predecessors in a series of new parts. The driving advantages of silicon were low-cost through wafer level fabrication (typically an order of magnitude), small size and manufacturability. In the 1970s, a typical non-silicon pressure sensor was fabricated from strain gauges bonded to a metal diaphragm and a typical price was in the $100 range. As a result, applications were limited to permanent sensors and volumes were relatively small.

Initial silicon pressure sensor elements had a very similar structure, with a pressure diaphragm that was photolithographically defined and etched into the silicon wafer. In place of bonded strain gauges, strain-sensitive elements were diffused into the silicon diaphragm using IC techniques. The resulting sensors were mounted in an IC package, e.g., a 'TO' (Fig. 2.5). This reduced the price to about $30 in the early 1980s and opened up new industrial and medical markets.

At the same time, ceramic hybrid substrates with laser-trimmed thick-film resistors for calibration and compensation were developed. The resulting automotive modules started in the $10 range and fell to around $5. The automotive pressure sensor was the largest single market driver in the MEMS pressure business and rapidly grew to several million parts per year. For much of this period, the pri-

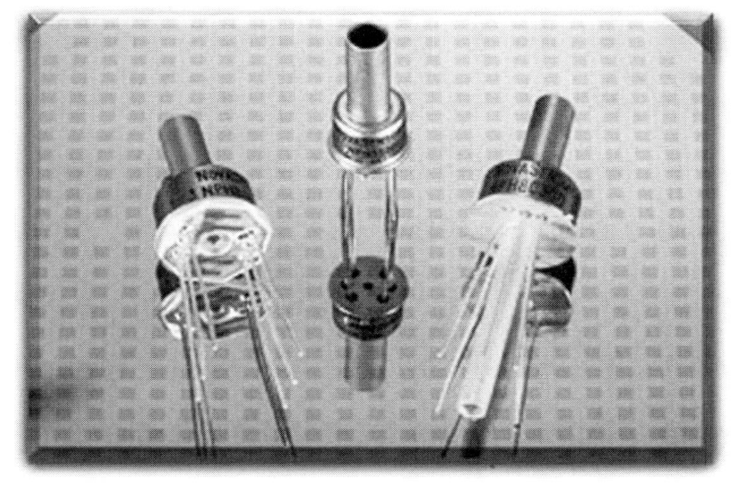

Fig. 2.5 NovaSensor TO-8 IC package. Photograph courtesy of NovaSensor Inc.

mary manufacturers were captive wafer fabs within vertically integrated automotive companies. As a result, the impact on the stand-alone MEMS industry was minimal.

Medical disposable modules also emerged during this period, enabled by silicon. Prior to the mid-1980s, medical pressure sensors typically cost about $80–100 and were re-used. Blood pressure was monitored by connecting a pressure sensor to intravenous (iv) tubes in hospital patients. Consequently, they could come in contact with blood and any infections could be transmitted to the next patient if the sensor was not properly cleaned and sterilized. In addition, cleaning and sterilizing sensors for re-use were very expensive and often could damage the sensor. Silicon technology provided a path to reduce the sensor price to the $5 range and enabled this application (Fig. 2.6). While not identified in the market surveys of the early 1980's, the medical disposable business was a major revenue source for the early MEMS fabs of the 1980's.

Similar low-cost markets emerged in diverse areas such personal diving depth monitors and hand-held tire pressure gauges, where low cost was critical.

In the mid-1980s, tire manufacturers developed a run-flat tire, which eliminated the need for a spare tire (reducing car costs and saving trunk space). However, a pressure monitor was required to warn the driver if tire pressure was lost, so that the driver would reduce speed and go to a repair center within the limited range of the run-flat tire. Responding to this requirement, R&D departments developed MEMS-based tire pressure monitors in the mid-1980s that transmitted pressure from the moving tire to an on-board display. Several systems were developed, but the run-flat tire did not prove popular and the market for tire-pressure monitoring did not develop over the next 15 years. This was a classical example of a miss on market timing. Periodically there were several additional cycles of renewed inter-

Fig. 2.6 IC Sensors Piezoresistive Airbag Accelerometer. Photograph Courtesy of IC Sensors Division of Measurement Specialties, Inc.

est from automotive manufacturers that rekindled interest in this product. The latest cycle was started with the well-known under-inflation/roll-over problems in the late 1990s and possible government-mandated systems. This may finally revitalize the long dormant market and help it take off.

During this period, pressure sensor evolution continued. Very low-pressure devices with contoured stress-concentrating diaphragms opened markets that were previously served by non-silicon parts. Fully integrated single chip sensors with signal processing have also evolved to reduce system costs further. These devices provide amplification, calibration and compensation on a single chip to reduce handling and programming time. The technology has evolved from laser-trimmed adjustment to EEPROM programming. At the same time, better modeling and MEMS manufacturing technology have allowed further shrinks in die size, with additional cost reductions, opening up additional markets.

The pressure sensor applications of MEMS are a good example of a market that has evolved over a 25-year period in a series of steps and facets that has generate an infrastructure of successful suppliers.

2.3.5 Evolution of Inertial Markets and Milestones

While also evolutionary, the inertial MEMS markets have had more discrete steps. The primary element has been the MEMS accelerometer. As with pressure sensors, mechanical accelerometers preceded the silicon MEMS. These were devices hand-assembled from metal parts and were typically a suspended mass on a suspension with some detection mechanism. The first silicon parts migrated into the industrial parts of the market using a silicon mass and springs to emulate the components of the earlier metal parts. Targeted at the same industrial markets, these parts typically were individually assembled and remained relatively expensive. Many of the early parts were silicon cantilever structures and suffered from off-axis sensitivity, parasitic resonance and low-shock protection.

By the mid-1980s, initial bulk silicon accelerometers were emerging that had more complex suspensions and structures to provide frequency control (damping) and overshock protection. This started an evolution of new lower cost silicon accelerometers into industrial and consumer applications. Prior to the MEMS parts, accelerometer prices were typically over $500. The new silicon parts lowered these to under $100. At the same time, automotive air-bag and suspension requirements drove MEMS manufacturing and packaging technology such that the price point came in under $10 by the early 1990s.

Surface micromachined accelerometers represented another discrete step in accelerometer evolution. Combining amplifier circuitry on the same chip as the suspended mass further simplified package and allowed further product evolution. This allowed prices for air-bag sensors to move to well under $5 (Fig. 2.7).

While the air-bag sensor was a major driver, silicon accelerometers have evolved into other markets including consumer applications, additional low-cost industrial monitoring and joy-sticks and other subsidiary markets. Other developments in-

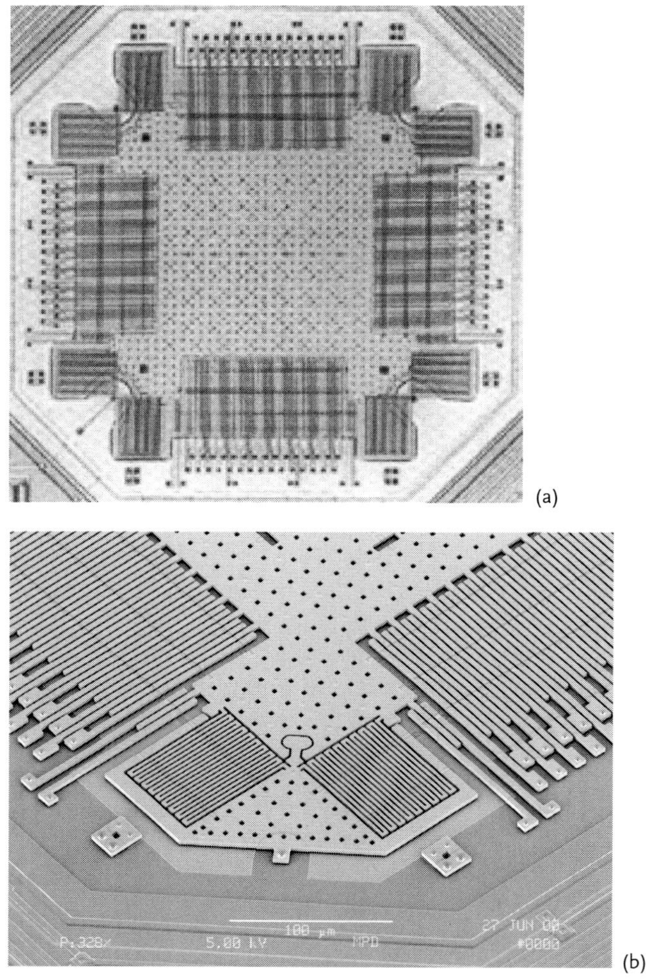

Fig. 2.7 Analog Devices Inc. accelerometer: (a) top view and (b) scanning electron micrograph of structure. Photographs courtesy of Analog Devices Inc.

clude multi-axis sensing and evolution into silicon gyroscopes in the current phase of product development (Fig. 2.8).

Hence the inertial MEMS markets of suspended mass structures has also evolved into a self-sustaining industry over a period of many years.

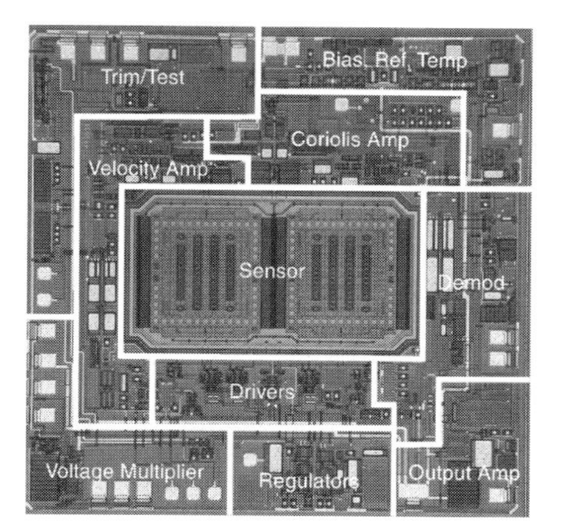

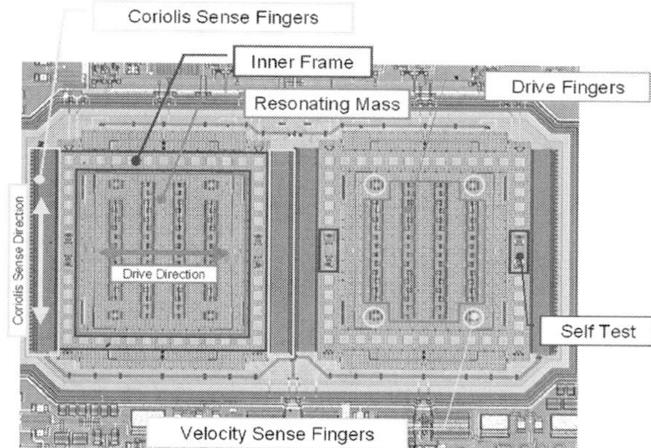

Fig. 2.8 Analog Devices Inc. gyroscope die. Photographs courtesy of Analog Devices Inc.

2.3.6
Other Markets and the Drivers

Several other products that developed in the 1980s have not yet generated the level of sustainability as the pressure and inertial parts. Mass flow sensors were reported in the mid-1980s, but packaging problems and no clear advantage over traditional technologies prevented commercial success. Many microvalve designs have also been reported. Again packaging and valve performance showed no clear advantage to justify investment in continued development. Gas analysis, biochemical, microrelays, pressure and accelerometer switches and several other devices also fall into the category where fabrication using MEMS is technically feasible. However, there was insufficient clear-cut competitive advantage to justify investment in additional commercial development.

2.3.7
The Optical MEMS Era Discontinuity

Deflectable optical mirrors were reported in the early 1980s, including arrays of single hinged parts. Several applications were proposed, including displays and light steering. Unlike the pressure and accelerometer markets, there were no substantial equivalent parts made with other technologies to serve as initial applications. The net result was that, with exception of the DLD display applications by Texas Instruments (Fig. 2.9), very little investment was made to generate products. Similarly, in the mid 1980s, Fabry-Perot structures suitable for optical filters were also reported using MEMS. Again, no immediate market emerged as the driver to sustain development.

By the late 1990s, wideband communications requirements to service the emerging growth of the Internet pointed toward the need to expand optical fiber communications networks. Existing mechanical switches were expensive and

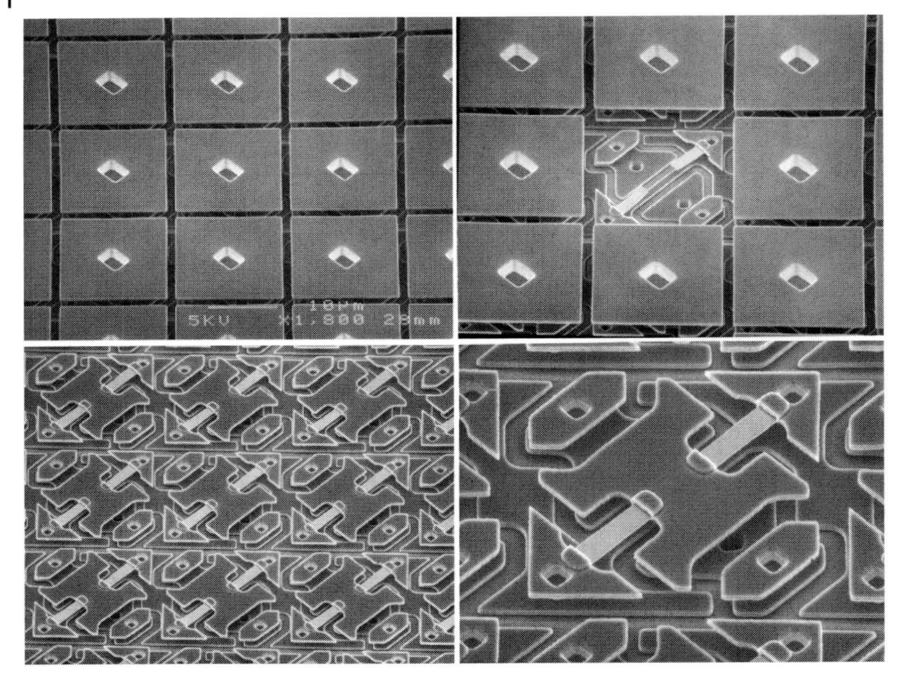

Fig. 2.9 Texas Instruments DLD devices. The top-left view shows nine mirrors. In the top-right view, one central mirror was removed to expose the underlying hidden-hinge structure. The bottom-right view shows a close-up of the mirror substucture. The mirror post, which connects to the mirror, sits directly on the center of this underlying surface. The bottom-left view shows several pixels with the mirror removed. Photographs courtesy of Texas Instruments

large. Huge markets were forecast needing optical switches and MEMS devices were seen as the best commercial solution and a wide variety of approaches were proposed by an equally large number of companies.

The subsequent economic slowdown, high competition and substantial overbuilding with non-MEMS parts resulted in a sudden reduction and delay in the market. There were no sustained existing sockets with traditional technologies and demand for traditional parts also slowed. The net result was a limited market and oversupply of companies. This led to shut-downs and very long delays where products have been put on hold. Several devices continue to be developed with more limited markets. The huge markets that were originally forecast are at best several years out. Commercialization was not successful because of a faulty market analysis.

The notable exception in the optical arena is the digital light device of Texas Instruments. After many years of development, a commercial product has been developed. The timing coincides very nicely with current demand for high-performance displays. For many years, the DLD existed as an investment in future markets and a drain on profits.

2.3.8 The Mobile Handset Market

By the end of the 1990s, cell phones had become a pervasive device with a huge market. A major cost of traditional cell phones is focused on the mechanical assembly and mechanically assembled components such as the microphone, speaker and coils. MEMS structures have been reported that can provide a competitive advantage. This market is promising because, as with the successful pressure and accelerometer markets, there is an existing market to replace and the MEMS part has clear cost and performance advantages.

2.3.9 The Aerospace and Government Factor

Aerospace and government influences have also had a direct hand in shaping MEMS product directions. Unlike other IC technologies at the time, early MEMS markets did not quickly generate a high sustained level of revenue such that product development could be completely funded by the industry. Government support worldwide has been a major influence on MEMS product directions. In the early phases, government support shaped the device development directions. In the USA, medical device research and NASA requirements provided the seed for many of the initial devices. Subsequent development was strongly supported by defense requirements, which drove the funding directions.

In Europe and Japan, similar government funded programs and research laboratories proved a focal point for MEMS research. This was a sharp contrast to other IC technologies, where the revenue stream is solid and the businesses have expanded to the level that development can be funded internally.

One of the early influences on product direction has also been driven by aerospace requirements. In particular, much of the early accelerometer work was funded by aerospace contracts to provide very specific accelerometer devices. Similarly, much of the current silicon gyroscope development has been funded in much the same way. Although the initial funding is usually for a very specific and high set of specifications, typically demand in other high-volume areas dictates modifications to reach the lower cost markets. The development of many of the early bulk microstructure elements was funded by high-end aerospace requirements followed by modifications to reduce costs in the commercial applications.

Government initiatives in Europe and Asia have also influenced technology directions both in the emphasis of various technologies and in industrial subsidies for specific devices.

2.4
Funding and Corporate Structures

2.4.1
Captive Early Fabs

During the 1970s, many of the commercial MEMS activities were carried out by special groups or divisions within larger companies. However, it soon became obvious that the infrastructure requirements of the newly emerging technology may have been better suited for a more entrepreneurial environment. Sales channels, customers, wafer processing, packaging and testing and also market size and growth rate were different from those of the other (typically bipolar) circuits that were being sold by these companies.

2.4.2
External Investment, Phase One

The early 1980s saw a wave of new start-ups focused in the pressure area (such as IC Sensors and Sensym), in dedicated focus products (such as Microsensor Technology in the gas chromatography market) and in general MEMS commercialization (such as Transensory Devices, later part of IC Sensors, and Novasensor). These companies were very vertically integrated. At this stage, there were very few support industries. Most equipment, tooling and fixtures were adapted from semiconductor equipment or were internally designed and constructed. Even test systems were primarily internally developed. Very little process standardization or modeling and design tools existed. Packaging was usually accomplished by adapting standard IC housings and assembly techniques and quality systems were also varied and internally generated. Microstructures (as they were typically known) generated a lot of interest with forecasts of very fast growth in an emerging technology. Returns were expected to meet the typical requirements of the VC industry and substantial investments were made.

During this period, the start-ups quickly found that there was not the single part 'killer application'. Instead, even standard parts required a lot of customization to meet each individual application with a very fragmented market. In many cases, the MEMS structure enabled the market, but was a very small percentage of the overall cost of the finished system. So the MEMS parts were leveraged into large markets by a wide variety of customers, but the growth of the MEMS supplier was limited.

Similarly, the vertically integrated systems supplier with captive MEMS fabs found that the fabs were very expensive to maintain and difficult to control. Hence these companies also looked to outsource the MEMS fabrication.

The MEMS dilemma was that while being the key element of new systems, in most cases it was only a small part of the whole system.

2.4.3
Bootstrap and Customer-funded Development

Owing to slower than expected growth and returns, by the late 1980s and early 1990s, it became obvious that the growth rates and profitability required to be attractive to venture investors were not being met. This led to a period of very limited additional venture investment. Many of the early start-ups were acquired by larger companies (allowing early investors to 'cash-out') and growth was through customer-funded development, strategic partnerships or investments by parent companies.

At the same time, MEMS companies were starting to become a little less vertically integrated. New start-ups were focusing on strategic partnerships that would allow the venture to deliver on its area of strength without the full infrastructure capital needs of a full vertical systems company. As a result, companies emerged that were focused on a very specific aspects of the MEMS industry such as structure design, modeling and modeling software or packaging and sales. Investments were kept low and there was a period of steady growth of both revenue and infrastructure. A number of companies consolidated their market positions with solid products. This was particularly true in various pressure, accelerometer and several other special vertical MEMS markets (such as DMD displays).

2.4.4
External Investment, Phase Two (Optical Communications)

By the late 1990s, there had been very little large-scale investment in new ventures when the boom of the communications industry spread to the MEMS area. This was fueled by recognition that MEMS structures had a potential advantage in providing switching and signal processing components for wide-band communications.

The result was both a radical shift in business directions by many of the small companies in addition to a near-frantic set of investments in new MEMS ventures. Much publicity was generated by the money invested on the upturn and also some of the more spectacular failures when the market did not materialize.

During this phase, most of the established MEMS manufacturers remained quietly focused on their core business. However, the increased publicity opened a new awareness of the MEMS industry. One of the more significant results of this attention has been a renewed interest in a MEMS infrastructure and segmented support industries. For established suppliers, this provided next-generation options to be less vertically integrated and, for ventures in new applications, reduced cost barriers to entry. It has become easier to enter specific applications without the heavy investments in wafer fabs and equipment so that several application-specific MEMS system companies have emerged in a variety of industries.

2.5
Commercial Infrastructure

Over the approximately 25 years of MEMS industry evolution, the commercial infrastructure has evolved from adapting and borrowing from the parallel IC industry to a number of infrastructure elements specifically targeted to support the MEMS industry. At the same time, the IC industry infrastructure has also progressed. Some of the evolution has been related while special MEMS requirements have determined other aspects.

2.5.1
Captive Fabs Versus Fabless and Process Standardization

Early MEMS structures had very little standardization between devices. Fabs were usually captive with a single goal of manufacturing the MEMS structure needed with a dedicated process.

The advantages of total flexibility on process and device structure in captive fabs are similar to those of the IC industry. The disadvantage that hindered early MEMS commercialization often resulted from inadequate initial volume and very long development times. During these extended ramp-ups, the captive fab was running way below capacity, with continued cash drain and inadequate volume for process control.

By tightening internal procedures, consolidating processes and designs and running several processes through the same line, current MEMS dedicated foundries have evolved such that a fabless model is a viable approach. The results is a new business model, where a fabless MEMS company can focus on its role to create products and markets for its target industry and the MEMS foundry will supply manufactured parts.

2.5.2
Similarities and Differences with Standard IC Process

Drawing comparisons between the MEMS industry and standard IC business models can be very dangerous. While MEMS processes use silicon and some processes similar to the IC industry, there are a significant number of different characteristics that change the way in which a MEMS foundry must operate:

1. Standard silicon structure – Most circuit or system requirements in a given technology can easily be customized with a photoplate change. Processing occurs following the same process sequence to create the underlying devices. In comparison, a MEMS structure typically uses the underlying structures as part of the mechanical device. As result, standardizing the process sequence at best will compromise the performance and often needs to be modified to build the desired structure. Consequently, process steps may follow a standard guideline or recipe, but dimensions are often modified and the process sequence may also be customized. Several surface processes have been proposed to standard-

ize the sequence, but the typical high-volume device will follow a custom flow. For the manufacturer, this means major differences in areas such as material flow and planning, control charts and statistical process control, recipe control, wafer handling, in-process monitoring and testing.

2. Double-sided processing – Many MEMS processes require processing on both sides of the wafer instead of the single-sided processing of a standard IC. Handling, edge and back-side protection, inspection and test are all technical differences. For a foundry, this means extra planning on automated handlers, often some manual handling and differences in tolerances.
3. Contamination – Another key difference is in contamination, cleaning and surface handling. A standard IC must have no lithography defects or impurity contamination in the electrically active areas. The MEMS foundry must also consider any particle contamination in later steps, surface conditions to prevent sticking, cleaning irregular or undercut surfaces and handling wafers that become fragile as material is removed.
4. Parasitic effects – Many MEMS devices include steps that are problems for standard circuits. MEMS foundries have been forced to develop procedures and work-arounds to minimize these problems. Standard wet etches such as KOH can contaminate semiconductor junctions at high temperatures. Noble metals such as gold are often used in MEMS and can be fatal to signal processing. Similarly, plasma, rf effects and passivation layer requirements are all considerations.
5. Stress – MEMS devices often depend on small mechanically isolated structures. While built-in stress and strain typically has little effect on a circuit IC, it can prevent a MEMS structure from working. Similarly, built-in stress also can impact etch rates and cause a MEMS process to fail.

2.5.3
Packaging and System Integration

Packaging, assembly and test have evolved into a separate industry for the standard IC industry. In most cases, it has evolved to reduce cost and size and also increase pin-count and reliability. A few standards have emerged based on the requirements of several high-volume segments. These typically are adapted across a wide range of products.

For the MEMS industry, it has not been as simple. The MEMS device usually has to interact with the physical environment (for example, to sense, actuate, steer light or control the flow of a gas or liquid). Early MEMS devices usually used modified IC packages and this remains a preferred approach. In addition to leveraging low-cost and materials with the standard ICs, it permits adapting standard handlers to assemble MEMS parts at the same time as other system components. However, standard packages need to be modified to meet this need.

2.5.4
Effect of Calibration and Compensation Performance on Commercial Success

Calibration and compensation have been one of the major drivers on packaging and system integration cost and complexity. Because of the mechanical nature of many MEMS, parasitic stress introduced while packaging may change the performance of the device. As a result, calibration and compensation must often occur after packaging. Approaches have often used laser trimming of thin- or thick-film resistors. The resulting packages have been bulky and relatively expensive. Handling during assembly has also added cost. Other devices have used fused links. More recently, co-integrated MEMS with built-in EEPROM or electrical programming have provided a very attractive alternative.

In many cases, the calibration and compensation cost exceeds the cost of the MEMS device. It can also be the cause of major yield loss and reduced reliability. The overall goal is to accomplish calibration and compensation using low-cost, reliable standard IC assembly techniques. Most current MEMS packaging and assembly have evolved to meet these requirements.

2.5.5
Design and Modeling

Design and modeling requirements for MEMS structures have substantially different requirements to those of the IC industry. The major requirements of the latter are to use design tools to generate complex signal processing circuits using standard cells and to develop the cells at the device physics level. In comparison, the MEMS designer needs to predict the performance of a mechanical device that is often a single device or possibly an array, where each element still operates independently. Hence the design tool depends less on huge arrays and more on the dynamic performance of a single structural element.

One of the greatest challenges of MEMS design tools is to match the wide variety of structures being developed. The second is to determine the mechanical and stress characteristics of a compound structure with different layers of materials on a very small scale. Often parasitic effects and built-in stress can be larger than the primary design. The most effective results have been a series of very general-purpose platforms and models that can be used by MEMS designers to create basic structures. These have emerged in the later stages of the industry evolution. A general rule of the MEMS industry remains the need to link designs back to a matrix of test structures or related designs as a first step to a complex design.

2.5.6
Technology Drivers: Surface, Bulk, DRIE and Co-integration

Along with design and modeling tools, several technology drivers have had a substantial impact on the commercialization process. In addition to changing the design or structure and performance of the devices, these technologies have made a mark on the structure of MEMS companies.

Surface micromachining provided a different angle to making mechanical structures from the early bulk structures. More recently, combination bulk and surface parts promise more flexibility in device design and better performance. The bulk characteristics can be used to increase signal level and structural integrity and surface processes to achieve tight control of small-dimensional structures and compound suspensions or elements.

Deep reactive ion etching (DRIE) has also added another level of complexity and capability to MEMS fabrication. The resulting vertical etch walls and dry processing have been incorporated in bulk, surface and combination parts. Low-stress nitrides, polysilicon and PSG films, along with dry etching, have also enhanced recent MEMS structures.

Finally, co-integrated structures combine circuitry and mechanical devices on the same chip. The advantages of such structures include lower cost, smaller size, reduced assembly complexity and improved performance and reliability.

One impact of these new technologies on the industry has been increased design and modeling complexity and cost. A far greater effect has been a substantial increase in the capital structure of the foundry. This led to an increase in the base load requirement of the foundry for profitability in addition to wafer lot sizes and volume requirements. The nature of these tools also requires more wafers to keep the processes in control. The industry has evolved from single wafer handling to multiple cassettes. Increased requirements of these new technologies on MEMS manufacturing facilities have been the principal factor in forcing companies away from vertical MEMS infrastructures and toward a segmented industry structure.

2.5.7 Quality and Reliability

MEMS quality can be directly linked to the business systems of the suppliers, system integrators and customers. Early companies were heavily driven by researchers and technologists. Processes were not well controlled and companies had few quality systems in place. Increased medical activity in the 1980's drove the first wave of increased quality systems. A second push was required by the automotive industry during MEMS air-bag sensor development in the late 1980s.

Current suppliers in commercial production typically have well-established design and production control. Companies supplying the automotive industry typically follow appropriate standards such as QS9000 and TS16949. In contrast, for many of the start-up MEMS companies focused on device development, the transition to series manufacturing has been a major block. This can be a particular problem for companies suddenly attempting to move from device research into foundry business as a way to absorb equipment capacity.

Whereas a major part of the MEMS industry is following quality guidelines, new product development phases within an established industry can generate highly visible start-ups that may not have as well-established business systems. Standard quality systems can be effectively applied to the MEMS industry as demonstrated by the many parts produced each day without quality issues.

Reliability of MEMS structures can also be addressed using the tools of other industries. The technology overall is not inherently more or less reliable than other technologies. Specific aspects must be analyzed on a very focused basis. For example, one early analysis of silicon pressure sensors showed that the silicon diaphragm was more reliable than previous metal structures. The silicon part did not show the fatigue characteristics of metal. In another example, possible particles in parallel-plate capacitive accelerometers were found to lead to a reliability problem. However, a modification to the structure and process provided a solution.

Both quality and reliability need to be addressed by each MEMS supplier according to the needs and requirements of their market. Although some parts of the industry have lagged in reviewing these issues, there is nothing inherent in the technology to prevent this from occurring.

2.6 Summary and Forecast for the Future

2.6.1 Market

The overall market for applying MEMS structures keeps growing with many new applications. Continued production shows a wide variety of companies solidly commercializing the technology. There is no question that the potential markets keep emerging. The technology continues to both replace alternative devices and generate new markets that are enabled by MEMS costs and performance. Much real growth in the industry has historically resulted from these hidden markets. Many examples of enabled markets such as medical disposable sensors, air-bag sensors or hand-held consumer gauges were not even mentioned in market surveys a few years before the applications emerged.

At the same time, many overly optimistic predictions have over-inflated the expected market size and growth rate. Coupled with a number of misplaced investments, the MEMS industry image has suffered.

2.6.2 Infrastructure

During its over 20-year lifetime, a considerable commercial infrastructure has emerged in the MEMS industry. Contrasting the early vertically integrated companies, the industry has now developed a well-defined structure. Fabless companies can now successfully focus on their charter to create products and markets and foundries exist to supply MEMS products reliably and cost-effectively. Design support, software, equipment suppliers and package test suppliers exist as dedicated entities.

2.7 References

1 C. S. Smith, Piezoresistance effect in germanium and silicon. *Phys. Rev.* **1954**, *94*, 42–49.

2 O. N. Tufte, P. W. Chapman, D. Long, Silicon-diffused-element piezoresistive diaphragms. *J. Appl. Phys.*, **1962**, *33*, 3322–3327.

3 A. C. M. Gieles, Subminiature silicon pressure transducer. *Digest IEEE ISSCC*, Philadelphia, **1969**, 108–109.

4 P. Bergveld, Development of an ion-sensitive solid-state device for neurophysical measurements. *IEEE Trans. Biomed. Eng.*, **1970**, *BME-19*, 70–71.

5 L. M. Roylance, J. B. Angell, A batch-fabricated silicon accelerometer. *IEEE Trans. Electron Devices*, **1979**, *ED-26*, 1911–1917

6 C. S. Sander, J. W. Knutti, J. D. Meindl. A monolithic capacitive pressure sensor with pulse-period output. *IEEE Trans. Electron Devices*, **1980**, *ED-27*, 927–930.

7 Source: IC Sensors, Sensonor

3
Capacitive Interfaces for MEMS

V. P. Petkov, B. E. Boser, University of California, Berkeley, CA, USA

Abstract
Micromachined sensors and actuators have the advantages of low cost, low power and small size, which make them particularly attractive as a replacement for macro-scale devices in a number of consumer applications. Most resolution specifications for microelectromechanical systems translate into stringent requirements for the operation of the electronic interface on very small signals in the presence of large parasitics. Capacitive sensing is often employed in such applications owing to its high resolution combined with low power dissipation, low temperature coefficient and compatibility with most fabrication processes. This chapter presents circuit and system design techniques for capacitive sensing interfaces. Chopper stabilization and correlated double-sampling are introduced as approaches to implement the interface front-end. Discrete-component and monolithic design examples are provided. System design considerations explore open-loop sensing issues and the applicability of force feedback to certain types of sensors. Sigma–delta modulation is introduced as an approach combining intrinsically linear, two-level force feedback with analog-to-digital conversion.

Keywords
capacitive sensing; chopper stabilization; correlated double-sampling; sigma–delta.

Advanced Micro and Nanosystems. Vol. 1
Edited by H. Baltes, O. Brand, G. K. Fedder, C. Hierold, J. Korvink, O. Tabata

ISBN: 3-527-30746-X

3.1 Introduction

Capacitive interfaces find widespread use in a large variety of sensors including accelerometers, gyroscopes, strain and pressure sensors, fluid-level detectors, biosensors and a number of other applications needing low-cost and low-power detection in a small form factor [1]. In applications such as optical mirrors, capacitive interfaces serve for both detection and actuation.

A number of techniques have been developed for signal transduction in interface front-ends, including piezoresistive, capacitive, tunneling and resonant sensing [2]. Capacitive sensing is often chosen in industrial applications owing to its robustness and compatibility with most fabrication processes. Comparison of resolutions achievable by different transduction techniques demonstrates that capacitive sensing can achieve performance comparable to or better than those of the alternative approaches [3]. For example, capacitive displacement detectors with resolution down to 10^{-14} m in a 1 Hz bandwidth have been demonstrated [4]. This figure is comparable to the classical radius of an electron and orders of magnitude smaller than the spacing of atoms in a crystal. The capacitance resolution of such sensors is similarly impressive: as low as 10^{-20} F, many orders of magnitude smaller than the gate capacitance of the transistor detecting the change. Achieving these results depends on many factors in the capacitive interface of the sensor itself and the associated electronics that are the subject of this chapter.

One reason for the popularity of capacitive interfaces is their pervasiveness: any two isolated conductors form a capacitor that may be used in the interface. Unlike piezoresistive, optical or tunneling displacement sensors, capacitive interfaces usually require no special processing and may not even need special features added to the sensor. Because of this, a practical capacitive sensor interface usually includes several capacitors with only few actually participating in the sensing function. Other capacitances, for example to the substrate or shield electrodes, reduce the signal amplitude and may even couple unwanted interference into the sensing element. In sensors with capacitive interfaces that are not carefully designed, parasitics can reduce the resolution by an order of magnitude or more [5].

The possibility of a capacitive interface to act as both a sensor and an actuator constitutes another potential advantage. For example, electrostatic force-feedback can be employed in inertial sensors to broaden the bandwidth and lower the temperature coefficient of the sensor. Mirrors for optical beam steering also often make dual use of the capacitive interface for sensing and actuation. However, this versatility has potential drawbacks in cases where the actuation is undesired. The most common of these problems is so-called electrostatic pull-in: if the voltage across a sense or actuation capacitor is increased beyond a critical value that depends on the geometry and size of the capacitance and the spring constant of the suspension, the capacitor snaps. For many micromachined sensors with capacitive interfaces and nominal gaps in the micron range, pull-in occurs for a few volts, often well within supply voltage limitations [6, 7].

Many sensor applications require functionality that goes beyond sensing, such as calibration, temperature compensation, self-test or even analog-to-digital conversion. To reduce cost, power consumption and size, these functions are often implemented in the electronic sense interfaces [8, 9]. By shifting requirements such as linearity or matching to the electronic interface, the manufacturing requirements for the mechanical elements can be relaxed. These requirements have brought about a variety of interface implementations. As a minimum, an electronic front-end is required in order to convert the information coming from the mechanical element to a voltage. Additional components such as filters, gain stages and demodulators are often included after the front-end for further signal processing. An analog-to-digital converter (ADC) can provide digital output from the sense interface. Depending on the system specifications and the characteristics of the mechanical element, feedback control may also be required.

Capacitive interfaces are employed in a broad range of applications and numerous examples of architectural and circuit solutions can be found in the literature. Humidity and gas sensors [10, 11] detect a measured quantity through variation in the dielectric constant of a micromachined capacitor. Pressure and inertial sensors convert the signal to deflection, which is sensed capacitively by the electronics [4, 9, 12]. Resonators and resonant sensors employ both capacitive sensing and actuation to excite oscillations in a micromachined element and evaluate the oscillation frequency [13, 14]. Capacitive sensing and actuation have also been combined in force-rebalanced feedback systems [9, 15–17]. The principle of capacitive actuation is also suitable for positioning applications such as magnetic disk drive microactuators [18] and micromirrors [19].

Capacitive interfaces are certainly not limited to the above examples. Their robustness, power efficiency and excellent resolution give interface designers the opportunity to target an increasing number of applications emerging with the advancement of micromachined technology.

This chapter concentrates on circuits for capacitive sensing with applications introduced as examples. Section 3.2 reviews the principles of capacitive interfaces and their application to sensing and actuation. Practical circuit implementations are introduced in Section 3.3. Sections 3.4 and 3.5 look at applications of these interfaces to closed-loop sensing and sigma–delta interfaces that combine electrostatic force-feedback with analog-to-digital conversion.

3.2
Capacitive Sensing and Actuation

The value of a capacitor depends on the geometry and the dielectric constant. Both mechanisms are used in capacitive sensors. Many sensors, such as accelerometers, use a capacitive interface to detect a displacement that is proportional to the sensor input. The physical motion changes the spacing between the capacitor electrodes, hence introducing a change in capacitance that can be detected electronically. In other cases, for example humidity or fingerprint sensors, variations of the dielectric constant are responsible for changing the nominal value of the capacitor. The same electronic circuits can be used in both cases.

Fig. 3.1 a shows a possible arrangement of a capacitor for displacement sensing. A movable electrode M is part of or attached to the sensor, for example the proof mass of an accelerometer. Together with two fixed electrodes S+ and S– it forms two capacitors C_{S+} and C_{S-} that are normally balanced. A displacement Δx of the movable electrode results in imbalance ΔC in the sense capacitors. For a parallel-plate configuration and for small displacements $\Delta x << x_0$, the value of ΔC is approximately equal for C_{S+} and C_{S-} (Fig. 3.1 b) and is proportional to Δx:

$$\Delta C = \frac{\partial C_S}{\partial x} \Delta x \approx \frac{C_{S0}}{x_0} \Delta x \tag{1}$$

where C_{S0} is the value of C_{S+} and C_{S-} for zero deflection.

An electronic evaluation circuit detects the imbalance ΔC. Depending on the type of sensor, ΔC can be a low-frequency signal, for example in accelerometers and pressure sensors, or it can be located at high frequency as in gyroscopes and resonators. The frequency content of the signal is one factor which influences the design decision for the sensing scheme used in the electronic interface.

The electrical model in Fig. 3.1 b splits the sense capacitors into a constant part and the signal ΔC. While the constant capacitor C_{S0} is an inherent part of the system, only ΔC is used for sensing. For high sensitivity it is advantageous to maximize the ratio between ΔC and the fixed capacitance at the sensing node. The motivation

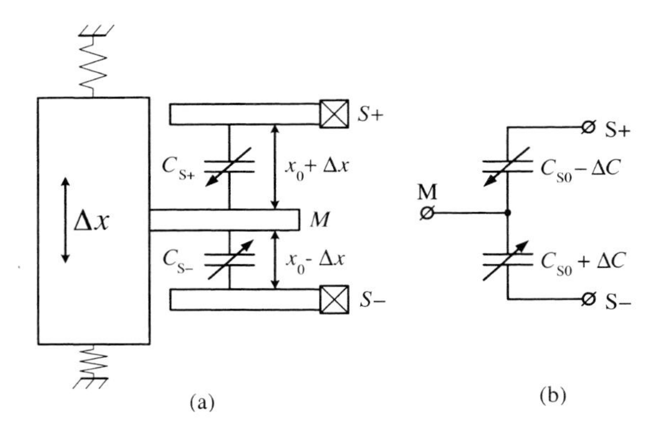

Fig. 3.1 Differential capacitive position sensing. (a) Mechanical configuration; (b) equivalent electrical model

for this choice will become obvious during the noise analysis of the interface. For simplicity, only the desired circuit elements, the sense capacitors, are shown in Fig. 3.1 b. Any actual realization includes additional stray capacitors, for example between the fixed electrodes or to the substrate. Care must be taken when designing the electromechanical interface to keep these strays small, typically comparable or smaller than the sense capacitors, C_{S0}. In addition to parasitic capacitance, the complete model of the micromachined element must include any resistance added by the interconnect lines between the sensor and the electronics. A series resistance of several kilo-ohms can contribute comparable noise to that of the amplifier in the interface front-end. The parasitic resistance also creates undesirable time-constants.

Fig. 3.2 shows the electrical model of a capacitive transducer with the parasitic components taken into account. It illustrates the case of a stand-alone mechanical element, which uses an electronic interface implemented on a separate chip. Reference [20] gives an example of such a device. Two-chip solutions bring increased parasitic capacitance due to the need for bonding pads and longer interconnects. The value C_p can exceed significantly the nominal sense capacitance C_S.

In an integrated MEMS process the mechanical element and the electronic interface are implemented on the same die, which allows for short interconnect wires and therefore lower parasitics. This however comes at the expense of smaller transducer size and increased process complexity. Reference [3] describes an integrated accelerometer and its electrical model. The value of C_p in this case is comparable to the sense capacitance.

The sensing configuration in Fig. 3.1 is differential. Simpler arrangements use only one of the two stationary electrodes. In this case the single sense capacitor is compared with some reference capacitor. This demands good matching between the sense and reference capacitors to avoid systematic offset. In general, single-ended arrangements are prone to additional errors, such as a poor temperature coefficient.

In addition to sensing, capacitive interfaces are often used for electrostatic actuation. Electrostatic force is generated when voltage is applied between the plates of a micromachined air-gap capacitor. The schematic in Fig. 3.3a shows the capacitor connected to a voltage source through an interconnect resistance R_p. If the plates are allowed to move relative to one another, capacitance becomes a function

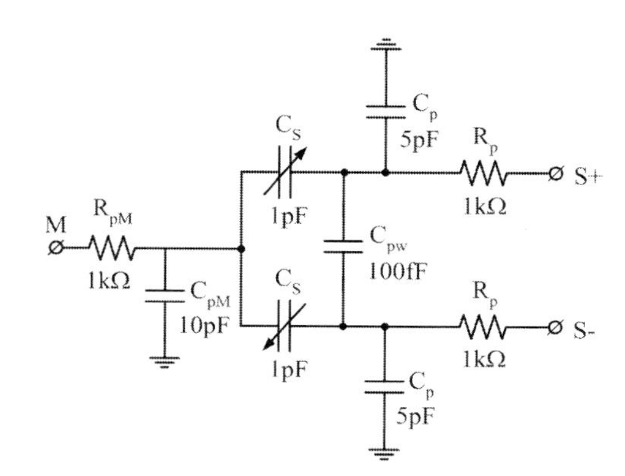

Fig. 3.2 Complete electrical model of a capacitive transducer

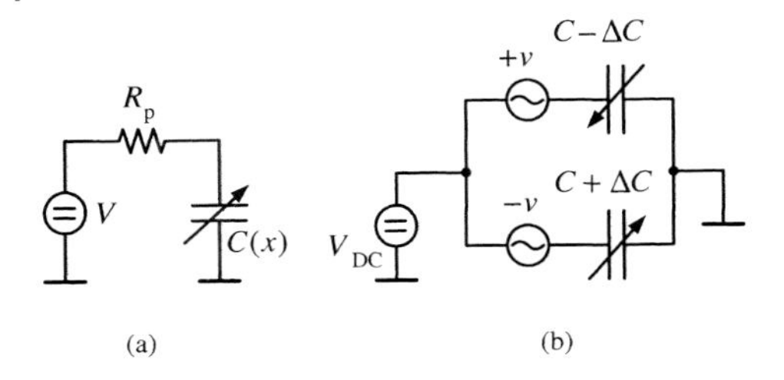

Fig. 3.3 Electrostatic force generation. (a) Single-ended; (b) differential

of displacement. A change in capacitance causes charge redistribution between the capacitor plates and the voltage source leading to a net change in the potential energy of the system:

$$\delta E = \frac{\delta C(x)\,V^2}{2} - \delta q V = -\frac{\delta C(x)\,V^2}{2} \tag{2}$$

where the first term represents the change in potential energy stored on the capacitor. The second term gives the energy due to flow of charge $\delta q = \delta C(x)\,V$ through the voltage source. The electrostatic force is calculated based on Equation (2):

$$F_{\mathrm{el}} = -\frac{\partial E}{\partial x} = \frac{\partial C(x)}{\partial x}\frac{V^2}{2} \tag{3}$$

Equation (3) can be extended to the case when C is a function of more than one coordinate.

An important property of electrostatic force is its proportionality to the square of the applied voltage. This quadratic dependence is inconvenient for linear actuation since an electronic circuit, which accurately computes the square root of the actuation voltage, is difficult to implement. Differential actuation is an alternative technique allowing electrostatic force to be linearized. The principle is illustrated in Fig. 3.3 b. The schematic corresponds to the differential, parallel-plate configuration from Fig. 3.1. V_{DC} is a large DC bias and v is a variable actuation voltage. The net force on the mobile electrode of the structure is equal to the difference between the forces applied by the two stationary combs. For small deflections the quantity $\partial C(x)/\partial x$ is approximately equal on both sides and the differential force is

$$F_{\mathrm{el\_dif}} \approx 2\frac{\partial C(x)}{\partial x} V_{\mathrm{DC}} v \tag{4}$$

Equation (4) shows a linear relation between force and actuation voltage.

Systems with large displacements can use a differential, lateral-comb configuration, in which the capacitor plates move parallel to one another and ΔC is obtained through a change in capacitor area [5]. Lateral combs have a constant $\partial C(x)/\partial x$ over a large displacement range. They also have higher pull-in voltage than parallel-plate structures. The penalty for using lateral combs is their lower sensitivity compared with parallel-plate designs.

So far we have introduced a mechanism relating displacement to capacitance. The next section will describe circuit techniques for measuring the capacitance and transforming it to a voltage signal.

3.3 Capacitance-to-Voltage Converters

The most common approach for measuring capacitance is analogous to an ohmmeter and realized by applying an AC voltage excitation to the capacitor and detecting the resulting current. A charge integrator or a voltage buffer can be used to convert the signal to voltage [5]. Charge integration is usually preferred in capacitive front-ends owing to its lower susceptibility to parasitics. Fig. 3.4 shows the simplified diagram of a differential capacitance-to-voltage (C/V) converter based on charge integration. An AC waveform V_M is applied to the center electrode M causing charge from the sense capacitors of the mechanical element to be transferred and integrated on the feedback capacitors (C_I) of the amplifier. A variation of $\pm\Delta C$ in the sense capacitors will result in a differential output voltage:

$$V_0 = \frac{2\Delta C}{C_I} V_M \tag{5}$$

Note that the fully differential amplifier operates with a large input common-mode signal. If the common-mode range of the amplifier cannot accommodate the voltage step caused by V_M, input common-mode feedback (ICMFB) or a feed-forward path must be added [9].

One of the challenges in capacitive sensor interfaces is the small signal level. Capacitive interfaces are not only capable of detecting minute displacements at the sub-atomic level or very small capacitance changes; the majority of applications in fact require such sensitivity. Electronic amplification readily can boost these signals to manageable amplitudes, but in doing so it also amplifies noise and interference. The fact that many sensors have low-frequency signal compounds this problem. At low frequency, all physical systems suffer from an elevated noise floor owing to so-called flicker noise. Because of its density that is ap-

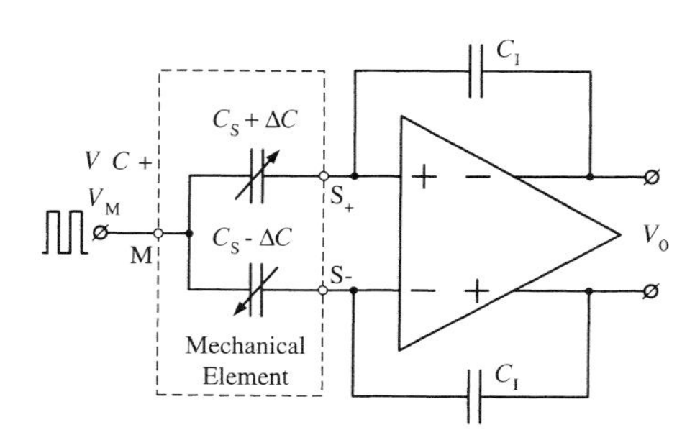

Fig. 3.4 Differential capacitance-to-voltage conversion

proximately inversely proportional to frequency, flicker noise is often referred to as $1/f$ noise. A good measure of the level of flicker noise of a particular device is its noise corner – the frequency above which broadband thermal noise dominates. For MOS transistors this frequency is usually above 100 kHz and can be as high as several MHz, depending on device geometry and bias, and consequently has the potential of significantly degrading the performance of a sensor operating at lower frequency.

Other types of transistors, most importantly JFETs, have significantly lower flicker noise. Unfortunately, because of limited availability of processes with JFETs, these devices are used mostly in discrete implementations. Most integrated solutions use CMOS devices only and employ circuit solutions to mitigate flicker noise problems. Two alternative techniques, chopper stabilization [21] and correlated double-sampling (CDS) [9, 17], are used for this purpose. Both achieve comparable performance and therefore system and component constraints are usually the factors determining the choice of one solution over the other.

Chopper stabilization is a continuous-time technique, suitable for systems with analog output. It provides intrinsically high resolution because signal modulation does not introduce noise aliasing [22]. The use of chopper stabilization is particularly convenient in printed circuit board implementations since all the necessary components are readily available. Chopper-stabilized amplifiers must be followed by a filter, which may introduce extra delay and also add to the power consumption and especially area of an integrated solution.

Correlated double-sampling (CDS) is a discrete-time approach, appropriate for systems in which a central clock is available. An advantage of CDS is that its output can be digitized directly by an analog-to-digital converter without the need for additional pre-filtering. As a sampled-data technique, however, correlated double-sampling inevitably suffers from noise folding. Section 3.3.2.2 discusses this in detail.

3.3.1 C/V Conversion Using Chopper Stabilization

3.3.1.1 Design and Implementation

Chopper stabilization is a technique which reduces the effect of amplifier offset and low-frequency noise. Fig. 3.5 a shows the principle. The voltage V_M is sinusoidal with frequency f_M. A mismatch ΔC results in a current difference Δi_x flowing into the integrating capacitors C_I, which in turn produces an output voltage at frequency f_M with amplitude proportional to ΔC. If the modulation frequency f_M is chosen higher than the $1/f$ noise corner frequency of the amplifier, flicker noise will not corrupt the modulated signal. After amplification, the signal is demodulated back to DC. Low-frequency noise, on the other hand, is shifted to a higher frequency where it can be removed with a low-pass filter. This process is akin to the modulation in an AM transmitter and receiver. Fig. 3.5 b shows the spectra of the signal in the circuit, where the input ΔC is assumed to occupy low frequencies, as is typical for many sensors such as accelerometers and pressure transduc-

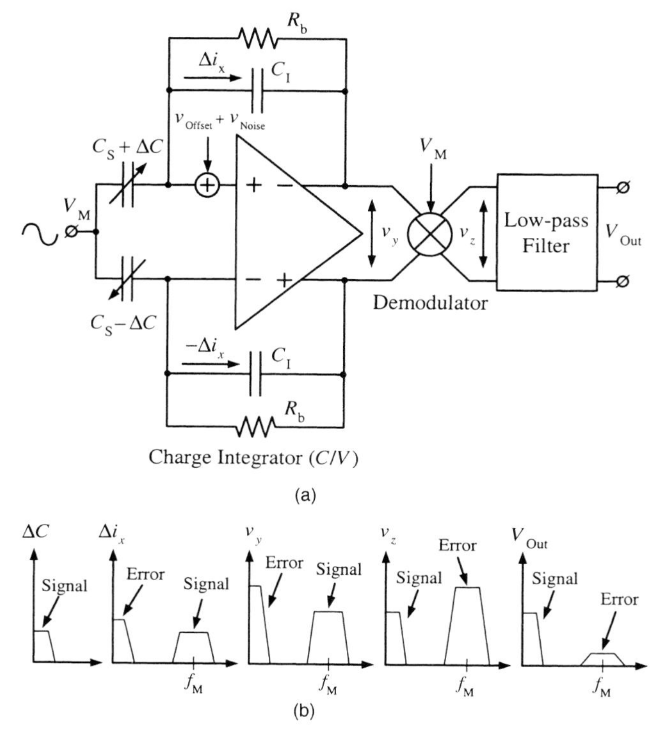

Fig. 3.5 Chopper stabilization. (a) Schematic diagram; (b) signal spectrum

ers. In many practical implementations a square wave is used for modulation. In this case, some of the signal power appears at multiples of f_M.

We will describe the design of the chopper-stabilized interface starting with noise analysis of the capacitive front-end. Fig. 3.6 shows an equivalent half-circuit noise model of the C/V converter. Several components have been added to the conceptual schematic in Fig. 3.4. R_b is a biasing resistor. R_P, C_{PS} and C_{PA} represent the interconnect resistance and parasitic capacitance of the mechanical structure and the capacitance at the amplifier input accordingly. We will also use the notation $C_P = C_{PS} + C_{PA}$ for the lumped parasitic capacitance. The block *preamp* stands for all gain stages following the charge integrator.

The main components contributing noise in the first stage are R_P, R_b and the amplifier itself. Noise from the first preamplifier stage should also be taken into account. The noise contributions in Fig. 3.6 are indicated as follows: $\overline{v_{nRp}^2} = 4k_BTR_P\Delta f$ is the noise voltage of the interconnect resistance, $\overline{i_{nb}^2} = (4k_BT/R_b)\Delta f$ is the noise current of the biasing resistor, v_{na}^2 is the input-referred noise voltage of the charge integrator amplifier and $v_{n\_pre}^2$ is the input referred noise voltage of the preamplifier. The components attributed to the micromachined element are indicated in the figure. In the noise expressions T is the absolute temperature and

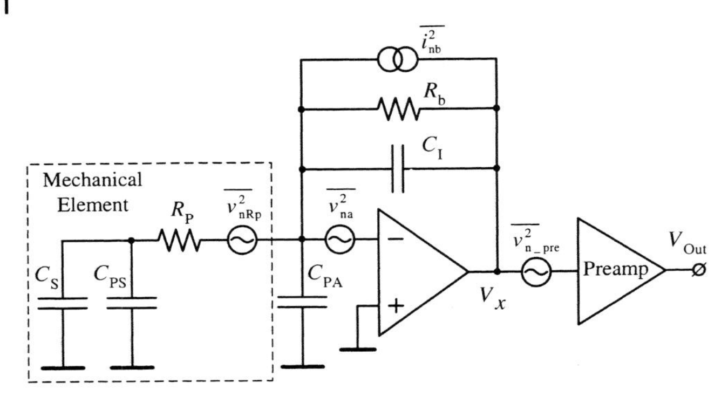

Fig. 3.6 *C/V* converter noise model

k_B is the Boltzmann constant. The term k_BT is approximately 4×10^{-21} J at room temperature.

To analyze and compare the noise contributions we will first refer them to the same node V_X. Since the amplifier of the charge integrator is the core component of the front-end its noise contribution will then be taken as a base for comparison between the different noise sources.

The amplifier of the charge integrator introduces noise, which depends on its equivalent noise resistance R_{na} and on the feedback factor of the stage:

$$\frac{\overline{v^2_{xna}}}{\Delta f} = \left(\frac{C_T}{C_I}\right)^2 \overline{v^2_{na}} = \left(\frac{1}{F_{CV}}\right)^2 4k_B T R_{na} \tag{6}$$

where $C_T = C_S + C_I + C_P$ is the total capacitance at the input node and F_{CV} is the feedback factor. Equation (6) shows that a small value of C_T results in lower output-referred noise. This implies low parasitic capacitance and maximized ratio between the signal ΔC and the constant part of the sense capacitance C_S. A large value of C_I also reduces the noise, but it also reduces proportionally the signal. Therefore, C_I should in fact be kept small since it is also a term in C_T. Reducing C_I comes as a tradeoff with amplifier speed. The closed-loop bandwidth of the charge integrator is:

$$f_{-3\,\mathrm{dB}} = F_{CV} f_u \tag{7}$$

where f_u is the unity-gain frequency of the amplifier. Equation (7) shows that lower C_I and lower F_{CV} accordingly need to be compensated with higher f_u and therefore higher power dissipation. $f_{-3\,\mathrm{dB}}$ is typically chosen several times higher than the modulation frequency since the gain of the chopper stabilized front-end is sensitive to delay introduced by its amplifiers [22]. The application of the overall system may pose additional constraints for the phase-lag of the front-end.

The term C_P in the expression for C_T combines the parasitic capacitance of the mechanical transducer C_{PS} and the input capacitance C_{PA} of the amplifier. While

C_{PS} should be minimized under any circumstances, C_{PA} is indirectly related to the input-referred noise of the amplifier. Increasing the size of the input transistors results in higher transconductance and lower input-referred noise, but it also increases C_{PA}. An exact analysis shows that an optimum value of $C_{PA} = C_S + C_{PS} + C_I$ exists [23]. This optimum may not be achievable for sensors with large parasitic capacitance, but it should serve as a design guideline.

Equation (6) assumes that there is no flicker noise contribution, which is true in the frequency region above the $1/f$ corner of the amplifier. This condition sets a criterion for the choice of the modulation frequency – the entire signal band should be modulated above the flicker noise corner.

A second significant contributor to the interface noise is the interconnect resistance R_P. The noise at V_X due to this component is:

$$\frac{\overline{v_{xnR}^2}}{\Delta f} = \left(\frac{C_S + C_{PS}}{C_I}\right)^2 4k_B T R_P \tag{8}$$

Equation (8) agrees with Equation (6) in the requirement for a low value of C_{PS} and maximum ΔC to C_S ratio. The value of C_I was already constrained by the requirement for maximum signal and by the closed-loop bandwidth of the charge integrator. R_p should always be minimized.

We find the noise from the biasing resistor R_b using its equivalent noise current. The noise current causes a frequency-dependent voltage drop across the feedback capacitor C_I. Since the signal bandwidth is located at the modulating frequency f_M, we calculate the noise contribution from R_b at f_M using the equation

$$\frac{\overline{v_{xnRb}^2}}{\Delta f} = \left(\frac{1}{2\pi f_M C_I}\right)^2 \frac{4k_B T}{R_b} \tag{9}$$

As we will show, R_b should be typically in the megaohm range in order to avoid a significant penalty from the noise in Equation (9). Megaohm resistors are readily available in discrete-component designs. In monolithic implementation, however, they require prohibitive chip area. Therefore, a variety of circuits with MOS transistors are used to simulate large-resistor behavior. A detailed design example of one such circuit can be found elsewhere [4].

Finally, the input-referred noise of the preamplifier is:

$$\frac{\overline{v_{n\_pre}^2}}{\Delta f} = 4k_B T R_{npre} \tag{10}$$

where R_{npre} is the equivalent noise resistance.

So far we have introduced the individual noise sources. Now, using equations (6), (8)–(10), we can express the total noise at V_X in the form:

$$\frac{\overline{v_{xn}^2}}{\Delta F} = \left(\frac{1}{F_{CV}}\right)^2 4k_B T R_{na}\left[1 + \left(\frac{C_S + C_{PS}}{C_T}\right)^2 \frac{R_P}{R_{na}} + \left(\frac{1}{2\pi f_M C_T}\right)^2 \frac{1}{R_{na}}\frac{1}{R_b} + F_{CV}^2 \frac{R_{npre}}{R_{na}}\right] \tag{11}$$

We further simplify Equation (11) by combining some of its terms and introducing them as noise factors:

$$\frac{\overline{v_{xn}^2}}{\Delta f} = \left(\frac{1}{F_{CV}}\right)^2 4k_B T R_{na}\left[1 + N_P \frac{R_P}{R_{na}} + N_b \frac{1}{R_b} + N_{pre}\frac{R_{npre}}{R_{na}}\right] \tag{12}$$

Equation (12) represents the interface noise by a main term due to the amplifier of the charge integrator and noise overhead due to parasitic resistance, biasing resistance and the preamplifier. With typical values for the transducer parameters we can compute the noise factors and compare the amplifier noise contribution with each of the overhead terms. As an example we will consider a capacitive transducer with $C_S = 1$ pF and $C_{PS} = 5$ pF and a charge integrator with $R_{na} = 5$ kΩ, $C_{PA} = 1$ pF and $C_I = 2$ pF. We will also use a modulating frequency of 200 kHz. For the above values we find that $N_P = 0.45$, which shows that an interconnect resistance of 10 kΩ will contribute approximately the same noise as the main amplifier. We also find that this is true for the noise from R_b if the resistor is 1.5 MΩ. For the chosen parameters N_{pre} is smaller than 0.1, which makes the noise requirement for the preamplifier less stringent. Its equivalent noise resistance can be as large as 50 kΩ. For transducers with low parasitic capacitance, in which the charge integrator has large feedback factor, the noise from the preamplifier may not be negligible.

So far we performed the noise analysis using the equivalent half-circuit of the interface. A fully differential configuration would double the total noise in Equation (12). The signal power, however, is four times larger in this case, which results in a 3 dB improvement in the resolution.

We can find the capacitance resolution of a fully differential interface from the total voltage noise using Equations (12) and (5):

$$\frac{\overline{\Delta C_n^2}}{\Delta f} = \frac{C_I}{4V_M^2}\left(2\frac{\overline{v_{xn}^2}}{\Delta f}\right) \tag{13}$$

The factor of two in front of the noise term and the factor of four in the denominator are due to the differential configuration. Equation (13) introduces the amplitude of the modulating waveform V_M as an additional design parameter. The maximum amplitude of V_M is limited by the supply rails or by the electrostatic pull-in voltage of the mechanical element.

This concludes the noise analysis of the chopper stabilized interface. Typically the signal after the charge integrator is in the microvolt range. The preamplifier is designed to provide additional gain before demodulation is performed. In this way the noise from the demodulator and the following stages does not degrade the resolution.

The demodulator can be either a linear multiplier or a chopper. The latter solution is preferable in monolithic implementations since non-linear mixing is easier to realize in a CMOS technology. Linear multipliers are readily available as discrete components and are therefore convenient in printed circuit board implementations. Circuit-level design of mixers and multipliers is covered elsewhere [24–26].

The low-pass filter, which follows the demodulator, attenuates the upconverted error components (Fig. 3.5 b). The order of the filter depends on the ratio between the modulating frequency and the bandwidth. A higher order filter will be needed if this ratio is small. Circuit-level design of continuous-time analog filters has been reported [27].

Chopper-stabilized front-ends are particularly suitable for prototyping with discrete components since all parts for the implementation are available off-the-shelf. Below we present the design of a printed circuit board for evaluation of inertial sensor elements.

A lateral micromachined accelerometer is used as an example. Tab. 3.1 lists the parameters of the sensor element. Our design goal will be to implement an interface which contributes comparable or lower noise than the intrinsic noise of the sensor element. When referred to capacitance, the intrinsic mechanical noise of the accelerometer is $0.058\ \mathrm{aF}/\sqrt{\mathrm{Hz}}$ [23, 28].

The circuit topology of the interface is based on the architecture in Fig. 3.5. Most amplifiers available as discrete components have a single-ended output. Therefore, a pseudo-differential architecture is appropriate for the charge integrator (Fig. 3.7). The first stage requires low-noise amplifiers with low input current. The use of dual amplifiers (two in a package) ensures better matching between the two channels and a dual power supply eliminates the need for analog ground reference on the board. Several commercially available parts which satisfy the above requirements have been described [29–31]. In this example the charge integrator is implemented with OPA2107 [30]. Tab. 3.2 lists some of the amplifier parameters.

Tab. 3.1 Parameters of a micromachined accelerometer

Parameter	*Value*	*Units*
Resonant frequency (f_{res})	5	kHz
Mass (m)	5	μg
Nominal gap (x_0)	2.5	μm
Quality factor (Q)	5	–
Sense capacitance (C_s)	1	pF
Parasitic capacitance (C_p)	5	pF
Parasitic resistance (R_p)	1	kΩ

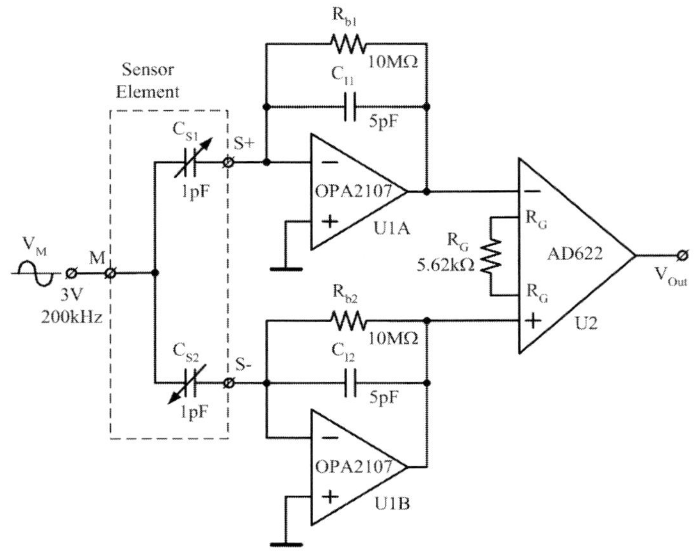

Fig. 3.7 Schematic of a discrete-component charge integrator

Tab. 3.2 Amplifier parameters

Parameter	***OPA2107***	***AD622***	***Units***
Input-referred noise voltage (v_{na})	8	12	$nV/\sqrt{Hz}$
Flicker noise corner (f_{co})	10	<1	kHz
Input bias current (I_{ib})	10	2000	pA
Input capacitance:			
Differential (C_{id})	2	2	pF
Common-mode (C_{icm})	4	2	pF
Gain bandwidth product (GBW)	4.5	–	MHz
Closed-loop bandwidth	–	0.8 ($G=10$)	MHz
Supply voltage	±15	±15	V
Channels	2	1	–

The prototyping board is designed for characterizing the resolution and obtaining the transfer function of the sensor element. To accommodate the entire frequency region of interest, the frequency of the modulating waveform is chosen to be 200 kHz.

The choice of V_M depends on the pull-in voltage of the mechanical element. Assuming a differential comb structure the critical voltage is calculated from [23]

$$V_{pi} = \sqrt{\frac{1}{2}\frac{x_0^2}{C_s} m\omega_{res}^2} \tag{14}$$

where $\omega_{res} = 2\pi f_{res}$. V_{pi} is 3.9 V for this design. Therefore with a certain margin the amplitude of V_M can be 3 V.

Next, the value of the feedback capacitors C_I is chosen to be 5 pF, which results in a feedback factor of 0.29 for the charge integrator. The closed-loop amplifier bandwidth calculated from Equation (7) is 1.32 MHz. $f_{-3\,dB}$ is therefore several times higher than the modulating frequency f_M as required. A faster amplifier would allow for a smaller value of C_I, which is required by Equation (13). We should note, however, that in a discrete-component implementation, board parasitics increasingly affect capacitor matching in the sub-picofarad range. Also, the matching of the discrete capacitors per se deteriorates as their value decreases. Therefore, 5 pF is chosen as a tradeoff.

The preamplifier is built as a differential to single-ended converter. The stage is implemented with an instrumentation amplifier AD622 [32] (U2 in Fig. 3.7) and provides a gain of 10. The bandwidth of the amplifier for this gain is 800 kHz. After the amplification the signal path proceeds as single-ended.

The noise performance of the interface is estimated from Equation (12) and the values in Tabs. 3.1 and 3.2 and Fig. 3.7. The input-referred noise of AD622 is specified in the datasheet for closed-loop operation, hence it is used directly as $2v_{n\_pre}^2/\Delta f$.

The total noise of the interface is 42 nV/$\sqrt{\text{Hz}}$. The dominant contribution is 38 nV/$\sqrt{\text{Hz}}$ from the amplifiers of the charge integrator, hence the use of low-noise parts in this stage is critical. The interconnect resistors, biasing resistors and the preamplifier contribute 3, 5 and 10% of the total noise, respectively.

Equation (13) gives 0.035 aF/$\sqrt{\text{Hz}}$ for the capacitance resolution of the interface. This value is lower than the intrinsic noise floor of the sensor.

Demodulation is the last operation required in the front-end. The demodulator is implemented with the analog multiplier AD734 [33] (Fig. 3.8). Since the AD734 has output-referred noise of 1 μV/$\sqrt{\text{Hz}}$, additional low-noise amplification is needed after the differential-to-single-ended converter in order to reduce the effect of demodulator noise on the interface resolution. OPA627 [34] is a low-noise amplifier, suitable for implementing the additional gain.

If the output is evaluated with a spectrum or a network analyzer, further filtering is not necessary. Otherwise, a filter with sharp cutoff at the signal bandwidth is needed. High-order analog filters are available as discrete components [35]. Their design will not be covered in this example.

Several practical considerations are given below. Symmetry between the two channels is a primary concern in the layout of the charge integrator. A mismatch in the signal paths can lead to gain error, offset and lower rejection of common-mode disturbances.

The parasitic capacitance of the lines which connect the mechanical element to the charge integrator is added to C_p. The tested element can be mounted directly on the board or in a package. In both cases the interconnect lines should have minimal length.

An RC filter at each power-supply pin of the amplifiers is recommended. Typical values can be 10 Ω for the resistor and 10 μF for the capacitor.

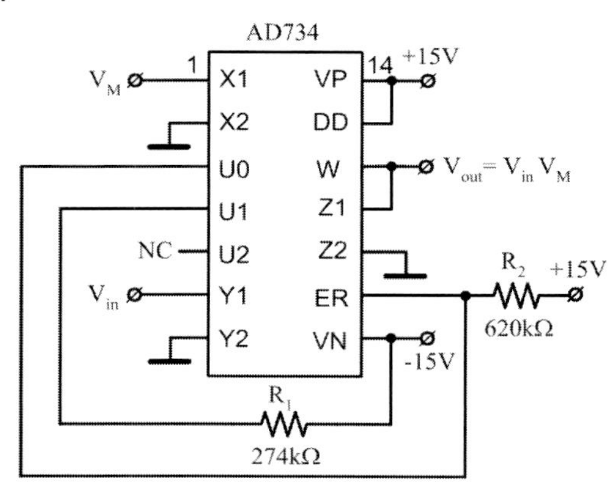

Fig. 3.8 Schematic of discrete-component demodulator

Some issues become obvious during testing. An option to bypass the mechanical element with a pair of discrete capacitors is very helpful in the initial process of assembling and testing the electronics. Sometimes the fact that the board needs to be mounted is overlooked. Ground pins and probe points are also very useful.

Fig. 3.9 shows a photograph of a printed circuit board which has been used for the evaluation of several micromachined gyroscopes and accelerometers. It contains the stages described in this example and also some additional electronics necessary for the operation of the gyroscope sensors. The board has proven to be a valuable tool for characterizing the frequency response of mechanical structures

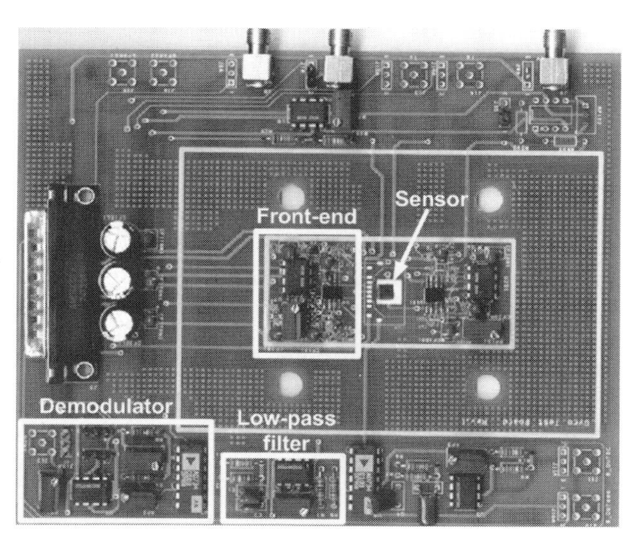

Fig. 3.9 Printed circuit board for evaluation of inertial sensors

using the electromechanical amplitude modulation technique presented in Section 3.3.1.2.

Continuous-time, capacitive front-ends have also found their place in a number of monolithic devices. Reported gyroscopes [36, 37] and an accelerometer [38] use single-ended interfaces. In the single-ended implementation the charge integrator is connected to the middle electrode of the structure (the sensor proof-mass), while the carrier is applied to the sense capacitors as two out-of-phase waveforms (Fig. 3.10a). In this case the amplifier input is at virtual ground and input common-mode feedback is not needed. The large parasitic capacitance of the proof-mass, however, lowers the feedback factor of the amplifier and degrades its speed and noise performance. Also, the single-ended implementation is more susceptible to common-mode disturbances. The reported accelerometer [21] uses a differential chopper-stabilized interface, which essentially duplicates the structure from the single-ended design and provides two isolated sensing nodes with a pair of sense capacitors connected to each one (Fig. 3.10b). Although this approach still suffers from the large parasitic capacitance of the proof-mass, it has the advantages of a differential circuit and does not require ICMFB. The solution in Fig. 3.10b requires careful design of the mechanical element in order to prevent differential disturbance from appearing across the two center nodes.

In systems whose signal has no DC component, sensing can be realized by applying a large constant bias V_{DC} to the sense capacitors and detecting the flow of charge $\Delta Q = (\Delta C) V_{DC}$ due to the change of capacitance ΔC. This is an example of continuous-time interface, which does not use chopper stabilization. Micromachined resonators [13] and gyroscopes [4] have been built based on this technique. The resonator [13] and the rate detection interface of the gyroscope [4] are implemented using charge integration. An alternative front-end architecture is the trans-resistance amplifier, which can be obtained from the circuit in Fig. 3.4 by replacing the integrating capacitors with large resistors (typically several hundred kilo-ohms). Trans-resistance front-ends have been employed in resonators and gyroscope drive circuits [4, 36, 39]. They provide the required signal phase for oscillation directly, without the need for additional 90° phase shift in the following

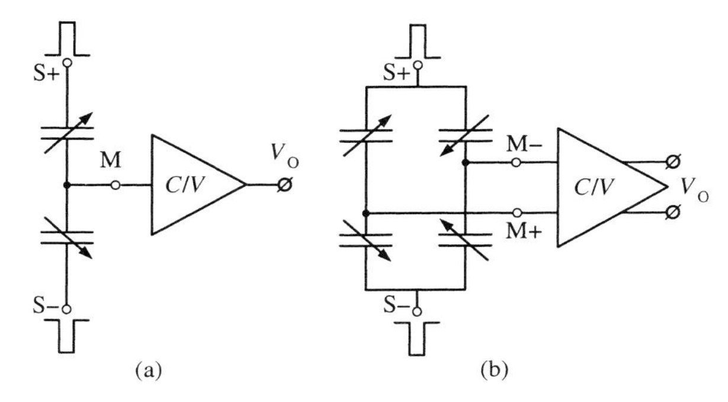

Fig. 3.10 Capacitive sensing from the center node of the mechanical element. (a) Single-ended; (b) differential

stages. Charge integrating interfaces however are preferable in high-resolution applications owing to their superior noise performance [13].

A disadvantage of continuous-time interfaces, which do not use chopper stabilization, is that the signal is not separated in frequency from flicker noise and also from feedthrough in the case of actuation (see Section 3.3.1.2). On the other hand, the need for generating a carrier signal with low phase noise is avoided.

In summary, continuous-time capacitive sensing has been used in both monolithic and discrete-component interfaces. Since continuous-time interfaces do not experience noise folding, they potentially have a resolution advantage over sampled implementations for a given power consumption [21]. The resolution and power consumption tradeoffs will be discussed further in Section 3.3.2.2. One disadvantage of the continuous-time implementation is the need for additional analog filtering and a dedicated analog-to-digital converter (ADC) in systems with digital output. A technique known as *sigma–delta modulation* will be introduced in Section 3.5 as an approach which incorporates the ADC in the sense interface.

3.3.1.2 Electromechanical Amplitude Modulation

Many applications of electromechanical systems combine capacitive sensing and electrostatic actuation. The relation between the sensed and the actuating signal in the micromachined element depends on the mechanical properties of the device and is a function of frequency. We will refer to this relation as frequency response. In this section we describe a method for implementing systems with sensing and actuation. It is generally applicable to force-feedback systems, which are the topic of Sections 3.4 and 3.5. Here the concept will be illustrated through the design of a testbench for obtaining the frequency response of micromachined capacitive transducers.

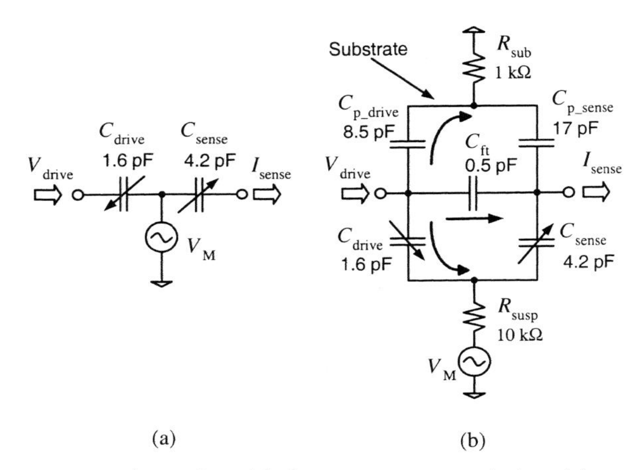

Fig. 3.11 Electrical model of a micromirror. (a) Ideal model; (b) model with parasitic components. Courtesy of B. Cagdaser, UC Berkeley

Fig. 3.11 a shows the simplified model of a micromirror. V_{drive} actuates the device and C_{sense} is used to detect the motion. To first order, the actuating and the sensed signals are related only through the motion of the mirror, hence the overall frequency response of the system will coincide with that of the mechanical element. The interface design in this case is straightforward. In practice, however, capacitive feedthrough paths from the point of actuation to the sensing node can cause V_{drive} to couple directly into the sense interface. Several coupling paths are easily identified in the complete model of the mirror, shown in Fig. 3.11 b. C_{ft} causes the dominant feedthrough contribution. The finite resistances of the substrate (R_{sub}) and the suspension (R_{susp}) lead to additional coupling paths through the main capacitors and the parasitics C_{p_sense} and C_{p_drive}. Capacitive feedthrough increases with frequency while the magnitude of the mechanical response decreases. Therefore, at high frequencies the detected response will be dominated entirely by the feedthrough components. This is illustrated in Fig. 3.12, which shows the measured frequency response of the micromirror with large feedthrough present in one case and significantly attenuated in the other. The difference between the two transfer functions can have a considerable impact on the overall system behavior, especially in cases when the micromirror is used in a wideband feedback control loop. Below we introduce several circuit techniques which address this problem by attenuating and separating the feedthrough from the signal.

To obtain the frequency response of a micromachined device we can actuate (drive) the mechanical structure electrostatically at various frequencies and detect its motion with a capacitive interface. The interface can separate the signal from the feedthrough either in time or in the frequency domain. If chopper stabilization is used in the front-end, the signal is modulated at the carrier frequency. On the other hand, feedthrough paths couple the driving signal directly into the charge integrator. If the carrier frequency is chosen to be higher than twice the

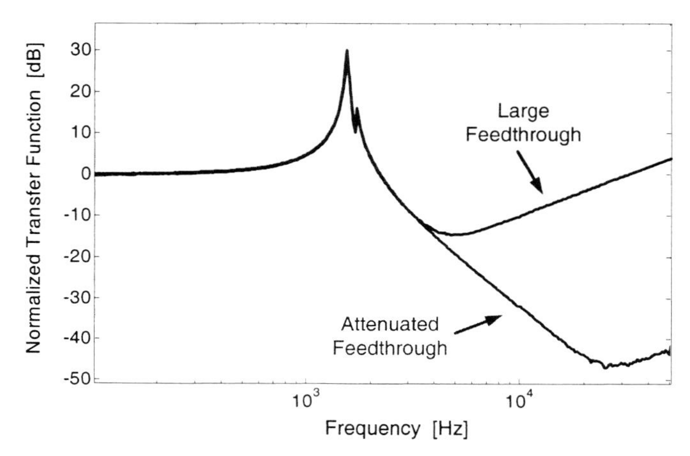

Fig. 3.12 Measured frequency response of a micromirror. Courtesy of B. Cagdaser, UC Berkeley

frequency range of the driving waveform, frequency separation is inherently present in the chopper-stabilized interface. This effect is known as *electromechanical amplitude modulation* (EAM) [39].

A system which uses EAM for frequency response measurement is shown in Fig. 3.13. A single-ended case is presented for simplicity. The generalized model of the mechanical element includes a sense capacitor C_S, drive capacitor C_D and feedthrough capacitors C_{PD} and C_{PC}, which couple the drive voltage v_d and the carrier v_m to the sense interface. The mechanical structure is actuated by v_d, obtained from the source terminal of a network analyzer. The motion is detected capacitively through C_S. After bandpass filtering (BPF) and demodulation the network analyzer extracts the signal component at the drive frequency f_d and obtains the frequency response of the tested element.

The voltage at the output of the charge integrator is

$$V_X = v_d(f_d)\frac{C_{PD}}{C_I} + v_m(f_m)\frac{(C_{S0} + C_{PC})}{C_I} + v_m(f_m)\frac{\Delta C(f_d)}{C_I} \tag{15}$$

where C_{S0} is the sense capacitance for zero deflection. The first term in Equation (15) represents the feedthrough from the drive node. The second term contains the component at the carrier frequency due to the feedthrough capacitance C_{PC} and the constant part of C_S. The last term contains the signal $\Delta C(f_d)$, modulated by v_m.

The voltage $V_{DC} + v_d$ on the drive capacitor applies electrostatic force to the mechanical structure. V_{DC} is a large DC bias on which the small sinusoidal signal v_d is superimposed. Electrostatic force is proportional to the square of the applied voltage [40] and therefore has frequency components at DC, f_d and $2f_d$ (see Section 3.2). The actuation causes deflection, which results in capacitance change $\Delta C(f_d)$ with the same frequency content as the applied force. $\Delta C(f_d)$ is the component in Equation (15) which is a function of the frequency characteristics of the micromachined element.

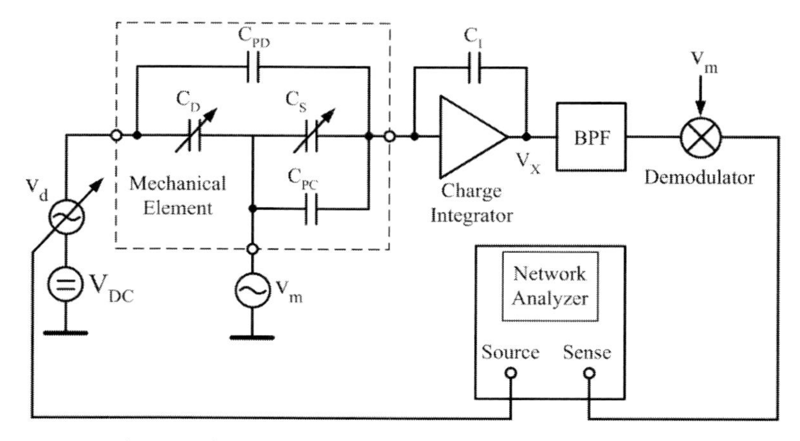

Fig. 3.13 Electromechanical amplitude modulation (EAM)

The overall frequency spectrum of V_X is shown in Fig. 3.14. The signal terms at f_d and $2f_d$ are recovered through demodulation, which also up-converts the feedthrough. The network analyzer evaluates the signal at f_d and filters all other components.

An ideal demodulator preserves the frequency separation between signal and feedthrough and does not require pre-filtering. Offset in the multiplier, however, will cause a fraction of the feedthrough to remain at f_d. Nonlinearity can also corrupt the signal band with interferers at other frequencies. To overcome the impact of nonidealities, a bandpass filter with a center frequency at the carrier (f_m) is used to attenuate all out-of-band components before demodulation is carried out. In practice, even a simple high-pass RC filter can successfully attenuate the feedthrough. For better rejection a higher order bandpass filter is recommended. An example of a particular circuit implementation has been presented [41]. Choosing a higher ratio between f_m and f_d further improves the attenuation of the feedthrough component.

With single-ended sensing and actuation the first two terms in Equation (15) can be many orders of magnitude higher than the signal. A differential configuration is advantageous from this point of view since waveforms with opposite phases are applied to the differential capacitors. This prevents most of the feedthrough charge from flowing into the charge integrator. In vertically moving structures or other devices where differential drive and sense are not an option, matching reference capacitors should be used to implement a pseudo-differential design. The reference capacitors can be a fixed replica of the moving structures. A substantial part of the residual feedthrough in Fig. 3.12 has been canceled using a reference capacitor.

The system in Fig. 3.13 measures the response of mechanical elements which have separate ports for driving and sensing. EAM can also be implemented for single-port devices [39]. Extending the concept to differential drive and sense is also straightforward based on the circuit board example from the previous section. It is helpful if the circuit board layout allows different configurations to be assembled.

Obtaining accurate information about the frequency behavior of a micromachined element is particularly important if the device needs to be incorporated in

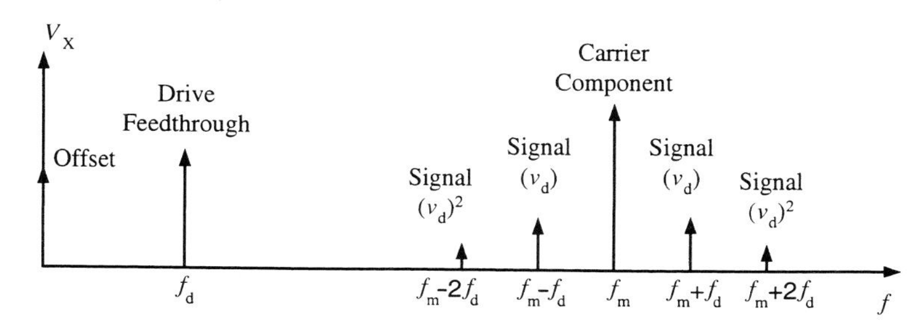

Fig. 3.14 Output spectrum of the charge integrator for EAM

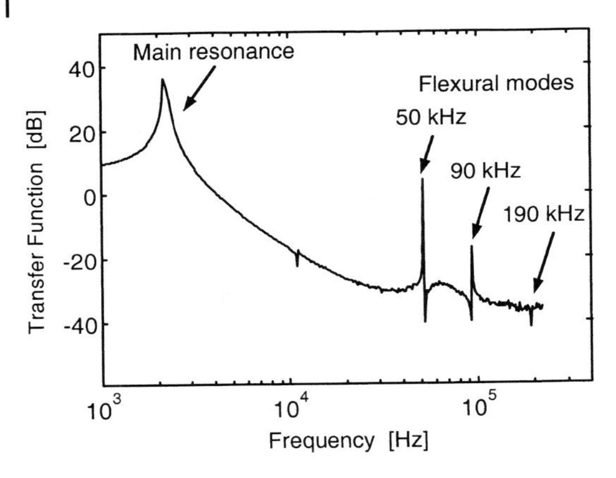

Fig. 3.15 Frequency response of a micromachined gyroscope measured with EAM

a feedback loop. The measured amplitude response of a micromachined gyroscope in Fig. 3.15 illustrates some of the effects which can be characterized using EAM. The frequency and quality factor of the main resonance can be determined as a function of temperature, pressure, electrostatic tuning and other factors. Also in the particular device several high-frequency flexural modes are observed in addition to the main resonance. The frequency and quality factor of such modes will affect the stability of a feedback system and have to be taken into account in the design procedure at an early stage [42, 43].

3.3.2
C/V Conversion Using Correlated Double-sampling (CDS)

3.3.2.1 Design and Implementation

Whereas chopper stabilization separates the error from the signal in frequency, correlated double-sampling is a technique which performs the separation in the time domain. The basic idea in CDS is to subtract two samples of the input, one of which contains both signal and error, while the second sample contains only the error. Thus, if the error has strongly correlated values in the two samples, it is cancelled or highly attenuated by the subtraction. The signal appears unchanged at the output. The condition for strong correlation between the two error samples suggests that only error components whose frequencies are significantly lower than the sampling frequency f_S are effectively cancelled by this technique. Therefore, in analogy with chopper stabilization, CDS is used to eliminate amplifier offset and to attenuate $1/f$ noise.

In the context of capacitive sensing, we will introduce the concept of correlated double-sampling with the differential charge integrator shown in Fig. 3.16 [9]. The operation is divided in three phases (Fig. 3.17). In Phase 1 all nodes, except the center node (M) of the mechanical element, are reset to analog ground (AGND). A voltage V_1 relative to AGND is applied to node M and the sense capacitors are

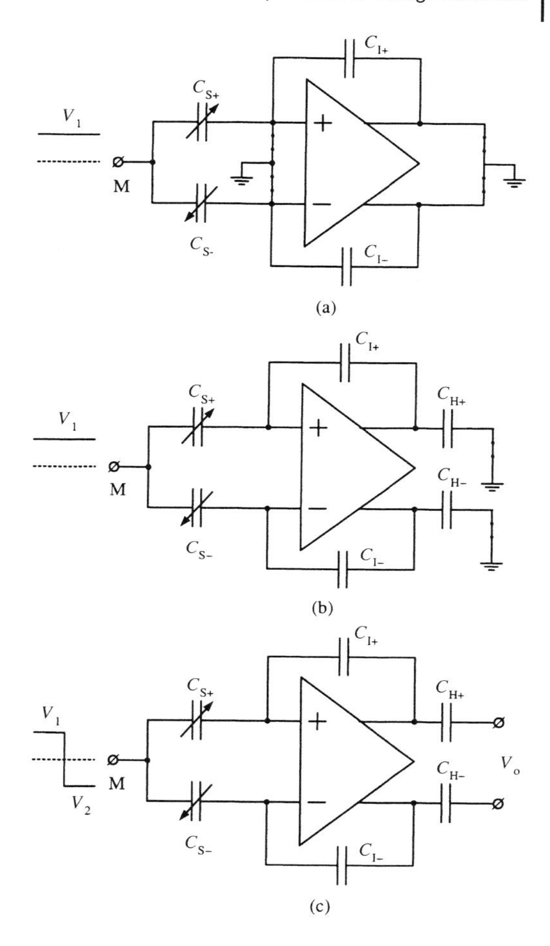

Fig. 3.16 Principle of correlated double-sampling (CDS). (a) Phase 1 – reset; (b) phase 2 – noise cancellation; (c) phase 3 – sensing

charged to the difference between V_1 and AGND. In Phase 2 the amplifier is active and error components including offset, $1/f$ noise, switch kT/C noise and charge injection are stored on capacitors C_{H-} and C_{H+}. At the beginning of Phase 3 a pulse of magnitude $V_M = V_1 - V_2$ is applied to node M, bringing its potential to a new value V_2 relative to AGND. C_{H-} and C_{H+}, which hold the error from Phase 2, are now connected in series with the signal path. Their value is subtracted from the output of the amplifier, which at the end of Phase 3 contains the error and the signal due to ΔC. In this way, the slowly varying error components are cancelled by the subtraction. The output voltage V_0 contains the signal and the kT/C noise, which has been sampled on C_{H-} and C_{H+} at the end of Phase 2.

We notice the similarity between Figs. 3.16c and 3.4. Indeed, for a constant ΔC the signal transfer function of the CDS stage is given by Equation (5). For signals other than DC, however, ΔC varies between Phase 2 and Phase 3 and the signal in V_0 at the end of Phase 3 becomes

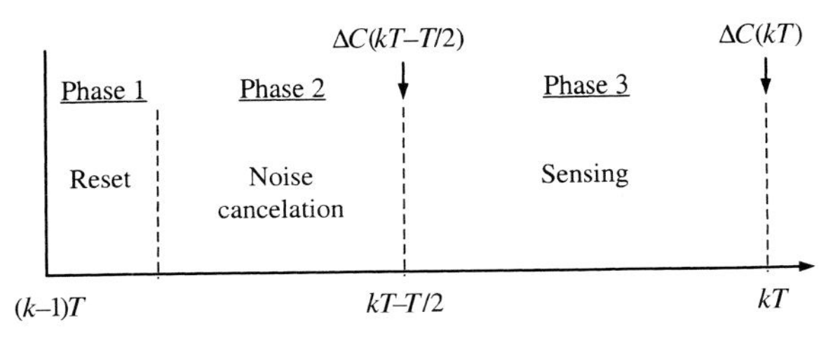

Fig. 3.17 CDS timing diagram

$$V_0(kT) = 2\frac{V_2\Delta C(kT) - V_1\Delta C\left(kT - \frac{T}{2}\right)}{C_I} \tag{16}$$

In the general case the circuit samples the signal during both CDS phases 2 and 3. When $V_1 = V_{AGND}$ the signal is sampled only once at the end of Phase 3. In this case Equation (16) is also identical with Equation (5).

We will present the design of CDS interfaces [3, 9] with an example of a front-end for the acceleration sensor element introduced in Section 3.3.1.1 (Tab. 3.1). The design considerations and the overall procedure can also be adapted easily to other types of sensors.

In addition to a sense interface, most electromechanical systems need some form of actuation, which might require voltages higher than the supply voltage of standard CMOS processes. For example, gyroscopes usually use driving voltages of 10–15 V. To accommodate the actuation voltage requirement, a 5 V, 0.5 μm CMOS process with a high-voltage (20 V) option is chosen for this design. An example of such process is AMIS C5F, available through MOSIS.

CDS stages have been used predominantly in sigma–delta interfaces [9, 17, 44, 45]. Therefore, our design example takes into account the requirements posed by this particular type of feedback system (see Section 3.5). These requirements will be indicated along the way in order to allow a designer to adapt the design to other applications. Fig. 3.18 shows a schematic and timing diagram of the front-end. A feedback phase is allocated in the timing diagram. The choice of sampling frequency in sigma–delta loops is dictated by resolution and stability requirements. For the purpose of this example a typical value of 500 kHz will be used. Open-loop designs can operate at lower sampling rates.

One design choice to be made at this point is whether input common-mode feedback should be included in the first stage. Implementing ICMFB inevitably loads the input node of the C/V converter with additional capacitance, which reduces the feedback factor. In addition, there is a power penalty from the ICMFB amplifier. Analysis of the overall circuit [23] shows that interaction between the two amplifiers leads to an underdamped response, which is also undesirable. If the input stage can tolerate the common-mode step caused by V_M, ICMFB can be omitted. The value of V_M will be set to 3 V in order to avoid pull-in (see Section 3.3.1.1 and Equation (14)). Since C_S and the total capacitance at the sense node

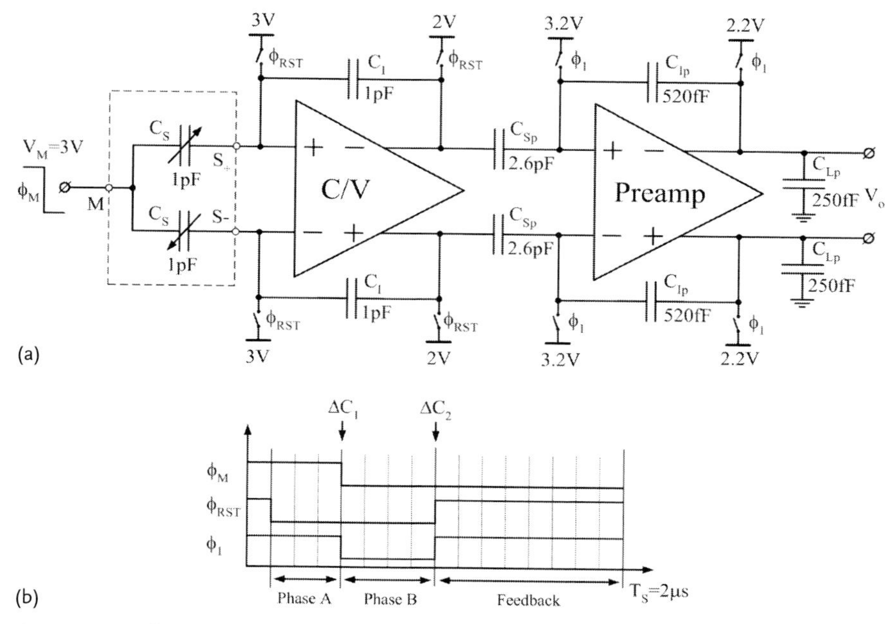

Fig. 3.18 CDS front-end. (a) Schematic; (b) clock diagram

form a capacitive divider, V_M will be attenuated at the amplifier input by approximately 1/7 for the given sensor parameters and a C_I value of 1 pF. Therefore, without ICMFB the amplifier will need to tolerate an input common-mode variation of approximately 400 mV, which is feasible in a 5 V process. A disadvantage of a *C/V* converter without ICMFB is that the common-mode step at the amplifier input reduces the effective amplitude of the sensing pulse. In the case when ICMFB is needed (for example, in a low-voltage interface) the architecture of Lemkin and Boser [9] can be implemented. The value of 1 pF for C_I has been chosen based on a tradeoff between amplifier speed and input-referred noise (see Section 3.3.1.1).

As a next step, the circuit topology of the amplifiers is determined. The signal swing at the output of the *C/V* converter is small compared with the supply voltage. For example, a 5 g maximum signal with the given sensor element will correspond to 120 mV swing for $C_I = 1$ pF in an open-loop design. A closed-loop system reduces the signal swing further by the loopgain. The amplifier in a sigma–delta loop also needs to accommodate the wide-band quantization noise. In any case, the output swing is in the low millivolt range. Under such conditions a telescopic topology is preferable owing to its simple architecture and fewer noise-contributing transistors. A double-cascode amplifier is chosen for its high DC gain.

The signal is sensed on the negative edge of V_M (Fig. 3.18 b). In this way, V_M will be at ground during $(3/4)T_S$, which will reduce the average voltage between the sensor proof-mass and the shield of the mechanical chip and therefore will minimize electrostatic tuning. At the negative edge of V_M the input common-mode level of the amplifier will be lowered by the voltage step. Therefore, PMOS transistors are chosen for the amplifier input in order to ensure that the differential pair will remain in the forward active region during the transient. Also, a PMOS input will provide larger headroom for the tail current source after the input common-mode level decreases. A schematic of the charge integrator amplifier is shown in Fig. 3.19. Note that instead of a common value for V_{AGND}, different voltages are chosen for different nodes in order to provide optimum biasing conditions. The input common-mode level of the first stage (V_{ICM1}) is chosen equal to the amplitude of V_M. In this way, $V_1=0$ and $V_2=-V_M$. The advantage of this design decision is that the output is proportional only to the signal ΔC_2 at the end of the sampling period (Fig. 3.18 b), which minimizes the delay caused by the stage. In contrast, in the hypothetical case when V_{ICM1} is set to zero the output will be determined entirely by ΔC_1, which occurs $(1/4)T_S$ before ΔC_2. As an intermediate case, choosing V_{ICM1} between zero and V_M will cause the output signal

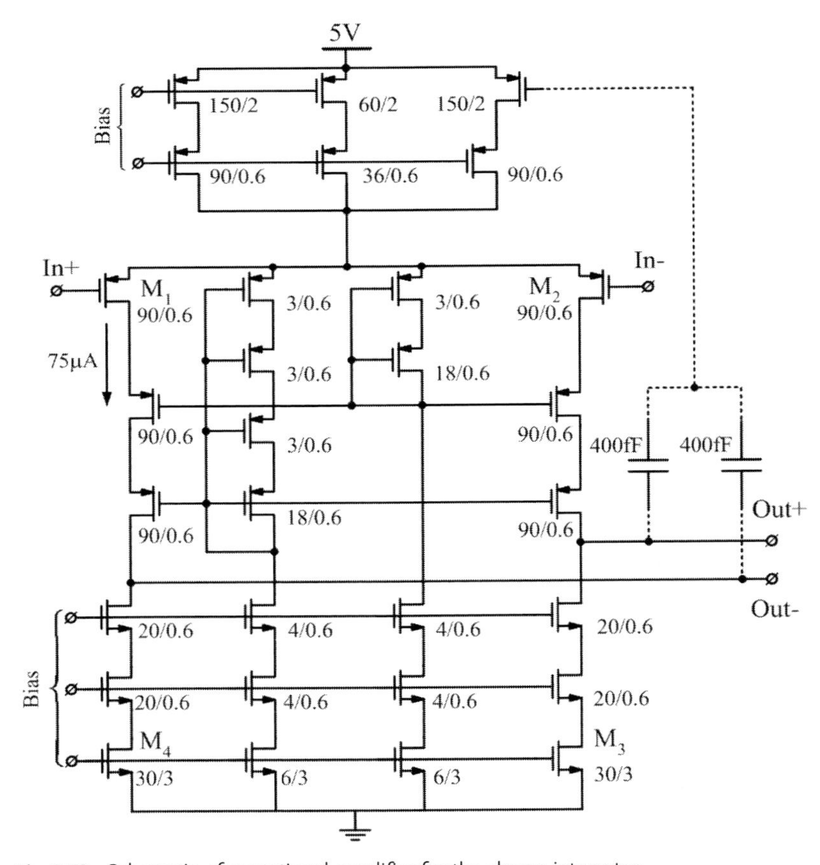

Fig. 3.19 Schematic of operational amplifier for the charge integrator

to be a weighted sum of ΔC_1 and ΔC_2. If the mechanical element allows V_M to be close to or higher than the supply voltage the condition $V_{ICM1} = V_M$ cannot be satisfied. In this case the difference between V_{ICM1} and V_M should be minimized in order to give maximum weight to ΔC_2.

The preamplifier will be designed with a gain of five. This value is sufficient for the noise from the charge integrator and the preamplifier to dominate the overall electronic noise floor. The topology in Fig. 3.19 is appropriate for this stage also. If the maximum signal amplitude from a particular mechanical element cannot be accommodated by the double-cascode architecture, higher swing topologies such as cascoded telescopic (single cascode transistor) or a folded-cascode amplifier should be used. In a feedback design the DC gain variation of the preamplifier and the subsequent stages is less critical than that of the first stage.

The amplifier design on transistor level will begin with the settling conditions. The settling error is given by the ratio of the minimum to the maximum signal in the sensor. The maximum acceleration in this example will be set to $a_{max} = 5$ g. The minimum value is set by the intrinsic noise floor of the sensor, which for a typical bandwidth of 100 Hz, is 150 μg. The value of the settling error is therefore $\varepsilon = 3 \times 10^{-5}$.

The charge integrator settles as a single amplifier during Phase A (see Fig. 3.18 b). The settling time for $f_S = 500$ kHz is $t_A = (3/16)T_S = 375$ ns. Owing to the small differential signal, no slewing occurs in the front end. Therefore, the amplifier closed-loop bandwidth is calculated based on the exponential settling of a single-pole stage for the accuracy given above. The required bandwidth is found from

$$f_{BW\_CV} = \frac{1}{2\pi\tau_{CV}} = \frac{1}{2\pi\left(\dfrac{t_A}{\ln(1/\varepsilon)}\right)} \tag{17}$$

where τ_{CV} is the amplifier time constant of the charge integrator. Equation (17) gives 4.5 MHz.

During Phase B, the charge integrator is connected in series with the preamplifier. In the complete interface one or several gain stages can follow the front-end. For the purpose of this example one additional gain stage will be added, which results in three single-pole stages settling in series. The settling time in this case is $t_B = (1/4)T_S = 500$ ns. The bandwidth of the charge integrator found from Phase A will be preserved and the bandwidth of the subsequent stages will be calculated based on the new settling requirement. This design decision is taken since the first stage operates under the most stringent conditions and therefore its operating conditions need to be maximally relaxed. The preamplifier and the third gain stage will be designed to have the same bandwidth. In this case the settling error can be found from the inverse Laplace transform of the transfer function formed by the three single-pole stages in series. A transcendental equation is obtained and solved numerically, giving a value of 5.3 MHz for the bandwidth of the second and third gain stages.

Once the amplifier bandwidth is known, the noise analysis of the sampled system can be performed. As in the chopper stabilization example, the interface will be designed to provide resolution comparable to the intrinsic noise floor of the sensor.

The amplifier noise model from Fig. 3.6 can be used if R_b is set to zero and load capacitances for the charge integrator and the preamplifier are included. Both amplifiers are implemented according to Fig. 3.19 with capacitive output common-mode feedback (OCMFB). All noise sources are referred to V_X.

An expression for the sampled noise density from the amplifier of the fully differential charge integrator is found using Fig. 3.6:

$$\frac{\overline{v_{n\_CV}^2}}{\Delta f} = \frac{16}{3}\frac{1}{F_{CV}}\frac{k_B T}{C_{Leff\_CV}} n_{CV}\frac{1}{f_S} \tag{18}$$

where F_{CV} is the feedback factor, $C_{Leff\_CV} = C_L + C_I(1 - F_{CV})$ is the effective load capacitance of the amplifier loaded with C_L and $n_{CV} = 1.35$ is the noise factor of the topology in Fig. 3.19. Since broadband white noise is sampled twice during CDS, its power is doubled and therefore an additional factor of two is included in the scale factor 16/3. Similarly to the continuous-time case, Equation (18) indicates that the feedback factor should be large in order to minimize the noise contribution from the amplifier. An important difference between the two cases, however, is that the noise resistance of the amplifier is not part of the equation in the discrete-time circuit. The noise is inversely proportional to the load capacitance and the sampling frequency. This sets a direct tradeoff between resolution and power consumption. C_L will be set to 3 pF. The value of C_L is distributed between the input sampling capacitance of the preamplifier ($C_{Sp} = 2.6$ pF – see Fig. 3.18) and the output common-mode feedback capacitors, which are chosen to be 400 fF (Fig. 3.19).

The sampled noise density from the wiring resistance R_p is

$$\frac{\overline{v_{n\_Rp}^2}}{\Delta f} = 4(4k_B T R_p)\left(\frac{C_S + C_{PS}}{C_I}\right)^2\left(\frac{\pi f_{BW\_CV}}{f_S}\right) \tag{19}$$

where once again a factor of two is included to account for the effect of CDS and a fully differential case is considered. Equation (19) is in fact identical with Equation (8) with the addition of the term $\pi f_{BW\_CV}/f_S$, which accounts for the noise folding in the sampled-data circuit (see Section 3.3.2.2).

The input-referred noise of the preamplifier is calculated next. The noise model is similar to that of the first stage with R_p and C_p neglected. The input sampling capacitor is $C_{Sp} = 2.6$ pF, the feedback capacitor C_{Ip} is therefore 520 fF for a gain of five and the load capacitor C_{Lpre} is chosen to be 0.3 pF, which includes the load from the next stage (C_{Lp}) and an OCMFB capacitor (50 fF). The expression for the noise density referred to V_X is

$$\frac{\overline{v_{\mathrm{n\_p}}^2}}{\Delta f} = \frac{8}{3}\frac{1}{F_{\mathrm{p}}}\frac{k_{\mathrm{B}}T}{C_{\mathrm{Leff\_p}}}n_{\mathrm{p}}\frac{1}{A_{\mathrm{p}}^2}\frac{1}{f_{\mathrm{S}}} \tag{20}$$

where F_{p} is the preamplifier feedback factor, $C_{\mathrm{Leff\_p}} = C_{\mathrm{Lpre}} + C_{\mathrm{Ip}}(1 - F_{\mathrm{p}})$ is the effective load capacitance, $n_{\mathrm{p}} = 1.35$ is the noise factor and $A_{\mathrm{p}} = 5$ is the preamplifier gain. Since the correlated double-sampling is carried out at the input of the preamplifier it does not double the noise contribution of this stage. In analogy with the charge integrator, the preamplifier noise is inversely proportional to the feedback factor, the load capacitance and the sampling frequency.

We will express the total noise in $\mathrm{V_X}$ in a form similar to Equations (11) and (12):

$$\frac{\overline{v_{\mathrm{xn}}^2}}{\Delta f} = \frac{16}{3}\frac{1}{F_{\mathrm{CV}}}\frac{k_{\mathrm{B}}T}{C_{\mathrm{Leff\_CV}}}n_{\mathrm{CV}}\frac{1}{f_{\mathrm{S}}}\left[1 + 3R_{\mathrm{p}}\frac{F_{\mathrm{CV}}C_{\mathrm{Leff\_CV}}}{n_{\mathrm{CV}}}\left(\frac{C_{\mathrm{S}} + C_{\mathrm{PS}}}{C_{\mathrm{T}}}\right)^2 \pi f_{\mathrm{BW\_CV}} + \frac{1}{2}\frac{F_{\mathrm{CV}}}{F_{\mathrm{p}}}\frac{n_{\mathrm{p}}}{n_{\mathrm{CV}}}\frac{1}{A_{\mathrm{pre}}^2}\frac{C_{\mathrm{Leff\_CV}}}{C_{\mathrm{Leff\_P}}}\right] \tag{21}$$

Introducing the noise factor coefficients in Equation (21) we arrive at

$$\frac{\overline{v_{\mathrm{xn}}^2}}{\Delta f} = \frac{16}{3}\frac{1}{F_{\mathrm{CV}}}\frac{k_{\mathrm{B}}T}{C_{\mathrm{Leff\_CV}}}n_{\mathrm{CV}}\frac{1}{f_{\mathrm{S}}}\left[1 + N_{\mathrm{p}}R_{\mathrm{p}} + N_{\mathrm{Pre}}\frac{C_{\mathrm{Leff\_CV}}}{C_{\mathrm{Leff\_P}}}\right] \tag{22}$$

By substituting the numerical values in Equation (22), we find that the noise contribution from the amplifier of the charge integrator is $330\ \mathrm{nV}/\sqrt{\mathrm{Hz}}$. The value of N_{P} is 6×10^{-4}, therefore an interconnect resistance of $1.6\ \mathrm{k\Omega}$ will contribute comparable noise to the main amplifier. The noise factor of the preamplifier (N_{Pre}) is 0.02, which allows for a small loading capacitor C_{Lp}. The total noise at V_{X} is $430\ \mathrm{nV}/\sqrt{\mathrm{Hz}}$. The dominant noise source is the amplifier of the charge integrator followed by the parasitic resistance, which contributes 35% and the preamplifier with 5%.

The capacitance resolution for our example is $0.08\ \mathrm{aF}/\sqrt{\mathrm{Hz}}$. The intrinsic noise floor of the sensor element ($0.058\ \mathrm{aF}/\sqrt{\mathrm{Hz}}$) is comparable to that value.

Designing the amplifiers in the AMIS C5F process gives the device sizes shown in Fig. 3.19 for the charge integrator. The bias current of M1 is 75 μA. The preamplifier requires 20 μA and is designed by scaling the device sizes accordingly.

The differential CDS interface presented in this section has been used in the front-ends of both open-loop [3] and closed-loop [9, 44] systems, achieving resolution better than $0.1\ \mathrm{aF}/\sqrt{\mathrm{Hz}}$ [3]. An alternative architecture can be found elsewhere [17]. The principle of operation of this front-end is illustrated in Fig. 3.20. The amplifier consists of a main and an auxiliary stage whose outputs are summed to produce the overall output signal. The circuit operates in two phases. The switch positions in the figure correspond to Phase 1 during which the amplifier is disconnected from the signal path and the offset of both stages is stored on

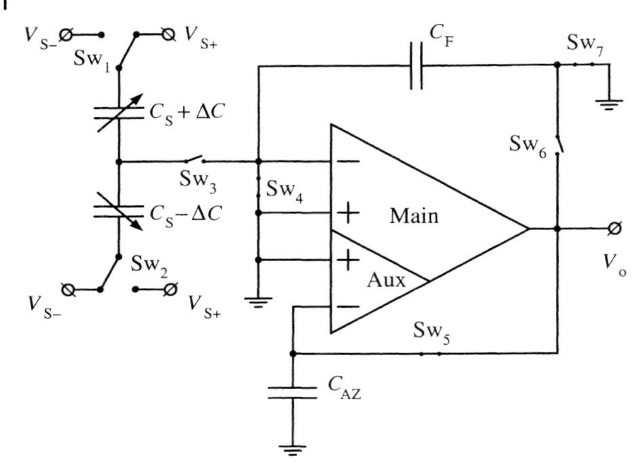

Fig. 3.20 Single-ended discrete-time charge integrator

C_{AZ}. The stored value is used for offset cancellation during Phase 2 when the input signal is detected. Switch charge injection from Sw_5 and low-frequency noise are also cancelled by this technique. A detailed analysis of the circuit can be found elsewhere [22, 46]. In analogy with the continuous-time implementations, comparison between the two approaches gives the differential solution the advantage of better disturbance rejection and lower parasitic capacitance at the charge integrator input. Also in the single-ended approach the additional noise contribution from the auxiliary gain stage should be minimized. On the other hand, the single-ended architecture operates without input common-mode variation. Other discrete-time front-end topologies can also be found in the literature [47, 48].

In general, discrete-time interfaces are suitable for systems with digital output since the sampling is intrinsically performed in the front-end. The tradeoffs involve noise folding, which will be discussed next.

3.3.2.2 **Discussion on Sampling and Noise Folding**

Sampling a continuous-time signal at a rate f_S causes all spectral components above the Nyquist frequency ($f_N = f_S/2$) to appear in the band from DC to f_N. To avoid corrupting the low-frequency signal band with aliased components, a low-pass anti-aliasing filter with sharp cutoff at f_N must be introduced before the point where sampling is performed.

In a correlated double-sampling capacitive interface the noise is filtered only by the roll-off of the amplifier open-loop transfer functions. In general, settling requirements result in an amplifier bandwidth which is several times higher than the sampling rate. The restrictions are even more severe if the system requires a large fraction of the sampling period to be allocated for feedback. In any case, noise folding (aliasing) results in an increased noise floor in the signal band. The effect is illustrated in Fig. 3.21.

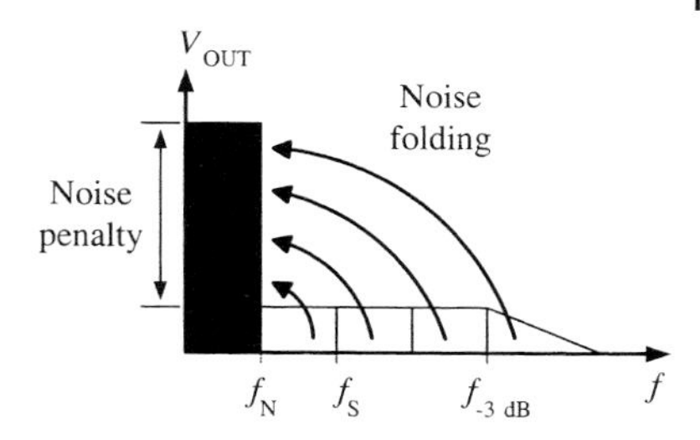

Fig. 3.21 Amplifier noise folding

Equations (18) and (20) show that the spectral density of the sampled amplifier noise is inversely proportional to the load capacitance. An increase in the load capacitance requires higher power consumption in order for the bandwidth to be preserved. Therefore, there is a direct tradeoff between noise and power consumption. For a given resolution the appropriate value of the load capacitor is chosen and the amplifier bias current is adjusted to provide the required bandwidth.

The need for a small phase lag in continuous-time systems leads to bandwidth requirements comparable to those in the discrete-time implementations. The noise density in the continuous-time case, however, does not depend on the load capacitance. This implies that a continuous-time front-end can meet certain resolution requirements without the need for a large load capacitance and therefore can operate with lower power than its discrete-time equivalent [21]. This claim, however, is justified only if the output in either case can be utilized directly by the rest of the system. In a system with digital output the signal from a discrete-time front-end can be supplied directly to an analog-to-digital converter. On the other hand, the intrinsically high resolution of the continuous-time interface can be preserved only if a high-order anti-aliasing filter is included before the ADC. In this case a continuous-time design may not have significant advantage in power consumption.

3.3.3 C/V Converters in Open-loop Systems

The circuit content of a capacitive interface is strongly dependent on the specifications of the overall system in which it is embedded. In general, the interfaces can be open-loop or feedback systems and can have analog or digital output. Open-loop interfaces have a simple implementation with minimum circuit overhead, intrinsically high resolution and are naturally stable. Capacitive sensors based on dielectric constant variation operate in an open-loop configuration [10, 11]. An open-loop interface is also suitable for displacement sensing in the case when the signal band is located away from the mechanical resonance or when the mechanical element has low Q. Applications such as crash and rollover detection typically use open-loop inertial sensors as threshold devices. On the other hand, inertial navigation specifications require high-resolution sensors with large dynamic range and low drift over time and temperature. Under such conditions feedback control is more appropriate.

The simplest open-loop interface converts the measured quantity to voltage and provides a band-limited analog output. Such system consists of capacitive front-end and a sharp-cutoff analog filter. A continuous-time open-loop interface was presented in Section 3.3.1. This approach is particularly suitable for purely analog systems where the voltage output can be used directly.

Alternatively, in a sampled-data implementation the front-end from Section 3.2 can be applied. The analog output in this case is a sampled signal, which is convenient for further discrete-time processing such as analog-to-digital conversion, without the need for pre-filtering. The sharp-cutoff band-limiting filter can be implemented digitally.

When continuous-time output is also required from the sampled-data interface, a direct charge-transfer low-pass filter [49] can be used as a buffer for discrete-time to continuous-time conversion. Fig. 3.22 shows a schematic of the stage. Since the input and the feedback capacitors are connected in parallel during ϕ_2, charge is transferred directly from input to output and ideally the amplifier does not have to provide current for charge redistribution. Therefore, the amplifier speed requirements are significantly relaxed and linear settling is ensured. The stage is suitable for driving large off-chip capacitive loads. A direct charge-transfer low-pass filter was used in the switched-capacitor interface of the monolithic open-loop accelerometer presented by Jiang et al. [3].

The simplicity and high resolution of open-loop interfaces unfortunately come at the price of reduced bandwidth and dynamic range and also high sensitivity to fabrication tolerances and ambient variations. Several examples of undesirable effects in open-loop system performance can be given. Low dynamic range allows unwanted out-of-band interferers with large amplitude to overload the interface. Low bandwidth can cause excessive phase lag in applications where the signal is located close to the natural frequency of the mechanical element. In high-Q systems both gain and phase characteristics at resonance can vary significantly with temperature and pressure. An example is a micromachined gyroscope, where the signal is purposely kept away from the resonance of the sensing mode to avoid

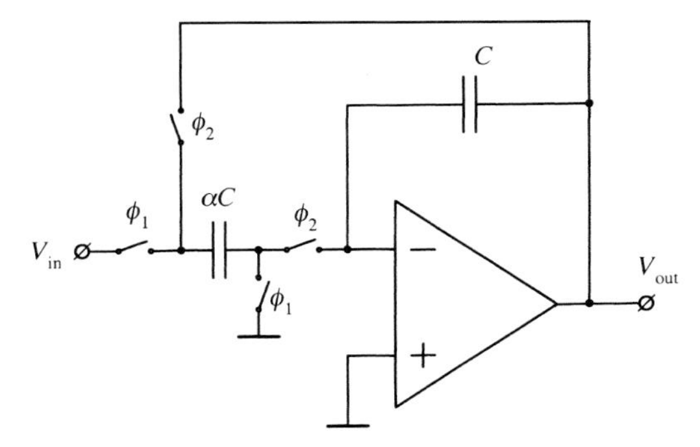

Fig. 3.22 Direct charge-transfer low-pass filter

large gain and phase variations, although the increased gain near resonance could provide a significant sensitivity advantage. Another undesirable effect appears in positioning applications such as micromirror steering, where an underdamped response of the mirror can increase considerably the settling time of the system.

Feedback is generally employed to address the above problems. Feedback control is used predominantly in displacement-sensing systems or positioning applications. Including the mechanical element in the forward path of a feedback loop attenuates any parameter variations by the loop gain. Also by the loop gain are scaled the bandwidth and dynamic range of the system. In an electromechanical system closed-loop operation is realized through force rebalancing of the mechanical structure. Several techniques for implementing force-feedback will be presented in the following sections.

3.4 Electrostatic Force-feedback

Electrostatic actuation was introduced Section 3.2. In the context of feedback it can be used to control the position of a mechanical element. Applications such as inertial sensors employ force-feedback to maintain the proof-mass close to its center position. In micromirror positioning a feedback loop can control the coordinates of the mirror according to a reference value.

The differential technique shown in Fig. 3.3 b is particularly suitable for implementing continuous, linear feedback in force-rebalanced electromechanical systems due to the linear relation between feedback voltage and electrostatic force (see Equation (4)).

A block diagram of a force-feedback system is shown in Fig. 3.23. The front-end (*C*/*V*) design was presented in Section 3.3. The differential feedback force can be generated with two summing amplifiers, which receive the voltages V_{DC} and v at the input and apply their sum and difference to the feedback capacitors as shown in Fig. 3.3 b. The compensator will be described next.

A micromachined element is generally modeled as a second-order system with transfer function from force to displacement given by

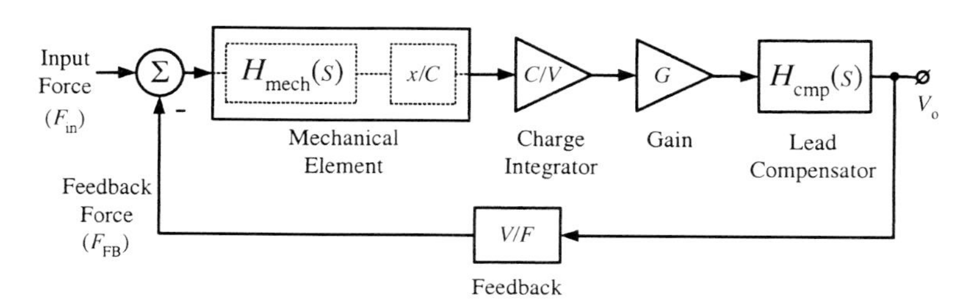

Fig. 3.23 Linear force-feedback system diagram

$$H_{\mathrm{mech}}(s) = \frac{x(s)}{F(s)} = \frac{1}{m\omega_0^2} \frac{1}{1 + \dfrac{s}{Q\omega_0} + \dfrac{s^2}{\omega_0^2}} \tag{23}$$

where m is the mass of the moving structure, ω_0 is the resonant frequency and Q is the quality factor. In most implementations the mechanical element is underdamped having a pair of complex poles at the resonant frequency. Since the two poles cause a phase lag of 180°, closing a feedback loop around such an element results in an unstable system. The compensator in Fig. 3.23 solves the instability problem by adding phase lead to the transfer function at the unity gain frequency of the loop. A phase-lead compensator has transfer function with a low-frequency zero and a high-frequency pole:

$$H_{\mathrm{cmp}}(s) = \frac{s + z}{s + p} \tag{24}$$

The value of the zero (z) is chosen to give the desired amount of phase lead. A zero in the transfer function inevitably leads to increased gain at high frequencies hence high-frequency noise is amplified by the compensator. To minimize this effect the pole (p) is positioned as low as possible such that the maximum phase lead occurs at the unity-gain frequency of the system. Additional phase lag due to finite amplifier bandwidth and nondominant poles should also be considered during the compensator design.

As an example, the system in Fig. 3.23 is designed based on the acceleration sensor introduced in Section 3.3. The mechanical element needs an additional set of feedback capacitors for the closed-loop design. Their value is chosen to be 300 fF as a tradeoff between area and feedback force. The system is designed for a DC loopgain of 33 dB, unity-gain bandwidth of 55 kHz and 50° phase margin. These parameters result in a lead compensator with a zero at 20 kHz and a pole at 150 kHz. A circuit which implements the compensator transfer function is shown in Fig. 3.24 [50]. The differential voltage-to-force converter is designed with 2.5 V DC bias assuming 5 V power supply. The gain G depends on the transfer function of the C/V stage. For a chopper stabilized C/V converter with 3 V carrier amplitude and 1 pF feedback capacitor, a gain G of 58 dB is needed in order to achieve an overall loopgain of 33 dB.

Ideally, a closed-loop system should achieve the same signal-to-noise ratio as the underlying open-loop design obtained from the same circuit blocks. Therefore, the resolution and power consumption of the feedback interface are determined mainly by the C/V front-end. Additional noise sources in the feedback system should be avoided or their contribution to the in-band noise floor should be minimized.

In our design example, the mechanical element was modeled as a second-order system with a single pair of poles at its main resonant frequency. Complex mechanical structures, however, can exhibit a number of additional resonances at higher frequencies due to flexural modes in the suspension, the electrostatic

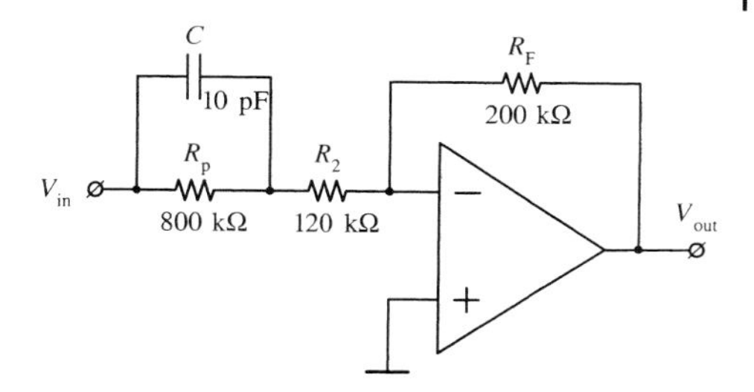

Fig. 3.24 Schematic of lead compensator

comb fingers and other sub-components. High-frequency modes tend to destabilize the feedback system and should be maximally damped when the operating conditions of the mechanical element allow this. Mechanical structures operating in vacuum require special attention in this case. Design techniques for stabilizing systems with high-Q modes can be found elsewhere [43]. In general, collocated control (i.e., using the same capacitors for sense and feedback) is preferable in such cases. A feedback interface with a discrete-time C/V front-end can implement collocated control by time-multiplexing the sense and feedback phases [9]. An approach which allows this type of control for chopper stabilized designs can be found in the literature [16]. It relies on frequency separation between the sensed signal, modulated at the carrier and the feedback signal, which remains at low frequencies.

An implementation of electromechanical linear-feedback system using lead compensation has been reported [18]. Alternative approaches for achieving stability have been proposed [15, 16]. A reported accelerometer [15] uses an overdamped (pressurized) mechanical element to achieve pole splitting. A disadvantage of this approach is a substantially increased Brownian noise of the sensor. The other implementation [16] relies on a low-Q mechanical element and a feedback system with low loopgain and narrow bandwidth in order to ensure the required phase lead at the unity-gain frequency. The bandwidth and loopgain restrictions make this solution unattractive. In addition, the phase margin in this case depends on the quality factor of the mechanical resonance.

The linear-feedback interfaces presented in this section provide a voltage output. In applications which require digital post-processing of the interface data, an anti-aliasing filter and a dedicated analog-to-digital converter are added to the system. An alternative feedback technique known as sigma–delta modulation allows the analog-to-digital conversion to be performed inherently by the feedback loop and therefore avoids the overhead from the filter and the ADC. Sigma–delta force-feedback is the subject of the next section.

3.5 Electromechanical Sigma–Delta Interfaces

The basic idea of sigma–delta modulation is to obtain higher effective resolution from a coarse quantizer by including it in an oversampled feedback loop [51]. Fig. 3.25 a shows a block diagram of a sigma–delta modulator. The difference between the input and the feedback signal is filtered and digitized by a quantizer (A/D), which has a resolution of one to several bits and operates at a sampling rate significantly higher than the bandwidth of the input signal. The oversampled digital sequence generated by the quantizer represents the output of the modulator. It contains the input signal and the error from quantization. The feedback is derived from the output after digital-to-analog conversion.

To understand how the above configuration can produce a high-resolution digital output, the quantizer, which is a highly non-linear element, is replaced by an additive error source modeling the quantization operation. Quantization error is typically represented as white noise [52]. The resulting linear feedback loop (Fig. 3.25 b) can be analyzed using conventional linear system theory. For high in-band loopgain, the transfer function for the input signal is unity. On the other hand, quantization error appears at the output multiplied by the inverse of the loop-filter transfer function. The loop-filter is always designed to have high gain in the signal band, which in turn results in high attenuation for the quantization error in that frequency region. In other words, output quantization error is shaped by the loop filter such that most of its variance is located out-of-band. Fig. 3.26 illustrates the principle qualitatively for a sigma–delta modulator using a low-pass loop filter. The quantization noise without shaping is shown for comparison.

The oversampled output of the sigma–delta modulator is processed by a digital filter and downsampled to the Nyquist rate. The filtering removes out-of-band quantization noise and therefore ensures high signal-to-quantization-noise ratio (SQNR).

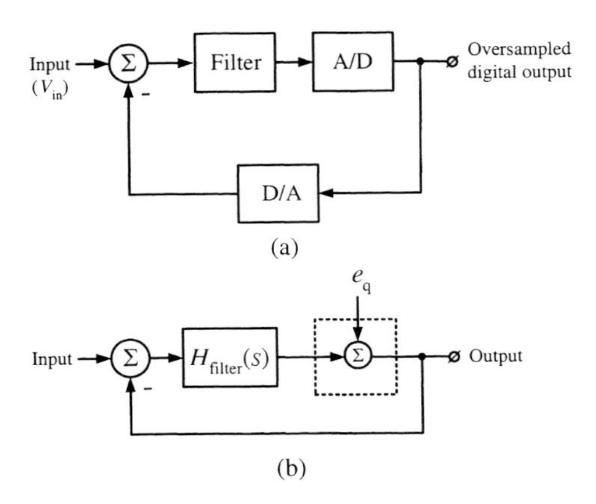

Fig. 3.25 Sigma–delta modulator. (a) Block diagram; (b) linear model

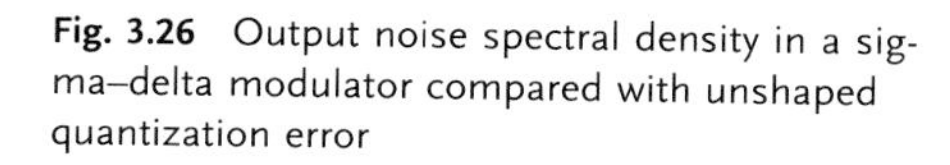

Fig. 3.26 Output noise spectral density in a sigma–delta modulator compared with unshaped quantization error

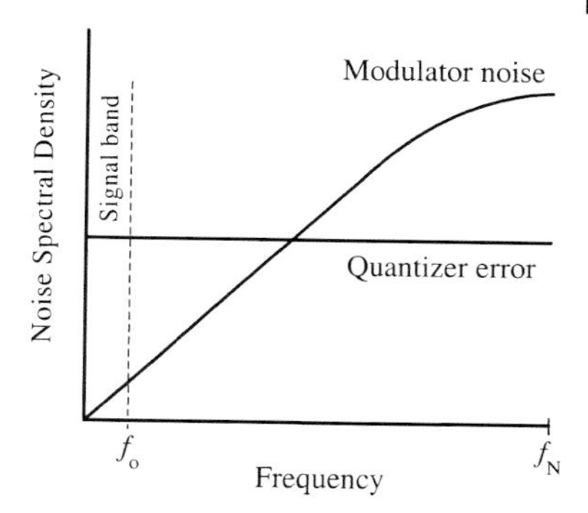

From an electromechanical system perspective, the sigma–delta technique can be particularly convenient if the inherent low-pass transfer function of the mechanical element is employed as a loop-filter. This approach has been proposed for the first time by Henrion et al. [53] and has seen a number of implementations since. So far the technique has been applied predominantly to inertial sensors although it is not limited only to this field. Other implementations such as a thermoelectrical sigma–delta wind sensor have also been reported [54].

Fig. 3.27 shows a block diagram of a single-bit electromechanical sigma–delta modulator. The system architecture resembles that of the linear loop in Fig. 3.23 with the addition of a binary quantizer and a modified feedback circuit, which implements two-level feedback. The intrinsic linearity of two-level feedback makes a binary quantizer particularly attractive in an electromechanical application owing to the quadratic relation between voltage and electrostatic feedback force.

Owing to the continuous-time nature of the mechanical element, the signal has to be quantized in both time and amplitude. Although the two operations could be performed by the quantizer, time quantization (sampling) can be done in the front-end instead. In this case the C/V converter is a sampled circuit (see Section 3.3.2) and the compensator operates in discrete time. A discrete-time front-end allows time multiplexing between signal sensing and feedback. This approach avoids the need for dedicated feedback electrodes in the mechanical element.

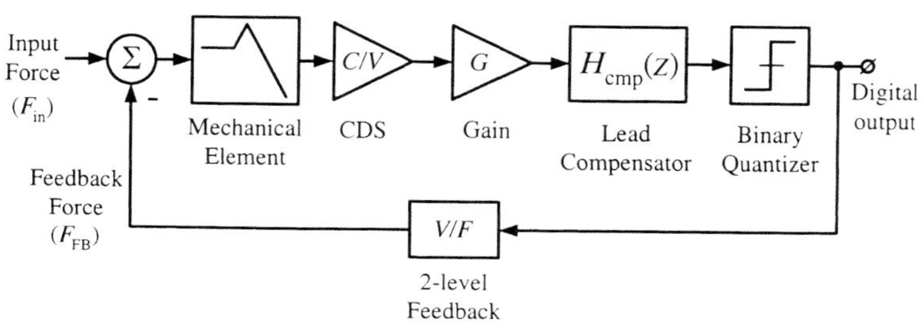

Fig. 3.27 Electromechanical sigma–delta modulator

Although the sigma–delta modulator appears similar to the linear feedback system in Fig. 3.23, some aspects of its behavior are considerably different owing to the non-linear nature of the binary quantizer. Stability in the sigma–delta loop cannot be defined in the conventional terms known from linear system theory since the output of the quantizer never converges to a static value even for constant input signals. Instead, it forms limit cycles, which can impair the system resolution significantly if they occur at frequencies inside the signal band. In particular, an improperly compensated loop tends to exhibit low-frequency limit cycles and can even produce a sustained oscillation as a large single tone. The role of the compensator from that point of view is to ensure stability in a sense that the modulator would achieve high in-band SQNR.

Another particular feature of the sigma–delta loop is the presence of quantization error as additional noise source. System design should aim at minimizing the contribution of quantization error to the in-band noise floor.

Also, owing to the continuous-time nature of the mechanical element, an effective loop-delay is contributed by the time interval between the instant when displacement is detected and the center of the finite-length feedback pulse. This delay requires additional phase compensation.

For a given mechanical element and a C/V front-end, optimized for noise and delay (see Section 3.3.2), the interface designer has control over the behavior of the sigma–delta loop only through the compensator transfer function. In analogy with linear feedback designs, second-order electromechanical sigma–delta modulators use lead compensation in this stage. The general transfer function of a discrete-time lead compensator is

$$H_{\mathrm{cmp}}(z) = \frac{az - 1}{z} \tag{25}$$

The design parameter a determines the position of the compensating zero. In a linear feedback system the position of the zero is chosen to provide sufficient phase margin at the unity gain frequency of the loop. In sigma–delta loops, however, the presence of quantization noise poses additional restrictions on the compensator transfer function. Since $H_{\mathrm{cmp}}(z)$ is part of the loop-filter, the contribution of quantization noise to the signal band is inversely proportional to the in-band gain of the compensator. Equation (25) shows that larger phase lead (smaller value of a) leads to higher attenuation at low frequencies and therefore higher in-band quantization noise level. This argument suggests that the amount of compensation should be reduced. Insufficient compensation, on the other hand, impairs the loop stability, resulting in low-frequency limit cycles and degraded quantization noise shaping. Based on the above discussion, it can be expected that an optimum value of a can be found. A quantitative analysis as described [23, 55] results in an optimum of $a \approx 2$.

A circuit implementation of the compensator is shown in Fig. 3.28. A capacitive FIR filter without active elements was chosen in order to minimize the settling time and power consumption of this stage. A single-ended case is shown for simplicity since it is straightforward to extend the circuit to a differential design.

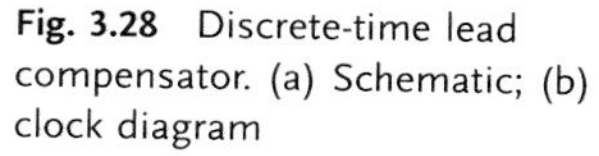

Fig. 3.28 Discrete-time lead compensator. (a) Schematic; (b) clock diagram

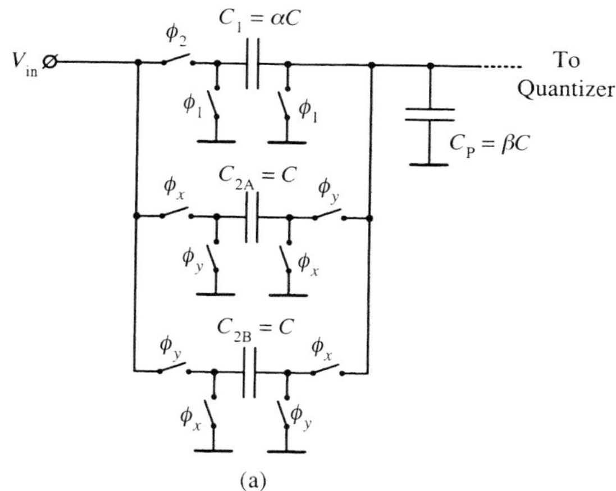

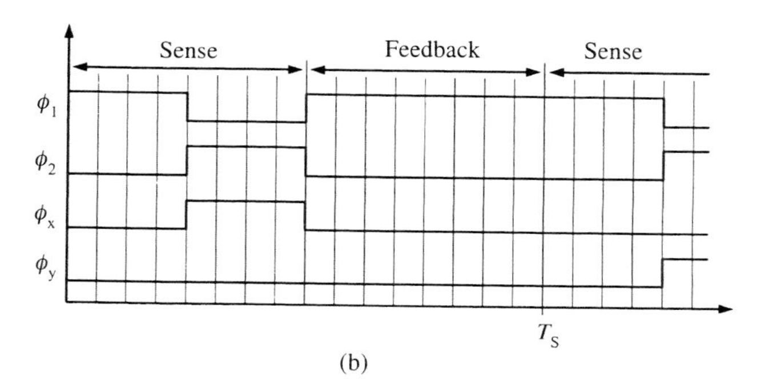

The operation of the circuit will be explained from its timing diagram. Phase ϕ_1 in the diagram is the same as the one used by the CDS front-end (see Fig. 3.18 b). During ϕ_1 capacitor C_1 is discharged, C_{2B} holds the input value from the previous cycle and the charge on C_{2A} is irrelevant. During ϕ_2, which coincides with ϕ_x on this cycle, charge redistribution between C_1, C_{2B} and C_p occurs at the output while the input is sampled on C_{2A}. C_p is parasitic capacitance, formed by the top-plate parasitics of the compensator capacitors and the input capacitance of the comparator pre-amplifier. In the following cycle ϕ_2 occurs together with ϕ_y and the output is determined by C_1, C_{2A} and C_p. The input in this case is sampled on C_{2B}. In other words, C_1 provides the current value of the input while the alternating capacitors C_{2A} and C_{2B} give the values with a one-sample delay. The transfer function of the stage is

$$H_{\text{cmp}} = \frac{1}{1+\alpha+\beta}\left(\frac{\alpha z - 1}{z}\right) \tag{26}$$

Equation (26) shows that the circuit implements a lead compensator with a scaling factor, which depends on the value of α and the size of the parasitic capacitance at the output. The parasitic capacitance does not affect the position of the zero. The compensator capacitors should be sized such that they do not present significant load to the preceding gain stage and also do not result in large attenuation (large value of β). Since the compensator is placed at the end of the amplifier chain, immediately before the quantizer, sampled noise is not critical in this case. Therefore, capacitor values of the order of several hundred femptofarads are reasonable for this design.

The quantizer can be implemented as a regenerative feedback comparator [51]. The comparator output provides the digital bit-stream to the decimation filter and controls the binary DAC. Depending on the value of the output, the DAC applies a constant voltage level to the appropriate feedback capacitor, while the complementary capacitor remains at the potential of the proof-mass. A time-multiplexed implementation, which uses the same electrodes for sense and feedback, has been described [9].

In analogy with linear-feedback loops, the stability of the sigma–delta interface is also affected by high-frequency modes in the mechanical element. The additional delay in the sigma–delta modulator, however, results in different stability criteria for the loop under collocated and non-collocated control depending on the frequency location of the modes. A detailed discussion of the problem has been presented [42]. In general, damping of the high-frequency resonances is the preferred solution in this case also.

Variations of the sigma–delta architecture in Fig. 3.27 have been used in a number of inertial sensors [9, 15, 17, 44, 45 56. Sigma–delta accelerometers achieved noise floors down to 1 $\mu g/\sqrt{Hz}$ [15]. A gyroscope with a resolution of $3^{\circ}/s/\sqrt{Hz}$ has been described [44]. The interface resolves displacement of 2×10^{-12} m, which translates into 0.1 aF capacitance in 1 Hz bandwidth.

The operation of three monolithic sigma–delta accelerometers integrated on the same substrate has been demonstrated [9]. Fig. 3.29 shows the die micrograph of this implementation [9]. This solution achieves resolution of 110 $\mu g/\sqrt{Hz}$ (or 0.085 $aF/\sqrt{Hz}$) with 100 fF sense capacitors. Owing to the small value of the proof-mass, the noise floor in this design is dominated by Brownian noise.

An accelerometer which uses a multi-bit quantizer and a digital PID controller to implement the compensator stage has been presented [17]. The PID controller provides additional shaping for the quantization noise and improves the SQNR of the modulator. The multi-bit feedback DAC in this design uses pulse-width modulation (PWM) for improved linearity of the voltage-to-force transducer. The resolution of the device is 150 $\mu g/\sqrt{Hz}$.

In summary, sigma–delta modulation has the advantage of closed-loop operation and inherent analog-to-digital conversion, performed by the feedback loop. The additional noise from quantization, however, can introduce a significant resolution penalty over an open-loop implementation. Minimizing the quantization noise overhead is a major challenge in the design of electromechanical sigma–delta modulators. This problem is addressed with an optimum compensator transfer function in second-order modulators or by increasing the order of the loop through introducing additional filtering in the electronic interface.

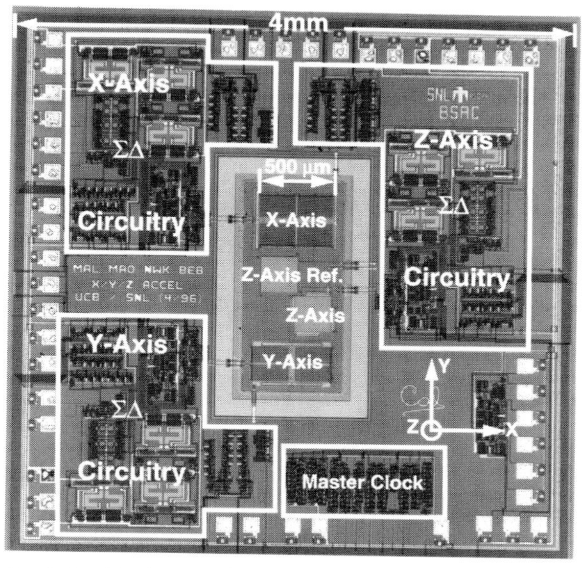

Fig. 3.29 Die photograph of a monolithic three-axis accelerometer [9]

3.6 Conclusions

The high resolution and robustness of capacitive sensing have led to the development of high-performance interfaces in a broad range of applications. This chapter has presented an overview of different approaches for implementing capacitive interfaces and provided guidelines on choosing and designing the appropriate architecture.

Noise analysis of the interface front-end showed that low interconnect resistance and parasitic capacitance of the transducer improve the resolution of the system. A further resolution advantage is achieved by devices which produce large capacitance variation with a small nominal value of the sense capacitors.

Owing to the low-frequency nature of the signal in most transducers, the electronic interfaces employ circuit techniques, which attenuate flicker noise. Chopper stabilization and correlated double-sampling were introduced as two alternative approaches. Chopper stabilization is a continuous-time technique, suitable for systems with analog output. Correlated double-sampling is a discrete-time approach, appropriate for systems in which analog-to-digital conversion is applied.

The resolution in both approaches is determined mainly by the noise contribution of the amplifier in the charge integrator. Therefore, there is a general tradeoff between noise performance and power consumption of the interface. Maximizing the signal is also critical for the resolution since the typical signals obtained from the transducer are in the sub-atofarad range. Therefore, micromachined devices,

Tab. 3.3 Comparison of capacitive sensing architectures

Interface type	*Advantages*	*Disadvantages*
Chopper-stabilized, open-loop	• High resolution – no aliasing, minimal number of noise sources • Low front-end power – SNR not limited by capacitor size • Suitable for discrete-component implementation	• Low bandwidth and dynamic range, sensitive to process and ambient variations (open-loop) • Requires additional filtering and ADC for digital output • Requires large biasing resistors
Discrete-time (CDS), open-loop	• Compatible with standard VLSI CMOS process • Output can be digitized directly	• Large capacitors needed for low kT/C noise • Low bandwidth and dynamic range, sensitive to process and ambient variations (open-loop)
Continuous-time force-feedback	• High resolution • Large bandwidth and dynamic range – scaled by the loop-gain • Reduced sensitivity to process and ambient variations	• Requires additional filtering and ADC for digital output • Requires linearization of voltage-to-force feedback • System stability affected by high-order dynamics in the mechanical element
Digital force-feedback (sigma–delta modulation)	• Analog-to-digital conversion performed by the feedback loop • Intrinsically linear with two-level feedback • Large bandwidth and dynamic range, low sensitivity to process and ambient variations	• Quantization noise affects resolution in second-order modulators • System stability affected by loop delay and high-order dynamics in the mechanical element

which allow large voltages to be used for extracting the signal, have a resolution advantage. The value of the sensing voltage is limited by the supply rails or by the pull-in voltage in compliant structures.

The interface design on a system level introduced closed-loop operation as an approach for increasing the bandwidth and dynamic range and reducing its sensitivity to parameter and ambient variations. Sigma–delta feedback is a further development of the closed-loop concept, which allows analog-to-digital conversion to be performed intrinsically by the feedback loop.

Tab. 3.3 summarizes the key characteristics of the interfaces presented in this chapter.

3.7 References

1 G. Kovacs, *Micromachined Transducers Sourcebook*; New York: McGraw-Hill, **1998**.

2 N. Yazdi, F. Ayazi, K. Najafi, *Proc. IEEE* **1998**, *86*, 1640–1659.

3 X. Jiang, F. Wang, M. Kraft, B.E. Boser, in: *Solid-State Sensor and Actuator Workshop, Hilton Head, SC*; **2002**, pp. 202–205.

4 J.A. Geen, S.J. Sherman, J.F. Chang, S.R. Lewis, *IEEE J. Solid-State Circuits* **2002**, *37*, 1860–1866.

5 B.E. Boser in *Sensors Applications*, J. Marek, H. Trah, Y. Suzuki, I. Yokomori (eds.); New York: Wiley, **2003**, Vol. 4, Chapter 6.1.

6 B.E. Boser, R.T. Howe, *IEEE J. Solid-State Circuits* **1996**, *31*, 366–375.

7 J.I. Seeger, B.E. Boser, *IEEE J. Microelectromech. Syst.* **2003**, *12*, 656–671.

8 K.D. Wise, K. Najafi, in: *IEEE Solid-State Sensor Conference, New York*; **1984**, pp. 12–16.

9 M. Lemkin, B.E. Boser, *IEEE J. Solid-State Circuits* **1999**, *34*, 456–468.

10 P. Malcovati, A. Haberli, F. Mayer, O. Paul, F. Maloberti, H. Baltes, in: *Symposium on VLSI Circuits, Tokyo*; **1995**, pp. 45–46.

11 C. Hagleitner, D. Lange, A. Hierlemann, O. Brand, H. Baltes, *IEEE J. Solid-State Circuits* **2002**, *37*, 1867–1878.

12 K. Kasten, N. Kordas, H. Kappert, W. Mokwa, in: *TRANSDUCERS '01, EUROSENSORS XV, Munich*; **2001**, pp. 510–513.

13 T.A. Roessig, R.T. Howe, A.P. Pisano, J.H. Smith, in: *Solid-State Sensor and Actuator Workshop, Hilton Head, SC*; **1998**, pp. 328–332.

14 A.A. Seshia, M. Palaniapan, T.A. Roessig, R.T. Howe, R. W. Gooch, T.R. Schimert, S. Montague, *IEEE J. Microelectromech. Syst.* **2002**, *11*, 784–793.

15 T. Smith, O. Nys, M. Chevroulet, Y. DeCoulon, M. Degrauwe, in: *IEEE Int. Solid-State Circuits Conf. Dig. Tech. Papers, San Francisco, CA*; **1994**, pp. 160–161.

16 S.J. Shermann, A.P. Brokaw, R. W.K. Tsang, T. Core, *Monolithic Accelerometer*, US Patent 5345824, **1993**.

17 C. Lang, R. Tielert, in: *Proceedings of the 25th European Solid-State Circuits Conference, Neuilly sur Seine*; **1999**, pp. 250–253.

18 D.A. Horsley, N. Wongkomet, R. Horowitz, A. Pisano, in: *Solid-State Sensor and Actuator Workshop, Hilton Head, SC*; **1998**, pp. 120–123.

19 V. Milanovic, M. Last, K.S.J. Pister, in: *TRANSDUCERS '01, EUROSENSORS XV, Munich*; **2001**, pp. 1298–1301.

20 K. Funk, H. Emmerich, A. Schilp, M. Offenberg, R. Neul, F. Larmer, *A surface micromachined silicon gyroscope using a thick polysilicon layer*, presented at Twelfth IEEE International Conference on Micro Electro Mechnical Systems, Orlando, FL, **1999**, pp. 57–61

21 J. Wu, G.K. Fedder, L.R. Carley, in: *IEEE Int. Solid-State Circuits Conf. Dig. Tech. Papers, San Francisco, CA*; **2002**, pp. 428–429.

22 C.C. Enz, G.C. Temes, *Proc. IEEE* **1996**, *84*, 1584–1614.

23 M. Lemkin, *Micro Accelerometer Design with Digital Feedback Control*; Doctoral Dissertation, UC Berkeley, Berkeley, CA, **1997**.

24 T.H. Lee, *The Design of CMOS Radio-Frequency Integrated Circuits*; Cambridge: Cambridge University Press, **1998**.

25 P.R. Gray, P.J. Hurst, S.H. Lewis, R.G. Meyer, *Analysis and Design of Analog Integrated Circuits*, 4th edn.; New York: Wiley, **2001**.

26 C.C. Enz, E.A. Vittoz, F. Krummenacher, *IEEE J. Solid-State Circuits* **1987**, *22*, 335–342.

27 M.S. Ghausi, K.R. Laker, *Modern Filter Design*; Noble Publishing, Norcross, GA, **2003**.

28 T.B. Gabrielson, *IEEE Trans. Electron Devices* **1993**, *40*, 903–909.

29 Analog Devices, *Data Sheet: AD8610/AD8620 Precision, Very Low Noise, Low Input Bias Current, Wide Bandwidth JFET Operational Amplifier*, Analog Devices, Wilmington, MA, **2002**.

30 Texas Instruments, *Data Sheet: OPA2107 Precision Dual Difet Operational Amplifier*, Texas Instruments, Dallas, TX, **2003**.

31 Texas Instruments, *Data Sheet: OPA2132 FET-Input Operational Amplifiers*, Texas Instruments, Dallas, TX, **1995**.

32 Analog Devices, *Data Sheet: AD622 – Low-Cost Instrumentation Amplifier*, Analog Devices, Wilmington, MA, **1999**.

33 Analog Devices, *Data Sheet: AD734–10 MHz, 4-Quadrant Multiplier/Divider*, Analog Devices, Wilmington, MA, **1999**.

34 Texas Instruments, *Data Sheet: OPA627/OPA637 Precision High-Speed Difet Operational Amplifiers*; Texas Instruments, Dallas, TX, **1998**.

35 Maxim, *Data Sheet: MAX274/275 4th and 8th-Order Continuous-Time Active Filters*; Maxim, Sunnyvale, CA, **1996**.

36 T. Juneau, A. P. Pisano, in: *Solid-State Sensor and Actuator Workshop, Hilton Head, SC*; **1996**, pp. 299–302.

37 W. A. Clark, R. T. Howe, R. Horowitz, in: *Solid-State Sensor and Actuator Workshop, Hilton Head, SC*; **1996**, pp. 283–287.

38 Analog Devices. *Data Sheet: ADXL 105 – High Accuracy ± 1g to ± 5g Single Axis IMEMS Accelerometer with Analog Input*; Analog Devices, **1999**.

39 C. Nguyen, T.-C., *Micromechanical Signal Processors*; Doctoral Dissertation, UC Berkeley, Berkeley, CA, **1994**.

40 W. C. Tang, C. Nguyen, T.-C., R. T. Howe, *Sens. Actuators* **1989**, *20*, 25–32.

41 J. Cao, C. T.-C. Nguyen, in: *10th International Conference on Solid-State Sensors and Actuators, Sendai*, **1999**, pp. 1826–1829.

42 J. I. Seeger, X. Jiang, M. Kraft, B. E. Boser, in: *Solid-State Sensor and Actuator Workshop, Hilton Head, SC*, **2000**, pp. 296–299.

43 A. Preumont, *Vibration Control of Active Strctures: an Introduction*, 2nd edn.; Dordrecht: Kluwer, **2002**.

44 X. Jiang, J. I. Seeger, M. Kraft, B. E. Boser, in: *Symposium on VLSI Circuits, Honolulu, HI*, **2000**, pp. 16–19.

45 H. Kulah, A. Salian, N. Yazdi, K. Najafi, in: *Solid-State Sensor, Actuator and Microsystems Workshop, Hilton Head, SC*, **2002**, pp. 219–222.

46 M. Degrauwe, E. Vittoz, I. Verbauwhede, *IEEE J. Solid-State Circuits* **1985**, *20*, 805–807.

47 N. Yazdi, K. Najafi, in: *IEEE Int. Solid-State Circuits Conf. Dig. Tech. Papers, San Francisco, CA*; **1999**, pp. 132–133.

48 T. Kajita, M. Un-Ku, G. C. Temes, in: *2000 IEEE International Symposium on Circuits and Systems. Emerging Technologies for the 21st Century. Proceedings (IEEE Cat No.00CH36353)*; Lausanne: Presses Polytech. Univ. Romandes, 2000, Vol. 4, pp. 337–340.

49 J. A. C. Bingham, *IEEE Trans. Circuits Syst.* **1984**, *CAS-31*, 419–420.

50 G. F. Franklin, J. D. Powell, A. Emami-Naeini, *Feedback Control of Dynamic Systems*, 3rd edn.; Addison-Wesley, New York, NY, **1994**.

51 S. R. Norsworthy, R. Schreier, G. C. Temes, *Delta-Sigma Data Converters: Theory, Design and Simulation*; IEEE Press, New York, NY, **1997**.

52 B. Widrow, *IRE Trans. Circuit Theory* **1956**, *CT-3*, 266–276.

53 W. Henrion, L. DiSanza, M. Ip, S. Terry, H. Jerman, in: *Solid-State Sensor and Actuator Workshop, Hilton Head, SC*, **1990**, pp. 153–157.

54 K. A. A. Makinwa, J. H. Huijsing, in: *International Solid-State Circuits Conference, San Francisco, CA*; **2002**, pp. 432–433.

55 X. Jiang, *Capacitive Position-Sensing Interface for Micromachined Inertial Sensors*; Doctoral Dissertation, UC Berkeley, Berkeley, CA, **2003**.

56 C. Lu, M. Lemkin, B. E. Boser, *IEEE J. Solid-State Circuits* **1995**, *30*, 1367–1373.

4
Packaging of Advanced Micro- and Nanosystems

V. M. Bright, C. R. Stoldt, NSF Center for Advanced Manufacturing and Packaging of Microwave, Optical and Digital Electronics, Department of Mechanical Engineering, University of Colorado, Boulder, CO, USA
D. J. Monk, M. Chapman, A. Salian, Motorola Sensor Products Division, Phoenix, AZ, USA

Abstract
Packaging of micromachined transducers plays a vital functional role in the production of a microsystem. The package both protects the microsystem and provides an interface to the environment. However, it can affect the output of the microsystem, dictate the assembly and testing methodology of the device, and represent a considerable amount of the cost of a product. This chapter reviews current methods of MEMS and nanosystems packaging. First, general packaging definitions and associated technologies are presented. Then specific processes are reviewed including protective coatings, directed and self-assembly, and device-level packaging. The focus of the chapter is on the zero- and first-level packaging. Finally, two specific case studies of transducer packaging are presented: tire pressure sensor and accelerometer from Motorola.

Keywords
micro; nano; MEMS; packaging; pressure; sensor; accelerometer; microsystem; nanosystem.

Advanced Micro and Nanosystems. Vol. 1
Edited by H. Baltes, O. Brand, G. K. Fedder, C. Hierold, J. Korvink, O. Tabata

ISBN: 3-527-30746-X

4.1 Introduction

Microelectromechanical systems (MEMS) are integrated microdevices that combine electrical and mechanical components. These systems can sense, control and actuate on the nano- and microscale and function individually or in arrays to generate effects on the macroscale. They are fabricated using batch processing techniques similar to the integrated circuit (IC) industry and can range in size from micrometers to millimeters, with much attention given these days to nanoscale-sized device components. A number of standardized processes are commercially available for the fabrication of MEMS devices. Bulk micromachining processes involve the removal of significant amounts of substrate material in order to form the desired structure. In contrast, surface micromachining processes take place on the surface of the wafer. In surface micromachining, thin-film structural elements are commonly deposited using techniques such as chemical vapor deposition (CVD) and electroplating.

Most MEMS act as sensors or actuators, that is, devices that interact directly with the physical environment. Sensors transduce such physical signals as pressure, temperature, acceleration, displacement and light into electrical signals to be processed by electronics. Moreover, electrical signals can be transduced in actuators to perform work on the environment, such as pumping and valve operation in microfluidics and switching in optics and/or electronics. It is possible that the physical signals induced by the package can be misinterpreted as signals from the environment [1, 2]. For example, in pressure sensors, packaging stress can adversely affect the sensor inside, thus altering the output of the device. Vibrational modes in the package can influence the operation of an accelerometer. On the other hand, the package can play a vital functional role in the production of a microsystem. For example, the package can isolate the induced stress from the mounting of the device, it can make the electronics compatible with the harsh environment and it can even be an integral part of the microsystem (e.g. mechatronics packaging).

Besides the fact that packaging can affect the output of a microsystem, assembly and testing of these devices represents a considerable amount of the cost of a product, with a typical cost breakdown roughly 33% silicon content, 33% package and 33% test. In some cases, >70% packaging cost vs. silicon content has been reported [3–10]. Furthermore, of this silicon content, often less than half of it is the microsensor or actuator; the rest is the control electronics cost.

Packaging for MEMS has been significantly under-represented in academic publications. Much of the grant money, academic research and publications are devoted to the micromachined device itself and not the rest of the microsystem. However, packaging for MEMS is an important challenge to the commercialization and long-term reliability of microsystems.

This chapter summarizes many of the practical issues in MEMS packaging. We start by describing general packaging definitions and associated techniques. This is followed by specific processes for MEMS packaging, including coatings, directed and self-assembly and device-level packaging. The focus of the chapter is on the zero- and first-level packaging. We conclude the discussion with specific practical examples (case studies) of MEMS packaging: pressure sensors and accelerometers from Motorola.

4.2 MEMS Packaging

In general, packaging of MEMS is an outgrowth of the IC packaging industry and the hybrid packaging industry, but with the following requirements [1, 5, 11–15] (see Fig. 4.1):

- interaction with the environment (e.g. media compatibility or hermetic, vacuum sealing to protect accelerometers, resonators, etc.);
- low cost;
- small size;

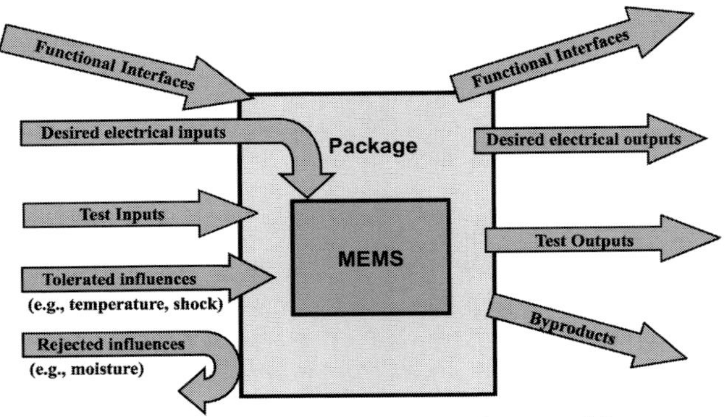

Fig. 4.1 MEMS packaging requirements [16]. Courtesy of Professor Y. C. Lee, University of Colorado at Boulder

- high reliability and quality;
- standardization (although custom packaging for MEMS seems to be the rule);
- acceptable electrical interconnection (e.g. minimum power supply voltage drop, self-inductance, cross-talk, capacitive loading, adequate signal redistribution and, perhaps, electrical feedthroughs);
- thermal management (i.e. power dissipation and matching CTEs with substrates to minimize package-induced stress);
- acceptable mechanical interconnection (e.g. porting for pressure sensors and mounting techniques that do not apply undue stress to the device yet support the device appropriately);
- protection from electromagnetic interference;
- testability and trimability (e.g. internal test/trim nodes);
- precision alignment (especially for optical MEMS);
- mechanical protection and stress isolation.

4.2.1
Package Partition Definitions

A commonly accepted hierarchy of MEMS packaging is (see Fig. 4.2):

- zero-level packaging: device encapsulation (wafer level);
- first-level packaging: die-to-package;
- second-level packaging: package-to-substrate (board);
- third-level packaging: board-to-module.

The discussion in this chapter is focused on the zero and first packaging levels. An additional level, the sub-zero level, is introduced to describe nano- and microstructure assembly using self- and directed assembly processes, and also two-dimensional nano- and microstructural coatings as the initial steps to device reliability and protection prior to device encapsulation at level zero.

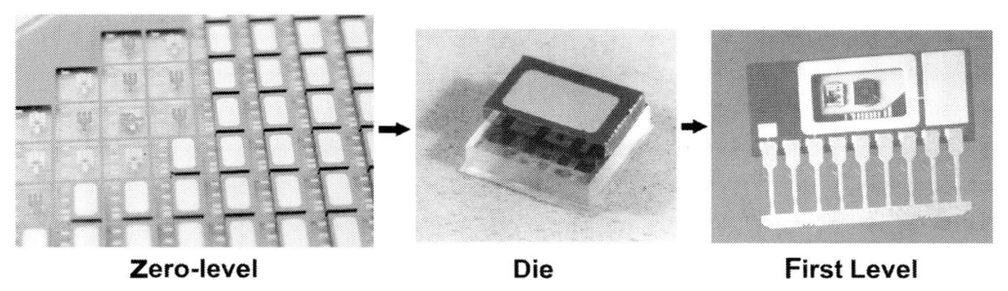

Fig. 4.2 Package partitioning. Courtesy of Dr. L. Spangler, Aspen Technologies

4.2.2 Sub-zero Level Packaging

4.2.2.1 Nanoscale Bottom-up Assembly

Reducing the scale of electronic, photonic and magnetic systems below 50 nm presents many technical challenges in device construction, the foremost of which is the inability of conventional microfabrication processes to mass produce nanoscale devices and systems efficiently in a cost-effective manner. The natural alternative to this approach is bottom-up fabrication, where the nanoscale device or system is built from atomic or molecular components by a self-assembly process. A recent review details many of the emerging techniques now employed for the self-assembly of nanoscale components, as well as many of the recent applications for nanofabrication in the development of new devices and systems [17].

Perhaps the greatest challenge facing the future development of nanoscale devices lies in the difficulties of component integration. Issues such as nanoscale manipulation and interconnectivity remain largely unsolved for large-scale nanodevice manufacturing and, as a result, most of the viable nanodevices demonstrated thus far are the hybridization of nanoscale functional components with relatively straightforward microfabricated device platforms. To date, a wide range of functional components have been pursued by bottom-up self-assembly and include but are not limited to nanotubes, nanoparticles, quantum dots, nanowires and biotemplated nanostructures. Three of the more prominent types of nanoscale functional components and their nanofabricated devices are briefly reviewed next.

Carbon is the most familiar and pursued form of the nanotube for engineering applications owing to its superior physical properties. One of the most promising routes to carbon nanotube (CNT) fabrication for future integration into nano- and microscale devices utilizes metal-catalyzed vapor-phase processing. An example of vertically oriented CNTs grown by this process is shown in Fig. 4.3a [18]. Through the lithographic patterning of the metal catalyst layer, CNT bundles can be selectively grown and integrated into microfabricated devices that utilize the nanostructures as the key functional components, such as integrated cold cathode field emitters shown in Fig. 4.3b [19]. Alternatively, individual CNTs can be utilized as the functional elements in nanodevices and are typically integrated through me-

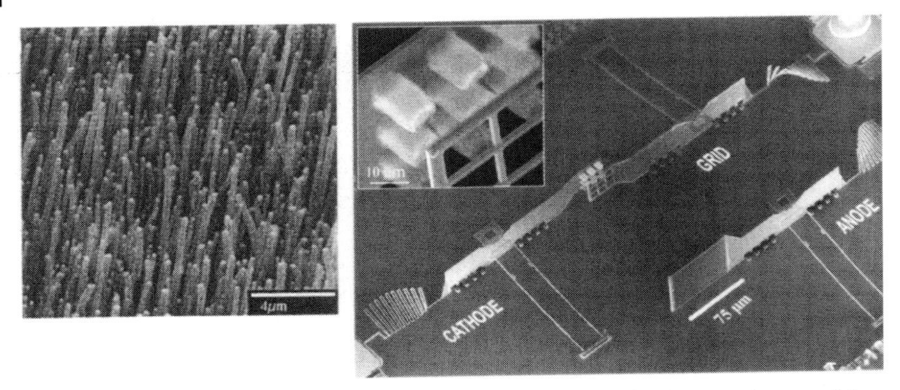

Fig. 4.3 Electron micrographs showing (a) self-assembled carbon nanotubes [18] and (b) a MEMS-based cold cathode device that utilizes CNT assemblies as the field emitters [19]. The inset in (b) shows a close-up of the CNT field emitter assemblies that are selectively deposited using metal-catalyzed vapor phase deposition

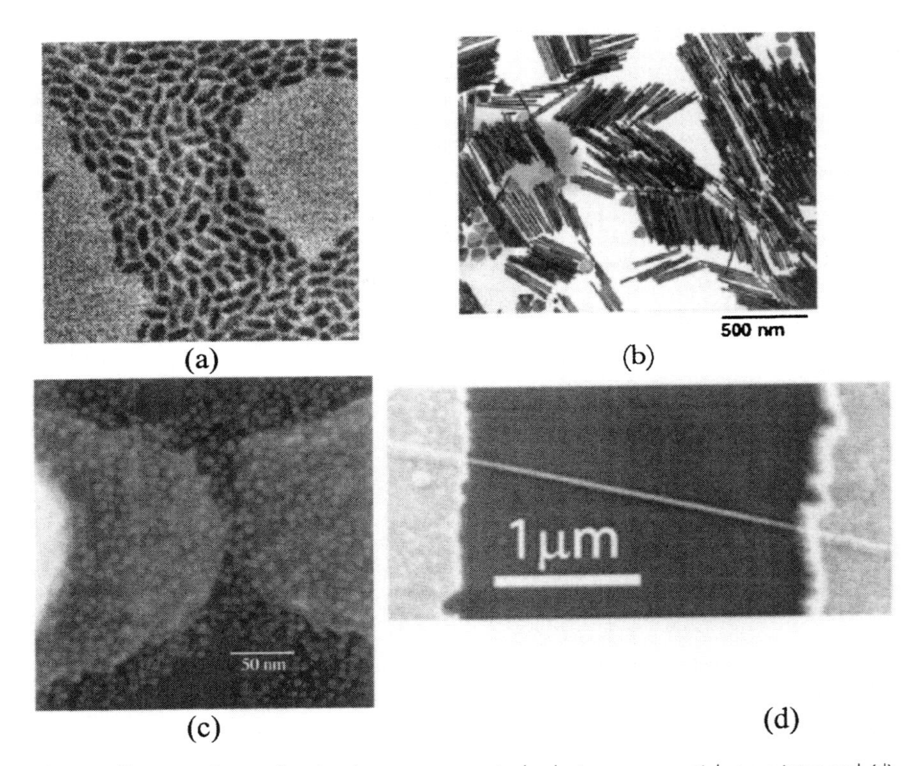

Fig. 4.4 Electron micrographs showing assemblies of (a) CdSe nanoparticles [22] and (b) Au nanorods [27], (c) a microfabricated single-electron nanoparticle transistor and (d) an Si nanowire FET sensor [30]

chanical or liquid-phase assembly. Notable examples include a CNT-based field-effect transistor [20] and a CNT-based gas sensor [21].

Through both vapor-phase and liquid-phase assembly routes, a wide range of solid-state nanowires and nanoparticles have been fabricated [22–27]. Various metal, ceramic and semiconductor compositions have been synthesized in the form of isotropic nanoparticles to highly anisotropic nanowires (nanorods), with the primary goal thus far being the determination of their collective physical properties. Examples of two-dimensional assemblies of semiconducting CdSe nanoparticles [22] and metallic Au nanorods [27] are shown in Fig. 4.4 a and b, respectively. As with CNTs, small-scale devices have been successfully integrated with both assemblies and single numbers of nanoparticles and nanowires. For instance, Sn oxide nanoparticle assemblies have been successfully integrated into microscale hotplate platforms for residual gas detection [28], and single-electron transistors and (bio)chemical sensors have been developed that utilize single CdSe nanoparticles (Fig. 4.4 c) [29] and individual Si nanowires (Fig. 4.4 d) [30], respectively.

In all nanodevice instances, the state-of-the-art remains primitive and, as a result, packaging issues are yet to be adequately defined. Before an issue such as nanoscale packaging can be dealt with, the more immediate challenges of component manipulation and integration must be overcome for nanodevices to find mass production and the commercial market.

4.2.2.2 MEMS-directed Self-assembly

Assembly of Surface Micromachined MEMS Using Pre-stressed Microstructures

Surface micromachined designs are planar through fabrication and may require some assembly after release. A great number of MEMS are based on 3D assembly of planar microstructures. The primary components that allow 3D assembly of planar parts are microhinges and microlatches, originally proposed by Pister et al. [31]. Microhinges enable surface micromachined parts to be rotated out of the plane of the substrate. This allows fabrication of flip-up components composed of plates of a structural material. There are several types of microhinges: the substrate hinge, the scissors hinge and a flexure hinge. A substrate hinge is used to hinge released structures to the substrate. It consists of a pivot arm and a staple. The staple, which is anchored to the substrate, allows free rotation of the pivot arm. Scissors hinges are fabricated by interlocking fingers of two structural layers and allow two plates to be connected together while still allowing them to pivot at the connection. The flexure hinge is the flexible beam or spring connected between two structural layers. Microlatches are used to lock flip-up structures into specific positions [32]. When the flip-up plate is raised into position, the latch moves along the surface of the plate until it drops into position in the locking slot cut in the plate. The advantage of the self-engaging lock is that it is only necessary to flip up one plate. This is an important feature in the design of self-assembled microsystems.

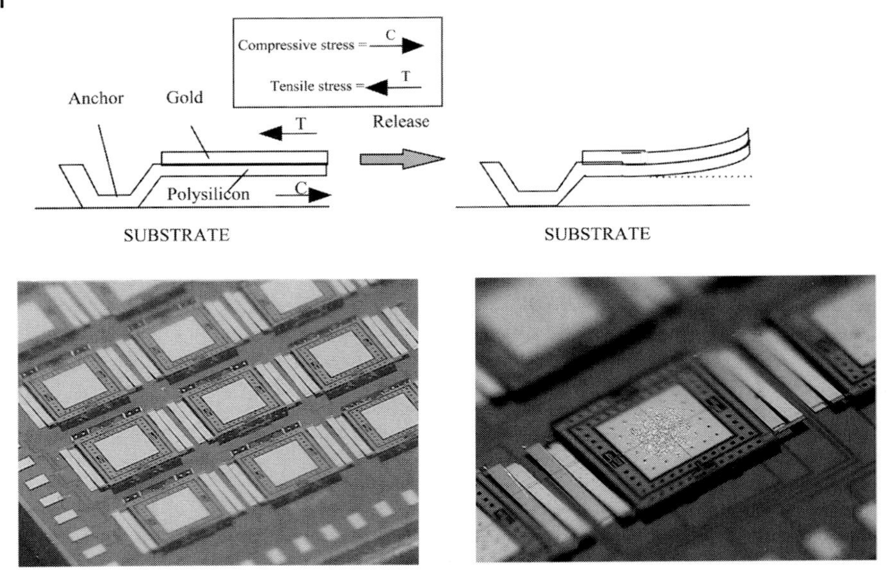

Fig. 4.5 Illustration of cantilever curling due to residual material stress and an array of micromirrors assembled using pre-stressed cantilevers (V. Bright group, University of Colorado)

The suspended surface micromachined structures may curl owing to internal material or thermo-mechanical stresses. Although typically undesirable in many MEMS, the stress-induced curling can be used as a lift mechanism during directed MEMS assembly. Consider the example in Fig. 4.5. Prior to release, the cantilever is held flat owing to the surrounding oxide. After release, the residual material stresses in the gold (tensile) and polysilicon (compressive) layers cause the free end of the cantilever to curl upwards. The amount of curl or deflection is dependent on the amount of stress and the physical dimensions of the material layers; more on this is discussed in the thermomechanical behavior of microstructures section below.

MEMS Surface Tension-driven Assembly

Another method of assembling MEMS has been proposed that uses the surface tension properties of molten material such as solder or glass as the assembly mechanism [33]. The solder method uses hinged micromachined plates with specific areas metalized as solder wettable pads. A fluxless soldering process can be used [34]. Once the solder is deposited, the substrates are placed in a chamber filled with nitrogen and formic acid gas. The nitrogen gas prevents the solder from oxidizing at high temperatures and the formic acid gas removes any existing oxide. The substrates are then heated to melt the solder and the force produced

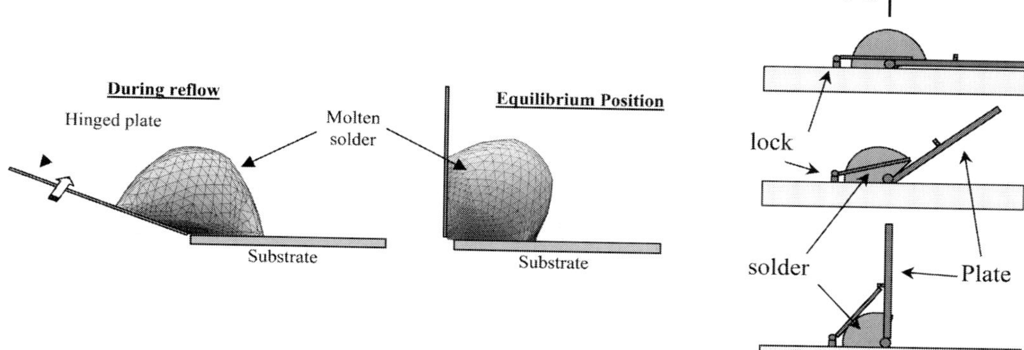

Fig. 4.6 Illustration of 3D surface micromachined MEMS assembly using surface tension of molten solder: (left) side view of the assembly in progress and (right) using a micro-latch mechanism to define the final position of the assembly. (V. Bright group, University of Colorado)

by the natural tendency of liquids to minimize their surface energy pulls the hinged plate away from the substrate (see Fig. 4.6). One way to control the precision of solder assembly is through accurate solder volume control. The assembly precision in this case is dependent on the amount of solder deposited. An alternative approach for precision-demanding applications is to use a self-locking design integrated with solder technology. In this case the assembly precision is determined by the design geometry of the assembled microstructure and the self-engaging microlatches [34].

Using a surface-tension assembly mechanism, arrays of precision assembled MEMS can be accomplished with a single batch process and the cost/assembly can be reduced by orders of magnitude. Moreover, using solder provides solid mechanical, thermal and electrical connections to the assembled device. Several surface-tension assembly processes exist [35], with Fig. 4.7 illustrating some examples.

Flip-chip Transfer and Assembly of Microstructures

Common surface micromachining processes provide few choices of materials or number of structural layers. A flip-chip transfer and assembly followed by removal of the initial silicon substrate can be used to integrate MEMS with other devices or circuits on substrates other than silicon [36–39]. Such an approach is particularly beneficial to rf or optical components that require alternative substrates other then silicon. The flip-chip silicon substrate is removed to produce a microsystem where specific material properties are paramount or where more structural layers are required. Challenges in this technology are the repeatable flip-chip bonding with sub-micron gaps, post-assembly release of MEMS and removal of the silicon host substrate. Fig. 4.8 illustrates the processes based on indium bumps:

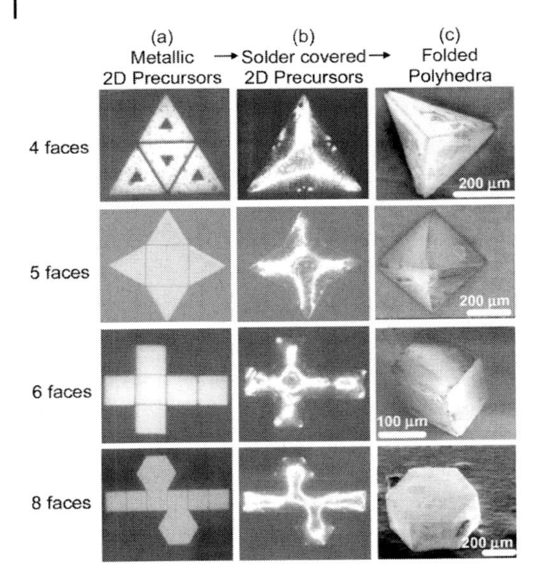

Fig. 4.7 Examples of solder assembled microstructures [35]. (a–c) Folded metallic polyhedra; (d) rotary fan with blades assembled on top of a scratch drive actuator motor; (e) electroplated inductor; (f) self-assembled hexagonal MEMS mirror

1. Design and fabricate MEMS with a layer of SiO_2 between the MEMS and the host silicon.
2. Deposit indium bumps on the bonding pads.
3. Bond MEMS to a target substrate through thermocompression or thermosonic processes.
4. Glue silicon substrate to the target substrate using epoxy. An alternative method uses breakable tethers to support the substrate (see Fig. 4.9).
5. Remove SiO_2 in HF to separate the host silicon from the MEMS.
6. Remove the host silicon by breaking the epoxy bond or the support tethers.

The bonding pads are only connected to the host silicon substrate with the sacrificial SiO_2 layer. When the SiO_2 is dissolved in HF, the bonding pads and the MEMS device are completely disconnected from the host silicon substrate. The

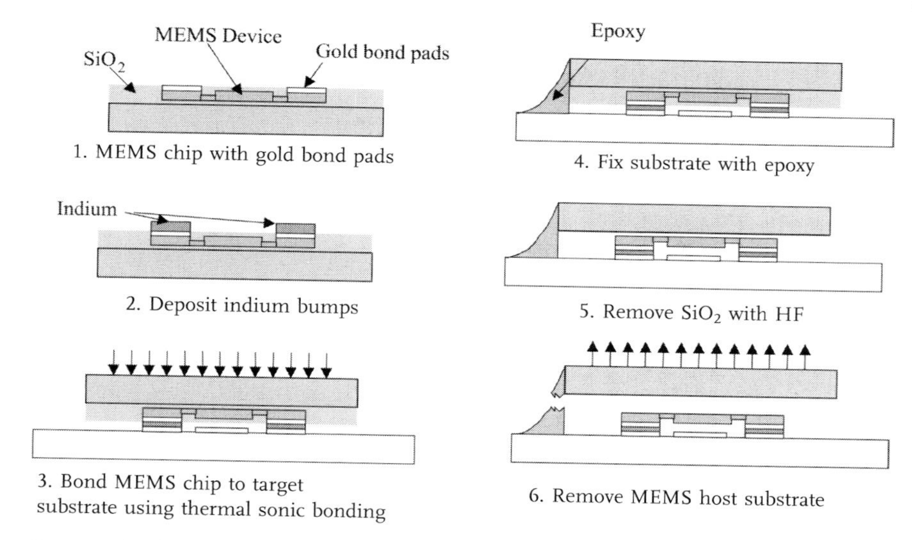

Fig. 4.8 Flip-chip assembly of MEMS

epoxy bond holds the silicon substrate in place during the HF release process to protect the transferred MEMS. A breakable tethers approach can be used instead of epoxy (Fig. 4.9).

Among the processes illustrated in Fig. 4.8, the bonding and the release are the two critical steps. Sub-micron height variations are required to assure repeatable MEMS performance that is sensitive to the height, e.g. air gap that affects the capacitance for rf applications or applied electrostatic voltage for driving actuators. With a special bonding tool designed for precision force control, it is possible to control height to within ± 0.1 μm. The second critical step is the post-assembly release of the MEMS structure from the host silicon substrate. If the release is incomplete, the MEMS structure can be damaged or destroyed when the host substrate is removed. On the other hand, if the MEMS structure is left too long in the etchant, the structure can be over-etched and damaged. A particular challenge for flip-chip MEMS is to remove the SiO_2 when the flow of the etchant is constrained by the narrow gap between the silicon and the target substrates. Therefore, a long etching time is needed for effective release, but a long release process can over-etch thin microstructures. A solution to this problem is to include a partial pre-release step before bonding, leaving the MEMS structure connected to the host silicon by breakable mechanical tethers. Essentialy, the majority of the silicon dioxide etching is done before bonding. After the MEMS structure is transferred, the support tethers are broken off during host silicon removal (Fig. 4.9).

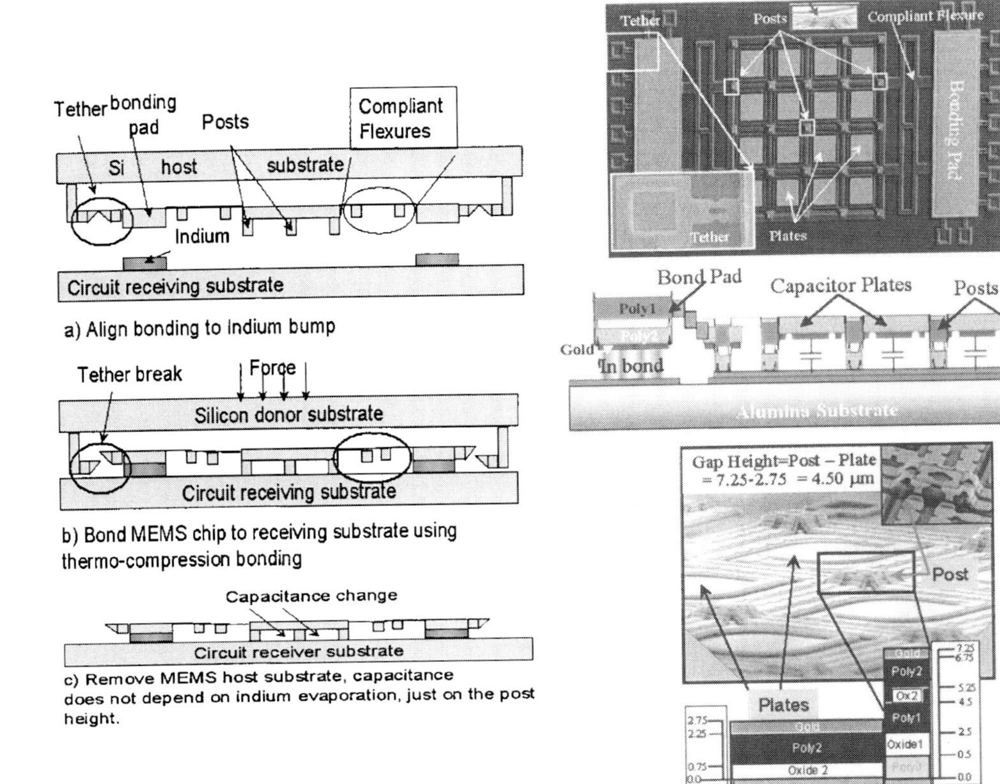

Fig. 4.9 Polysilicon MEMS voltage-tunable variable capacitor transferred to a ceramic substrate: process of transfer using breakable tethers, SEM of the transferred device and illustration of the flip-chip assembly height control posts. The posts are used to contol the gap height precisely to within 0.25 µm. (Courtesy of F. Faheem and Professor Y. C. Lee, University of Colorado at Boulder)

4.2.2.3 **MEMS Reliability**

Reliability of MEMS is still a relatively unexplored field. An early study performed by Brown et al. examined the reliability of silicon and other micromachining materials through the observation of slow crack growth [40, 41]. Typical MEMS reliability qualifications include humidity exposure, temperature cycling, high- and low-temperature storage, vibration testing and electromagnetic interference exposure. The following discussion summarizes some of the more notable reliability issues of MEMS and the work that has been done.

It is widely accepted that MEMS reliability is related to nanoscale interface phenomena. One approach to tackle this problem is through novel materials and/or fabrication methods. An alternative approach is to take into account interface phenomena such as adhesion or charging and compensate through innovative design.

Additionally, a detailed understanding of manufacturing and final application failures is required. For automotive safety critical applications (e.g. airbag accelerometers), acquiring a deep understanding of the root cause of product failures throughout a product's entire life cycle [e.g. MEMS fabrication, assembly, test, shipping, module manufacturing, car assembly and in use (warranty)] has been ongoing since the mid-1990s. Implementation of corrective actions has led to orders of magnitude reductions in product and system failures.

Primary micro and nanoelectro-mechanical device reliability challenges include the following:

1. Thin-film microstructures are susceptible to adhesion due to large contact surface area, chemical and capillary forces and/or dielectric charging.
2. Microstructures exhibit a short life-time due to significant friction and wear on the microscale.
3. Multilayered microstructures are susceptible to deformation during fabrication, assembly and use.
4. Multilayered MEMS structures experience changes in curvature during thermomechanical loading or fabrication/packaging processes that require temperature cycling.
5. Multilayered MEMS structures experience stress relaxation with time, thus changing device functionality.

In MEMS, surface forces are more significant than gravity and inertia, which are the dominant forces at the macroscale. This is due to the large increase in surface-to-volume ratio at the microscale, which makes interfacial adhesion a significant issue. The structural elements in MEMS are only a few micrometers thick and a few micrometers apart and are often constructed of polycrystalline materials that exhibit nanometer-scale roughness. Moreover, the surface properties of MEMS devices are strongly related to their processing and to their environmental exposure. For example, polysilicon roughness depends on deposition temperature. Likewise, the surface energy of the polysilicon depends on the final surface treatment process. If supercritical drying is performed to release the parts, the surfaces are hydrophilic. The surfaces can instead be rendered hydrophobic by applying a self-assembled monolayer (SAM) [42]. The SAMs have a number of limitations, however. One limitation is that the coatings are organic and have a very limited range of properties. Another limitation is that the coatings are limited to a single monolayer and as such do not provide sufficient electrical insulation or sufficient protection against frictional wear. Moreover, SAMs can result in nonuniform films, are affected by temperature variations during device operation and have also proven not to last for extended periods of time. Another approach is to use chemical vapor deposition (CVD) to coat MEMS. However, CVD also has several limitations. One limitation is that the high deposition temperatures for CVD are often not suitable for post-processing of MEMS. For example, the temperature for tungsten deposition by CVD is generally not less than 450 °C and may be even higher (e.g. 650 °C). Differences in coefficients of thermal expansion between the deposited layer and underlying materials, over the large temperature range between deposition temperature and ambient, may create significant material stress gradients between the deposited layer and the underlying materials. Another limitation is that the deposited layers do not have uniform thickness. In particular, CVD primarily deposits on line-of-sight surfaces, which means that bottom surfaces and small void regions receive less or almost no coating. Another technique that is being developed for coating MEMS is atomic layer deposition (ALD). The technique allows uniform device coating with one atomic layer at a time. Although ALD shows much promise, it is a relatively novel method of coating for MEMS that requires further studies. Thus, much work remains to be done to find inorganic, tough coatings that can be used as lubricating layers and also adhesion prevention and mass transfer barriers in MEMS surface contacts.

Because surface forces are not well characterized and are often difficult to reproduce in MEMS structures, most commercial applications, such as accelerometers, avoid contact between structural members. In general, however, many MEMS devices may experience contact between structural elements. A representative example is the two contacts of a microswitch that are intentionally brought together to complete an electric circuit. When the applied force is removed, the contact may stick together and not separate. MEMS switch problems in general include metal contact switches (contact degradation due to adhesion, mass transfer between contacts, other contamination such as oxidation) and dielectric switches (charge build-up, contamination). Besides operational contact, surfaces in MEMS devices may also come into contact while in storage owing to shock, electrostatics or capillary action. Adhesive forces can then develop which prevent device subsequent function [43]. To address the effect of prolonged storage and environment (such as humidity), a brute-force solution has been to develop hermetic packaging for MEMS. However, hermetic packaging is very difficult to implement, expensive and application specific. It is common to have at least 70% of the packaged device cost to be embedded in the package itself. Therefore, solving the fundamental adhesion problem at the interface on the microscale may result in a relaxed overall device packaging requirement, thus lowering the cost of manufacturing, increased performance, increased repeatability, longer device lifetime and overall improved reliability of the system.

Surface adhesion (or stiction as it is commonly referred to in the MEMS community) can be described as the phenomenon where microstructures become stuck to the neighboring surfaces. Several authors have provided reviews of the stiction phenomenon [44–48] and comparative studies of various stiction reduction techniques [49]. Typically, stiction occurs at one of two times (particularly for surface micromachined structures): post-sacrificial layer etch and during usage. The former is called 'release stiction' and the latter 'in-use stiction'. Packaging engineers are most often concerned with in-use stiction because of the potential reliability problems that it can cause; however, a phenomenon similar to release stiction can be a problem if devices are exposed to liquids during packaging and assembly – particularly during the wafer sawing operation.

Release stiction has been recognized as a problem, especially for surface micromachined structures, since the late 1980s. During drying, surface tension from the liquid-vapor interface causes a downward force on the structural layer. If the layer touches the substrate, it is prone to stick to the surface. It is hypothesized that etch products and/or contaminants in the rinse water can then precipitate out of solution during drying and cause a bond that is stronger (e.g. a chemical bond) than electrostatic bonding between the two semiconductors [50–55]. To prevent microstructures from sticking, the surface area of the contact interface must be minimized, thus the technique of using dimples or standoff bumps [56–60] has been developed. Alternatively, breakable tethers can be used to hold the device in place temporarily during sacrificial layer etching and drying. These tethers can be broken or melted following release [61]. Efforts have been aimed at modifications to the etch process to include: the use of gas-phase HF to eliminate liquid totally from the sacrificial etch process, use of a rinse solution that exhibits a lower surface tension force during drying (e.g. IPA) and freeze-drying and sublimation to eliminate the surface tension forces. The most successful, however, seems to be the use of the supercritical carbon dioxide drying technique [62, 63].

In the case of in-use stiction, it is hypothesized that moisture from the environment comes in contact with the MEMS structural surfaces. If, during operation, these structures come in contact, the moisture can cause a temporary bond, which, like release stiction, can then become permanent with time. To reduce in-use stiction, three basic techniques have been attempted. The first is to use a hermetic seal around the microstructure to eliminate the possibility of moisture encountering the structure. An example of this is the Motorola accelerometer, which uses a glass frit wafer bond that hermetically seals the microstructure from the environment, to eliminate the possibility of moisture affecting the device. Second, the use of techniques (such as ammonium fluoride) to minimize the work of adhesion has been employed [64, 65]. Lastly, a variety of coatings and/or surface treatments have been used on the microstructure to eliminate the chance of contact between two surfaces that have a prevalence to stick [46, 66–70].

Thermomechanical Behavior of Multilayer Microstructures

Multilayer material systems abound in MEMS applications, serving both active and passive structural roles. In these many applications dimensional control is a critical issue. Surface micromachined mirrors, for example, require optically flat surfaces, with $<10\%$ of a wavelength variation across the mirror's surface. This level of flatness is difficult to achieve owing to curvature commonly seen in micromachined mirrors composed of several different material layers. For other applications, such as rf MEMS, cantilever and bridge-like structures are used to make electrostatic switches. It is important to control warpage of these large-area, thin actuator structures in order to achieve desired deflection versus voltage relationships, on/off switching times and rf frequency response. An inherent characteristic of such multilayer material structures is that misfit strains between the layers (for example, due to intrinsic processing stresses or thermal expansion mismatch between the materials upon a temperature change) lead to stresses in the layers and deformation of the structures. To obtain the necessary thermal expansion mismatch in bilayer actuators, one film is typically a metal and the other a material with a much lower thermal expansion coefficient such as a glass or ceramic. Innovative designs and layering arrangements have led to microactuators capable of deforming both in and out of the plane of the bilayer, providing three-dimensional motion. Such actuators have been used for many purposes, including rf switches, optical positioning, accelerometers and 3D microassembly [71–77]. Irrespective of the intended application, MEMS multilayer material systems are susceptible to substantial deformation upon temperature changes and their implementation relies on a solid understanding of this behavior, either to use it productively or to compensate for it. In either case, the most straightforward phenomenon to deal with from a design viewpoint is linear thermoelastic behavior where the multilayer deforms linearly with a temperature change. Metal films, however, do not typically exhibit a stable microstructure in their as-deposited state and exposing them to temperature excursions results in highly nonlinear deformation behavior during the first few cycles due to microstructural evolution such as annihilation of excess vacancies, void coalescence and grain growth. This phenomenon has been known for some time in the microelectronics materials community [78–81], where the primary interest has been in understanding the evolution of stresses in deposited metal lines over a few thermal cycles. It exists in the context of MEMS multilayers also and is important for short- and long-term dimensional stability issues due to inelastic deformation mechanisms in the metal. Relevant examples include rf switch arrays [82], tunneling accelerometers [75] and resonant mass sensors [83]. In MEMS applications the film thicknesses are often, but not always, comparable, unlike in microelectronics where the metal film is typically orders of magnitude thinner than the substrate. The latter case can be referred to as the thin-film limit of the case of two arbitrary thickness films. In the cases where the film thicknesses in MEMS multilayers are not comparable, i.e. one layer is orders of magnitude thinner than the other layer, the studies in microelectronics are directly applicable. In cases where the film thicknesses are comparable, i.e. prestressed cantilever structures used for MEMS assembly, this leads to different stress states in the films because the large stresses that would exist in the thin-film limit are substantially relaxed by bending of the multilayer. Furthermore, there is a through-thickness stress gradient, unlike in the thin-film limit where the stress through the thickness of the film is essentially constant since the substrate is so thick.

The thermoelastic deformation experienced by layered MEMS structures has been the subject of numerous studies and is fairly well understood, including both linear and geometrically nonlinear deformation [84]. Consider the basic thermomechanical response of a layered plate when subjected to temperature changes or other sources of misfit strains between the layers (see Fig. 4.10). When such a layered plate is subjected to a temperature change, two key aspects of deformation occur: straining of the midplane and bending. When the transverse deflections due to bending are of prime importance, as is often the case, one way broadly to characterize the deformation response, especially for plates with relatively large in-plane dimensions compared with their thickness, is in terms of the average

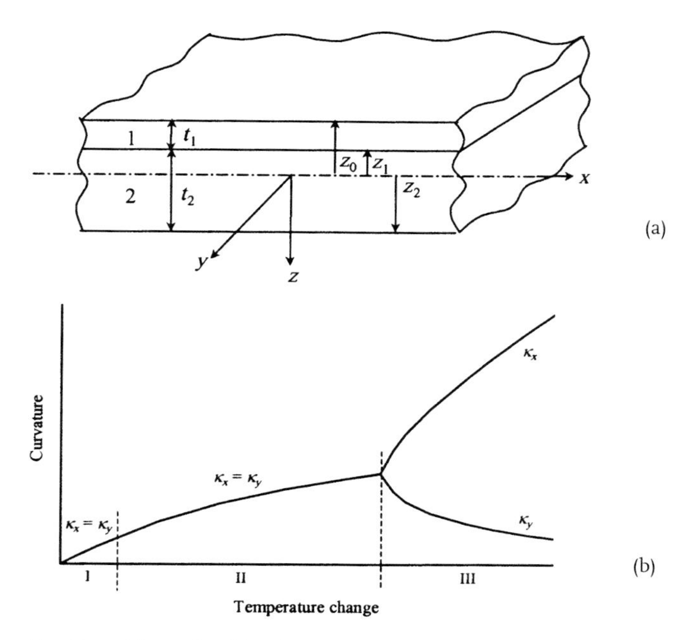

Fig. 4.10 Schematic of (a) the geometry of the two-layer plate microstructure showing relevant dimensions and (b) the general characteristics of the average curvature vs. temperature change of a two-layer plate microstructure. (Courtesy of Professor M. L. Dunn, University of Colorado at Boulder)

curvature developed as a function of temperature change. Formally, the curvature is a second-rank tensor and for the problem at hand it can be wholly described by the two principal curvature components, e.g. in the *x*- and *y*-directions, κ_x and κ_y. The curvature is a pointwise quantity, meaning it varies from point to point over the in-plane dimensions of the plate. To illustrate the nature of deformation, consider the seemingly simple case of a plate with total thickness much less than the in-plane dimensions of the plate composed of two isotropic layers with different material properties (elastic modulus and thermal expansion) subjected to a temperature change (Fig. 4.10a). In terms of the average curvature variation as a function of temperature change, three deformation regimes are possible, as illustrated in Fig. 4.10b [84]. The first regime, *I*, consists of a linear relation between the average curvature and temperature change where $\kappa_x = \kappa_y$, i.e. the average curvature is spherically symmetric. This symmetric deformation would not exist if the material properties were anisotropic. This deformation regime is characterized by both small transverse displacements and rotations and so conventional thin-plate theory adequately describes the deformation. The second regime, *II*, consists of a nonlinear relation between the average curvature and temperature, but again $\kappa_x = \kappa_y$. The behavior is due to geometric nonlinearity that results when the deflections become excessively large relative to the plate thickness and they contribute significantly to the in-plane strains. It has been shown that in these two regimes the symmetric deformation modes are stable. The second regime ends at a point when the deformation response bifurcates from a spherical to ellipsoidal deformation, i.e. $\kappa_x \neq \kappa_y$. At this point, the beginning of regime *III*, it becomes energetically favorable for the plate to assume an ellipsoidal shape because to retain the spherical deformation under an increasing temperature change requires increased midplane straining. After the bifurcation, the curvature in one direction increases whereas that perpendicular to it decreases; the plate tends towards a state of cylindrical curvature. This observation helps to explain the energetic argument as unlimited cylindrical curvature can be obtained with no midplane straining, whereas spherical curvature cannot. This discussion was based on the assumption of linear material behavior. Additional deformation regimes result if material nonlinearity is present, e.g. yielding [85].

Most work regarding the deformation of layered MEMS systems has focused on the first linear regime. This includes most of the understanding developed in the context of microelectronics applications where the thin film limit of this behavior is applicable. Indeed, much of the understanding of these issues in MEMS applications is built upon this knowledge base. In this case one layer (the film) is much thinner than the other (the substrate). A 0.5 μm metal film on a 500 μm thick, 100 mm diameter silicon substrate is a reasonable example. If subjected to a 100 °C temperature change, the maximum deflection would be about 2% of the thickness if the film fully covered the substrate and even less if it were patterned discontinuously. The deformation falls into the linear regime *I* of Fig. 4.10 b. In fact, the most common application of this behavior in microelectronics is the use of the Stoney equation [86] to determine thin-film stresses (which are typically biaxial and spatially uniform) from measured wafer curvature. In MEMS applications the layer thicknesses are not only small (on the order of micrometers) relative to in-plane dimensions, but are often comparable to each other. An example that is not unreasonable is a 0.5 μm gold film on a 1.5 μm thick, 400 μm diameter polysilicon plate. If subjected to a 100 °C temperature change, the maximum deflection would be about six times the thickness. This falls into the nonlinear second regime *II* of Fig. 4.10 b and perhaps even into regime *III*. Although not as heavily studied as the first, the second and third deformation regimes have been observed in structural composites cured at elevated temperatures [87] and more recently in microelectronics thin-film systems [88] and MEMS microstructures intended for rf applications [89, 90].

Examples of the deformation behavior of a series of square gold/polysilicon plate microstructures fabricated through polyMUMPS [91] are shown in Fig. 4.11. The plates are subjected to uniform temperature changes which generate internal stresses and deformation via thermal expansion mismatch of the gold and polysilicon. Linear and geometrically nonlinear deformations are observed, in addition to bifurcations in the equilibrium deformed shapes [84]. The nonlinear deformation and bifurcations depend strongly on the size of the plate. Moreover, the deformations and onset of bifurcations are a function of temperature. Similar analysis and results can be considered in the context of some MEMS applications, providing guidelines to the design of layered microstructures with controlled curvature. De-

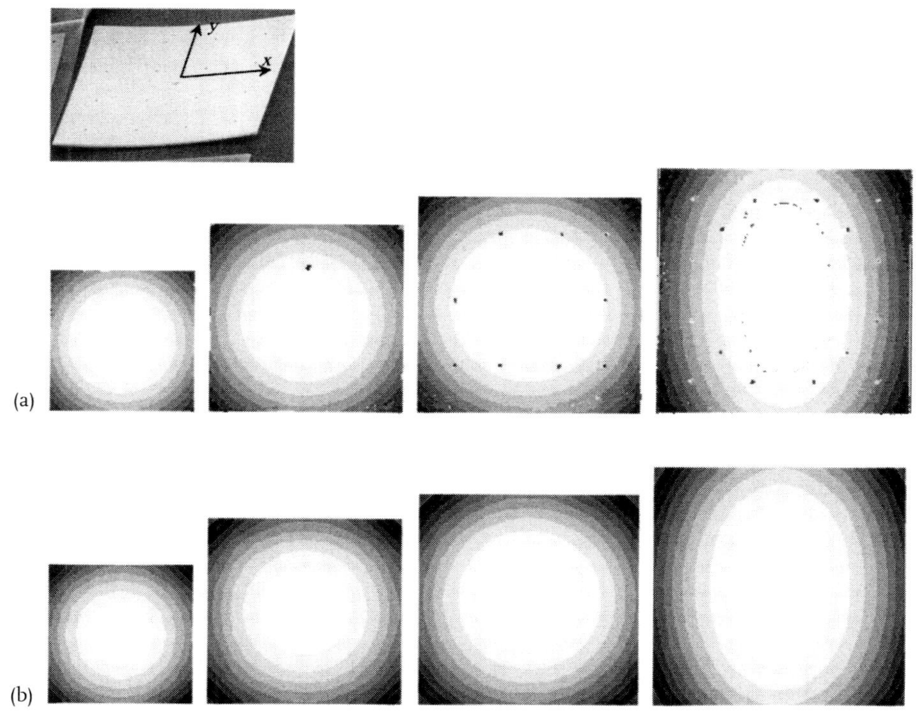

Fig. 4.11 Contour plots of the (a) measured and (b) predicted transverse displacements at room temperature for the four gold/polysilicon square plates of side L: $L = 150$, 200, 250 and 300 μm from left to right

pending on the application, tailoring the curvature might entail minimizing it (e.g. micromirrors) or maximizing it (e.g. microactuators).

Very little work has been focused on inelastic deformation of thin-film microstructures subjected to thermal loading [82, 92, 93]. As an example consider the deformation response of thin-film bilayer beams during the first few thermal cycles to elevated temperatures (of 100–275 °C) following processing and release from the substrate [92] (see Fig. 4.12). While this temperature range is modest, it is relevant for practical microsystems applications [94], including post-fabrication packaging processes. Suppose the bilayer gold/polysilicon cantilever beam in Fig. 4.12 serves as the active mechanical element in a switch array. During operation there are two key parameters that are strongly impacted by the cyclic thermal behavior: the linearity of the curvature developed as a function of temperature and the curvature at room temperature (or any other specified temperature) which sets the reference state for the device. For example, in an electrostatically actuated switch, the room-temperature deflection sets the open or closed configuration, depending on the design and linear response with a temperature change desired to facilitate robust temperature compensation schemes. As shown in Fig. 4.12, the

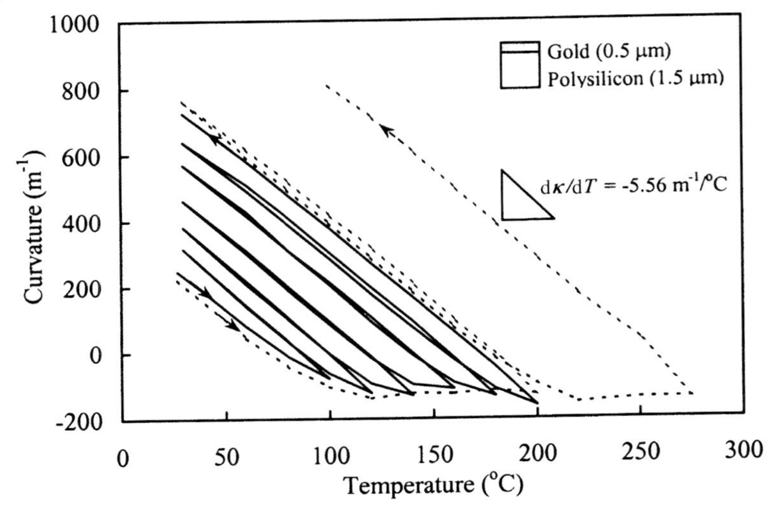

Fig. 4.12 Curvature vs. temperature for gold (0.5 μm thick)/polysilicon (1.5 μm thick), 300×50 μm, fully covered by gold beam microstructures. The solid lines are tests with six cycles from room temperature to 200 °C with the maximum temperature in each successive cycle increasing by 20 °C. The dashed lines are tests with two cycles: the first from room temperature to 200 °C and then back to 30 °C and the second to 275 °C and then back to 30 °C. The arrows indicate the direction at the beginning and end for each set of tests (courtesy of Professor M. L. Dunn, University of Colorado at Boulder)

linearity can be ensured over a certain temperature range by cycling the microstructure to a temperature higher than the maximum intended use temperature, i.e. by annealing it. For example, if the microstructure has been cycled to 200 °C, it can be expected to deform linearly over a temperature range with a maximum temperature less than 200 °C. Again, the complete characterization of the deformation-temperature behavior requires the consideration of the rate of temperature change and, in particular, the length of time the microstructure is held at the elevated temperature. In addition, the thermomechanical behavior under cyclic loading can also be important in practice.

The room-temperature (or any other temperature) deformation can be tailored by a combination of the structural geometry and the annealing process parameters. Assume that a device application requires that after packaging, the beam in Fig. 4.12 is expected to have a curvature of 300 m^{-1} to yield a desired tip deflection of a cantilever. In addition, assume that a packaging process, e.g. wire bonding, exposes a microstructure to a temperature of 120 °C. If one simply carried out this process, upon return to room temperature the curvature would be about 400 m^{-1}, rendering the device unusable. Instead, if an anneal step were performed at 180 °C, for example, the curvature at room temperature after the anneal would be about 700 m^{-1}. Subjecting it to a temperature of 120 °C during the packaging process would induce no further deformation upon return to room temperature since the deformation during that temperature change would be linear thermoelastic. Hence the proper anneal process would make the deformed shape of the microstructure stable during the packaging process. However, the curvature of about 700 m^{-1}, although stable, would still not be acceptable. The challenge is then to design the geometry, e.g. the layout of the gold film on the polysilicon, so that the room-temperature curvature after the anneal is the desired curvature. In the present example this can be accomplished by patterning the polysilicon beam with a gold line. This example is intended to demonstrate that if one understands the deformation response, both material and structural, it can be used in the design process to facilitate high device yield. Furthermore, it can be taken advantage of by a clever designer to increase design flexibility.

Finally, while much of the understanding regarding the thermomechanical behavior of layered systems derives from experiences in microelectronics, it is important to point out that significant differences exist for many MEMS applications and these must be well understood when optimizing the design of reliable MEMS.

Coatings for MEMS Reliability

Self-assembled Monolayers (SAMs) A serious problem in MEMS reliability is the capillary condensation of water (H_2O) in small enclosed MEMS structures. One method to avoid capillary condensation is to coat microstructural surfaces with thin organic films that repel H_2O and prevent capillary condensation.

MEMS surfaces can be made hydrophobic by deposition of nonpolar molecules on the surface, the current method of choice being SAMs. SAMs provide an effective means for anti-stiction treatment when applied to MEMS devices through liquid-phase processing steps [95, 96]. In general, a properly integrated SAM layer is shown to eliminate release stiction, reduce in-use stiction on horizontal surfaces, reduce friction between contacting microstructures and survive some packaging environments. Several single-step, liquid-phase monolayer systems have been developed for these purposes and include alkyl- and perfluoroalkyltrichlorosilanes, alkene-based molecular films and dichlorosilane monolayer films. Physical property data for a number of these films are summarized in Tab. 4.1.

With the ability to fabricate complex microstructures such as the microgear transmission shown in Fig. 4.13, the contact of vertical microstructure surfaces such as gear hubs and intermeshing gear teeth during operation has become an increasingly important issue in MEMS reliability. Recently, the successful reduction of in-use stiction on microstructure sidewalls (vertical surfaces) has been demonstrated through the incorporation of a hydrophobic SAM system [97].

The standard processing steps for the inclusion of a SAM on a micromechanical device invariably require a microstructure release step, a wafer cleaning step and subsequent exposure to the monolayer system. Typically, one or more of these steps use liquid reactants and therefore pose technical challenges when integration with full-wafer level, vapor-phase processing methodologies is desired. As a result, the scalability of standard SAM processing steps is limited. To address this problem, Ashurst et al. have developed a novel processing method that per-

Tab. 4.1 Physical property data for several surface preparations [96]: octadecyltrichlorosilane (OTS); dimethyldichlorosilane (DDMS); 1-octadecene; and untreated oxidized silicon

Surface treatment	*Water contact angle (°)*	*Work of adhesion (mJ/m^2)*	*Static friction coefficient*	*Thermal stability in air (°C)*	*Particulate formation*
OTS	110	0.012	0.07	225	High
FDTS	115	0.005	0.10	400	Very high
DDMS	103	0.045	0.28	400	Low
1-Octadecene	104	0.009	0.05	200	Negligible
Oxide	0–30	20	1.1	–	–

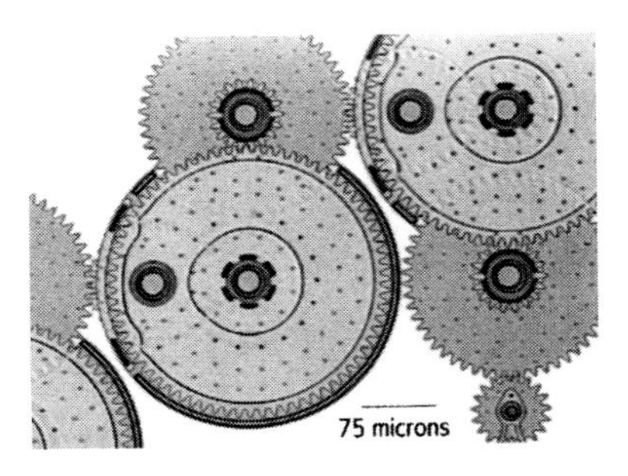

Fig. 4.13 A complex micromechanical system fabricated from the SUMMiT V^{TM} technology. Such a device contains a large number of sidewall contacts that can experience in-use adhesion [97]

mits in situ cleaning at the full-wafer level followed by the application of the DDMS coating system, all done using vapor-phase steps in a single reactor system [98]. Wafer-level, vapor-phase processing of this nature, although still technically challenging, is the most readily scalable of all available SAM processing strategies for eventual application to multi-wafer cassettes. This transition to the vapor phase promises integration of SAM processing into standard microfabrication routines and may result in the increased use of SAMs in future MEMS applications.

Chemical Vapor Deposition (CVD) A number of new materials with enhanced physical and chemical properties have been developed for the purpose of replacing Si as a primary structural layer in micromechanical devices, in turn enhancing reliability. These materials include, but are not limited to, cubic silicon carbide (c-SiC) [99], amorphous diamond [100, 101] and silicon–germanium [102]. New materials such as these pose a number of processing challenges that have made their incorporation into mainstream MEMS fabrication limited at best. For instance, MEMS designers have been faced with issues such as high residual stresses in released structures, excessive processing temperatures making back-end integration impossible and a lack of effective etch processes to pattern the microstructural elements. An attractive alternative to replacing Si structural layers altogether instead involves the modification of their surfaces with solid-state coatings in a single post-processing step using CVD.

CVD is a commonly employed processing method whereby one or more vapor-phase reactants are utilized in the growth of solid-state thin films. Typically, elevated reaction vessel temperatures are employed in vacuum to thermally activate the reactant molecules, producing highly mobile film precursors which react and subsequently bond with the heated surface. Although CVD is conventionally used in the growth of Si-based structural layers for integrated circuit and MEMS applications, it can also be employed as a post-processing step for surface modification in Si-based micromechanical structures. Similarly to using SAMs as a surface treatment for reduction of microstructure adhesion and friction, a thin, solid-state coating formed by CVD can also provide performance enhancement with additional packaging benefits arising from the processing methodology and solid-state film characteristics. In general, a post-processing deposition method for solid-state coatings should be:

1. Low temperature so as to not induce deformation of the poly-Si microstructures. In order to perform back-end processing, the growth temperature must be significantly low so as to be CMOS compatible.
2. Highly conformal, insuring that all surfaces of fully released microstructures are evenly coated. An uneven coating will yield preferential mass loading and unequal stresses across the structure, causing out-of-plane curvature. Other processes such as plasma deposition, laser ablation and rf sputtering are line-of-sight methods [103] and are therefore not applicable when conformal coatings are required.
3. Selective to Si only, thus preventing deposition on all device surfaces and preventing the need for extensive patterning of the post-processed thin film.
4. By nature self-limiting, ensuring necessary process control. This in turn will help insure the conformality of the deposited films across the entire microstructure surface.

Both c-SiC and tungsten thin films have been explored in this context. Both materials have been successfully integrated into poly-Si micromechanical structures using CVD in a post-processing step.

Post-processing of poly-Si micromechanical structures with undoped c-SiC has recently been reported [104]. This method utilizes the single-source CVD precursor 1,3-disilabutane at growth temperatures near 800 °C to produce high-quality, insulating polycrystalline films. Although the growth temperature is relatively high for a post-processing deposition step, it is lower than that of conventional high-temperature SiC MEMS processes in excess of 1000 °C [105]. In Fig. 4.14, fully released silicon microstructures are shown following the deposition of a relatively conformal SiC coating. Ultra-thin SiC films deposited on micromechanical structures by this process enhance device performance through the addition of a stiff, encapsulating outer shell having an elastic modulus of about 380 GPa, in addition to high physicochemical stability under aggressive etchant conditions [106]. SiC films grown using the 1,3-disilabutane precursor have been shown to grow on Si, SiO_2 and Si_3N_4 surfaces with varying rates and crystal structure.

Post-processing of poly-Si microstructures with a low temperature (<450 °C) W CVD process has also been successfully implemented [107]. Building upon a CVD

(a)

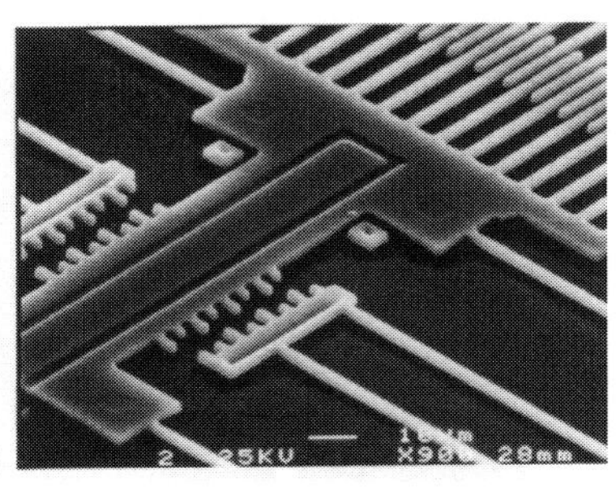

(b)

Fig. 4.14 Scanning electron micrographs showing (a) the cross-section of a 130 nm thick c-SiC film deposited on a Si cantilever and (b) the top view of a released poly-Si lateral resonator coated with a thin SiC film [104]. The resonator is viable after application of a 35 nm thick c-SiC film

process developed for the IC industry, metallic W deposition using the precursor WF_6 is self-limiting in nature, therefore yielding highly conformal films as evidenced in Fig. 4.15 a. The W CVD process is also selective to Si so no extraneous patterning steps are required. Wear resistance tests have been performed using

(a)

(b)

Fig. 4.15 Scanning electron micrographs showing (a) a cross-section of a 10 nm thick conformal W coating on a poly-Si microstructure and (b) uncoated and W-coated poly-Si gears. The uncoated gear shows wear debris (indicated by the arrows) after 1 million cycles whereas the W-coated gear shows no indication of wear after 1 billion stress cycles. The gear diameter is 76 μm

the Sandia microengine [108], a diagnostic MEMS device that experiences sliding friction during operation. Microengines coated with the selective W CVD process showed longer lifetimes than uncoated engines, exhibiting no failure when tested to 2 million cycles. Fig. 4.15 b shows uncoated and W-coated poly-Si gears following wear resistance testing.

In both of the aforementioned studies, thin solid-state coatings produced by CVD are shown to enhance the performance of poly-Si micromechanical devices. In either case, the reliability of micromechanical structures can be improved to varying extents with these processes. Not only are performance enhancements evidenced with both ceramic and conductive thin films, but also overall microstructure stability is realized, thus making CVD post-processing a viable tool for MEMS packaging methodologies. To date, CVD post-processing as a means of improving MEMS reliability remains relatively unexplored, but shows future promise as a potential packaging technology.

Atomic Layer Deposition (ALD) ALD is a specialized subset of CVD. It is different from other CVD techniques in that the source vapors are present in the chamber one at a time. This allows each exposure step to saturate the surface with a monomolecular layer of that step's precursor. The result is a self-limiting growth mechanism that is extremely conformal and uniform. It is a simple technique with accurate film thickness control and Ångstrom-level resolution. The ALD reactions typically take place at temperatures lower than 200 °C and can be run at temperatures as low as room temperature. Because the chemical reactions are self-limiting, the deposited layers are atomically flat resulting in atomic control of film thickness and optically flat surfaces. Materials that have been deposited by ALD include Al_2O_3, ZnO, TiO_2, W, TiN, SiO_2, WN, Si_3N_4, ZrO_2, Ta_2O_5, Ti and single-element metals [109]. Typical film thicknesses for ALD range from a few to hundreds of Ångstroms. It has been demonstrated that Al_2O_3 ALD films on electrostatically actuated beams prevent electrical shorting and increase the number of actuation cycles before device failure [110]. To date little research has been done towards coating MEMS devices with ALD films, although the technique may hold much promise.

ALD is a gas-phase deposition technique that relies on sequential self-limiting surface reactions (Fig. 4.16) to deposit ultrathin conformal films [111]. The notches in the starting substrate for reaction A in Fig. 4.16 represent discrete reactive sites. Exposing this surface to reactant A results in the self-terminating adsorption of a monolayer of A species. The resulting surface becomes the starting substrate for reaction B. Subsequent exposure to molecule B will cover the surface with a monolayer of B species. Consequently, one AB cycle deposits one monolayer of the compound AB and regenerates the initial substrate. By repeating the binary reaction sequence in an ABAB... fashion, a film of any thickness can be deposited.

For example, Al_2O_3 ALD films can be deposited using alternating trimethylaluminum (TMA) and H_2O exposures. The A and B surface reactions that define an AB cycle for Al_2O_3 ALD are as follows:

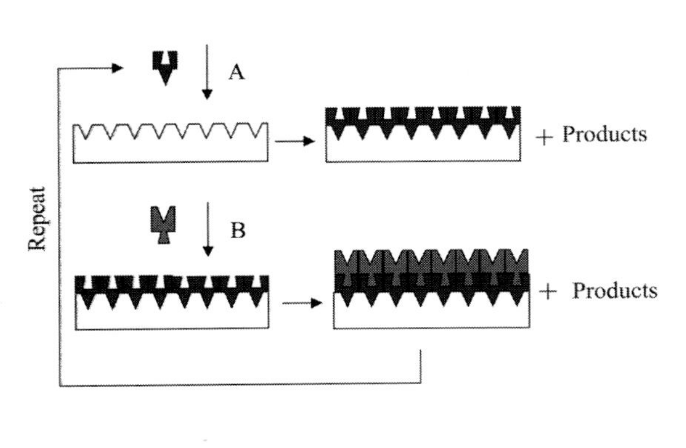

(a)

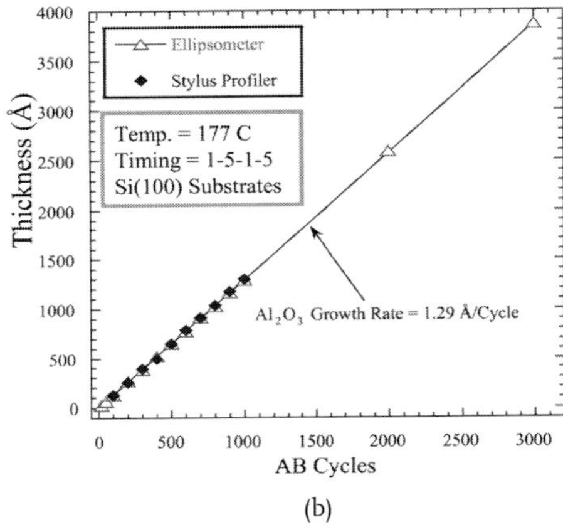

(b)

Fig. 4.16 (a) During ALD, the surface is first exposed to reactant A, which completely reacts with the initial surface sites. Next, the surface is exposed to reactant B, which regenerates the initial functional groups. (b) The film is then grown to the desired thickness by repeating this AB sequence

(A) $AlOH^* + Al(CH_3)_3 \rightarrow AlOAl(CH_3)_2^* + CH_4$
(B) $AlCH_3^* + H_2O \rightarrow AlOH^* + CH_4$

where the asterisks designate the surface species.

ALD films are extremely smooth and conformal to the underlying substrate. This superb conformality has allowed the successful coating of powders, nanoporous membranes and high aspect ratio trench structures [109]. Atomic force microscopic inspection of ALD Al_2O_3 films deposited on planar surfaces reveals pinhole-free coatings with nearly the same surface roughness as the underlying substrate. Smooth, dense films result from forcing each of the half-reactions to saturation. Fig. 4.16 also presents ellipsometric and stylus profilometric thickness measurements for ALD Al_2O_3 films deposited using alternating $Al(CH_3)_3$ and H_2O exposures. The ALD Al_2O_3 film growth is extremely linear with the number of AB cycles performed and the growth rate is 1.29 Å/cycle. Precise thickness control, pinhole-free coatings and superb conformality make ALD a very good candidate technique for coating MEMS devices (Fig. 4.17).

In general, ALD has the following unique features: nano-scale conformal film with atomic layer control; processing temperature is MEMS compatible and can be as low as the room temperature; Al_2O_3 layer, for example, can be coated on almost any material surface, including gold, silicon, silicon dioxide, silicon nitride and ceramics. Typical processing temperature for Al_2O_3 films is 177 °C [109, 110].

Interesting physical properties can be obtained by combining multilayers of different materials. For example, alloys of Al_2O_3 and ZnO can be grown by varying the relative amounts of Al_2O_3 and ZnO in the thin film. This can be accomplished by varying the number of individual $Al(CH_3)_3$ or $Zn(CH_2CH_3)_2$ exposures used with the alternating H_2O exposures. This alloy system allows the electrical conductivity

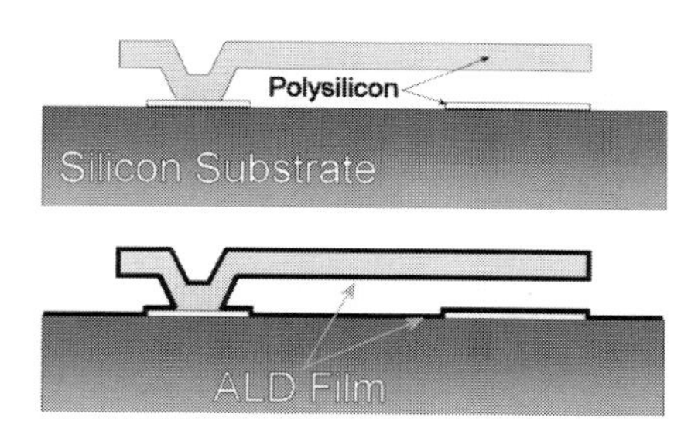

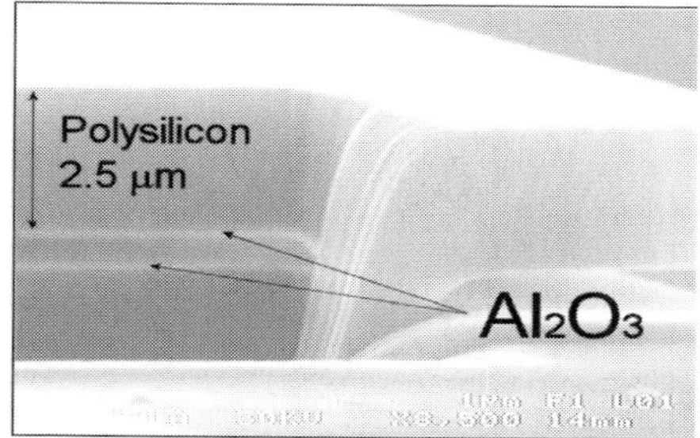

Fig. 4.17 MEMS cantilever coated with 60 nm thick, protective alumina using ALD

to be varied over 18 orders of magnitude in thin film coatings. Fig. 4.18 demonstrates a multilayer coating with Al_2O_3 and ZnO layers with a resistivity plot for the ZnO/Al_2O_3 system [109]. Other multilayer composite structures perhaps can be designed and fabricated to control other surface properties of MEMS structural layers.

ALD can be used for hydrophobic coatings [112]. The hydroxyl groups can be added to the MEMS surface using Al_2O_3 ALD with $Al(CH_3)_3$ and H_2O. After 5–10 cycles of Al_2O_3 ALD using $Al(CH_3)_3/H_2O$, a 5–10 Å layer of Al_2O_3 is added to the MEMS surface. This Al_2O_3 surface has a very high concentration of AlOH* groups after the last H_2O reaction. There is a wide range of chlorosilanes that may yield hydrophobic surfaces. A variety of R groups on chlorosilanes are available for attachment through the surface hydroxyl groups. By controlling the R groups and their coverage, the surface hydrophobicity can be tuned across the entire spectrum from very hydrophobic to very hydrophilic. Nonpolar $-(CH_2)_nCH_3$ and $-(CF_2)_nCF_3$ groups are very hydrophobic and polar OH and NH_2 groups are very hydrophilic. The coverage of these hydrophobic or hydrophilic groups can be

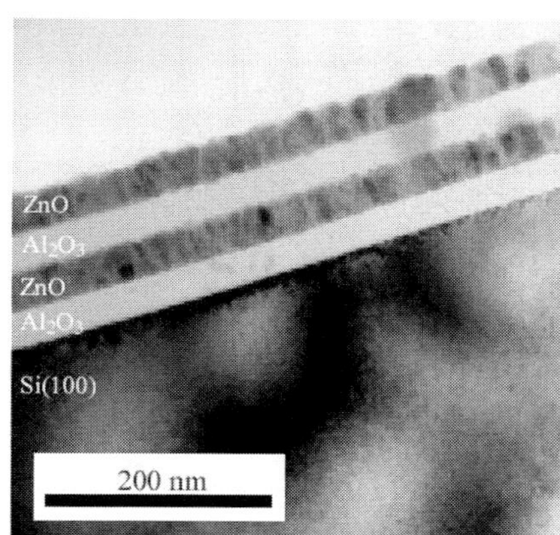

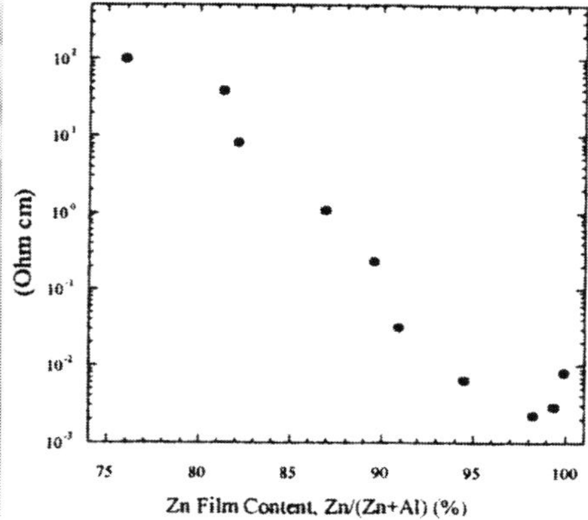

Fig. 4.18 ALD multilayer structure of alternate layers of Al_2O_3 and ZnO on an Si(100)substrate and its resistivity as a function of composition for the ZnO/Al_2O_3 system (courtesy of Prof. S. M. George, University of Colorado at Boulder)

controlled by the initial hydroxyl surface coverage and the extent that the chlorosilane reaction is allowed to reach completion.

The hydrophobic films can be deposited in a solution using chlorosilane attachment of alkylsilanes or perfluoroalkylsilanes to surface hydroxyl groups [95]. However, this deposition technique is challenging. Chlorosilanes are known to react first with small quantities of H_2O in solution to form silanols [113]. These silanols then polymerize, resulting in poorly ordered and weakly bonded films as illustrated in Fig. 4.19 A. An alternative method for the deposition of reliable and robust hydrophobic coatings on MEMS devices has been demonstrated [112]. First, a thin film of Al_2O_3 is deposited via ALD and is used as a seed layer to prepare and optimize the MEMS surface for the subsequent attachment of the hydrophobic precursors. Then the alkylsilanes are chemically bonded to the surface hydroxyl groups on the ALD seed layer. This technique results in a dense and ordered hydrophobic film as illustrated in Fig. 4.19 B. Al_2O_3 ALD is used to optimize the initial MEMS device for hydrophobic precursor attachment by (1) covering the MEMS surface uniformly with a continuous adhesion layer (2) providing a high surface coverage of hydroxyl groups for maximum precursor attachment and (3) smoothing and removing nanometer-sized capillaries that may otherwise lead to microcapillaries and thus moisture-induced stiction problems. Additionally, polymerization is avoided by using alternative precursors, such as alkylaminosilanes, instead of the traditional chlorosilanes. These alternative precursors react more completely and effectively with the surface hydroxyl groups without initial reaction with H_2O. The hydrophobic films on Al_2O_3 ALD adhesion layers on silicon wafers were observed to increase dramatically the contact angle to greater than 100° (Fig. 4.20). MEMS test structures consisting of polysilicon cantilever beams 600 μm long, 20 μm wide and 2 μm thick and suspended 2 μm off the substrate were coated with the hydrophobic ALD film to investigate any change in moisture-induced stiction. As shown in Fig. 4.21, the beams without the hydrophobic ALD coating are pinned to the substrate, while

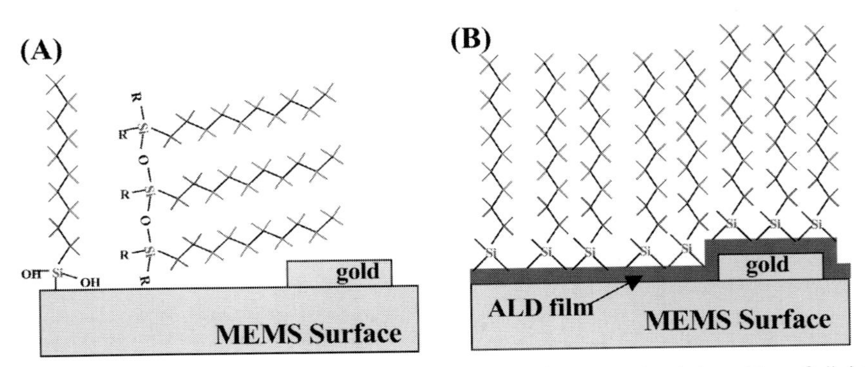

Fig. 4.19 A comparison between deposition techniques. The conventional deposition of alkylsilanes via a solution on bare MEMS surfaces (A) leads to a poorly ordered and weakly bonded film. Adding the ALD adhesion layer (B) provides a uniform reactive surface to which hydrophobic precursors can chemically bond

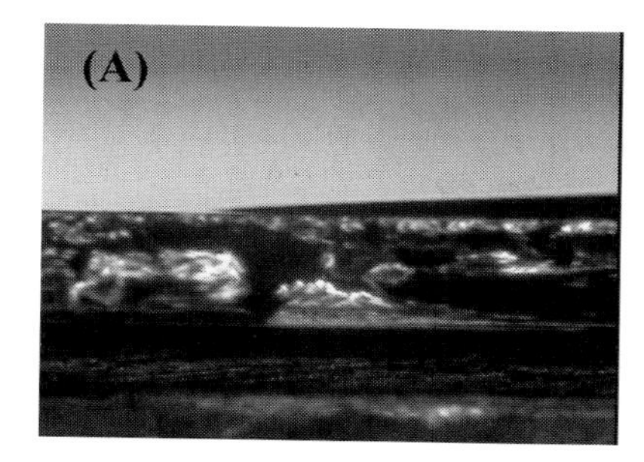

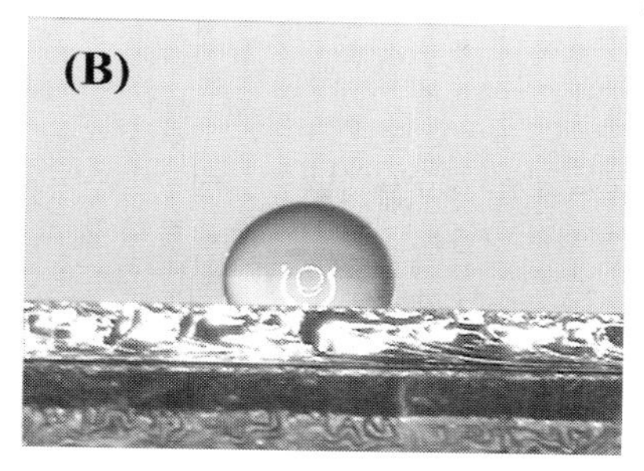

Fig. 4.20 Water droplets on silicon wafers. Without the ALD coating (A), the contact angle is ~5°. With the hydrophobic ALD coating (B), the contact angle is ~107°

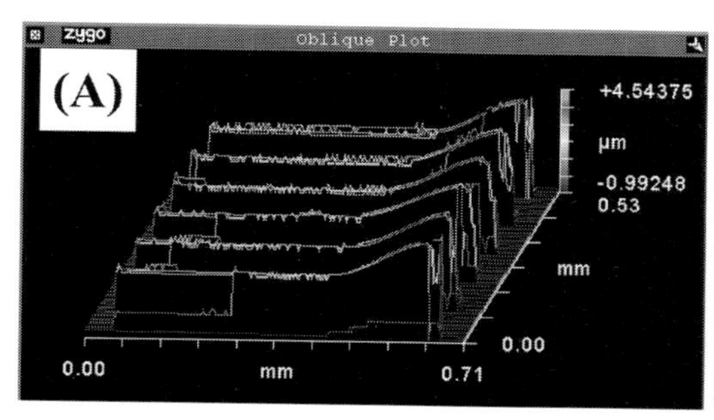

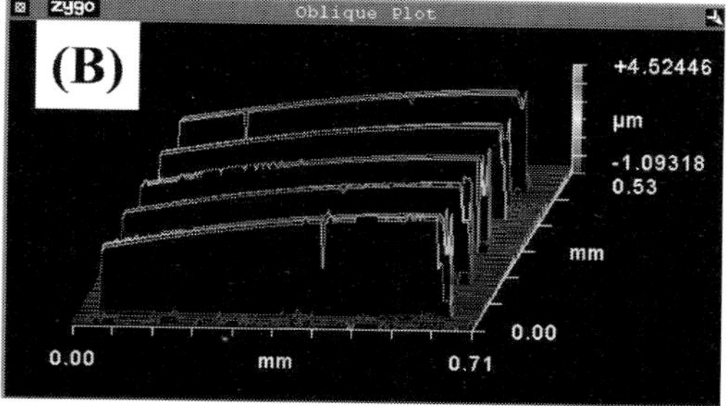

Fig. 4.21 Cantilever beams without hydrophobic ALD coating (A) are susceptible to stiction after exposure to water. The stiction of the beams is dramatically reduced with the addition of the hydrophobic ALD coating (B)

the coated beams exhibit much less stiction. Based on the adhered section of the beam, the adhesion energy of the hydrophobic ALD-coated beams was calculated using literature methods [114]. The ALD coated beams were determined to have an adhesion energy of 0.11 ± 0.03 mJ/m^2 compared with the same beams without a coating of 12 ± 1 mJ/m^2. Thus the ALD technique does seem to hold promise for hydrophobic surface engineering of MEMS.

4.2.3 Zero-level Packaging/Wafer Bonding

The zero-level (wafer-scale) packaging approaches (Figs. 4.22 and 4.23) can be classified as follows:

- *Integrated thin-film MEMS encapsulation processes.* These are highly process dependent, not versatile and not easy for post-processing. However, they offer minimum package size (high-density integration) at low cost. These are relatively new technology (initially developed in the early 1990s) with unknown reliability and, in particular, unproven hermeticity.

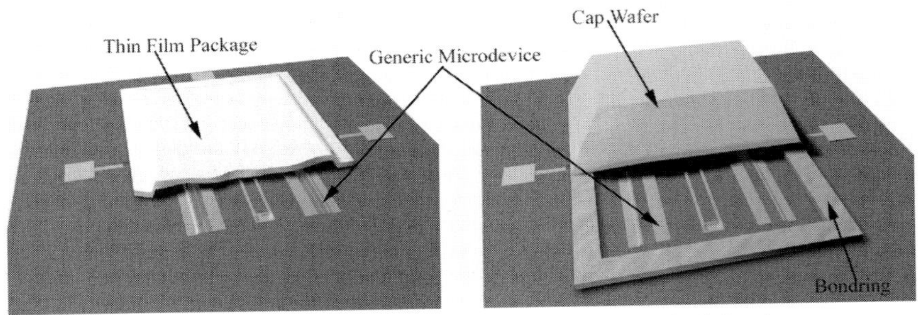

Fig. 4.22 Two methods of packaging a MEMS device: integrating the package into the process flow (left) and bonding a cap on top of it (right). Courtesy of Brian H. Stark, University of Michigan, 2003 [115]

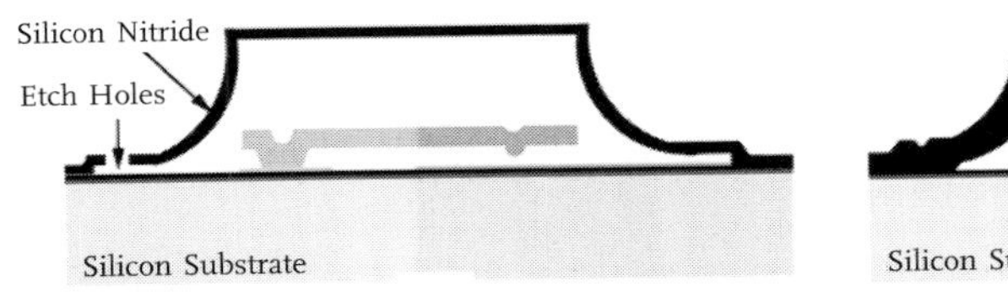

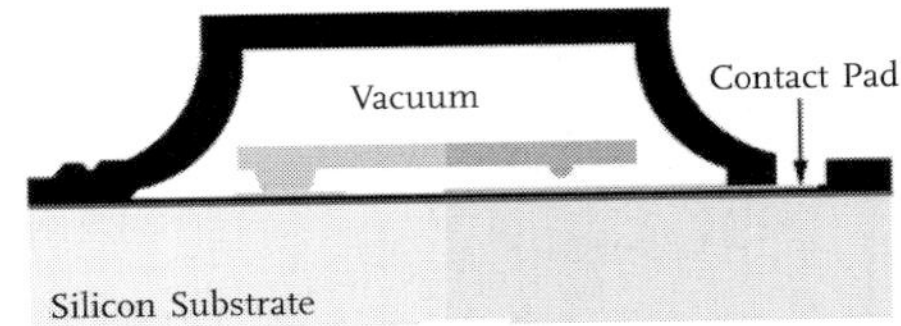

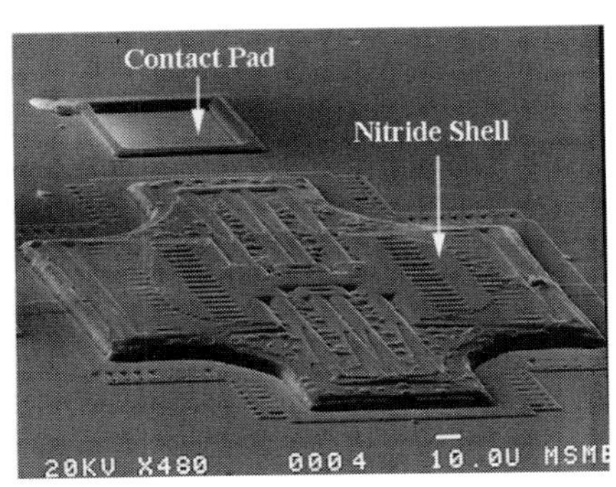

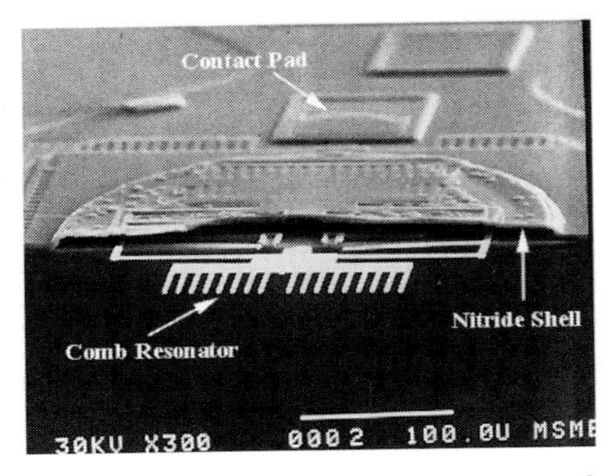

Fig. 4.23 Illustration of thin-film packaging: LPCVD encapsulation process. Nitride shell deposition followed by etch hole definition, and removal of all sacrificial material inside the shell followed by global LPCVD sealing. Courtesy of Professor Liwei Lin, University of California at Berkeley [116]. An alternative technique is being developed at the University of Michigan using electroplated thin-film caps [115]

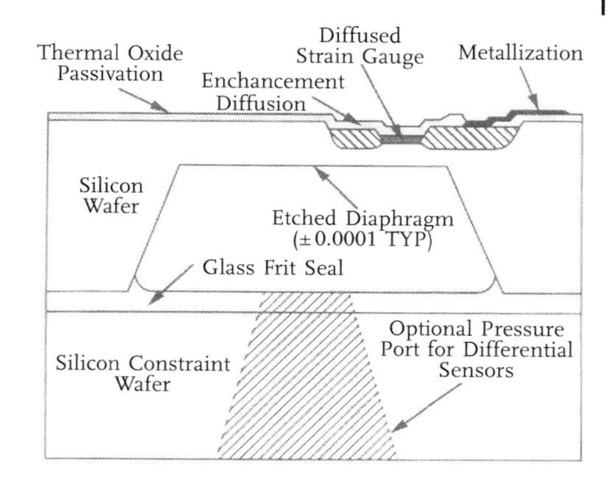

Fig. 4.24 Wafer bonding to create an absolute vacuum cavity for an absolute pressure sensor device [44]. The sensor is created by an adhesive layer bonding with low temperature (450–500 °C) glass frit. An optional port through the silicon wafer constraint can be formed with bulk micromachining to create a constraint wafer for a differential or gauge pressure sensor

- *Wafer bonding processes (anodic, fusion, eutectic).* These processes are high temperature (may damage microelectronics or temperature-sensitive MEMS materials) and require very smooth and flat surfaces for bonding. They are, however, proven technology (since the late 1960s), with demonstrated reliable hermetic packaging.
- *Localized thermal bonding processes.* These include eutectic, fusion, laser, inductive and solder bonding. A variety of epoxy and polymer bonding methods are also emerging. The issues with these technologies are manufacturability on a large scale and reliability over the lifetime of the device.

Wafer bonding or cavity sealing is present on many MEMS products. Wafer-level packaging is used in pressure sensors (Fig. 4.24) and in accelerometers or micromachined resonators to create a controlled atmosphere for the device (Fig. 4.25). Complete reviews of wafer-to-wafer bonding for MEMS were published by Schmidt [117] and Ko et al. [118].

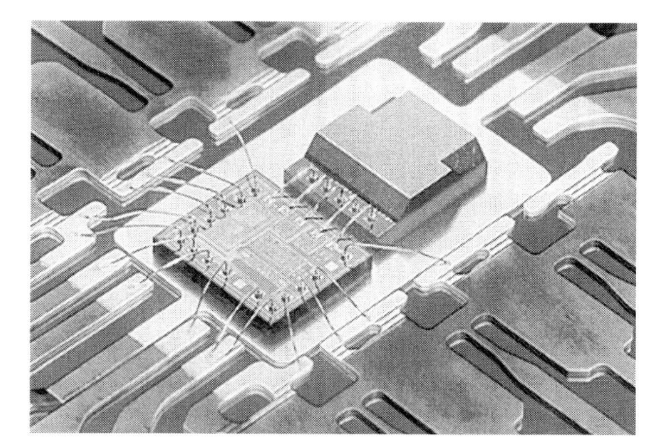

Fig. 4.25 The Motorola two-chip accelerometer, where the micromachined device is located within the wafer-bonded die (right side of the photograph). Wafer bonding, using an adhesive glass frit, for the hermetic wafer-level sealing of an accelerometer device is done prior to wafer saw and assembly [44]

Examples abound of MEMS that use wafer bonding. In general, wafer bonding can be categorized based on the following processes: (1) direct wafer bonding, (2) anodic bonding and (3) intermediate layer bonding.

4.2.3.1 Direct Wafer Bonding

Direct wafer bonding dates back to the mid-1960s [44, 119]. A complete review of the history, mechanism, processing considerations and applications for direct wafer bonding to micromachined devices was given by Ristic [44] and an overview of the applications for MEMS by Desmond and Abolghasem [120]. The key requirements for the wafer bonding process to occur are to have a specularly smooth surface, to maintain contact between the two wafers throughout the process and to elevate the temperature so that the chemical bonding process can occur. Typically, bond energies an order of magnitude higher than those obtained at room temperature are observed when bonding at temperatures in the 800–1200 °C range [117, 121]. The proposed mechanism for two oxidized wafers (silicon wafers will normally have a native oxide on them; hence the process for silicon wafers is similar to that described here for oxidized wafers) is that the silanol (SiOH) on the surface of either wafer reacts to form water and siloxane (Si–O–Si) [44, 122]. This occurs at relatively low temperature ($\sim$300 °C) [119]. At temperatures between 300 and 800 °C, water dissociates and silicon bonding begins to occur. Between 800 and 1400 °C, oxygen diffuses into the lattice and bonding is complete. Experimental evidence has shown that this process requires the presence of water. Key processing issues include interfacial integrity (including surface microroughness, surface hydrophilicity, particle contamination and surface morphology [122]), bond strength, wafer thinning/polishing and microdefects. Often scanning acoustic tomography [44], IR imaging [117, 123] and/or transmission electron microscopy are used for determining interfacial integrity and the presence of microdefects. A blade technique for separating wafers (using a knife blade) or a pressure burst test is used to measure bond strength [117, 123].

Provided that one can design around the high-temperature requirements for direct wafer bonding, it is a versatile technique that can be used for wafer-level packaging for a variety of MEMS. This works well when the bonding process is the first (or nearly the first) process step. However, often the bonding process is the last step and high-temperature processing is not possible at that point in the procedure. Alternatively, the use of surface activation bombardment of the surface with Ar energetic particle beams [124–126], the use of chemical cleaning [127] in ultra-high vacuum, surface activation for bonding using an ammonium fluoride etch mixture [128], the use of low vacuum ($\sim$700 Pa) followed by a 150 °C air anneal [129] or the storage of wafers at slightly elevated temperatures ($<$150 °C) for a long period [130] have shown promising low-temperature direct wafer bonding results.

Another concern with direct wafer bonding is the result of diaphragm deflection following bonding of a cavity. Cavities sealed by direct wafer bonding (or anodic bonding) have been observed to contain residual gas pressure (i.e. hydrogen, water, nitrogen and oxygen) higher than the bonding pressure [131]. Observation of the residual gas in a cavity can be deduced by watching the deflection of a diaphragm or by FTIR measurement of gases within the cavity [132]. Conclusions from work by Schmidt [117, 123] and Secco d'Aragona et al. [133] suggest that cavities must be sealed in a controlled oxygen environment (e.g. 20% O_2) to eliminate plastic deformation from the diaphragms. Sealing of the cavities in nitrogen resulted in observed plastic deformation of the single-crystal silicon diaphragms [133].

4.2.3.2 **Anodic Bonding**

Anodic, or electrostatic, bonding is accomplished by placing a polished silicon wafer against a glass wafer, heating to between 180 and 500 °C [134] and applying 200–1000 V. Originally, this process was used for bonding metal to glass, but the metal was replaced by silicon for application in the MEMS industry [117]. The glass wafers used are often Pyrex for MEMS applications [135]. These wafers have sodium ions that are mobile during the bonding process and aid in completion of the bond. With the application of high electric fields, the mobile sodium ions migrate away from the bonded interface. This electrically driven diffusion creates a fixed charge in the glass, thus inducing a high electric field across the interface. Although the bonding mechanism is not completely understood, it is assumed that a chemical bond forms between the silicon and the glass when they are under the combination of high electric field and high temperature [117]. More generally, glasses with positive alkali metal mobile ions can be used with conductors or semiconductors. With applied temperature, the alkali metal ions become more mobile and migrate away from the bonding interface with the application of high electrical bias. This leaves oxygen ions at the bonding surface available for chemical bonding [136].

The advantage of anodic bonding over direct wafer bonding is that the temperatures used are much lower. This allows anodic bonding to be used after the first step in a fabrication process. Anodic bonding is generally considered more reliable than the low-temperature wafer bonding processes. In fact, several companies have used anodic bonding in producing MEMS devices, including pressure sensors for automotive applications from Motorola [137], Honeywell and Bosch [138] (Fig. 4.26).

However, anodic bonding is not without its own limitations [138]: it requires nonstandard IC equipment; alkali metal ions are needed in the glass material, which can be a problem with on-chip electronics; high voltages are required for the bonding, which also can affect on-chip electronics; glass wafer structuring can be complicated; and the CTE of Pyrex and other glasses is not the same as that of silicon over a wide range of temperature, although it is close [118]. The following requirements are essential for this process [118]: the glass must be slightly conductive; the metal used as the anode cannot inject mobile ions into the glass; and the surface roughness must be small (<1 μm) for both the glass and the metal. Experiments have shown that a 20 nm groove under the bond will still allow adequate sealing whereas a 58 nm groove will not [138]; the surfaces must be free of

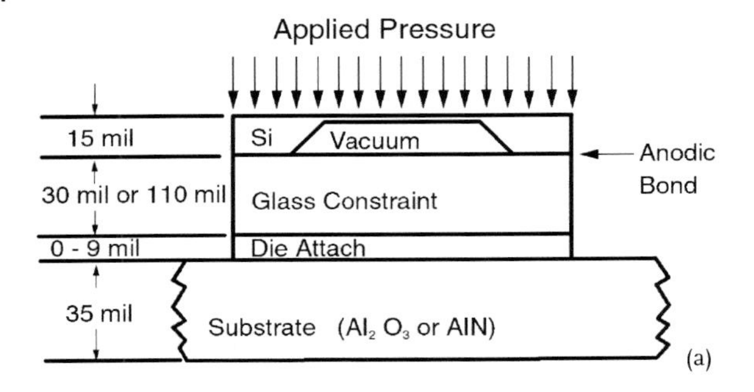

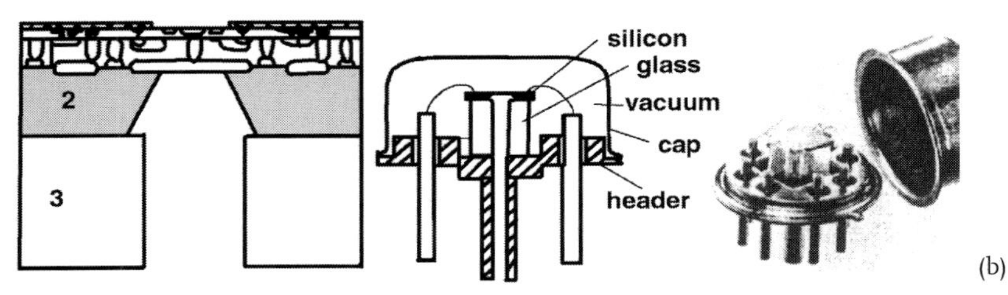

Fig. 4.26 Examples of anodic bonding being used in products. (a) An absolute pressure sensor used for automotive applications from Motorola [137]. Bosch has also presented a similar device [138]. (b) A backside exposure 'wet–dry' (i.e. harsh media from the backside of the device, benign media on the frontside of the device) differential pressure sensor from Bosch [138]

contamination; pressure is not required, but has been observed to improve the anodic bonding [139]; the CTE of the two materials being bonded should be closely matched; voltage requirements are dependent upon temperature used, glass type and thickness; and ambient atmosphere can be used for bonding although nitrogen, forming gas, argon and helium have also been used.

In addition, these techniques have been used with laser heating [118] and with a Pyrex thin film on silicon to create a silicon-to-silicon bond [118, 140]. Also, nonevaporable getters have been used with anodic bonding to eliminate the problem of residual gas in the bonded cavity [141].

4.2.3.3 **Other Bonding Techniques**

Several alternative bonding techniques have been developed for the MEMS industry. Many methods use an intermediate material such as epoxy and polyimide [118]. Eutectic bonding is another alternative [117, 118]. The eutectic point in a two-component phase diagram is where the composition is such that the melting temperature is at a minimum. For example, in the Sn/Pb system, this occurs at 61.9 wt% Sn and 38.1 wt% Pb. The melting temperature is 183 °C. In the Au/Si system it is 97.1 wt% Au and 2.85 wt% Si and the melting temperature is 363 °C. An interesting method based on eutectic bonding on localized areas using a poly-

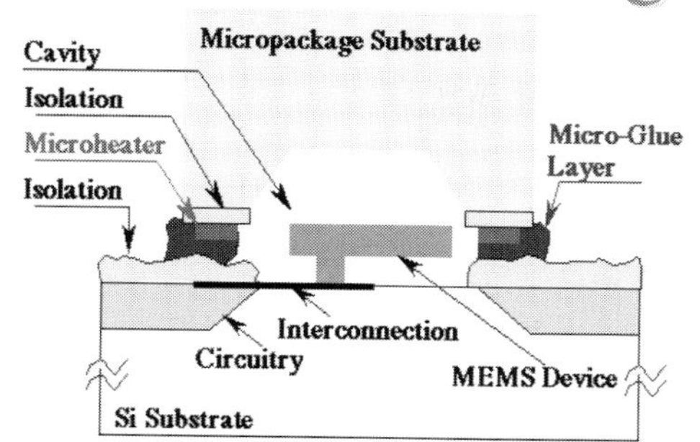

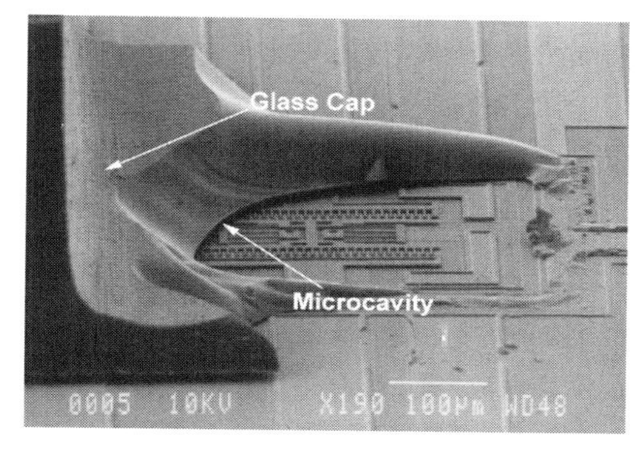

Fig. 4.27 Diagram of wafer bonding using localized heating [142] and an example of a vacuum encapsulated comb resonator. (Courtesy of Professor Liwei Lin, University of California at Berkeley)

silicon heater has been demonstrated (Fig. 4.27) [142]. This localizes the heat source so a material can form a hermetic bond directly above the heater. Variations on this method exist where instead of a embedded heater the heat is provided by laser heating, inductive heating or rapid thermal annealing. Another notable method has used permeable polysilicon as a shell for later sealing [143]. In this process, thin (≤1500 Å) polysilicon is permeable to HF. The sacrificial layer etching, which does not require etch access holes, etches the underlying glass (particularly phosphorus-doped glass), while leaving the thin polysilicon intact. The permeable polysilicon can then be sealed with a subsequent LPCVD deposition.

4.2.4 Package Assembly Processes – First Level Packaging

Package assembly processes consist of several basic steps (Fig. 4.28), e.g. dicing and die separation, die attach, interconnection, encapsulation.

4.2.4.1 Dicing and Die Separation

For microelectronic ICs, dicing and die separation are a mature process and the most widely used today. Typically, a wafer is placed on a Mylar film with light adhesive that holds the wafer in place during diamond sawing. The adhesive is used to keep the die on the Mylar film during sawing; however, the adhesive is mild enough to allow the 'pick and place' (followed by die attach) operation that occurs next in the assembly process. A diamond blade saw is used to cut partially or completely through a silicon wafer substrate. A continuous stream of water is used to cool the blade during operation. In production, a vision system with programmable spacing and rotation of a stage carrying the wafer is used to adjust

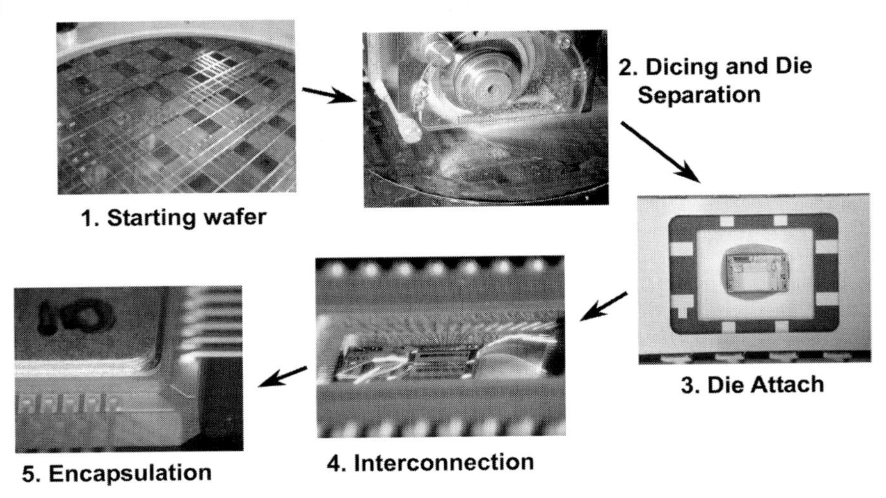

Fig. 4.28 Overview of package assembly processes. Courtesy of Dr. L. Spangler, Aspen Technologies [144]

the location of the wafer so that the saw cuts through the wafer 'streets' (i.e. areas in between die locations). There are design rules accounting for the roughness of the cut that extend beyond the saw cut width.

The traditional dicing techniques used in the semiconductor industry are not satisfactory for several classes of MEMS. Micromachined devices are more difficult to saw because of induced stress that affects the device performance: delicate microstructures can be broken during the sawing operation and/or stiction resulting from the water used to cool the wafer saw during operation. Surface micromachined devices after sacrificial layer release are easily damaged by moisture and particles and are incompatible with the substantial silicon debris, blade cooling water spray and vibration generated by diamond wheel dicing. Wafer-level caping of the devices eliminates the potential problems listed above. An ideal dicing process for MEMS would cut silicon in a dry fashion, without generating debris. Some candidate processes for improved die separation include dicing with wafer masking, wafer scribing, laser cutting, diamond wire cutting and abrasive jet machining.

Conventional dicing can be used if the sensitive portions of the micromachined device can be protected or masked in some manner. Spin-on polymer coatings, such as polyimide, can be used for this purpose, especially if the polymer can be removed in a plasma ash process afterwards (Fig. 4.29). However, a necessarily hard and thick polymer for device protection is usually difficult to remove after dicing without leaving a residue. Furthermore, the coating may be damaging to some kinds of devices.

When protecting the frontside is not practical or sufficient, wafer scribing is sometimes preferred using scribers that scribe and break the wafer, while only contacting the wafer backside. The scribing and fracturing process eliminates the

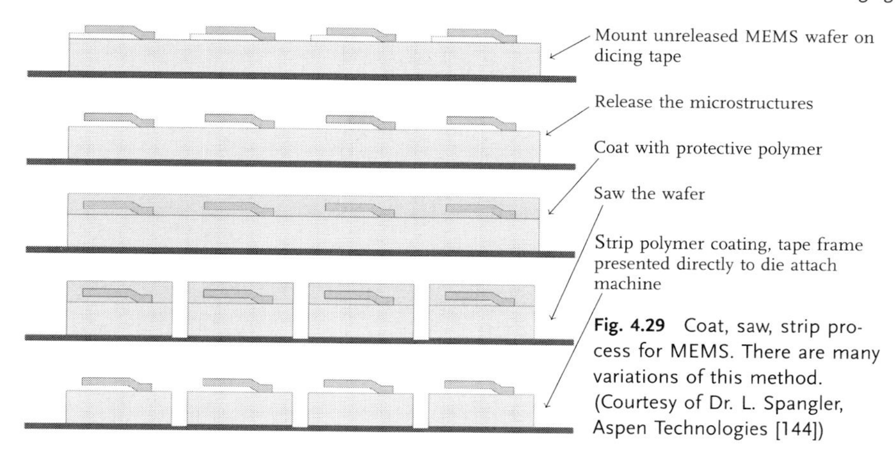

Fig. 4.29 Coat, saw, strip process for MEMS. There are many variations of this method. (Courtesy of Dr. L. Spangler, Aspen Technologies [144])

debris and moisture of traditional dicing; however, fracturing can still generate particulates that may be unacceptable depending on the sensitivity of the device to particles and manufacturing yield requirements.

Other cutting techniques which merit consideration include laser cutting, diamond wire cutting and abrasive jet machining. Laser cutting can be fast and precise, but tends to create a significant amount of splatter and debris that damage devices. Diamond wire cutting can be used for a wide variety of sensitive cutting applications. Diamond wire cutting uses a fine loop of diamond wire on a motor-driven capstan to cut into the wafer surface. This type of cutting produces very little surface damage, but may not have sufficient throughput for high-volume applications. Abrasive jets are commonly used in repair and rework of conventional electronic components. It is plausible that with a suitably small nozzle and dust collector, abrasive jets could be used to dice MEMS devices safely. Although currently no cutting technique meets all of the desired characteristics for MEMS dicing, as the market size for MEMS process equipment grows manufacturers will undoubtedly develop new and innovative dicing methods.

Finally, die separation can be achieved by using micromachining. For example, bulk micromachining can be used to create an opening through a wafer in a pattern such that individual die can be separated [5].

4.2.4.2 **Pick and Place**

Pick and place is the process in which dies from a wafer are mounted on a lead frame or into a package. A wafer is probed as one of the last steps during the fabrication process, where the wafer is marked to show the 'good' die from the 'bad' die. This is used to minimize cost in the assembly area by not working with the bad die. Typically, a bad die is marked with a drop of black ink. Once the dies have been marked, dies are picked using a vacuum tool that holds the die during mounting on a lead frame or into a package.

With micromachined devices, handling of delicate microstructures can be a problem throughout the assembly process. Pick and place is one operation that could cause a potential problem because of mechanical damage or damage to a delicate microstructure caused by the change in pressure (to vacuum). An additional advantage of the wafer-level packaging is that it provides protection to the microstructure during pick and place.

Some research has been performed to investigate the possibility of self-assembly of die. Yeh and Smith have observed trapping of semiconductor ICs in micromachined wells [145, 146]. Cohn et al. have modified this technique by adding electrostatic alignment of the die prior to settling into the cavities [147]. Although these techniques may not yet be ready for large-scale production, they represent research efforts to minimize the cost of pick and place through batch deposition of dies on a substrate.

4.2.4.3 Leadframe and Substrate Materials

Support materials for micromachined structures come in a variety of styles. MEMS dies are mounted on metal leadframes, ceramics, pre-molded packages, silicon or other wafer substrates and header package materials. A leadframe is often used as a mounting substrate for the MEMS die. For example, Motorola's two-chip accelerometer is shown in Fig. 4.25. In this case, both the sensor die and the CMOS control die are mounted on the leadframe prior to molding. Alternatively, Analog Devices uses a header substrate in many cases, as shown in Fig. 4.30. Many pressure sensor manufacturers also use header package designs [5]. Figs. 4.30 and 4.31 show many of these design variations.

Performance of MEMS can be affected by the package itself. For example, stress in the package can affect the device output for many physical sensors. Thus, thermal management is required when designing the package. In fact, modeling can be performed to show the impact of package and substrate materials on the device performance of a physical sensor. Tab. 4.2 shows typical thermal expansion coefficients for materials used to package MEMS. One method for minimizing the impact of the substrate is to use a very low modulus die attach material (e.g.

Fig. 4.30 (left) Pressure sensor (http://www.sensonor.com) and (right) MEMSIC, from website http://www.memsic.com [148]

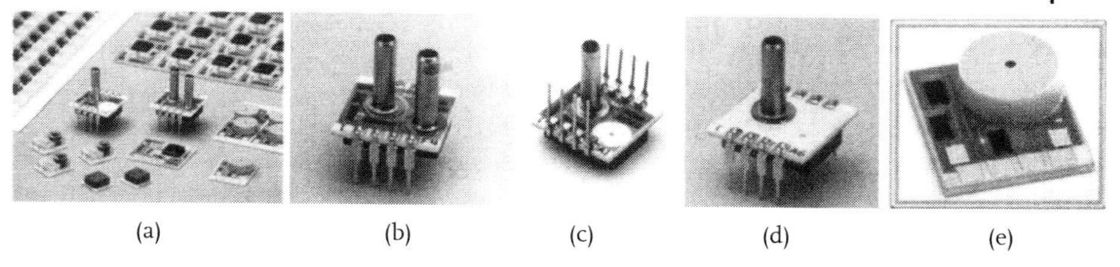

(a) (b) (c) (d) (e)

Fig. 4.31 Pressure sensors mounted in ceramic packages in various styles [149]: (a) a variety of package styles using ceramic supplied by Lucas NovaSensor; (b) a dual topside port package; (c) a backside port version; (d) a single topside port version; (e) a version used in angioplasty [149]

Tab. 4.2 Coefficients of thermal expansion for several typical packaging materials [150–152]

Application	*Material*	*Coefficient of thermal expansion (ppm/°C)*
Die	Silicon	2.6
	Gallium arsenide	9.7
Leadframes	Copper	17
	Alloy 42	4.3–6.0
	Kovar	4.9
	Invar	1.5
Substrates	Alumina (99%)	6.7
	AlN	4.1
	Beryllia (99.5%)	6.7
	Pyrex 8329	2.8
	Tempax	3.18
	Pyrex 7740	3.1 (at room temperature)
Adhesives	Lead glass	10
	Au-Si eutectics	14.2
	Pb-Sn solder	24.7
	Ag-filled epoxy	32 a)
	(Fluoro-)silicones	300–800 b)
Conformal coatings or encapsulants	Polyimide	40–50
	(Fluoro-)silicones	300–800 b)
	Silicone gel	300 b)
Epoxy molding compounds	Unfilled epoxy	60–80 a)
	Silica-filled epoxy	14–24 a)

a) Below the glass transition temperature (T_g) range.
b) Above the T_g range.

silicone). An alternative (or complementary) method is to use a substrate with the same CTE as the die material.

4.2.4.4 **Die Attach**

Once the die is separated from the wafer, it must be attached to the chosen substrate material. This process precedes interconnection. Generally, an adhesive bond layer is deposited on the substrate material prior to placing the die on the substrate. This is followed by a curing, annealing or firing step to secure the adhesive bond.

The stress induced by the packaging materials is a function of the leadframe, header or substrate material choice and the die attachment method. Stress is also a function of the material choice for the die attach. Often silicones are chosen for stress-sensitive MEMS devices, even though they have very high CTE values, because they are very low modulus materials (often six orders of magnitude lower than silicon). In general, stress caused by eutectic (e.g. Au–Si alloy) die attach is higher than epoxy, which is also higher than silicones [5].

One important challenge in designing with these types of materials is that they are often polymeric. Polymer materials differ from silicon, most MEMS fabrication materials (e.g. polysilicon, silicon nitride, silicon dioxide) and metals in that they are viscoelastic. This suggests that they exhibit both a viscous (or lossy term) and an elastic term for modulus. This property makes these materials more difficult to model. Also, material geometry is difficult to determine precisely, because these materials are often drop-dispensed and will flow readily prior to curing.

Location of the sensor die is another concern when die attach methods are considered. For example, often the micromachined die is not the same die as the circuit die. If the output required includes circuit calibration, amplification or other signal processing, the package may contain two or more die. Fig. 4.25 is an example of the Motorola accelerometer, which contains a micromachined, wafer-bonded accelerometer or 'g-cell' die and a CMOS control IC. This side-by-side bonding is one configuration for the die attach. An alternative configuration is 'vertical assembly'. Kelly et al. (after Lyke [153]) describe this as the 'Towers of Hanoi' approach to die attach [154]. In this approach, the sensor die is placed on the silicon control IC and the two are wirebonded together with chip-to-chip wirebonds.

4.2.4.5 **Interconnection**

Typical electrical interconnection for microelectronics includes wirebonding (ball/wedge and wedge/wedge), tape automated bonding and flip chip (Fig. 4.32). Most MEMS devices utilize wirebonding as the method for electrical interconnection. The wirebonding process is of two types: thermocompression bonding and ultrasonic bonding. Thermocompression bonding uses heat and pressure to create the metal-to-metal bond (usually Au wire). Ultrasonic bonding uses ultrasonic vibration to create the metal-to-metal bond (usually Al wire). Because of the stress isolation that is often essential with MEMS, soft die attach materials are used in

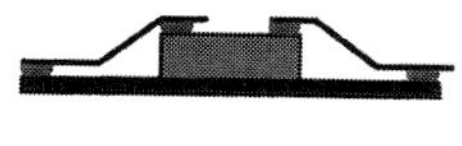

Fig. 4.32 Electrical interconnection methods for die inside a package

many cases. This requires that thermocompression bonding be used. On the other hand, if high temperatures are not allowed, ultrasonic bonding is necessary.

Wirebonding is typically a 'ball-wedge' bond. The ball is formed initially with the end of the existing wire on the die bondpad. After a stitching process, the wire is placed on the leadframe post with a wedge bond and is cut. Fig. 4.33 shows the process [150], which is fast and automated. However, several die and package properties affect this process: the die attach material, the die flatness (or tilt), the allowable wirebond loop height, the level of the bondpad versus the leadframe post and the cleanliness of the surfaces. In addition, wirebonding properties such as the temperature and load with a thermocompression bond or the energy applied with an ultrasonic bond affect the bond quality. A method for qualitatively evaluating the bond strength is the bond pull test. With this test, a hook is

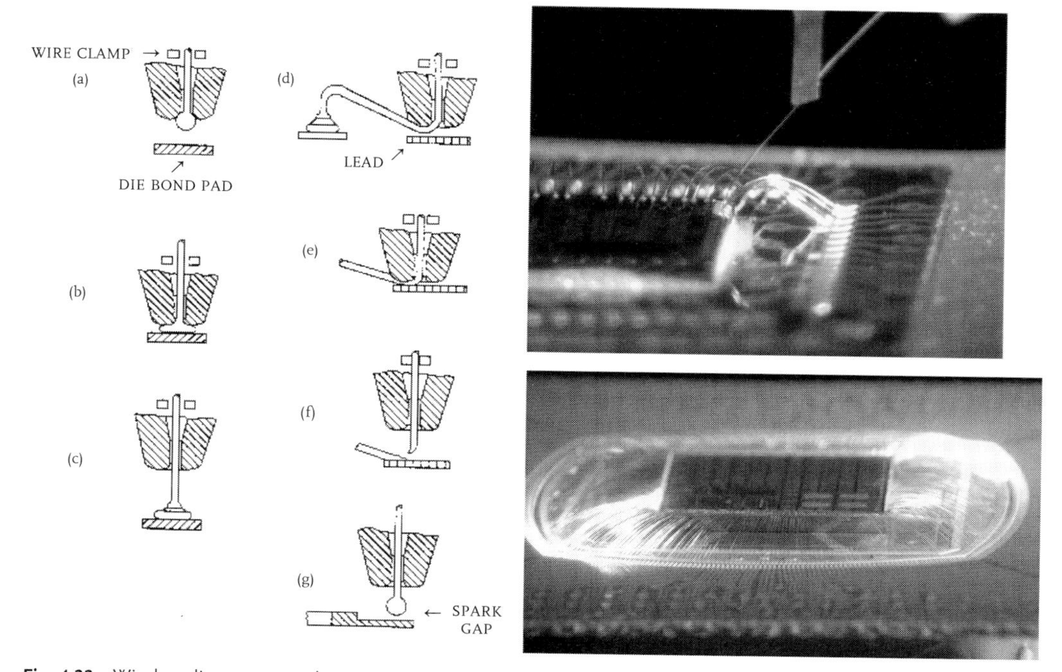

Fig. 4.33 Wirebonding process diagram [150]. Photographs courtesy of Dr. L. Spangler, Aspen Technologies [144]

placed under the wirebond and pulled with an increasing force upwards until the wirebond breaks. The location of the break also provides information concerning the quality of the wirebond. For example, if the bond ball or wedge pulls up, this often indicates that the surface of the metal was not clean and/or the wirebonder parameters were not selected properly. Alternatively, a break in the middle of the wire suggests a bond that is as strong as the strength of the wire itself. Failures in wirebonds include mechanical stress, interdiffusion, oxidation and interfacial contamination [155].

Tape-automated bonding (TAB) can increase the reliability of interconnections [155]. Although a significant amount of literature exists on TAB processing, little has been done with TAB and MEMS devices.

Flip chip interconnection provides superior electrical and density performance because it eliminates leads altogether. It also provides protection of the interconnection from the environment [156, 157] and may be more economically viable than monolithic integration [158]. From the research standpoint, it provides a method for rapid prototyping of microsystems [159]. The die or chip is 'flipped' over and connected to the substrate via solder bumps; the ability to attach (or remove) the die is simply a matter of locally heating the substrate to reflow the solder.

Finally, the use of micromachining to create interconnections, such as through-wafer interconnections, has been investigated [160]. Also, research has been pursued to create optical interconnections using micromachining [161–164].

4.2.4.6 Integration with Electronics

The miniaturization advantages of MEMS are realized only if they can be efficiently integrated with microelectronics. At first glance, it may seem that the most desirable approach to integration of MEMS and microelectronics would be to create a single or monolithic fabrication process capable of supporting both microelectronics and MEMS. This, however, is a difficult undertaking. For example, the removal of all oxide layers in a polysilicon surface micromachining process does not allow monolithic integration with CMOS VLSI circuits. Additionally, the high-temperature anneal of MEMS devices to relieve internal stress can be harmful to the carefully controlled diffusion budgets of microelectronic circuits. However, some custom processes, such as Sandia's Modular, Monolithic Micro-Electro-Mechanical Systems (M^3EMS) process, have been realized to allow monolithic integration of simple surface micromachined MEMS and electronics by fabricating the MEMS before the microelectronics [165].

Although several custom monolithic fabrication processes of MEMS and microelectronics have been demonstrated, the requirement to remove sacrificial layers in micromachining presents a set of unique problems. For example, the choice of sacrificial material for the MEMS device may be incompatible with the CMOS devices or microelectronic packaging. In CMOS, silicon dioxide is critical for the transistor gate insulation and for circuit passivation; but many MEMS processes use silicon dioxide as the sacrificial layer. Thus, the selection of structural and sacrificial materials in the MEMS devices impacts the integration with CMOS electronics. Moreover, some etchants for surface or bulk micromachining are not compatible with microelectronic materials or wiring. For example, KOH is often used in bulk micromachining, but it dissolves aluminum, which is commonly used in IC metallizations. Another bulk micromachining etchant, EDP, is not as aggressive in attacking aluminum but still requires masking the aluminum to preserve the integrated circuits or the package. Other important integration issues include the material properties of the films used (e.g. gate polysilicon may not be the best choice of mechanical polysilicon), the thickness of the thin films and thermal budget restrictions. As a result, the integration of MEMS with electronics usually requires additional processing steps and materials to protect the CMOS circuits and the package during the final MEMS release. Furthermore, electrical and environmental conditions also hamper monolithic integration of CMOS and MEMS. Many MEMS are designed to operate electrostatically with high voltages, which are challenging to implement with digital CMOS technologies. Other MEMS are designed to operate inside living organisms or are exposed to temperatures, radiation or chemicals that would be destructive to the CMOS integrated circuits.

All of these factors provide challenges for integration of MEMS and CMOS technology. Building monolithically integrated MEMS and electronic circuits in the same process may not always be cost-effective or realizable. However, it may be possible to use multichip module (MCM) technology to gain the benefits of MEMS and CMOS integration with minimum extra cost or additional technical challenges. MCMs offer an attractive integration approach because of the ability to support a variety of die types in a common substrate without requiring changes or compromises to either the MEMS or electronics fabrication processes. Furthermore, MCMs offer packaging alternatives for applications for which it is cost or time prohibitive to develop a monolithic integration solution. One of the main benefits of MCM packaging for MEMS and IC integration is the ability to combine die from incompatible processes in a common substrate. Other benefits of MCM technology are the electrical, size and weight performance improvement over conventional packaging techniques. The two common characteristics for MCM classification are the type of substrate used and the means of interconnecting signals between the dies [166]. The three dominant MCM substrate technologies are MCM-laminate, MCM-ceramic and MCM-deposited, but other substrate alternatives exist.

Of particular interest to wafer-scale integration of MEMS may be the MCM-silicon (MCM-Si) or 'silicon on silicon' technology. In MCM-Si, IC fabrication processes are used to deposit the interconnect and dielectric layers on a silicon substrate. MCM-Si has the highest signal interconnect density of any substrate choice and its coefficient of thermal expansion is an excellent match for any silicon die [166, 167]. The primary disadvantage of MCM-Si is that silicon is not a good base for the MCM package assembly because it is relatively fragile and consequently the MCM-Si substrate must be repackaged, causing additional cost. An advantage of MCM-Si technology, however, is that bulk micromachining tech-

NEC Laser Array

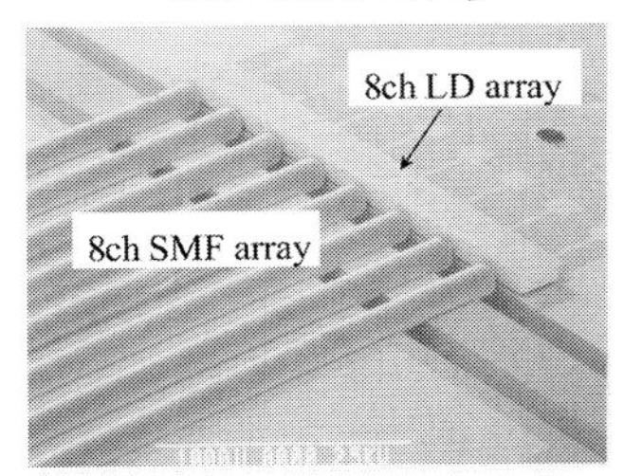

IBM Laser Module

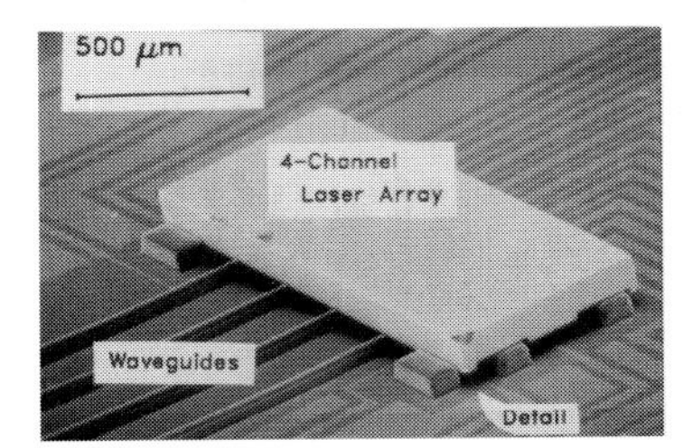

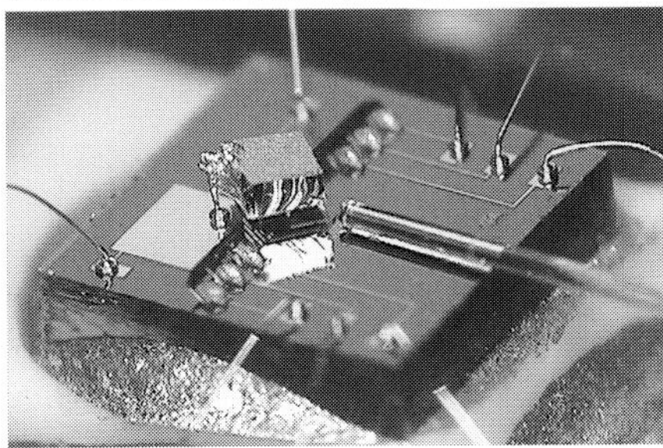

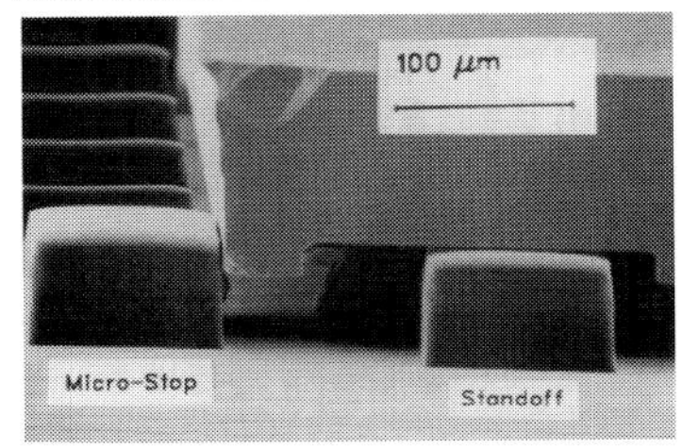

VCSEL to Fiber Module (Univ. of Colorado)

Fig. 4.34 Examples of micromachined silicon substrates for hybrid integration. Figures courtesy of Professor Y. C. Lee [169] and Koji Ishikawa [170], University of Colorado

niques can be used to pattern the silicon substrate. The substrate can be patterned using silicon bulk etchants and wafer bonding to form useful features such as embedded components or microchannels. Microchannels may be used to align optical fibers (Fig. 4.34) or carry fluids to and from MEMS mounted in the module. In addition, these microchannels can provide an efficient way to cool the MCM [168].

4.2.4.7 Encapsulation and Overmolding

Although overmolding of typical IC chips is common [171], there are more constraints placed on its application when packaging a MEMS device. First and foremost, package stress is often an important consideration for MEMS devices. The difference in coefficient of thermal expansion (CTE) is often the source of the package stress that is observed [172]. Tab. 4.2 shows typical CTE values for a variety of packaging materials. Stresses in MEMS devices can adversely affect the output of a MEMS device. For example, coatings on a pressure sensor have been shown to affect the device output [173]. Careful consideration must be paid to the package materials that come in contact with the MEMS device. Some solutions to this problem include the use of materials with similar CTE values (this is most often used with die attach materials), use of very thin coatings (e.g. thin-film polymer coatings, such as polyimide or parylene over a pressure, flow or chemical sensor) and/or the use of materials with very low moduli (e.g. silicone gels on a pressure sensor [174]). In addition, some techniques to isolate the MEMS device from a high-modulus material by using an intermediate low-modulus material have also been implemented [175].

Media compatibility is another important issue that constrains the choice of encapsulants or overmolds. For sensors or actuators, where the device must be in contact with an external environment, protection of the device from that environment, that is, media compatibility, is a very important consideration [150, 176]. In general, there are few techniques for low-cost media compatibility: thick, low-modulus film coatings (e.g. silicone gel [174]); thin, higher modulus film coatings (e.g. polyimide, parylene [177–179], other organics [180, 181] or wafer-level inorganics [182–185]); backside interconnection and typical overmolding processes [186–190]; selective encapsulation to avoid coverage of the sensor; secondary diaphragm and silicone oil [191].

Organic encapsulation may be divided into three basic categories [192]: (1) nonelastomeric thermoplastics, (2) nonelastomeric thermosetting polymers and (3) elastomers. Several types of polymers can be listed for each of these categories. The common IC encapsulant chemistries include nonelastomeric thermoplastics such as poly-*p*-xylylene (parylene) and pre-imidized silicone-modified polyimides, among other chemistries. Thermosetting polymers include silicones, polyimides, epoxies, silicone-modified polyimides, benzocyclobutenes and silicone-epoxies, and silicone gels and polyurethanes are examples of elastomers. Currently, many pressure sensor manufacturers use various types of silicone gel to encapsulate the device. A drawback of these gels is their limited chemical resistance. Alternative materials that have better chemical resistance are being considered for use in new media-resistant pressure sensor packaging.

4.2.4.8 **Package Housing Materials and Molding**

Package materials used to house MEMS are varied. MEMS devices are packaged in metal (Fig. 4.35), ceramics, thermoplastics (Fig. 4.36) and thermosets. In some cases, package housing materials are used to hermetically seal a device or to seal an atmosphere or fluid within the package.

Polymer materials are being used more often for packaging MEMS. Motorola has used both thermoplastics (Fig. 4.36) and thermosets to house MEMS devices. Thermoplastics are materials that are injection moldable. Typically, engineering thermoplastics are used because of their high-temperature tolerance properties. For example, polyesters, nylon, poly(phenylene sulfide) (PPS) and polysulfone have all been used. When molding a material with a metal insert (called 'insert molding'), as shown in Fig. 4.36 (a), CTE mismatch between the polymer material and the metal can cause leakage of gases or liquids from the inside of the package to the outside of the package (or vice versa). Package designers must consider this as a potential reliability problem when choosing package housing materials.

Thermoset plastics are materials that are transfer moldable. Typically, epoxy materials are used because of their moldability and high-temperature tolerance proper-

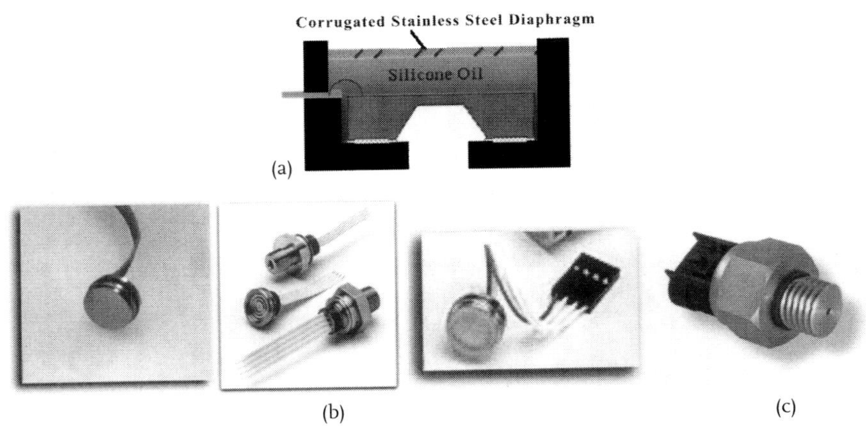

(a)

(b)

(c)

Fig. 4.35 A stainless-steel package used to house a media-compatible pressure sensor. (a) A typical cross-section of a stainless-steel diaphragm with silicone oil fill used to create media-compatible pressure sensor packages. (b) Examples of stainless-steel diaphragm, silicone oil-filled packages from Lucas NovaSensor [149]. (c) A high-pressure metal package for housing pressure sensors [193]

(a) (b)

Fig. 4.36 MEMS devices packaged in thermoplastics. (a) The Motorola unibody package using polyester thermoplastic housing material [150, 194]. (b) The Motorola manifold absolute pressure (MAP) module [193]

ties. Thermosets are cross-linked plastics. They require very high pressures for molding. Thermosets are generally low-viscosity materials when they enter the mold. Following heating, they cure and chemically react to create the 'set' material.

4.3
Case Studies – Packaging of Physical Sensors

4.3.1
Case Study 1: Tire Pressure Sensor Packaging

Commercial package development for MEMS pressure sensor devices is intimately linked to silicon development for (at least) three major reasons: (1) stress-induced performance effects caused by the packaging on the sensor, (2) environmental reliability factors of the sensor in the application and (3) the cost of the final product.

The emerging tire pressure automotive sensor application is an example of a case where all apply. The US government has mandated tire deflation monitoring for all US automobiles by 2006 through the TREAD Act [195]. Recently, a Court of Appeals ruling has called into question the use of the indirect tire pressure measurement technique that takes advantage of the differential wheel speed sensor capability within the ABS systems on many cars [196]. Without this approach, the so-called direct tire pressure measurement approach is the leading candidate to fill the requirement for 2006 automotive production.

4.3.1.1 Tire Pressure Monitoring Application Requirements

Sensor Performance – Accuracy

The tire pressure application has three very significant performance requirements that have impacted the design of the transducer itself, which ultimately affects the design of the package. First, the accuracy requirement for the pressure sensor is stringent: ~1.5% FSS in the nominal pressure, temperature and battery voltage range. While manifold absolute pressure sensors (MAP sensors) for engine management applications have become even more accurate than this, very few other applications for MEMS require this level of accuracy. Because this is total accuracy, it includes offset, sensitivity, linearity, temperature coefficient, pressure and temperature hysteresis and ratiometicity, among other errors that impact the device performance. MEMS packaging has been shown to impact device performance through CTE mismatch of thin-film passivation materials on pressure sensors [197], metal stress hysteresis [198], changes in strain caused by the package that are transmitted through anchor points in the transducer [199], etc. Therefore, careful consideration of the package design must be made while designing the transducers.

Sensor Performance – Power Consumption

The second performance requirement that is uniquely taxing for tire pressure sensors among current automotive sensor requirements is power consumption. In other applications, efforts are made to lower power consumption, but in the direct tire pressure measurement application within the tire, the most prominent approaches are battery-powered, so power consumption becomes an overriding requirement. Typical battery lifetime is specified in the 5–10 year range, depending on the automotive use; however, most passenger cars require a 10 year battery lifetime. Not only does this alter the design approach for current consumption, it also requires significant consideration of the signal-to-noise ratio for the device as filtering to minimize noise requires power [200] and the battery lifetime puts significant limits on circuit blocks that can be added.

Motorola already had an existing bulk micromachined, piezoresistive, bipolar integrated pressure sensor in its production portfolio when tire pressure application development began in 1998. However, the power consumption requirement motivated the development team to review alternative, surface micromachined, capacitive pressure sensor technology. The requirement for a digital output to allow data conversion to a bitstream rf output, the likelihood of additional features required for the application and the power consumption requirement pushed the development team to CMOS instead of bipolar circuit technology. A choice was made for cost reasons to integrate the standard digital output, power-saving multi-mode (sleep, pressure read, temperature read and data transmit) and capacitive pressure sensor on to a single chip. This provided a chip outline that could be used for this first-generation tire pressure monitoring system (TPMS). This device, because of its die size, could fit into an existing Motorola standard 'super-small outline package' (SSOP) for pressure sensors, which also minimized development costs and added volume to an existing production line allowing further depreciation of existing (not new) capital assets.

Tire Pressure Environmental Influence – Harsh Chemical Environment and Centrifugal Acceleration

There is a third significant performance/reliability requirement that was specified for the tire pressure application: performance of the pressure and temperature measurements in the tire environment, including harsh chemical media and the centrifugal forces that would be present in a rotating tire.

The harsh chemical environment required for reliability testing included not only the standard temperature cycling, pressure cycling and humidity exposure, but also exposure of the sensor to chemicals common to the roadway environment – salt water, ice, dust, chemicals common to the automotive environment such as fuels, oils and other automotive fluids (e.g. transmission fluid, battery acid), and chemicals common to the tire environment – tire mounting lubricant. Existing Motorola pressure sensor packages included a fluorosilicone or fluorocarbon polymeric low-modulus gel material that protected the silicon from the environment while causing minimal performance changes to the device and allowing an accurate pressure measurement at the diaphragm surface because of the incompressibility of the gel material.

However, in the case of the TPMS application, the tire would be rotating and the additional mass of the gel material on the diaphragm could cause an unwanted acceleration cross-sensitivity – an error in the pressure sensor measurement in what was already a very stringent accuracy requirement for the sensor. Analytical modeling was performed to show how this physical effect would induce an accuracy error to the pressure measurement [201].

4.3.1.2 Tire Pressure Monitoring Sensor Encapsulation Evaluation

An evaluation of various methods to protect the sensor from the tire environment while not inducing any additional performance errors to the device was performed. The fact that a new MEMS technology was being used provided new constraints and new possibilities. With a conventional bulk micromachined pressure sensor for tire pressure applications, a backside exposure approach could be used (see Fig. 4.37, environmental exposure on the opposite side of the silicon from where the circuitry resides). Backside exposure provides an effective barrier of the harsh chemical environment to the electronics that are typically on the frontside of the pressure sensor diaphragm (e.g. from corrosion). However, a surface micromachined device, without very sophisticated silicon and/or package processing, must be exposed on the topside (see Fig. 4.38 [202], same side of the silicon as the sensor electronics). This is a constraint for the surface micromachining technique. On the other hand, backside exposure of silicon in this environment has its own drawbacks. First, linearity of a pressure sensor is worse when exposed from the backside than when exposed from the frontside. Second, at the pressure required for these applications (2–6 atm or bar gauge), the die attach material is in tension during the lifetime of its exposure, which could result in material fatigue with time in the application.

The opportunity that is availed by using a surface micromachined pressure sensor is the result of the much smaller size of the transducer. The sensitivity of a pressure sensor goes as a ratio of the diaphragm length to thickness (raised to the third power). Hence, as the thickness decreases, the length (and area) of the pressure sensor can be reduced significantly for a given sensitivity. A typical bulk micromachined pressure sensor is 12–24 μm thick (single-crystal silicon). A surface micromachined pressure sensor can be anywhere from 0.5 to 3 μm thick (polysilicon). The result is a transducer that takes up a much smaller portion of the die itself. Hence the cost of the transducer can be reduced.

Because of the smaller transducer size, alternative environmental protection techniques that were dubbed 'selective encapsulation' were reviewed. This general approach allowed the evaluation of several chemically resistant materials that would be too stiff for the existing pressure sensor, provided that a technique could be found for depositing them around the electronics (wirebonds, package leadframe, etc.) and not on the pressure sensor diaphragm. Three types of approaches were attempted (Fig. 4.39):

1. 'Drawn-dam' selective encapsulation (Fig. 4.39 a), where a polymer gel material was drawn in the shape of a dam around the diaphragm to protect it from subsequent material deposition in the resulting annular ring around the diaphragm.

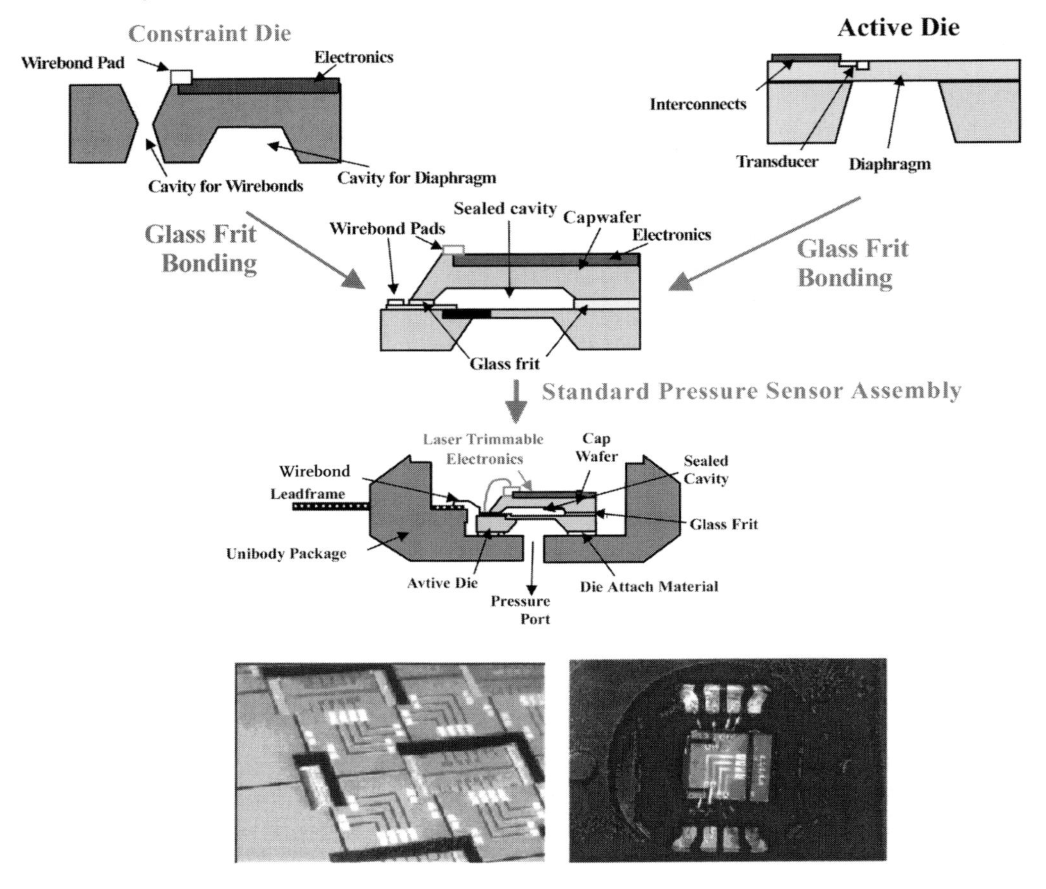

Fig. 4.37 Backside exposure for a pressure sensor device

2. 'Silicon-dam' selective encapsulation (Fig. 4.39 b), where a second bulk micromachined silicon wafer was wafer-bonded to the surface micromachined, CMOS integrated pressure sensor to create a dam around the diaphragm to protect it from subsequent material deposition in the resulting annular ring around the diaphragm.
3. 'Cap encapsulation' (Fig. 4.39 c), where a portion of the mechanical metal protective cap that covers the pressure sensor was used to create the barrier around the pressure sensor diaphragm. This was intended to allow pressure (and the associated chemical environment) only to the diaphragm and not to the electronics on the chip.

Furthermore, two alternatives were reviewed with higher modulus films that could be deposited in a thin, uniform and repeatable manner, thus minimizing variability in the performance effects that these films had on the devices. Many re-

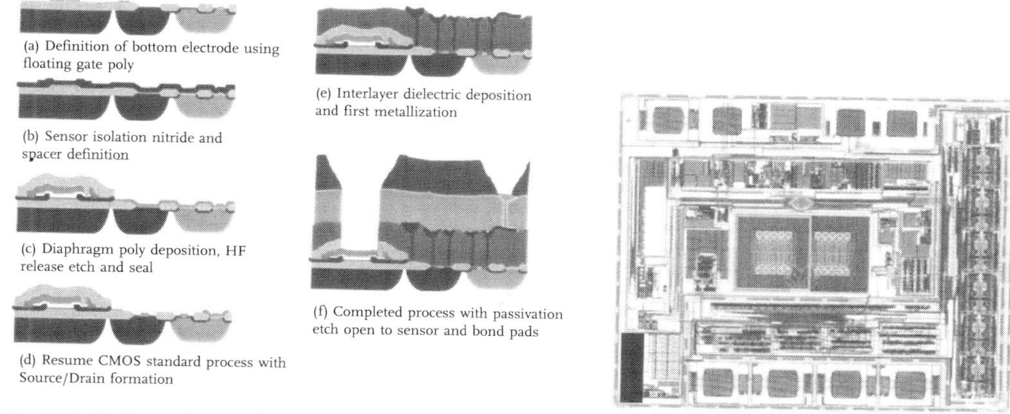

Fig. 4.38 Surface micromachined pressure sensor device [202]

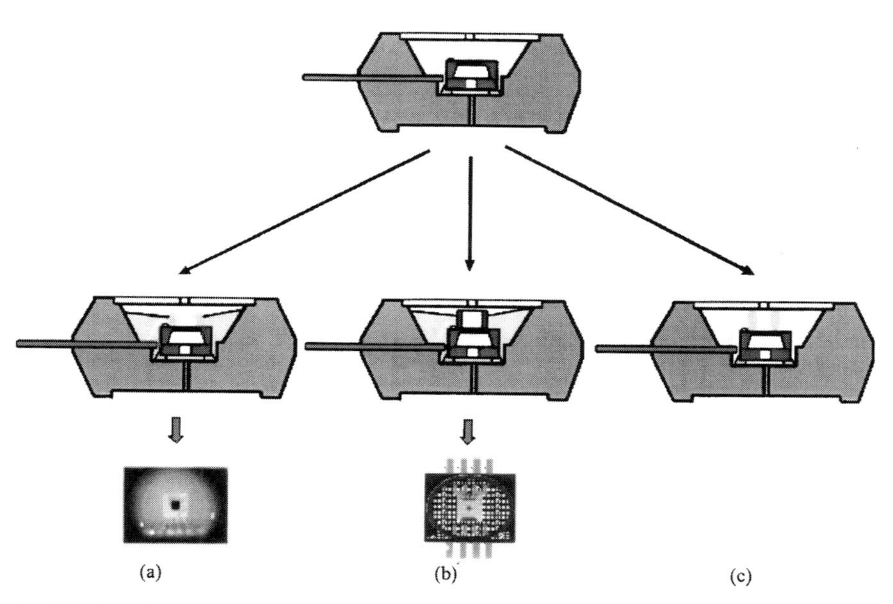

Fig. 4.39 Selective encapsulation approaches

searchers have suggested using films with slower diffusivity to harsh environments (especially water), including thermoset polymers and/or inorganic films [203]. The problem with doing this on pressure sensors is that these materials must coat the extremely high aspect ratio wirebonds and other packaging features; in other words, they need to be deposited at the package level and not the wafer level, as most users are accustomed to doing. The two films reviewed were Teflon and parylene because of their known chemical resistance. Moreover, parylene is deposited by a moderate pressure deposition technique that allows very conformal coating on these high aspect ratios in the package.

Finally, an approach was reviewed using various types of filters that would allow the passage of gases but not liquids [204]. A composite filter that included hydrophobic layers and oleophobic layers was found that could protect the device from the harsh environment and not impart any packaging stress or other performance issues on the device.

4.3.1.3 **Results and Conclusions**

The encapsulation evaluation yielded the following results and conclusions.

Drawn-dam selective encapsulation (Fig. 4.40) [205]. An assembly process was developed that could create this structure in the package repeatably using an existing die attach dispense machine. The wirebond loop height of 10 mil above the silicon, however, posed an additional constraint. The deposition of the polymeric gel yielded a resulting material with an aspect ratio (height to width) of ~0.8. A wider diameter deposition needle could create a tall enough 'dam' to prevent overflow of the subsequent 'fill' encapsulation material from the outside annulus into the diaphragm cavity. However, this wider diameter would cause a resulting dam material that impinged on the wirebonds, creating a situation where the wirebond went through two materials: the dam material and then the fill material. Heuristic knowledge suggests that this is a situation to be avoided because of the possible temperature cycling fatigue effects. An alternative approach to creating multilayer dams was also completed. Whereas the multilayer dam approach worked and allowed a dam to be developed that had a high enough aspect ratio to allow adequate fill material coating of the wirebonds and other package features, it also caused a throughput issue with this piece of assembly equipment. This work was discontinued when the cost versus wirebond reliability tradeoff was identified (and positive work on other techniques was progressing).

Silicon dam selective encapsulation (Fig. 4.40) [205]. A silicon wafer fabrication, wafer-bonding and assembly process was developed that could create the same type of structure. Two issues were identified with this approach that precluded adopting it for the tire pressure application. First, the bulk micromachined wafer that was used to bond on the surface micromachined wafer had a significant portion of the area removed to allow for two sets of holes per die: one for the pressure opening and one for the wirebonding opening. Although special care in handling this wafer resulted in insignificant wafer breakage, it was noted that this could be a breakage problem in manufacturing that could cause scrap material for this line of products, and, in the worst case, could also result in contamination (particulates) in other lines of products in the same manufacturing area. Furthermore, the cost of the additional wafer processing and the wafer bonding was higher than other alternatives that were being evaluated.

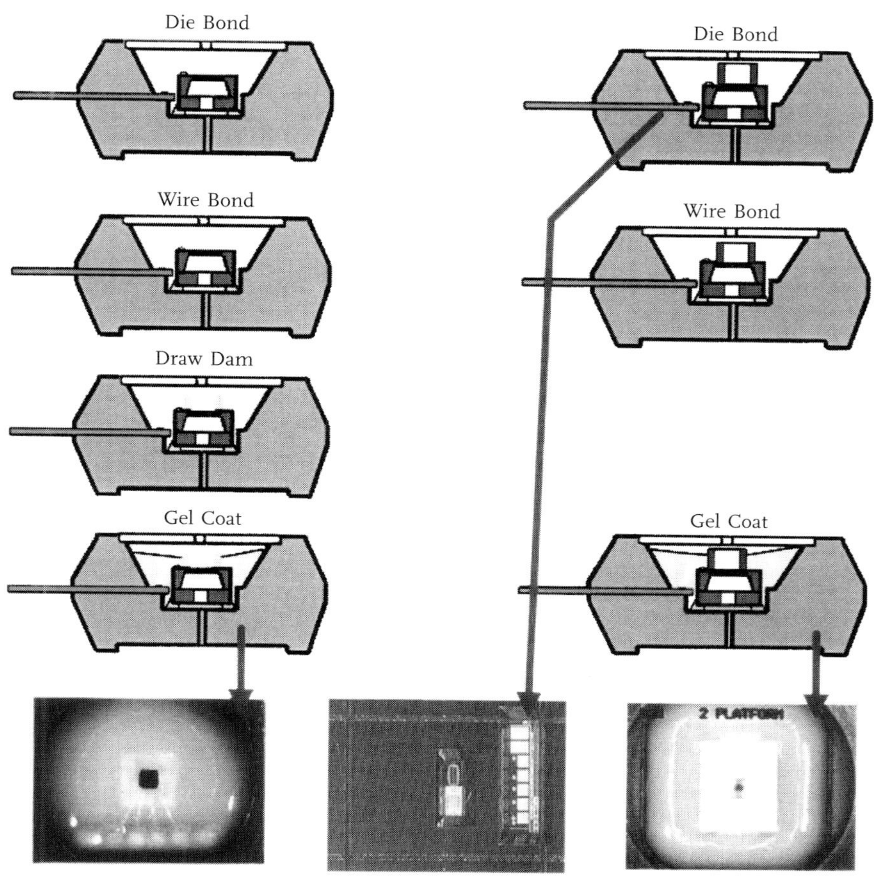

Fig. 4.40 Selective encapsulation assembly techniques [205]

Cap encapsulation. It was noted with this technique that a reliable seal between the cap and the silicon die was difficult to achieve, so this approach was abandoned early in the evaluation.

Teflon coatings. It was found that these coatings did not provide a reliable enough coverage of the various surfaces within the package to pass consistently the harsh media exposure evaluations. Coating of Teflon can be done in many ways, but these approaches are limited with packaged devices that have thermoplastic materials with softening temperatures between 180 and 220 °C. Because of this observation, these approaches for depositing Teflon were also abandoned as other techniques appeared successful.

Parylene films [206]. Motorola has evaluated parylene coatings three times in great detail to determine whether they could be used to provide a reliable coating material for harsh chemical media protection. In each case, the adhesion of the material was noted to be less reliable than needed for the particular pressure sensor applications. A large majority of the parylene-coated devices did pass the harsh chemical environmental testing, but not 100% of them. The observation of these failures was that delamination occurred, presumably following diffusion of an aqueous-based chemical to a site where corrosion was initiated. Corrosion products caused a propagation of the delamination, which opened more sites for corrosion, so the process continued until an open circuit was found. Significant work using corrosion techniques, such as electrochemical impedance spectroscopy, was performed on these films to determine the root cause of the failures. No definitive solution has yet been found, although it was noted that coating the parylene with a second film (such as a fluorosilicone gel) provided the reliability required for several applications in automotive and aqueous-based pressure sensor products. However, this solution also added the mass of the gel material over the diaphragm, which caused an acceleration cross-sensitivity error. And, two encapsulant materials were required, which increased the cost of the final solution.

Composite hydrophobic and oleophobic filters (Fig. 4.41) [204]. Initial results with these devices were very positive following significant work that was performed to seal the filters to the package so that no harsh chemical media could seep around the filter into the package. Devices with this protective approach passed all the tire pressure harsh media requirements repeatedly and did not impart unwanted packaging stress on the device. Furthermore, the manufacturing process for these devices allowed them to be produced in mass quantities at prices that were acceptable. A full assembly process was completed and qualified, including an evaluation of any response time delay. No significant response time delay was found with the filters attached on packages with the production tire pressure sensor device.

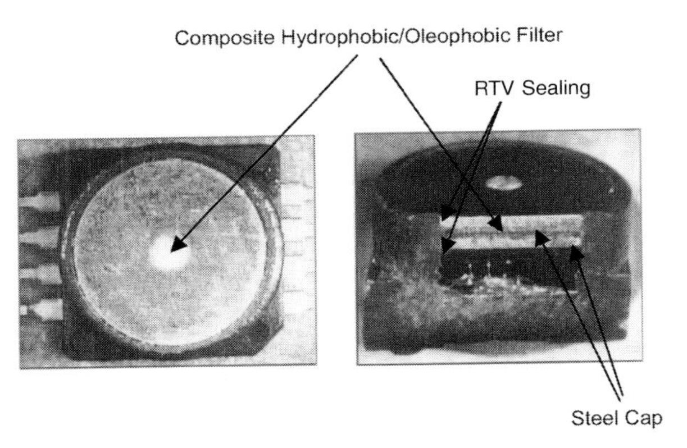

Fig. 4.41 Composite hydrophobic and oleophobic filter [204]

The final result is that the solution with the composite hydrophobic and oleophobic filter attached to the Motorola standard SSOP with the tire pressure sensor die was qualified and released to production in May 2003. This first-generation device is being followed quickly with a second-generation device that will integrate more of the tire pressure features into the package, so package development for this market will continue through the 2004 timeframe in preparation for the market that is slated to ramp to very high volume production in 2005–2006.

4.3.2 Case Study 2: Inertial Sensor Packaging – Second-generation SOIC Accelerometer Sensor Packaging

Another example of a commercialized, high-volume MEMS physical sensor is an accelerometer. The most prevalent use of accelerometers is in airbag deployment, so automotive safety specifications and reliability are required. As in the pressure sensor case, packaging for accelerometers is an integral part of the device function, reliability and product design. Designing a successful MEMS product is a collaborative concurrent team process involving the customer, transducer and ASIC designer(s), package designer, test development and manufacturing.

At the macro level, there are many technologies that must be combined in any successful MEMS products (Fig. 4.42). Each successful MEMS producer leverages their core competencies with the breadth of technologies that they possess. Motorola has chosen to partition the transducer from the ASIC and take advantage of 'combinational technologies' (or Systems in Packages: SiP). This was a key strategic business decision taken to provide the broad product portfolio for the markets they serve.

Motorola started to sell automotive-grade accelerometers in 1996 utilizing a straight tether z axis transducer in a family of plastic packages that was based on the 16 lead dual-in-line. The axis of sensitivity with reference to the PCB for this

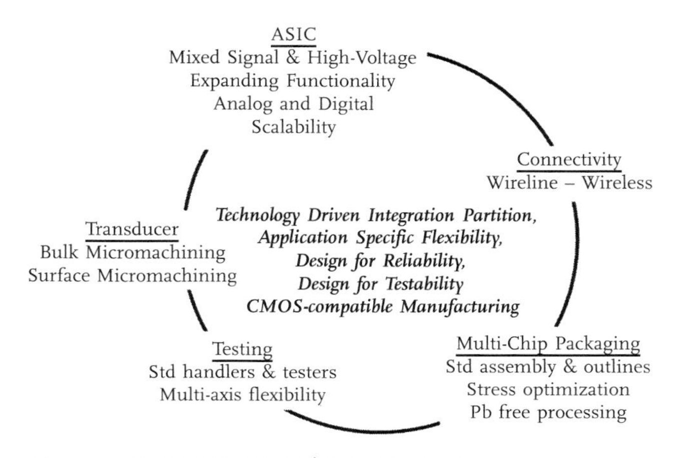

Fig. 4.42 Core competency and component partitioning

family was provided by the package. In 1999 Motorola introduced a family of z, x and x/y axis accelerometers with the axis of sensitivity independent of the package (Fig. 4.43). Fig. 4.43 shows the evolution from through-hole to surface-mount SOIC16/20 lead package.
The packaging strategy can be summarized by five goals [207]:

1. Maintain and execute a product/packaging road map that aligns with key customer inputs/requirements.
2. Reduce the overall system cost.
3. Provide differentiated package solutions.
4. Improve system robustness.
5. Reduce the time to qualify new packaging concepts.

Several requirements must be considered when designing an accelerometer package:

- Package size requirements
- Time to market
- Customer board attach and PCB attach reliability
- Die/dies dimensions and technology
  - Transducer hermetic seal
  - Passivation and metallization
  - Stiction
  - Level of silicon integration
- Mechanical and thermal stress management
- Automotive reliability requirements as per AEC-Q100
- Product liability
- Operating environment
  - Automotive temperature range (–40 to 125 °C)
  - Survive powered and unpowered high mechanical shock (to 2000 g and beyond);

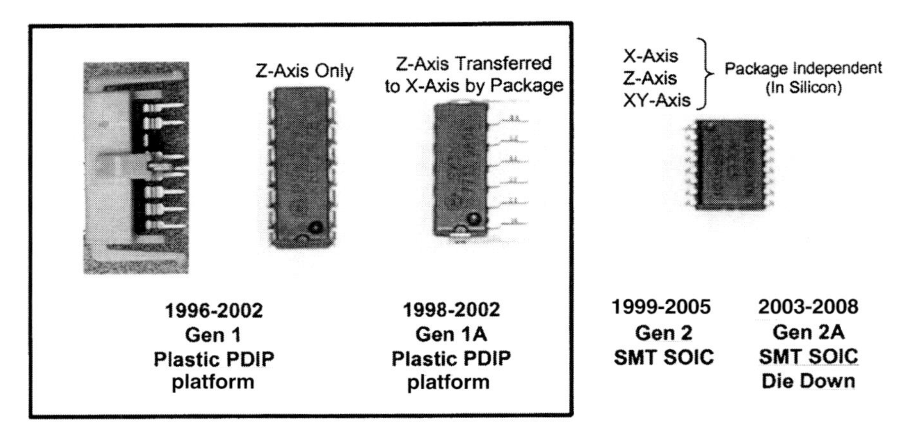

Fig. 4.43 The first two generations of Motorola's family of accelerometers

  - ESD
  - EMC
  - Package and transducer resonance
- Orientation of the transducer must be considered (for *x*-, *y*-, *z*- and rotational accelerometers packaging orientation may change depending upon the design of the transducer);
- Greater than 15 year lifetime
- Manufacturability and testability
- Designed for the environment
- Cost.

4.3.2.1 Typical Manufacturing Flow

Fig. 4.44 shows a simplified final manufacturing flow used for manufacturing Motorola's surface mount accelerometer. As can be seen from the requirements and the manufacturing flow, there are multiple challenges that need to be addressed when manufacturing MEMS products. In addition, the product and package designers have to include several other factors, such as cost, quality, reliability, time to market, economy of scale, global legislation and standards and competition. All add to the technology challenges of bringing a successful product to market.

The following discussion considers a few of the components from this flow and addresses some of the key manufacturing challenges.

In 1994, Motorola choose to develop its first-generation accelerometer in an over-molded plastic package taking advantage of its strength in high-volume leadframe technology and thus provided a low-cost advantage. In 1999, Motorola introduced a family of surface-mount accelerometers using the highly productive copper multistrand leadframe as shown in Fig. 4.45.

In the case of the Motorola accelerometer, glass frit bonding [208, 209] is used at the wafer level to protect the MEMS element during final packaging (e.g. handling/stiction/contamination/particles) and to create a stable pressure environment and provide a defined damping level for the accelerometer during operation, over the lifetime of the application (Fig. 4.46).

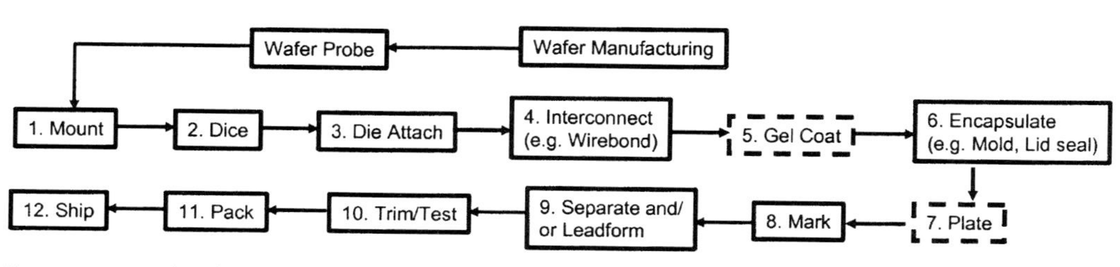

Fig. 4.44 Motorola's final manufacturing flow. Solid lines represent common processes utilized for most wirebonded packages (cavity packages, ceramic, plastic, arrays, etc.)

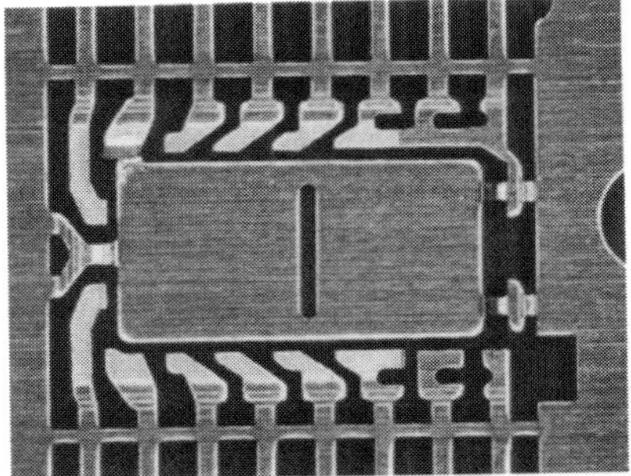

Fig. 4.45 Multistrand SOIC leadframe

(a)

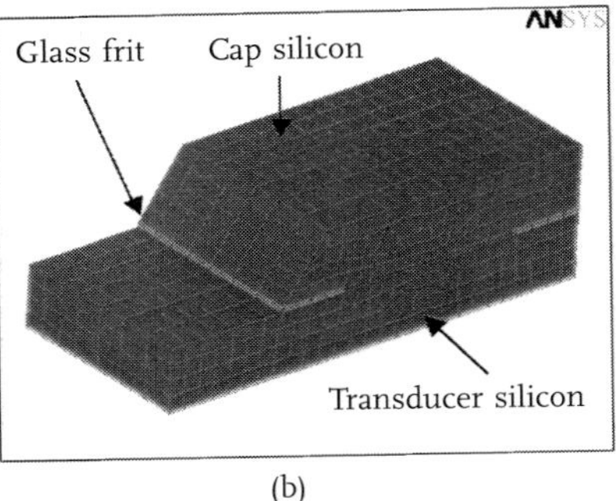

(b)

Fig. 4.46 (a) Motorola accelerometer with glass-frit bonded cap wafer to eliminate stiction and to protect the sensor from mechanical damage during assembly and to create a constant-pressure environment for the operation of the device during the lifetime of the application [210, 211]. (b) A wafer-level package model (a half symmetry model showing the microcavity for the transducer)

4.3.2.2 **Mechanical Stress Isolation**

In order to meet customers' need for zero defects and six-sigma quality, there are design challenges for a capable, well-controlled manufacturing process, for the transducer and the package. The type of the package will determine the strategy adopted to meet this challenge; for example, some companies have chosen open-cavity packages whereas Motorola has chosen over-molded plastic packages.

Transducer design challenges include mechanical element intrinsic stress and stress gradients (and resulting curvature), package-induced deformation and product standardization [212]. A comparison of a prior transducer design and its present successor illustrates some solutions to these challenges. Production data concretely demonstrate the actual improvements achieved.

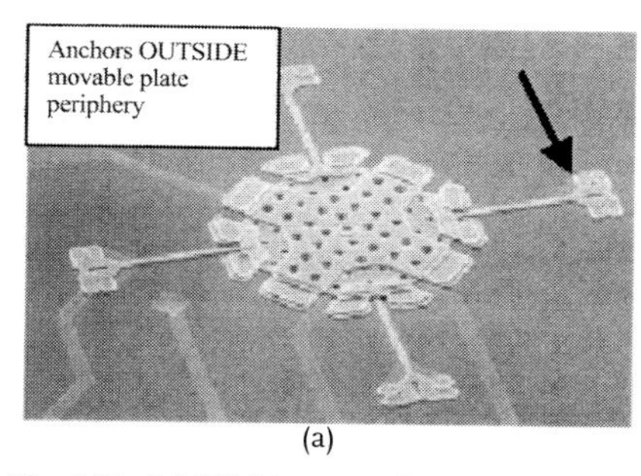

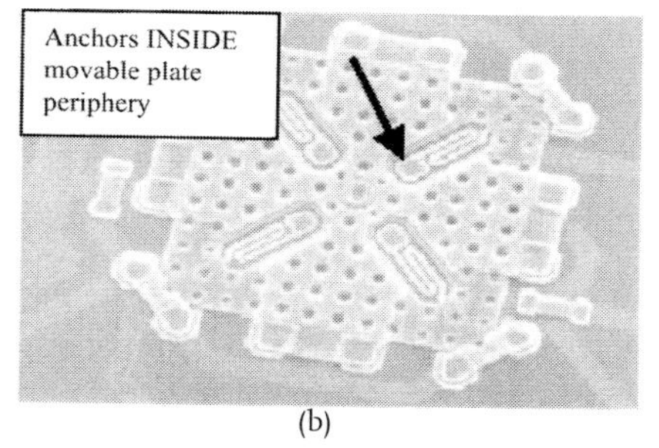

Fig. 4.47 (a) SEM images of Motorola's first-generation *z*-axis transducer design, Vega, and (b) improved design, Altair [212]

A simple approach for stress isolation was to replace the straight tensile tethers (Fig. 4.47 a) with their folded beam equivalents and leave the rest of the design unchanged (Fig. 4.47 b). Folded beams is a well-known approach [213, 214] in the industry for addressing stresses induced by the package. The folding and moving of the anchor position as shown in Fig. 4.47 b allows for anchor movement due to package-induced deformation with a minimal influence on the movable plate position. Fig. 4.48 a shows the impact of the folded beam versus straight tether design and optimized anchor position. Fig. 4.48 b shows the comparison between production and FEA models. This folding also allows for a more compact design (38% die size reduction).

Package stresses due to various materials that make up the package also cause curvature of the 'fixed' top and bottom plates. Therefore, the challenge in addition

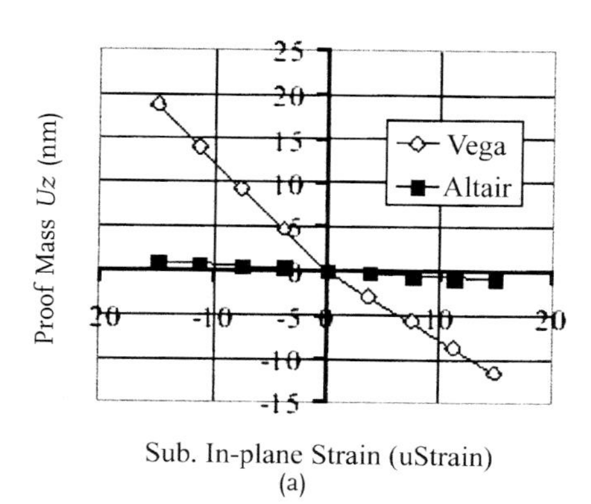

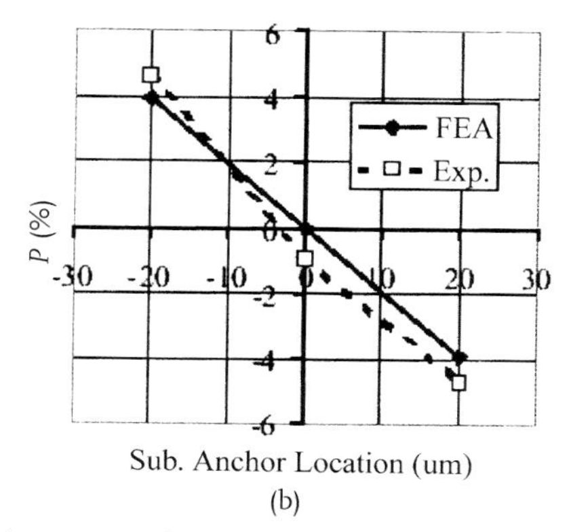

Fig. 4.48 (a) FEA on effects of substrate strain on transducer design performance. Package stress induced movable plate or proof mass movement. Vega (original design) is straight and Altair (improved design) is folded beams [212]. (b) Effect of anchor location on transducer capacitive matching. Comparison of FEA with experimental data. This plot shows designs with three different anchor locations (nominal +20 μm and −20 μm radial) [212]

to providing a mounting foundation to a PC board, stresses induced by mismatch in the thermal expansion coefficients of the materials used to fabricate the package and the external thermal loading of the package must be controlled and kept low enough to avoid impact on the sensors' electrical and reliability performance [215–217]. The selection criteria typically include the following: (1) the material must provide the required functions (i.e. conductive or nonconductive); (2) the resulting package must be reliable for the intended service life of the product; (3) it must meet environmental requirements; and (4) most important, the packaging stress due to mechanical constraints or thermal cycling must remain minimal or repeatable enough so that the characteristics of the transducer remain within the specification over its lifetime of service.

In order to reduce packaging stress on the transducer, Motorola developed [215–220] a process where the transducer is attached to the leadframe using RTV at its four corners instead of a single glob of RTV and a silicone gel coating process as illustrated in Fig. 4.49. On the top of the silicon, a thin layer of silicone gel is dispensed and cured before the final step of placing epoxy-molding compound (EMC). Fig. 4.50 illustrates an FEA model of the final package.

Fig. 4.50 a displays packaging stress/strain for a nominal SOIC-16 package for three different molding compounds (from three EMC vendors) [221]. Clearly, over a temperature range of –40 to 125 °C, the first molding compound exhibits the greatest stress/strain and the third compound shows the least. Although the stress is less, the stress/strain dependencies on temperature of EMC 2 and 3 are nonlinear (not straight). Therefore, it is difficult to compensate for this nonlinearity in the design of the accompanying control circuitry.

Another issue that needs to be considered early in the design stage is the dynamic characteristic of the package [215, 220]. For example, in certain types of automotive applications, the frequency of vibration signals can be as high as 20 kHz. If one or more natural modes of the package are at or near the frequency of a high-energy input signal, the package may exhibit large vibrations that could

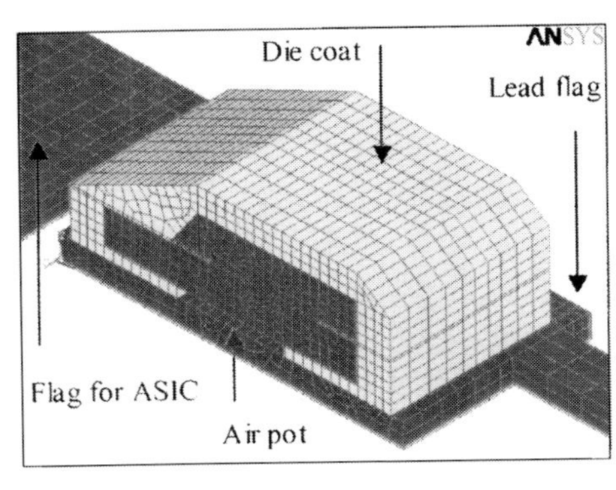

Fig. 4.49 The silicon is glued to the flag at only four corners and a soft silicone gel is dispensed to separate the silicon chip from the surrounding EMC [215]

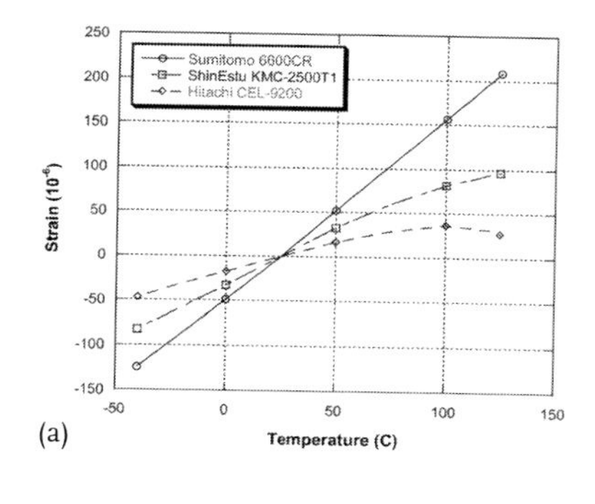

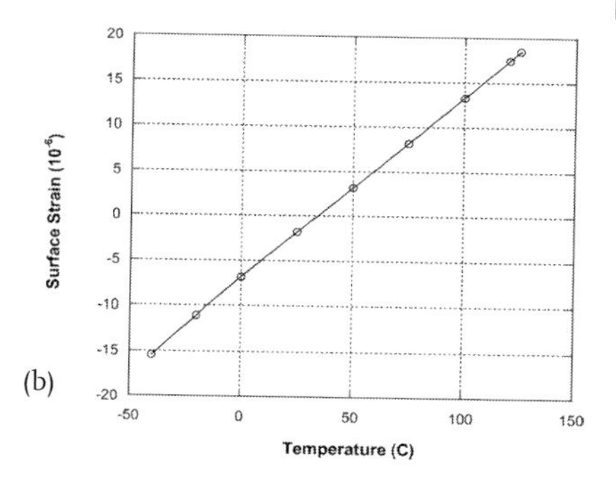

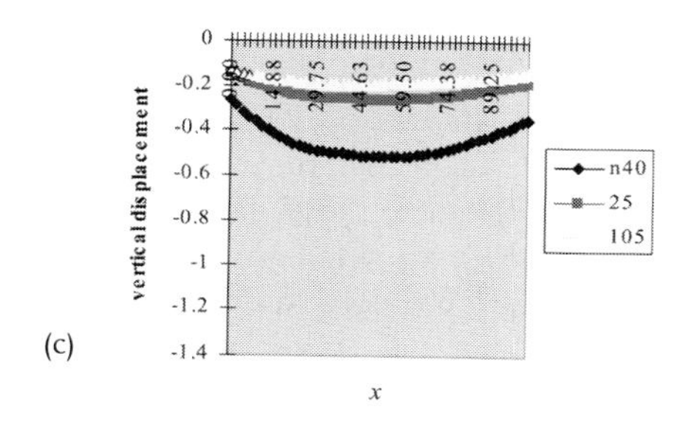

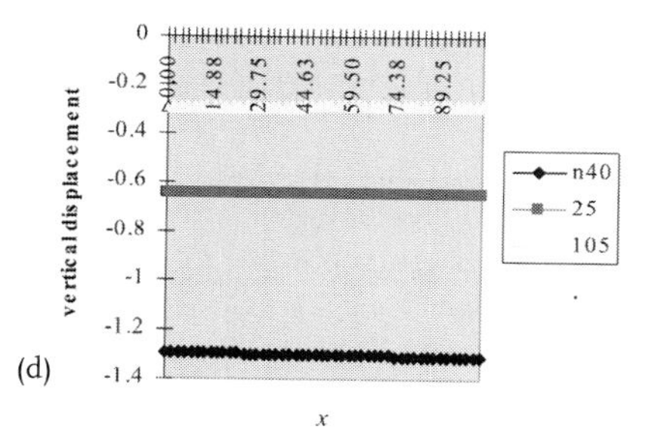

Fig. 4.50 (a) The surface strain on the transducer without any stress isolation built into the package for three different mold compounds [221]. (b) The surface strain on the transducer with RTV die attach and silicone gel covering the MEMS transducer for Sumitomo mold compound [221]. (c) The curvature across the transducer element at different temperatures without RTV die attach and silicone gel [222]. (d) No curvature across the transducer element with RTV die attach and silicone gel [222]

distort the output signal of the transducer or even damage the transducer mechanically. Fig. 4.51a displays a typical frequency response curve for a nominal leaded sensor package as shown in Fig. 4.51b. The severity of package resonance depends on the transducer technology (mechanically over-damped or under-damped), circuit design, package choice (leaded or unleaded) and module design (potted or conformal coated, etc.).

4.3.2.3 **Design for the Environment [223–226]**

The electronics industry actively supports the need for environmentally safe products to conform to community, customer and legislative requirements. This is being addressed by companies, industry associations, standards bodies and government legislation at a global level.

A Directive on the 'Impact on the Environment of Electrical and Electronic Equipment' (EEE) was proposed to the European Council in 2002. The Directive requires manufacturers to design electrical and electronic equipment in such a way that po-

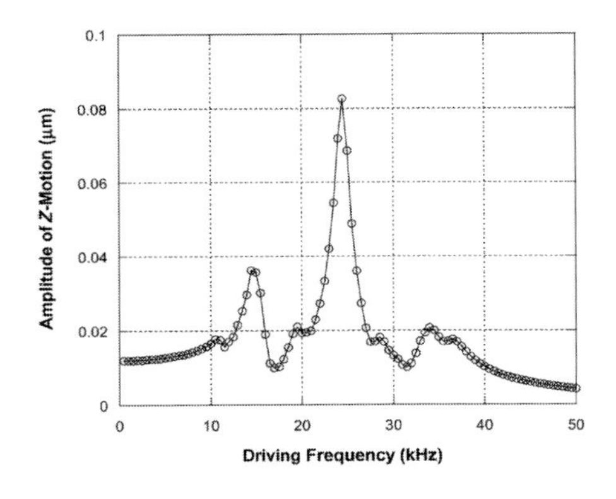

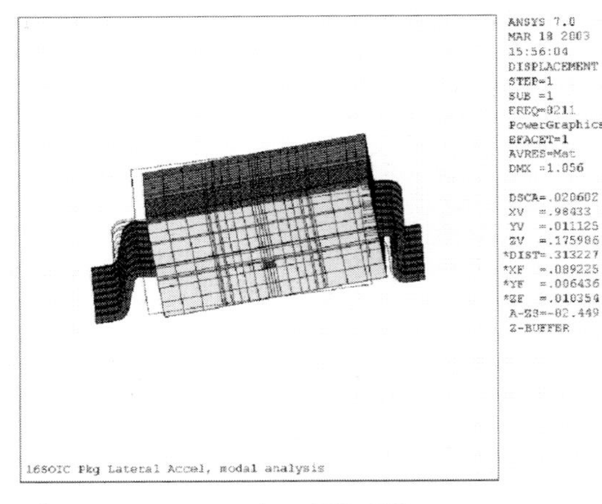

Fig. 4.51 A typical frequency response of a MEMS sensor package [215, 220]

tentially assesses and takes account of every environmental attribute in a product or component's life cycle, as a condition for products being marketed in the EU. Also, WEEE AND ROHS Directives were approved by the European Parliament in December 2002 and will bring significant challenges in the choice of materials. The ROHS requires the industry to eliminate lead, cadmium, mercury, hexavalent chromium and certain flame retardants (PBBs and PBDEs) from electrical and electronic products by 2006. Not only are these material substitutions costly to research and implement, technical challenges such as product reliability issues are still not well known. The WEEE Directive poses significant challenges since the industry will pay for the collection, treatment, recovery and recycling of all electrical and electronics products. The requirement to take back its end-of-life products is designed to drive the industry to design and build products that contain fewer hazardous materials and are easier to recycle. In the ELV (End of Life Vehicle) Directive, a maximum concentration of up to 0.1% by weight per homogeneous material for lead, hexavalent chromium and mercury and up to 0.01% by weight per homogeneous material for cadmium will be tolerated, provided that these substances are not intentionally introduced.

This raises multiple challenges for plastic leadframe packaging designer, such as

- removal of Pb from the lead finishes and the use of Pb free solders during board mount;
- higher reflow temperature required for Pb-free solders;
- increased susceptibility to moisture-related failures during PCB solder reflow;
- removal of fire-retardant halides used in mold compounds;
- increased cost of material.

To address these requirements, many new materials and/or processes are currently being offered or are being considered:

- replacement of Sn/Pb plating (leadfinish) with pure Sn or SnBi or pre-plated leadframes (e.g. NiPdAu, NiAu);
- new fire-retardant or self-extinguishing resins;
- higher filler content and lower moisture absorption mold compounds;
- stronger die attachment and improved adhesion between materials;
- solder alloys for high-conductivity die attachment and board mount.

4.3.2.4 Trends and Conclusions

Motorola has shipped more than 100 million accelerometers for automotive airbag applications since their introduction in 1996. Today the accelerometer portfolio spans from 1 to 250 *g* in all three axes, *z*, *x* and *x*/*y* with both analog and digital communications. In comparison with the integrated circuit industry, MEMS is still in a state of infancy.

In the immediate future, there are applications that are opening up in the 'low-g' (1–10 g) range. Fig. 4.52 shows some of these examples and their relative acceleration ranges. In the automotive market place, this is driven by the need for braking applications (electronic stabilization programs, ESP), roll-over (which is driven in the US by law defined in the TREAD Act) and navigation. Furthermore, several applications are appearing in the consumer and industrial markets for ga-

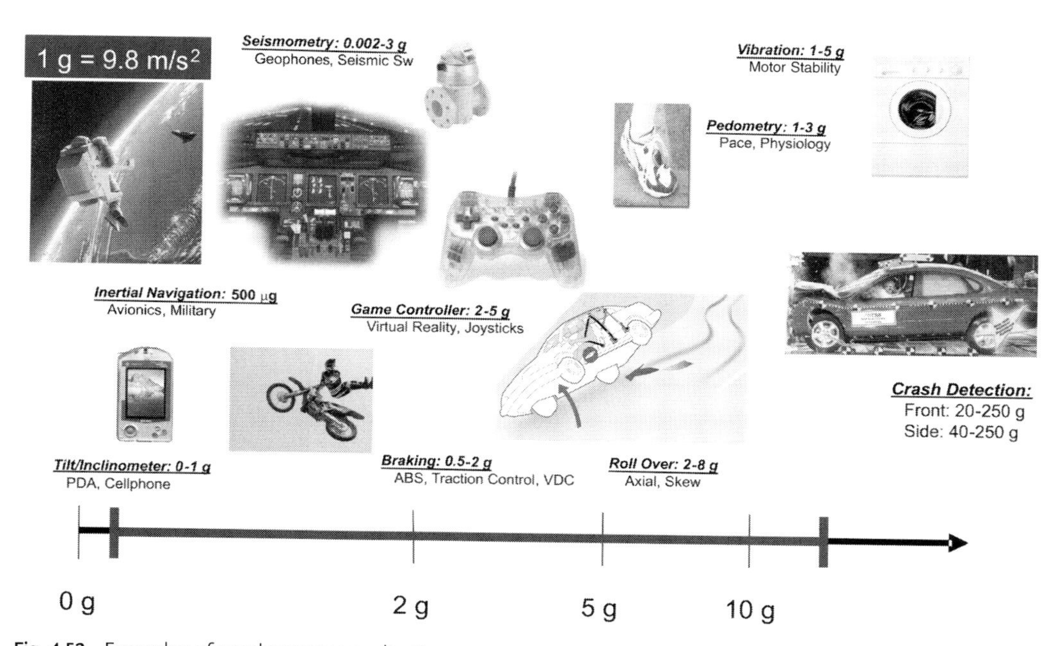

Fig. 4.52 Examples of accelerometer application

mepad/PDA/cellphone handset controls to motor-bearing health monitoring. As the cost, size and power consumption are driven down, these consumer applications will evolve and expand.

A second trend is the combination of additional technology into the product. For example, with the drive for more networking of accelerometers for crash detection, there is a requirement for more 'smarts' at the point of sensing. Sensors with ADC, logic and a network protocol are appearing on the market. It is likely, also, that this will expand into wireless networking protocols, as the driving market applications develop.

Third, combinations of sensory inputs will begin appearing in the same package. For example, *XY* acceleration sensors are now on the market in mass quantities. *XYZ* accelerometers have also been introduced. This will require significant care in the product design as the impact of package stress, for example, could be different depending on the axes. Furthermore, several applications are driving to smaller packages: not just in *X*- and *Y*-dimensions for the package, but also in the *Z*-dimension, as package thickness will be pushed below the 1.5 mm point. Smaller packages pose different problems for multiple axis sensors. Moreover, adding other sensor inputs into a single product has also been considered, such as pressure sensors (altimeters) and accelerometers (for navigation) and angular rate sensing and accelerometers (inertial measurement units).

Finally, the key point to these examples is to show that for product design and commercialization success of MEMS/microsystem devices, package design must be considered in concert with the rest of the product design. This is critical when considering the components/chips required to facilitate the system in package, the stress impact of the materials (i.e. now environmentally friendly materials) on device performance and reliability, the assembly flow itself and how it affects yield/quality (e.g. stiction) and the cost of the final product.

4.4 References

1 L. Ristic, M. Shah, in: *WESCON 96 Conference Proceedings, Anaheim, CA*; **1996**, pp. 64–72.

2 S. D. Senturia, R. L. Smith, in: *Microsensor Packaging and System Partitioning, Sensors and Actuators* **1988**, *15*, 221–234.

3 J. Swift, presented at the Solid-State Sensors and Actuators Workshop, Hilton Head Island, SC, 1998.

4 J. Schuster, presented at the Solid-State Sensors and Actuators Workshop, Hilton Head Island, SC, 1998.

5 M. Madou, *Fundamentals of Microfabrication*; Boca Raton, FL: CRC Press, **1997**.

6 R. Allan, *Electron. Des.* **1997**, January 20, 75–88.

7 H. Reichl, *Sens. Actuators A* **1991**, *25–27*, 63–71.

8 C. Song, in: *Transducers '97, International Conference on Solid-State Sensors and Actuators, Chicago, IL*; **1997**, pp. 839–842.

9 D. S. Eddy, D. R. Sparks, *Proc. IEEE*, **1998**, *86*, 1747–1755.

10 G. Beardmore, in: *IEEE Colloquium on Assembly and Connection in Microsystems*; **1997**, pp. 2/1–2/8.

11 R. R. Tummala, E. J. Rymaszewski, *Microelectronic Packaging Handbook*; New York: Van Nostrand Reinhold, **1989**.

12 W. H. Ko, in: *Microsystems Technologies '94: Proc. 4th International Conference and Exhibition on Micro Electro Mechanical*

Systems and Components; **1994**, pp. 477–480.

13 A. Morrissey, G. Kelly, J. Alderman, *Sens. Actuators A* **1998**, *68*, 404–409.

14 A. D. Romig Jr., P. V. Dressendorfer, D. W. Palmer, in: *High Performance Microsystem Packaging: A Perspective, Microelectron. Reliab.* **1997**, *37*, 1771–1781.

15 A. Bossche, C. V. B. Cotofana, J. R. Mollinger, in: *SPIE Conference on Smart Electronics and MEMS, San Diego, CA*, **1998**, 166–173.

16 *2000 NSF Workshop on Manufacturing of Micro-Electro-Mechanical Systems, Orlando, Florida*, 7 November **2000**.

17 B. A. Parvis, D. Ryan, G. M. Whitesides, in: *Using Self-assembly for the Fabrication of Nano-scale Electronic and Photonic Devices, IEEE Trans. Adv. Packaging 26(3)*, **2003**, 233–241.

18 Y. Y. Wei, G. Eres, V. I. Merkulov, D. H. Lowndes, *Appl. Phys. Lett.* **2001**, *78*, 1394.

19 L. H. Chen, S. Jin, in: *Packaging of Nanostructured MEMS Micro-Triode Devices, J. Electron. Mater.* **2003**, *32(12)*, 1360–1365.

20 V. Derycke, R. Martel, J. Appenzeller, Ph. Avouris, *Nano Lett.* **2001**, *1*, 453–456.

21 J. Kong, N. R. Franklin, C. Zhou, M. G. Chapline, S. Peng, K. Cho, H. Dai, *Science* **2000**, *287*, 622–625.

22 A. P. Alivisatos, *Science* **1996**, *271*, 933–937.

23 G. Markovich, C. P. Collier, S. E. Henrichs, F. Remacle, R. D. Levine, J. R. Heath, in: *Architectonic Quantum Dot Solids, Acc. Chem. Res.* **1999**, *32*, 415–423.

24 C. M. Lieber, *Solid State Commun.* **1998**, *107*, 607–616.

25 C. B. Murray, C. R. Kagan, M. G. Bawendi, *Annu. Rev. Mater. Sci.* **2000**, *30*, 545–610.

26 Y. N. Xia, et al., *Adv. Mater.* **2003**, *15*, 353–389.

27 C. J. Murphy, N. R. Jana, *Adv. Mater.* **2002**, *14*, 80–82.

28 R. E. Cavicchi, R. M. Walton, M. Aquino-Class, J. D. Allen, B. Panchapakesan, *Sens. Actuators B* **2001**, *77*, 145–154.

29 D. L. Klein, R. Roth, A. Lim, A. P. Alivisatos, P. L. McEuen, in: *A single-electron transistor made from a cadmium selenide nanocrystal, Nature* **1997**, *389*, 699–701.

30 Y. Cui, Q. Wei, H. Park, C. M. Lieber, *Science* **2001**, *293*, 1289–1292.

31 K. S. J. Pister, M. W. Judy, S. R. Burgett, R. S. Fearing, *Sens. Actuators A* **1992**, *33*, 249–256.

32 M. C. Wu, L. Y. Lin, S. S. Lee, in: *Micromachined Free-Space Integrated Optics, Proceedings of the SPIE, 2291, Integrated Optics and Microstructures II, San Diego, California*, July 28, **1994**, pp. 40–51.

33 P. W. Green, R. R. A. Syms, E. M. Yeatman, *IEEE J. Microelectromech. Syst.* **1995**, *4*, 170–176; R. R. A. Syms, *Sens. Actuators A* **1998**, *65*, 238–243.

34 K. F. Harsh, V. M. Bright, Y. C. Lee, *Sens. Actuators A* **1999**, *77*, 237–244.

35 R. R. A. Syms, E. M. Yeatman, V. M. Bright, G. M. Whitesides, *IEEE J. Microelectromech. Syst.* **2003**, *12*, 387–417.

36 A. Singh, D. A. Horsley, M. B. Cohn, A. P. Pisano, R. T. Howe, in: *Transducers '97. 1997 International Conference on Solid-State Sensors and Actuators, Chicago, IL*; **1997**, pp. 265–268.

37 M. B. Cohn, Y. Liang, R. T. Howe, A. P. Pisano, in: *Solid-State Sensor and Actuator Workshop, Hilton Head Island, SC*; **1996**, pp. 32–35.

38 P. Bell, N. Hoivik, V.M. Bright, Z. Popovic, presented at the 35th International Symposium on Microelectronics (IMAPS), 4–6 September **2002**, Denver, CO.

39 F. F. Faheem, K. C. Gupta, Y. C. Lee, in: *Flip-chip assembly and liquid crystal polymer encapsulation for variable MEMS capacitors, IEEE Transactions on Microwave Theory and Techniques* **2003**, *51(12)*, 2562–2567.

40 S. B. Brown, G. Povirk, J. Connally, in: *MEMS 93, Fort Lauderdale, FL*; **1993**, pp. 99–104.

41 S. B. Brown, W. Van Arsdell, C. L. Muhlstein, in: *1997 International Conference on Solid-State Sensors and Actuators, Chicago, IL*; **1997**, pp. 591–593.

42 R. Maboudian, W. R. Ashurst, C. Carraro, in: *Self-Assembled Monolayers as Anti-Stiction Coatings for MEMS: Characteristis and Recent Developments; Sensors and Actuators A (Physical)* **2000**, *A82(1–3)*, 219–223.

43 M. P. De Boer, et al., *Acta Mater.* **2000**, *48*, 4531–4541.

44 L. Ristic, *Sensor Technology and Devices*; Boston, MA: Artech House, **1994**.
45 R. Maboudian, R.T. Howe, *Tribol. Lett.* **1997**, 215–221.
46 R. Maboudian, R.T. Howe, in: *Critical review: Adhesion in surface micromechanical structures, J. Vac. Sci. Technol. B* **1997**, *15(1)*, 1–20.
47 R. Maboudian, *Surf. Sci. Rep.* **1998**, *30*, 207–269.
48 N. Tas, T. Sonnenberg, H. Jansen, R. Legtenberg, M. Elwenspoek, *J. Micromech. Microeng.* **1996**, *6*, 385–397.
49 J.Y. Kim, C.-J. Kim, in: *Comparative Study of Various Release Methods for Polysilicon Surface Micromachining, 10th IEEE International Conference on MicroElectro-Mechanical Systems,* **1997**, pp. 442–447.
50 R.L. Alley, G.J. Cuan, R.T. Hose, K. Komvopoulos, in: *Solid-State Sensor and Actuator Workshop, Hilton Head Island, SC*; **1992**, pp. 202–207.
51 H. Guckel, D.W. Burns, *Sens. Actuators A* **1989**, *20*, 117–122.
52 C.H. Mastrangelo, C.H. Hsu, *IEEE J. Microelectromech. Syst.* **1993**, *2*, 33–43.
53 C.H. Mastrangelo, C.H. Hsu, *IEEE J. Microelectromech. Syst.* **1993**, *2*, 44–55.
54 Y. Yee, M. Park, K. Chun, *IEEE J. Microelectromech. Syst.* **1998**, *7*, 339–344.
55 F.M. Serry, D. Walliser, G.J. Maclay, in: *The Role of the Casimir Effect in the Static Deflection and Stiction of Membrane Strips in Microelectromechanical Systems (MEMS), J. Appl. Phys.* **1998**, *84*, 2501–2506; W. Tang, in: *Electrostatic Comb Drive for Resonant Sensor and Actuator Applications*; PhD Thesis, University of California, Berkeley, CA, **1990**.
56 L.S. Fan, *Integrated Micromachinery: Moving Structures on Silicon Chips*; PhD Thesis, University of California, Berkeley, CA, **1989**.
57 Y.-C. Tai, R.S. Muller, in: *Solid-State Sensor and Actuator Workshop, Hilton Head Island, SC*; **1998**, p. 88.
58 R.C. Stouppe, *Surface Micromachining Process*; Norwood, MA, Analog Devices, **1997**.
59 T. Abe, W.C. Messner, M.L. Reed, in: *Proc. MEMS '95*; Amsterdam, **1995**, pp. 94–99.
60 R.T. Howe, H.J. Barber, M. Judy, *Apparatus to Minimize Stiction in Micromachined Structures*; Norwood, MA: Analog Devices, **1996**.
61 G.K. Fedder, R.T. Howe, in *Proc. MEMS '89*; **1989**, pp. 63–68.
62 G.T. Mulhern, D.S. Soane, R.T. Howe, in: *7th International Conference on Solid-State Sensors and Actuators, Yokohama;* **1993**, pp. 296–299.
63 C.W. Dyck, J.H. Smith, S.L. Miller, E.M. Russick, C.L.J. Adkins, in: *SPIE 1996 Symposium on Micromachining and Microfabrication, Austin, TX*; **1996**, pp. 225–235.
64 M.R. Houston, R.T. Howe, R. Maboudian, in: *Effect of hydrogen termination on the work of adhesion between rough polycrystalline silicon surface, J. Appl. Phys.* **1997**, *81(8)*, 3474–3483.
65 M.R. Houston, R. Maboudian, R.T. Howe, in: *The 8th International Conference on Solid-State Sensors and Actuators and Eurosensors IX, Stockholm*; **1995**, pp. 210–213.
66 R.L. Alley, G.J. Cuan, R.T. Hose, K. Komvopoulos, *Solid-State Sensor and Actuator Workshop, Hilton Head Island, SC*; **1992**, pp. 202–207.
67 P.F. Man, B.P. Gogoi, C.H. Mastrangelo, in: *Elimination of Post-Release Adhesion in Microstructures Using Conformal Fluorocarbon Coatings, J. MEMS,* **1997**, *6(1)*, 25–34.
68 M. Nishimura, Y. Matsumoto, M. Ishida, in: *Technical Digest of the 15th Sensor Symposium, Kawasaki*; **1997**, pp. 205–208.
69 Y. Matsumoto, K. Yoshida, M. Ishida, in: *1997 International Conference on Solid-State Sensors and Actuators, Chicago, IL;* **1997**, pp. 695–698.
70 J.R. Martin, Y. Zhao, *Micromachined Device Packaged to Reduce Stiction*; Norwood, MA: Analog Devices, **1997**.
71 J.S. Go, Y.H. Cho, B.M. Kwak, K. Park, *Sens. Actuators A* **1996**, *54*, 579–583.
72 C.L. Chang, P.Z. Chang, *Sens. Actuators A* **2000**, *79*, 71–75.
73 Y. Zhang, Y. Zhang, R.B. Marcus, *IEEE J. Microelectromech. Syst.* **1999**, *8*, 43–49.
74 D.C. Miller, W. Zhang, V.M. Bright, *Sens. Actuators A* **2001**, *89*, 76–87.

75 D. J. Vickers-Kirby, R. L. Kubena, F. P. Stratton, R. J. Joyce, D. T. Chang, J. Kim, *Mater. Res. Soc. Symp.* **2001**, *657*, EE2.5.1–EE2.5.6.

76 H. Sehr, A. G. R. Evans, A. Brunnschweiler, G. J. Ensell, T. E. G. Niblock, *J. Micromech. Microeng.* **2001**, *11*, 306 –310.

77 Y. Zhang, M. L. Dunn, *Proc. SPIE* **2002**, *4700*, 147–156.

78 W. D. Nix, *Metal. Trans. A* **1989**, *20*, 2217–2245.

79 Y. L. Shen, S. Suresh, *Acta Metall. Mater.* **1995**, *43*, 3915–3926.

80 J. Koike, S. Utsunomiya, Y. Shimoyama, K. Maruyama, H. Oikawa, *J. Mater. Res.* **1998**, *13*, 3256–3264.

81 S. P. Baker, A. Kretschmann, E. Arzt, in: *Thermomechanical Behavior of Different Texture Components in Chu Thin Films, Acta Materialia* **2001**, *49(12)*, 2145–2160.

82 D. C. Miller, M. L. Dunn, V. M. Bright, *Proc. SPIE* **2001**, *4558*, 32–44.

83 F. Shen, P. Lu, S. J. O'Shea, K. H. Lee, T. Y. Ng, in: *Thermal Effects on Coated Resonant Microcantilevers, Sens. Actuators A* **2001**, *95(1)*, 17–23.

84 M. L. Dunn, Y. Zhang, V. Bright, *IEEE J. Microelectromech. Syst.* **2002**, *11*, 372–384.

85 M. Finot, S. Suresh, *J. Mech. Phys. Solids* **1996**, *44*, 683–721.

86 G. G. Stoney, *Proc. R. Soc. London, Ser. A* **1909**, *82*, 172–175.

87 M. W. Hyer, *J. Compos. Mater.* **1981**, *15*, 175–194.

88 M. Finot, I. A. Blech, S. Suresh, H. Fujimoto, *J. Appl. Phys.* **1997**, *81*, 3457–3464.

89 K. F. Harsh, W. Zhang, V. M. Bright, Y. C. Lee, in: *Proc. 12th IEEE International Conference on Microelectromechanical Systems (MEMS '99)*; **1998**, pp. 273–278.

90 M. L. Dunn, Y. Zhang, J. Roy, P. E. W. Labossiere, V. M. Bright, in: *Nonlinear Deformation of Multilayer MEMS Structures, Proc. 1999 ASME IMECE MEMS Symposium*, MEMS-vol. 1, Nashville, TN; **1999**, pp. 75–79.

91 D. A. Koester, R. Mahadevan, B. Hardy, K. W. Markus, *MUMP^{TM} Design Rules*; Research Triangle Park, NC: Cronos Integrated Microsystems, **2001**; *http://www.memsrus.com/cronos/svcsrules.html*.

92 Y. Zhang, M. L. Dunn, in: *Proceedings of the MEMS Symposium, ASME International Mechanical Engineering Congress and Exposition* **2001**, *3*, 149–156.

93 K. Gall, M. L. Dunn, Y. Zhang, B. Corff in: *Thermal cycling response of layered gold/polysilicon MEMS structures, Mechanics of Material* **2004**, *36(1/2)*, 45–55.

94 D. M. Burns, V. M. Bright, *Sens. Actuators A* **1998**, *70*, 6–14.

95 R. Maboudian, W. R. Ashurst, C. Carraro, *Tribol. Lett.* **2002**, *12*, 95–100.

96 R. Maboudian, C. Carraro, *J. Adhesion Sci. Technol.* **2003**, *17*, 583–591.

97 W. R. Ashurst, M. P. de Boer, C. Carraro, R. Maboudian, *Appl. Surf. Sci.* **2003**, *212/213*, 735–741.

98 W.R. Ashurst, C. Carraro, R. Maboudian, W. Frey, in: *Solid-State Sensor, Actuator and Microsystems Workshop, Hilton Head Island, SC*; **2002**, pp. 142–145.

99 M. Mehregany, C. A. Zorman, N. Rajan, C. H. Wu, *Proc. IEEE* **1998**, *86*, 1594–1610.

100 J. P. Sullivan, T. A. Friedmann, M. P. de Boer, D. A. LaVan, R. J. Hohlfelder, C. I. H. Ashby, M. T. Dugger, M. Mitchell, R. G. Dunn, A. J. Magerkurth, in: *Developing a New Material for MEMS: Amorphous Diamond, Mater. Res. Soc. Symp. Proc.* **2001**, *657*, EE7.1.1–EE7.1.9.

101 I. S. Forbes, J. I. B. Wilson, *Thin Solid Films* **2002**, *420/421*, 508–514.

102 A. E. Franke, J. M. Heck, T. J. King, R. T. Howe, *IEEE J. Microelectromech. Syst.* **2003**, *12*, 160–171.

103 G. Radhakrishnan, R. E. Robertson, P. M. Adams, R. C. Cole, in: *Integrated TiC coatings for moving MEMS, Thin Solid Films* **2002**, *420/421*, 553–564.

104 C. R. Stoldt, C. Carraro, W. R. Ashurst, D. Gao, R. T. Howe, R. Maboudian, *Sens. Actuators A* **2002**, *97–98*, 410–415; C. R. Stoldt, C. Carraro, W. R. Ashurst, M. C. Fritz, D. Gao, R. Maboudian, in: *Proceedings of Tranducers '01/Eurosensors XV, Munich*; **2001**, pp. 984–987.

105 C. H. Wu, C. A. Zorman, M. Mehregany, *Mater. Sci. Forum* **2000**, *338*, 541–544.

106 C.R. Stoldt, M.C. Fritz, C. Carraro, R. Maboudian, *Appl. Phys. Lett.* **2001**, *79*, 347–349.
107 S.S. Mani, J.G. Fleming, J.J. Sniegowski, M.P. de Boer, L.W. Irwin, J.A. Walraven, D.M. Tanner, D.A. LaVan, *Mater. Res. Soc. Proc.* **2000**, *605*, 135–140.
108 D.M. Tanner, W.M. Miller, W.P. Eaton, L.W. Irwin, K.A. Peterson, M.T. Dugger, D.C. Senft, N.F. Smith, P. Tangyunyong, S.L. Miller, in *1998 IEEE International Reliability Physics Proceedings*; **1998**, p. 26.
109 Prof. S.M. George research group at the University of Colorado – Boulder. *http://www.colorado.edu/chemistry/GeorgeResearchGroup/index.html*, ca 20 April 2004.
110 N.D. Hoivik, J.W. Elam, R.J. Linderman, V.M. Bright, S.M. George, Y.C. Lee, *Sens. Actuators A* **2003**, *103*, 100.
111 S.M. George, A.W. Ott, J.W. Klaus, *J. Phys. Chem.* **1996**, *100*, 13121.
112 C.F. Herrmann, F.W. DelRio, V.M. Bright, S.M. George, in: *Hydrophobic Coatings using Atomic Layer Deposition and Non-Chlorinated Precursors, 17th IEEE International Conference on Micro Electro Mechanical Systems (MEMS2004)*, Maastricht, The Netherlands, January 25–29, **2004**.
113 M.L. Hair, C.P. Tripp, *Colloids Surf. A* **1995**, *105*, 95.
114 M.P. de Boer, T.A. Michalske, *J. Appl. Phys.* **1999**, *86*, 817.
115 B.H. Stark, K. Najafi, in: *An Integrated Process for Post-Packaging Release and Vacuum Sealing of Electroplated Nickel Packages, The 12th International Conference on Solid-State Sensors and Actuators (Transducers 03)*, Boston, MA, June 8–12, **2003**, pp. 1911–1914.
116 Prof. L. Lin, Research Group at the University of California at Berkeley, *http://www.me.berkeley.edu/faculty/lin/index.html*, ca 20 April 2004.
117 M.A. Schmidt, *Proc. IEEE* **1998**, *86*, 1575–1585.
118 W.H. Ko, J.T. Suminto, G.J. Yeh, in: *Micromachining and Micropackaging of Transducers*, C.D. Fung, P.W. Cheung, W.H. Ko, D.G. Fleming (eds.); Amsterdam: Elsevier Science, **1985**, pp. 41–61.
119 P.W. Barth, *Sens. Actuators A* **1990**, *21–23*, 919–926.
120 C.A. Desmond, P. Abolghasem, in: *Proceedings of the 4th International Symposium on Semiconductor Wafer Bonding*; **1997**, pp. 95–105.
121 M. Shimbo, K. Furukawa, K. Fukuda, K. Tanzawa, *J. Appl. Phys.* **1986**, *60*, 2987–2989.
122 C. Maleville, O. Rayssac, H. Moriceau, B. Biasse, L. Baroux, B. Aspar, M. Bruel, in: *Proceedings of the 4th International Symposium on Semiconductor Wafer Bonding*; **1997**, pp. 46–55.
123 M.A. Schmidt, in: *Solid-State Sensor and Actuator Workshop, Hilton Head Island, SC*; **1994**, pp. 127–131.
124 H. Takagi, R. Maeda, Y. Ando, T. Suga, in: *Room Temperature Silicon Wafer Direct Bonding in Vacuum by Ar Beam Irradiation, Proc. IEEE MEMS Workshop*, **1997**, pp. 191–196.
125 T. Suga, Y. Ishii, N. Hosoda, in: *Microassembly System for Integration of MEMS Using the Surface Activated Bonding Method, IEICE Transaction on Electronics* **1997**, *E80-C*, 297–302.
126 H. Takagi, R. Maeda, T.R. Chung, T. Suga, in: *Proceedings of the 4th International Symposium on Semiconductor Wafer Bonding*; **1997**, pp. 393–400.
127 A. Plössl, H. Stenzel, Q.-Y. Tong, M. Langenkamp, C. Schmidthals, U. Gösele, in: *Covalent Silicon Bonding at Room Temperature in Ultrahigh Vacuum, Mater. Res. Soc. Symp. Proc.* **1998**, *23*, 141–146.
128 M.J. Vellekoop, P.M. Sarro, in: *Technologies for integrated sensors and actuaturs, Proc. of SPIE's 1996 Symposium on Smart Materials, Structures and MEMS*, Bangalore, India **1996**, *3321*, 536–547.
129 Q.-Y. Tong, W.J. Kim, T.-H. Lee, U. Gösele, *Electrochem. Solid-State Lett.* **1998**, *1*, 52–53.
130 Q.-Y. Tong, G. Cha, R. Gafiteanu, U. Gösele, *IEEE J. Microelectromech. Syst.* **1994**, *3*, 29–35.
131 S. Mack, H. Baumann, U. Gösele, H. Werner, R. Schlögl, *J. Electrochem. Soc.* **1997**, *144*, 1106–1111.
132 M. Nese, R.W. Bernstein, I.-R. Johansen, R. Spooren, *Sens. Actuators A* **1996**, *53*, 349–352.
133 F. Secco d'Aragona, T. Iwamoto, H.-D. Chiou, A. Mirza, in: *A Study of Silicon*

Direct Wafer Bonding for MEMS Applications, Electrochem. Soc. Proc. **1997**, *97-36*, 127–134.

134 S. Shoji, H. Kikuchi, H. Torigoe, in: *Low-temperature anodic bonding using lithium aluminosilicate-β-quartz glass ceramic, Sensors and Actuators A*, **1998**, *64(1)*, 95–100.

135 T.A. Knecht, in: *Transducers '87*; **1987**, 95–98.

136 P.R. Younger, *J. Non-Cryst. Solids* **1980**, *38–39*, 909–914.

137 Y.-C. Lin, P.J. Hesketh, J.P. Schuster, *Sens. Actuators A* **1994**, *44*, 145–149.

138 H. Baumann, S. Mack, H. Münzel, in: *Proceedings of the Third International Symposium on Semiconductor Wafer Bonding: Physics and Applications*; **1995**, pp. 471–487.

139 K.B. Albaugh, P.E. Cade, D.H. Rasmussen, in: *Solid-State Sensors and Actuators, Hilton Head Island, SC*; **1988**, 109–110.

140 P. Krause, M. Sporys, E. Obermeier, K. Lange, S. Grigull, in: *The 8th International Conference on Solid-State Sensors and Actuators and Eurosensors IX, Stockholm*; **1995**, 228–231.

141 H. Henmi, S. Shoji, Y. Shoji, K. Yoshima, M. Esashi, *Sens. Actuators A* **1994**, *43*, 243–248.

142 Y.T. Cheng, L. Lin, K. Najafi, in: *LMEMS '99: Twelfth IEEE International Conference on Micro Electro Mechanical Systems, Orlando, FL*; **1999**, 285–289.

143 K.S. Lebouitz, R.T. Howe, A.P. Pisano, in: *The 8th International Conference on Solid-State Sensors and Actuators and Eurosensors IX, Stockholm*; **1995**, pp. 224–227.

144 Leland "Chip" Spangler, Ph.D., *http://www.aspentechnologies.net/*, ca 20 April 2004.

145 H.-J. Yeh, in: *MEMS '94, Oiso, Japan*; **1994**, pp. 279–284.

146 H.-J. Yeh, J.S. Smith, *Appl. Phys. Lett.* **1994**, *64*, 1466–1468.

147 M. Cohn, R.T. Howe, A.P. Pisano, in: *ASME 1995, San Francisco, CA*; **1995**, pp. 893–900.

148 *www.memsic.com*, ca 20 April 2004.

149 C. Bang, V. Bright, M.A. Mignardi, T. Kocian, D.J. Monk, in: *Assembly and Test for MEMS and Optical MEMS*, p. 370, Chapter 7 in *MEMS and MOEMS Technology and Applications*, P. Rai-Choudhury (ed.) ; SPIE Press, Bellingham, WA, **2000**.

150 L. Ristic, *Sensor Technology and Devices*; Boston, MA: Artech House, **1994**.

151 P. Krause, M. Sporys, E. Obermeier, K. Lange, S. Grigull, in: *The 8th International Conference on Solid-State Sensors and Actuators and Eurosensors IX, Stockholm*; **1995**, pp. 228–231.

152 W.H. Ko, J.T. Suminto, G.J. Yeh, in: *Micromachining and Micropackaging of Transducers*, C.D. Fung, P.W. Cheung, W.H. Ko, D.G. Fleming (eds.); Amsterdam: Elsevier Science, **1985**, pp. 41–61.

153 J.C. Lyke, in: *Microengineering Technologies for Space Systems*, H. Helvajian (ed.); El Segundo, CA: Aerospace Corp., **1995**, pp. 131–180.

154 G. Kelly, J. Alderman, C. Lyden, J. Barrett, A. Morrissey, in: *Micromachined Devices and Components III, Austin, TX*; **1997**, pp. 142–152.

155 H. Reichl, *Sens. Actuators A* **1991**, *25–27*, 63–71.

156 F. Mayer, G. Ofner, A. Koll, O. Paul, H. Baltes, in: *Smart Structures and Materials 1998: Smart Electronics and MEMS, San Diego, CA*; **1998**, pp. 183–193.

157 N. Najafi, S. Massoud-Ansari, in: *Method for Packaging Microsensors*; Integrated Sensing Systems, Inc., Ypsilanti, Michigan, **1998**, *www.mems-issys.com*.

158 R.F. Wolffenbuttel, in: *ISIE '97: Proceedings of the IEEE International Symposium on Industrial Electronics, Guimaraes, Portugal*; **1997**, pp. SS146–SS151.

159 R. Irwin, W. Zhang, K. Harsh, Y.C. Lee, in: *RAWCON '98: 1998 IEEE Radio and Wireless Conference, Colorado Springs, CO*; **1998**, pp. 293–296.

160 *http://acoustics.stanford.edu/group/cmut3.pdf*, ca 20 April 2004.

161 D.J. Warkentin, J.H. Haritonidis, M. Mehregany, S.D. Senturia, in: *Transducers '87*; **1987**, pp. 291–294.

162 L.Y. Lin, S.S. Lee, M.C. Wu, K.S.J. Pister, in: *MEMS95, Amsterdam*; **1995**, pp. 77–82.

163 S. Koh, C.H. Ahn, in: *Integrated Optics and Microstructures III, San Jose, CA*, **1996**, pp. 121–130.

164 M.A. Chan, S.D. Collins, R.L. Smith, in: *Transducers '93: 7th International Conference on Solid-State Sensors and Actuators, Yokohama*; **1993**, pp. 580–583.
165 J. Smith, S. Montague, J. Sniegowski, J. Murray, P. McWhorter, in: *Proceedings of the IEDM*; **1995**, pp. 609–612.
166 D. Doane, P. Franzon, *Multichip Module Technologies and Alternatives: the Basics*; New York: Van Nostrand Reinhold, **1993**.
167 J. Butler, *PhD Thesis*, Air Force Institute of Technology, **1998**.
168 L. Guérin, R. Sachot, M. Dutoit, in: *1996 IEEE Multi-Chip Module Conference, Santa Cruz, CA*; **1996**, pp. 73–77.
169 Y.C. Lee, University of Colorado, personal communication.
170 K. Ishikawa, J. Zhang, A. Tuantranont, V.M. Bright, Y.C. Lee, in: *An integrated micro-optical system for laser-to-fiber active alignment, 15th IEEE International Conference on Micro Electro Mechanical Systems (MEMS 2002)*, Las Vegas, NV, January 20–24, **2002**, pp. 491–494.
171 L.T. Manzione, *Plastic Packaging of Microelectronic Packages*; New York: Van Nostrand Reinhold, **1990**.
172 D.S. Soane, in: *Stresses in packaged semiconductor devices, Solid State Technol.* **1989**, *32(5)*, 165–168.
173 G. Bitko, R. Harries, J. Matkin, A.C. McNeil, D.J. Monk, M. Shah, in: *Thin film polymer stress measurement using piezoreisitive anisotropically etched pressure sensors, Thin Films: Stresses and Mechanical Properties VI. Symposium*, **1997**, pp. 365–371.
174 Motorola, *Pressure Sensor Device Data*; Phoenix, AZ: Motorola Literature Distribution Centers, **1995**.
175 J.B. Nysæther, A. Larsen, B. Liverød, P. Ohlckers, *Microelectron. Reliability* **1998**, *38*, 1271–1276.
176 T. Maudie, D.J. Monk, D. Zehrbach, D. Stanerson, in: *Sensors Expo, Anaheim, CA*; **1996**, pp. 215–229.
177 D.J. Monk, T. Maudie, D. Stanerson, J. Wertz, G. Bitko, J. Matkin, S. Petrovic, in: *Media compatible packaging and environmental testing of barrier coating encapsulated silicon pressure sensors, Digest. Solid-State Sensors and Actuators Workshop*, **1996**, pp. 36–41.
178 G. Bitko, D.J. Monk, T. Maudie, D. Stanerson, J. Wertz, J. Matkin, S. Petrovic, in: *Micromachined Devices and Components II, Austin, TX*; **1996**, pp. 248–258.
179 S. Petrovic, S. Brown, A. Ramirez, B. King, T. Maudie, D. Stanerson, G. Bitko, J. Matkin, J. Wertz, D.J. Monk, in: *Advances in Electronic Packaging, Kohala Coast, HI*; **1997**, pp. 455–462.
180 M.F. Nichols, *Biomed. Sci. Instrum.* **1993**, *29*, 77–83.
181 M.F. Nichols, *Crit. Rev. Biomed. Eng.* **1994**, *22*, 39–67.
182 G.F. Eriksen, K. Dyrbye, *J. Micromech. Microeng.* **1996**, *6*, 55–57.
183 K. Dyrbye, T. Romedahl Brown, G. Friis Eriksen, *J. Micromech. Microeng.* **1996**, *6*, 187–192.
184 R. De Reus, C. Christensen, S. Weichel, S. Bouwstra, J. Janting, G.F. Eriksen, K. Dyrbye, T. Romedahl Brown, J.P. Krog, O. Søndergård Jensen, P. Gravesen, *Microelectron. Reliability*, **1998**, *38*, 1251–1260.
185 T.A. Maudie, D.J. Monk, T.S. Savage, *Media Compatible Microsensor Structure and Methods of Manufacturing and Using the Same*; US Patent 5889211, **1999**.
186 H. Jakobsen, T. Kvisterøy, *Sealed Cavity Arrangement Method*; Sensonor A/S, Horten Norway, **1997**.
187 S. Bouwstra, in: *Micromachined Devices and Components II, Austin, TX*; **1996**, pp. 49–52.
188 S. Linder, H. Baltes, F. Gnaedinger, E. Doering, in: *The Ninth Annual International Workshop on Micro Electro Mechanical Systems, San Diego, CA*; **1996**, pp. 38–43.
189 M. Heschel, J.F. Kuhmann, S. Bouwstra, M. Amskov, in: *Smart Electronics and MEMS, San Diego, CA*; **1998**, pp. 344–352.
190 T. Maudie, D.J. Monk, *Pressure Sensor with Isolated Interconnections for Media Compatibility*, in: *Motorola*. Motorola, Inc., Schaumburg, IL, USA **1997**.
191 K. Ryan, J. Bryzek, in: *Sensor 95, Kongressband*; **1995**, pp. 685–690.
192 C.P. Wong, in: *Polymers for Electronic and Photonic Applications*, C.P. Wong

(ed.); Boston: Academic Press, **1993**, pp. 67–220

193 C. Bang, V. Bright, M. A. Mignardi, T. Kocian, D. J. Monk, in: *Assembly and Test for MEMS and Optical MEMS*, p. 396, Chapter 7 in: *MEMS and MOEMS Technology and Applications*, P. Rai-Choudhury (ed.); SPIE Press, Bellingham, WA, **2000**.

194 V. J. Adams, *Unibody Pressure Transducer Package*; in: *Motorola*, U. S. A.: Motorola, Inc., Schaumburg, IL, USA **1987**.

195 TREAD Act full text of FMVSS138 is available at: *http://www.nhtsa.dot.gov/cars/rules/rulings/TirePresFinal/index.html*

196 TREAD Act Court of Appeals Ruling full text of FMVSS138 is available at: *http://www.nhtsa.dot.gov/carsrules/rulings/TirePresFinal/index.html.*

197 D. J. Monk, M. Shah, in: *Thin Film Polymer Stress Measurements Using Piezoresistive Anisotropically Etched Pressure Sensors*, in: *Mat. Res. Soc. Symp. Proc.*, vol. 390, Electronic Packaging Materials Science VIII, R. C. Sundahl, K.-N. Tu, K. A. Jackson, P. Børgesen (eds.); MRS, San Francisco, CA, **1995**, pp. 103–109.

198 J. E. Vandemeer, G. Li, A. C. McNeil, in: *Analysis of thermal hysteresis on micromachined accelerometers, Proceedings of IEEE Sensors 2003 (IEEE Cat. No. 03CH37498)*, **2003**, *2*, 1235–1238. et al., IEEE Sensors.

199 A. C. McNeil, in: *A parametric method for liking MEMS package and device models, Technical Digest. Solid-State Sensor and Actuator Workshop*, **1998**, pp. 166–169.

200 Considerations to Improve Battery Life in Direct Tire Pressure Monitoring, SAE 2002-01-1078, Mark L. Shaw, Motorola Semiconductor Products Sector, US Patent 6472243: Method of forming an integrated CMOS capacitive pressure sensor, Motorola Inc., Motion Sensing Techniques and Analysis for Direct Tire Pressure Monitoring, SAE 2003-01-0202, Mark L. Shaw, Motorola Semiconductor Products Sector.

201 R. August, T. Maudie, T. F. Miller, E. Thompson, in: *Acceleration sensitivity of micromachined pressure sensors (for automotive use), Proceedings of the SPIE – The International Society for Optical Engineering*, **1999**, *3876*, 46–53.

202 B. P. Gogoi, S. Jo, R. August, A. McNeil, M. Fuhrmann, J. Torres, T. F. Miller, A. Reodique, M. Shaw, K. Neumann, D. Hughes Jr., D. J. Monk, in: *A 0.8 μm CMOS Integrated Surface Micromachined Capacitive Pressure Sensor with EEPROM Trimming and Digital Output for A Tire Pressure Monitoring System*, in *2002 Solid-State Sensors, Actuators and Microsystems Workshop* (Transducer Research Foundation, Hilton Head, SC), **2002**, pp. 181–184.

203 C. P. Wong, in: *Passivating Organic Coatings with Silicone Gels: The Correlation between the Material Cure & Its Electrical Reliability, Material Research Society Symposium*, San Diego, CA MRS, **1989**; C. P. Wong, in: *Recent Advances in the Application of High Performance Siloxanes Polymers in Electronic Packaging, 6th International SAMPE Electronics Conference, SAMPE*, **1992**; C. P. Wong, in: *Recent Adcances in IC Passivation and Encapsulation, Polymers for Electronic and Photonic Applications*, Academic Press, Inc., Boston, **1993**, pp. 167–220; C. P. Wong, J. M. Segelken, in: *Understanding the Use of Silicone Gels for Nonhermetic Plastic Packaging, IEEETrans. Comp., Hybrids, and Manuf. Tech.* **1989**, *12(4)*, 421–425.

204 S. Petrovic, D. J. Monk, H. J. Miller, in: *Teflon Filters for Media Compatible Pressure Sensors, Disclosure No. SC10838P, Disclosed on July 15*, **1999**, Status: Filed on October 25, **1999**.

205 G. Li, J. Schmiesing, A. McNeil, K. Neumann, B. Gogoi, G. Bitko, S. Petrovic, J. Torres, M. Fuhrmann, D. J. Monk, in: *Selective Encapsulation Using a Polymeric or Bonded Silicon Constraint Dam for Media Compatible Pressure Sensor Applications*, in *2001 International Conference on Solid-State Sensors and Actuators: Transducers '01*, Munich, Germany, June 11–14, **2001**, pp. 178–181.

206 S. Petrovic, A. Ramirez, T. Maudie, D. Stanerson, J. Wertz, G. Bitko, J. Matkin, D. J. Monk, in: *Reliability Test Methods for Media Compatible Pressure Sensors, IEEE Transactions on Industrial Electronics* **1998**, *45(6)*, 877–885; G. Bitko, D. J. Monk, T. Maudie, D. Stanerson, J. Wertz, J. Matkin, S. Petrovic, in: *Ana-*

lytical Techniques for Examining Reliability and Failure Mechanisms of Barrier Coating Encapsulated Silicon Pressure Sensors Exposed to Harsh Media, in *Proceedings of the SPIE 1996 Symposium on Micromachining and Microfabrication,* SPIE, Austing, TX, **1996**, 248–258; D.J. Monk, T. Maudie, D. Stanerson, J. Wertz, G. Bitko, J. Matkin, S. Petrovic, in: *MEdia Compatible Packaging and Environmental Testing of Barrier Coating Encapsulated Silicon Pressure Sensors,* in Transducer Research Foundation; Hilton Head, SC, **1996**, pp. 36–42.

207 P. Rai-Choudhury (ed.), *MEMS and MOEMS Technology and Applications*; Bellingham, WA: SPIE Press, **2000**.

208 L. Ristic, M. Shah, in: *WESCON 96 Conference Proceedings, Anaheim, CA*; **1996**, pp. 64–72.

209 L.A. Field, R.S. Muller, *Sens. Actuators A* **1990**, *21–23*, 935–938.

210 P.W. Barth, *Sens. Actuators A* **1990**, *21–23*, 919–926.

211 N. Koopman, S. Nangalia, in: *Proceedings of NEPCON WEST, Anaheim, CA*; **1995**, pp. 919–931.

212 J. Hammond, A. McNeil, R. August, D. Koury, in: *Transducers '03, International Conference on Solid-State Sensors and Actuators, Boston, MA*; **2003**, pp. 85–90.

213 H. Reichl, Sensors and Actuators, vol. A25- A27, pp. 63–71, **1991**.

214 C. Song, in: *Transducers '97. International Conference on Solid-State Sensors and Actuators, Chicago, IL*; **1997**, pp. 839–842.

215 G. Li, A. McNeil, D. Koury, D. Monk, *Curvature Study with Application in MEMS Packaging,* in: *Proceedings of IMECE'02, 2002 ASME International Mechanical Engineering Congress and Exposition, New Orleans, LA,* 17–22 November; **2002**, pp. 1–5.

216 G. Li, A.A. Tseng, *IEEE Trans. Electron. Packag. Manuf.* **2001**, *24*, 18–25.

217 G.X. Li, R.J. Gutteridge, D.N. Koury, Z.L. Zhang, R.M. Roop, *Proc. SPIE* **1996**, *2882*, 147–151.

218 D.S. Mahadevan, *Proc. SPIE* **1995**, *2642*, 265–272.

219 D. Dougherty, D. Mahadevan, M. Shah, R. Harries, V. Adams, in: *Proc. 1997 Winter Motorola AMT Symp., Schaumburg, IL*; **1997**, *2*, 1211–1219.

220 M. Kniffin, M. Shah, *ISHM Int. J. Microcircuits Electron. Packag.* **1996**, *19*, 75–86.

221 G. Li, D. Mahadevan, M. Chapman, *Packaging MEMS Inertial Sensors at Motorola,* Internal Report; Presented at LEOS 2003, paper WS2, Tucson, AZ, 26–30 Oct. **2003**.

222 M. Vujosevic, M. Shah, *Stress Isolation of an Accelerometer Die in a Post-molded Plastic Package – The Impact of Die Coat Coverage, SAE Technical Paper Series 1999-01-0157*; International Congress and Exposition, Detroit, MI, **1999**.

223 Directive 2002/95/EC of the European Parliament and of the Council, *Restriction of the Use of Certain Hazardous Substances in Electrical and Electronic Equipment*; Official Journal of the European Union EN, 13.2. 2003, L 37/19, 27 January **2003**.

224 Directive 2000/53/EC of the European Parliament and of the Council, *End-of Life Vehicles*; Official Journal of the European Union EN, 21.10.2000, L 269, pp. 0034–0043 (modified), 18 September **2000**.

225 Directive 2002/96/EC of the European Parliament and of the Council, *Waste Electrical and Electronic Equipment (WEEE)*; Official Journal of the European Union EN, 13.2. 2003, L 37/24, 27 January **2003**.

226 *Motorola's Commitment to the Environment; http://e-www.motorola.com/webapp/sps/site/homepage.jsp?nodeId=06j4R9,* accessed 05 April **2004**.

5
High-frequency Integrated Microelectromechanical Resonators and Filters

F. Ayazi, Georgia Institute of Technology, Atlanta, GA, USA

Abstract
Silicon-based micromachined electromechanical resonators with quality factors in the thousands and frequencies in the gigahertz range have become a reality. Microelectromechanical resonators are small in size, consume no power, have quartz-like characteristics and can be integrated in silicon for a variety of signal processing applications such as on-chip frequency references, bandpass filters and microsensors. Ultra-narrow bandwidth filtering can be achieved through strategic coupling of individual high-Q resonators. This chapter reviews recent development in silicon-based microelectromechanical resonator technologies and filter synthesis approaches. Research for implementation of silicon-based high-frequency microelectromechanical bandpass filters with acceptable impedance levels is under way.

Keywords
mechanical resonators; MEMS; RF MEMS; integrated filters; mechanical signal processing.

Advanced Micro and Nanosystems. Vol. 1
Edited by H. Baltes, O. Brand, G.K. Fedder, C. Hierold, J. Korvink, O. Tabata

ISBN: 3-527-30746-X

5.1 Introduction

Physical structures have mechanical resonances that occur at specific frequencies. The shape and frequency of these resonances are functions of the clamping conditions, as well as the effective mass (M_{eff}) and stiffness (K_{eff}) of the structure, which in turn relate to the Young's modulus and density of the resonator material:

$$\omega_{\text{resonance}} = 2\pi f_{\text{resonance}} = \sqrt{\frac{K_{\text{eff}}}{M_{\text{eff}}}} \tag{1}$$

Microelectromechanical resonators (MEMresonators) are comprised of a micro-scale mechanical element and *integrated* transducers that convert the motion of the micromechanical element into an electrical signal and vice versa. At least one high quality factor (Q) resonant mode of the micromechanical element can be excited using the transducers. The Q of a resonator, in the electronics world, is a measure of the purity of its resonance characteristic:

$$Q = \frac{f_{\text{resonance}}}{BW_{-3\,\text{dB}}} \tag{2}$$

The smaller the –3 dB bandwidth ($BW_{-3\,\text{dB}}$) of the resonance, the higher is the Q of the resonator. An equivalent universal definition of Q in a harmonic oscillator is given by

$$Q = 2\pi \cdot \frac{\text{peak energy stored}}{\text{energy dissipated per cycle}} \tag{3}$$

Therefore, to obtain high Q in micromechanical resonators, energy dissipation through various loss mechanisms should be minimized. Quartz crystals have numerous applications in electronic systems owing to their very high-Q mechanical resonance (in the range 10^4–10^6) and high stability of their resonance frequency against temperature and aging. Integrated MEMresonators have the potential to replace off-chip frequency-selective electromechanical components such as quartz crystals and surface acoustic wave (SAW) devices in a variety of microsystem applications (e.g. integrated filters and frequency sources for low-power wireless communication [1]). Silicon-based microresonators have shown very high Q values at very high frequencies. However, owing to the relatively large temperature sensitivity of Young's modulus, silicon microresonators have larger temperature coefficients of frequency (TCF) than quartz crystals and will most likely need some type of temperature compensation before they can be employed as frequency references. When a number of single degree of freedom MEMresonators are coupled to one another, higher order bandpass filters (with multiple degrees of freedom) can be realized. This chapter will review the operation and implementation of MEMresonators and high-order frequency filters built using these resonators and discuss the challenges faced in scaling the frequency of integrated microelectromechanical resonators and filters into ultra-high-frequency (UHF) range.

5.2 Microelectromechanical Resonators

Various electromechanical transduction mechanisms such as capacitive, piezoelectric, thermal and magnetostrictive can be used in MEMresonators. Here, we discuss only two types of such mechanisms that are more suitable for high-frequency integrated implementations: *capacitive* and *piezoelectric*. Fig. 5.1 shows a schematic diagram of a typical two-port capacitively transduced MEMresonator, in which the micromechanical resonator is a clamped-clamped silicon beam designed to operate in its in-plane bending flexural modes [2]. A DC bias voltage V_p, necessary for proper operation of the device, is applied to the body of the beam while the DC levels of the drive and sense electrodes are set to ground. In order to excite the beam into resonance, an AC drive signal v_d is applied to the drive electrode and the sense current i_s of the sense electrode is detected. The current i_s is a measure of the vibration amplitude of the micromechanical beam and is amplified by the Q of the resonator at its resonance frequency. If the resonator has

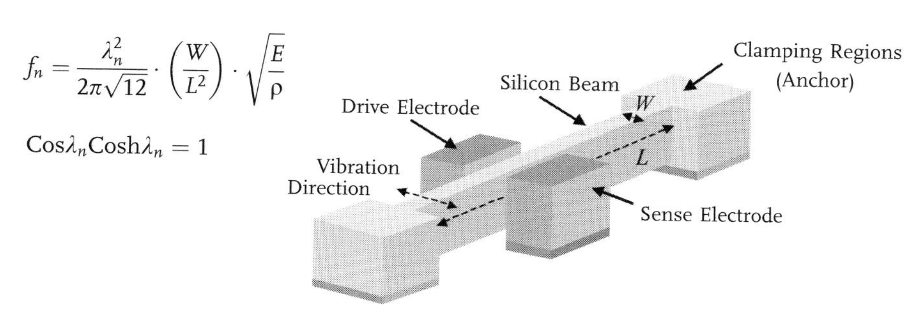

Fig. 5.1 A capacitively transduced clamped-clamped beam resonator with in-plane flexural vibrations

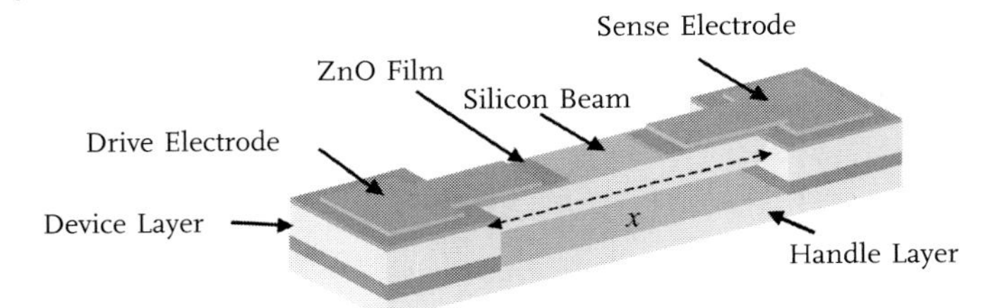

Fig. 5.2 A piezoelectrically transduced clamped-clamped beam resonator with out-of-plane flexural vibrations

high Q, the only frequency contents of the drive signal that can pass through the MEMresonator are those coinciding with detectable resonances of the micromechanical structure. Detectable resonances are a subset of the resonant modes of the micromechanical structure that can be simultaneously excited and detected by the drive and sense electrodes, respectively.

A schematic representation of a two-port piezoeletrically transduced MEMresonator is shown in Fig. 5.2. In this case, the micromechanical resonator is a clamped–clamped silicon beam with out-of-plane bending flexural modes. The electromechanical transducers are composed of a piezoelectric film sandwiched between two conductive electrodes [3]. In the MEMresonator of Fig. 5.2, the bottom electrode is the micromechanical beam resonator and the top electrodes are thin-film metal layers (such as Al). The piezoelectrically transduced resonators do not require a DC bias voltage for operation. When an AC voltage is applied to the drive electrode, it causes a corresponding strain in the piezoelectric film, which will be transferred to the micromechanical element. At resonance, the Q-amplified vibrations of the beam will be converted to a detectable voltage at the sense electrode of the MEMresonator.

An attractive feature of the MEMresonators in Figs. 5.1 and 5.2 is that their resonant frequencies are voltage-tunable. This is achieved by taking advantage of the nonlinearity of the electrostatic force with respect to the interelectrode gap spacing, which causes a reduction in the overall stiffness of the device (modeled as a negative electrostatic stiffness). The resonant frequency of the resonator in Fig. 5.1 is therefore a function of V_p [2] whereas that in Fig. 5.2 is a function of the DC potential applied between the handle layer and the device layer of the resonator [3].

Two-port MEMresonators can be modeled in the electrical domain using the admittance parameters (regardless of their transduction mechanism):

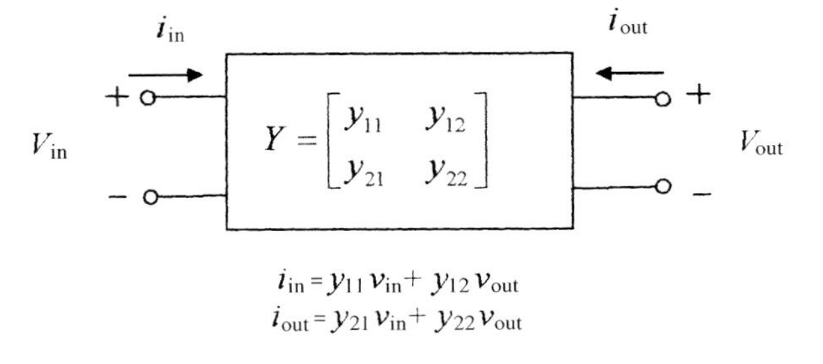

$$i_{in} = y_{11} v_{in} + y_{12} v_{out}$$
$$i_{out} = y_{21} v_{in} + y_{22} v_{out}$$

5.2.1
Resonant Modes

Mechanical resonant modes of a structure can have various shapes and stiffnesses. High aspect ratio structures have both low-stiffness bending flexural modes and high-stiffness bulk modes, depending on their clamping boundary conditions. As an example, let us consider the corresponding modes of a high aspect ratio silicon beam that is 10 µm in length, 0.5 µm in thickness and 2 µm in width. When the beam is clamped at both ends, the low-stiffness first bending flexural mode of the structure yields reasonably high Q owing to its large aspect ratio (20:1) at 43 MHz, as shown in Fig. 5.3 a. Higher frequencies can be achieved by using the higher order flexural modes that have lower Q due to increased mechanical clamping loss (support loss); e.g. the fifth flexural mode shown in Fig. 5.3 c has a frequency of 540 MHz. The fourth flexural mode, shown in Fig. 5.3 b, is an example of a mode that can neither be excited nor detected in a resonator with central and symmetric electrodes. If the support boundary conditions of this beam are now changed to what is shown in Fig. 5.4, i.e. to centrally clamp the beam at its side by two small tethers, we can then take advantage of high-Q, high-stiffness bulk modes of the resulting structure with frequencies into the gigahertz range. The first length extensional mode of this silicon structure shown in Fig. 5.4 a has a resonant frequency of ~430 MHz and the third length extensional mode shown in Fig. 5.4 b has a frequency of ~1.3 GHz. Both of these bulk modes will have high Q owing to symmetric force cancellation at the support area. The stiffness of the structure in its length extensional bulk modes is approximately two orders of magnitude larger than that of its flexural modes, resulting in ten times larger frequencies.

Bulk modes of *low aspect ratio* structures such as disks and blocks can also yield very high frequencies. As shown in Fig. 5.5, a side-supported silicon disk resonator that is 10 µm in diameter and 2 µm in thickness has high-order contour modes in the range 350 MHz and above. The disk is supported at a node of the bulk mode with a small tether to minimize clamping losses. Various examples of MEMresonators operating in the described modes are presented in the following section.

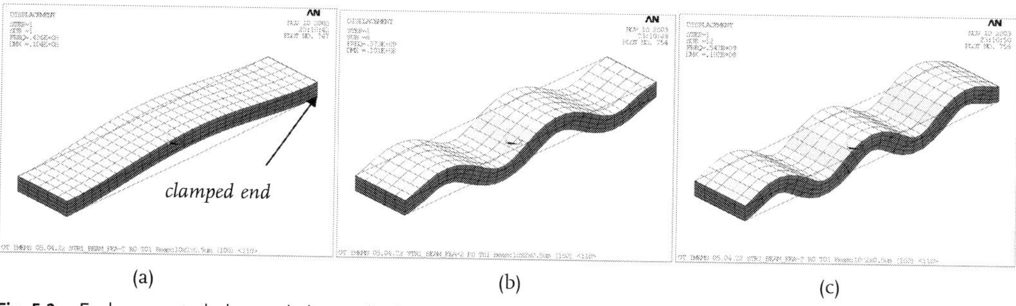

Fig. 5.3 End-supported clamped-clamped silicon beam resonator with length = 10 µm, thickness = 0.5 µm, width = 2 µm. Finite element analysis (FEA) results: (a) first flexural mode at 43 MHz; (b) fourth flexural mode at 373 MHz; (c) fifth flexural mode at 540 MHz

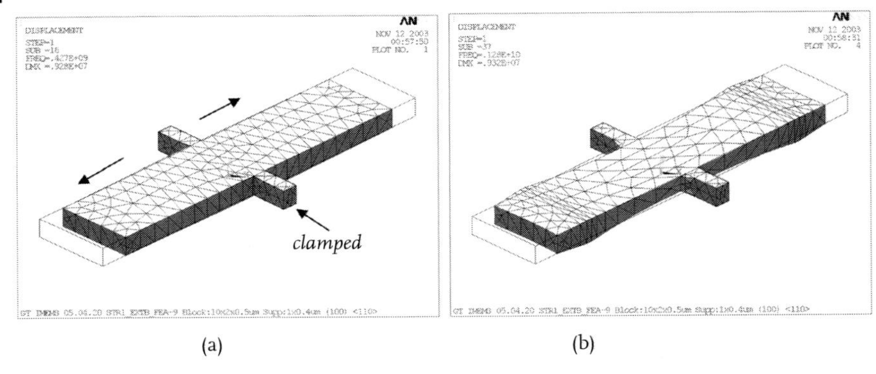

(a) (b)

Fig. 5.4 Centrally supported bulk-mode silicon beam resonator with length = 10 μm, thickness = 0.5 μm, width = 2 μm. FEA results: (a) first length extensional mode at ~430 MHz; (b) third length extensional mode at ~1.3 GHz

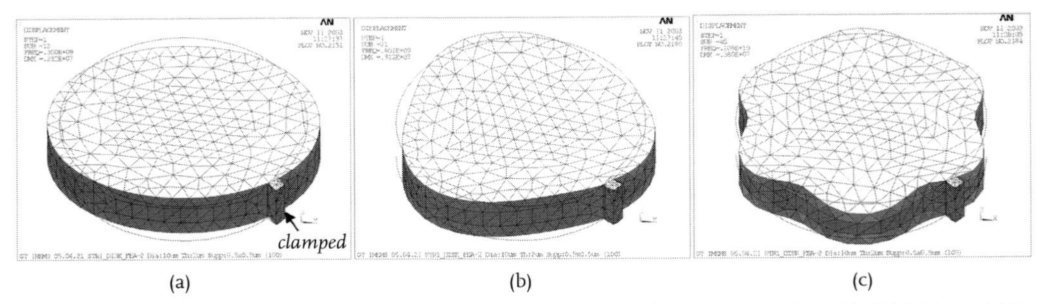

(a) (b) (c)

Fig. 5.5 (a) Side-supported bulk-mode silicon disk resonator with diameter = 10 μm and thickness = 2 μm has high-Q bulk contour modes at (a) 350 MHz, (b) 610 MHz and (c) ~1.1 GHz

5.3
Fabrication Technologies

A number of surface, bulk and mixed-mode (surface + bulk) micromachining technologies have been used to implement various types of MEMresonators. A few *silicon-based* representatives of the reported fabrication technologies that are scalable to high frequencies are discussed below.

5.3.1
Single-crystal Silicon Capacitive MEMresonators

Capacitive single-crystal silicon (SCS) resonators with sub-100 nm to sub-micron capacitive gaps and polysilicon electrodes have been implemented using the HARPSS process [2, 4]. High aspect ratio clamped-clamped SCS beam resonators operating in their first- and higher order flexural modes (up to the fifth mode)

Fig. 5.6 SEM picture of a 200 μm long, 5 μm wide and 30 μm thick clamped-clamped SCS beam resonator with capacitive gap of 200 nm fabricated in HARPSS

were fabricated on regular silicon substrates, while low aspect ratio high-Q silicon disk resonators have been implemented on SOI substrates [5]. Fig. 5.6 shows an SEM picture of a 200 μm long, 5 μm wide and 30 μm thick SCS beam resonator with 200 nm vertical capacitive gaps fabricated through the HARPSS process. Fig. 5.7 shows the measured first, third and fifth flexural resonant modes of this beam in vacuum. The second and fourth modes could not be excited owing to the midway position of the drive electrode. The fifth mode has a ~13 times larger

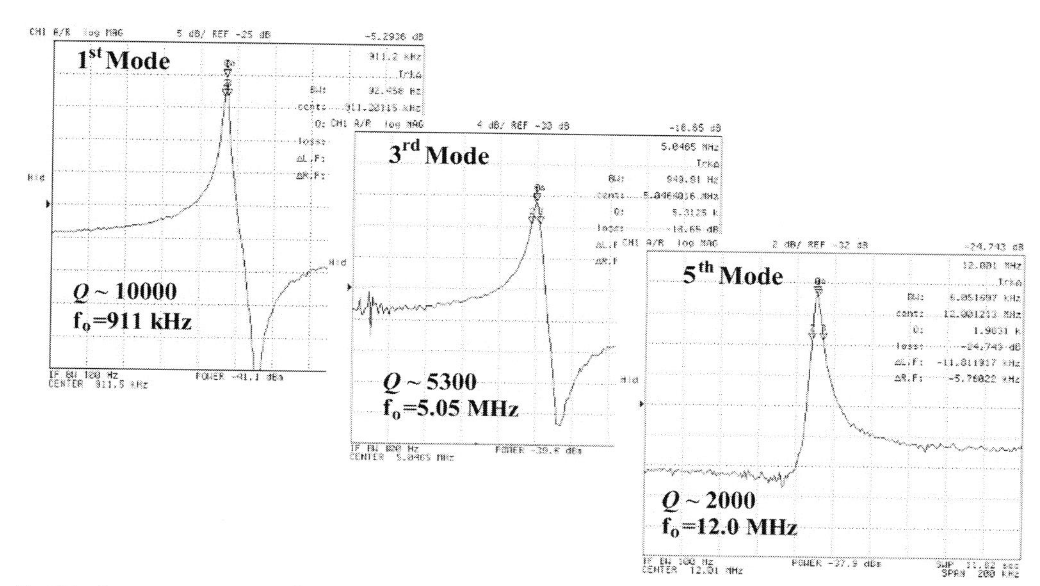

Fig. 5.7 Frequency response of a 200 μm long, 5 μm wide and 20 μm tall CC resonator with 200 nm capacitive gaps operating in its fundamental, third and fifth resonance modes

frequency than the first mode, but with a lower Q of 2000 (limited by the support loss) compared with the 10000 of the first mode. Low-frequency clamped-free silicon beams fabricated through the same technology have shown Q close to 200000 at 19 kHz [4].

The electrostatic frequency tuning characteristic of a 610 kHz clamped-clamped SCS beam resonator ($Q=6500$) with 80 nm capacitive gaps fabricated in HARPSS is shown in Fig. 5.8. A tuning range of 28% was obtained by changing the DC polarization voltage from 0.1 to 2.0 V [4]. Such a large frequency tuning over a small CMOS-compatible DC voltage range is a result of the ultra-narrow narrow gaps of the resonators.

The equivalent electrical output resistance of a capacitive micromechanical resonator (the motional resistance) is expressed by [5, 6]

$$R_{\text{io}} = \frac{\sqrt{KM}d^4}{Q\varepsilon_0^2 L^2 h^2 V_\text{p}^2} \propto \frac{d^4}{Qh} \tag{4}$$

where K and M are the effective stiffness and mass of the resonator, d is the capacitive gap size, Q is the resonator's quality factor, V_p is the DC polarization voltage and L and h are the electrodes' length and height, respectively. From this equation, it is evident that ultra-thin capacitive gaps, high Q and large electrode area are needed to reduce the equivalent output resistance of the capacitive resonators to reasonable values. Achieving smaller output resistance will facilitate the insertion of MEMS capacitive resonators in various high-frequency systems. HARPSS resonators provide all the necessary features to obtain reduced output resistance. First, the capacitive gaps of these resonators are determined in a self-aligned manner by the thickness of a deposited sacrificial oxide layer and can be reduced to their smallest physical limits (tens of nanometers and less) independent of lithography. Second, the thickness of the SCS HARPSS resonators can be increased to a few tens of microns while keeping the capacitive gaps in the nanometer scale, resulting in a lower equivalent motional resistance compared with thin-film fabrication technologies that have limited thickness. Finally, the ability of fabricating *all-silicon* resonators with high-Q single crystal silicon as the resonating element makes HARPSS a superior candidate for implementation of high-fre-

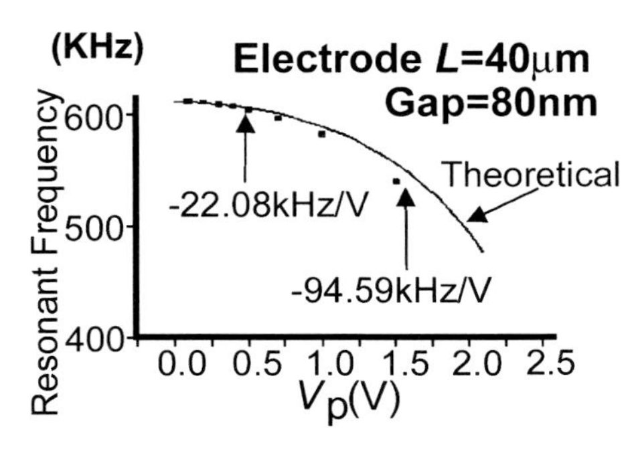

Fig. 5.8 Plots of the measured and calculated change of frequency versus polarization voltage for a 610 kHz clamped-clamped beam resonator with 80 nm capacitive gaps

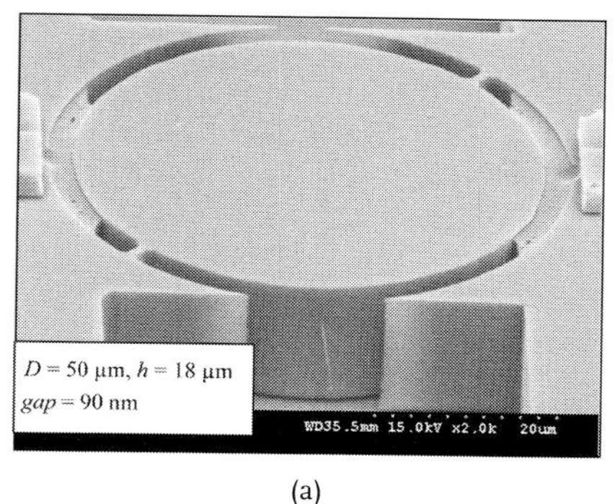

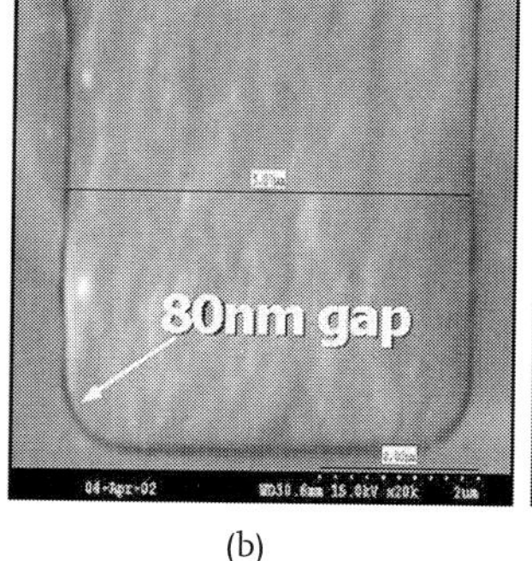

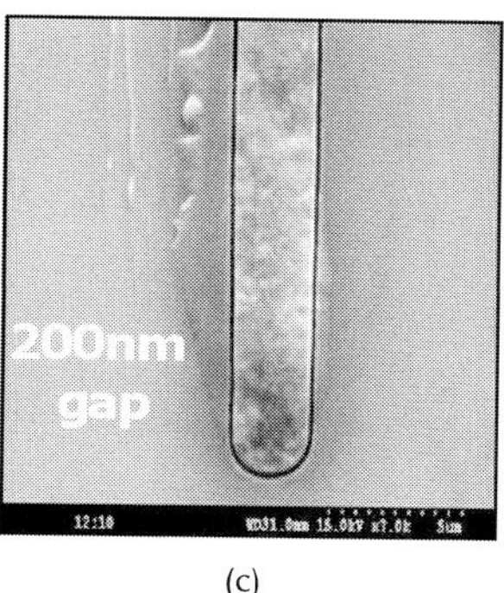

(a) (b) (c)

Fig. 5.9 (a) SEM picture of a 50 μm diameter, 18 μm thick side-supported SCS disk resonator with a gap spacing of 80 nm. Cross-sections of (b) 80 nm and (c) 200 nm uniform gaps between the trench-refilled polysilicon electrode and SCS substrate

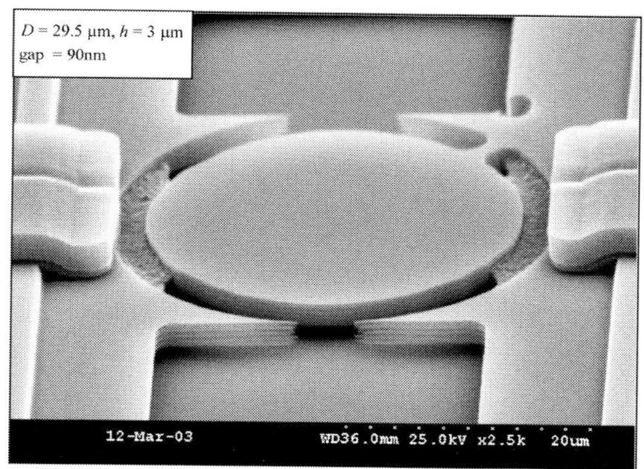

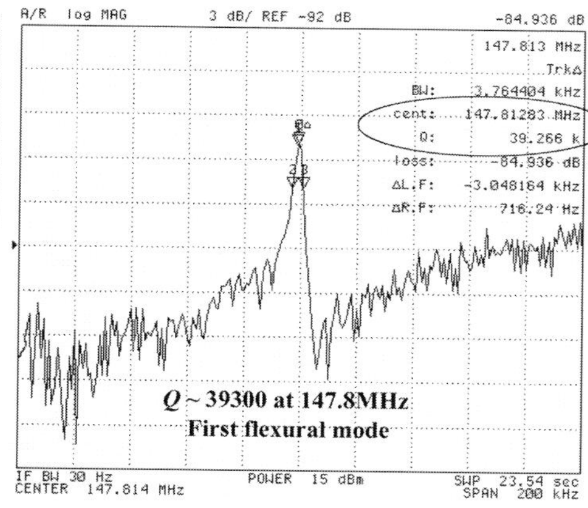

Fig. 5.10 SEM view of a 30 μm diameter side-supported SCS disk resonator fabricated in HARPSS, showing a Q of 40000 at 147.8 MHz in 1 mTorr vacuum

quency integrated MEMresonators. Fig. 5.9a shows an SEM view of a low aspect ratio, 18 μm thick side-supported silicon disk resonator with diameter of 50 μm and 90 nm capacitive gaps fabricated on SOI substrates. Fig. 5.9b and c show cross-sectional views of conformal 80 and 200 nm gaps between the trench-refilled polysilicon and the silicon substrate that continues uniformly all the way down to the bottom of the trench.

Fig. 5.10 illustrates the frequency response and an SEM view of a 30 μm diameter, 3 μm thick silicon disk resonator, supported by a 2.7 μm long support beam *at only one resonance node*. A quality factor of ~40000 has been measured in vacuum for the first elliptical bulk mode of this resonator (corresponding to Fig. 5.5a) at a frequency of 148 MHz, which is the highest reported quality factor for a single-crystal silicon device at such a frequency. Fig. 5.11 shows the fabrication process flow for SOI-based HARPSS resonators.

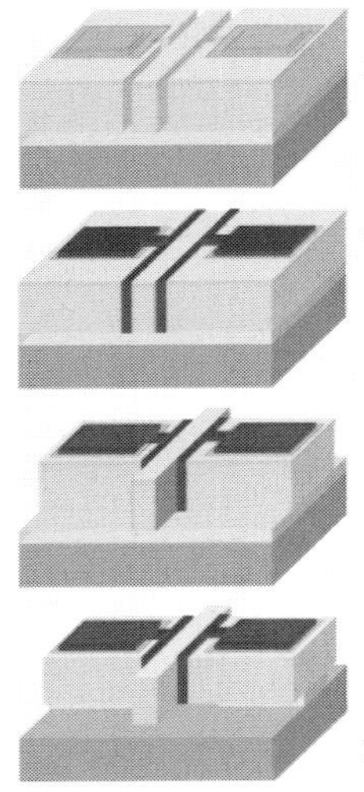

1. Grow and pattern initial oxide
2. Deposit and pattern LPCVD nitride
3. Etch trenches (Bosch process)
4. Grow and remove thin oxide (surface treatment)
5. Deposit and blanket etch sacrificial oxide
6. Deposit and pattern doped LPCVD polysilicon
7. Pattern initial oxide
8. Metallization
9. Etch release openings, pattern polysilicon for electrodes
10. HF release and undercut

Fig. 5.11 Fabrication process flow of single-crystal silicon resonators with sub-100 nm gaps on SOI

5.3.2
Piezo-on-Silicon MEMresonators

In order to reduce the motional resistance of bulk-mode capacitive MEMS resonators in the VHF and UHF range, an ultra-thin (sub-100 nm) electrode-to-resonator gap spacing is required, which can introduce complexities in the fabrication process. In addition, large DC bias voltages (in excess of 20 V) may be necessary in capacitive resonators to minimize R_m, which may become limited by either the transistor technology supporting the resonator or the available power supply.

In contrast to capacitive devices, the motional resistance in piezoelectric resonators is much smaller owing to their higher electromechanical coupling [7]. Examples of piezoelectric devices are surface and bulk acoustic wave resonators. The main drawback of surface acoustic wave (SAW) devices is their bulky size and incompatibility with silicon integration. Film bulk acoustic resonators (FBARs) overcome the size issue and have gigahertz frequencies with Q values of >1000 [1, 8]. However, since FBARs utilize the longitudinal thickness vibration of a piezoelectric film, accurate control of film thickness and quality across the wafer is critical for frequency stability and repeatability. In addition, multiple frequency devices/standards will need various piezo film thicknesses on a chip, complicating the manufacturing process. An effective trimming method is also not yet available for FBARs.

A new class of piezoelectrically transduced high-Q single-crystal silicon (SCS) resonators has been demonstrated recently [3]. The resonating element comprises the device layer of an SOI substrate, which has a higher inherent mechanical quality factor than bulk piezoelectrics and deposited thin films, whereas actuation and sensing are achieved by very thin films of piezoelectric materials (~ 3000 Å) such as zinc oxide (ZnO) deposited directly on the silicon resonator. Both high as-

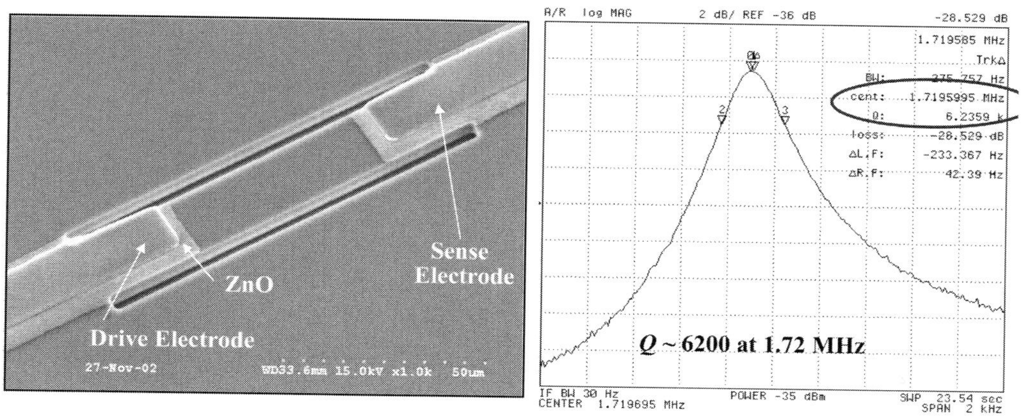

Fig. 5.12 Close-up SEM view of a 100 μm long, 4 μm thick piezo-on-silicon beam resonator, showing a Q of ~6200 at its fundamental out-of-plane flexural modes of 1.72 MHz

pect ratio beam resonators (low frequency) and low aspect ratio bulk-mode resonators (high frequency) have been implemented using this technique. Fig. 5.12 shows an SEM view of a 100 μm long, 20 μm wide and 4 μm thick SCS beam resonator, showing the ZnO film sandwiched between the low-resistivity silicon and the top aluminum electrodes. ZnO was etched away in the middle of the beam to increase the quality factor of the resonator. The first flexural mode of this out-of-plane beam resonator at 1.7 MHz has a Q of 6200 in vacuum.

A number of centrally supported blocks and beams operating in their in-plane length extensional bulk modes were also implemented using this technique to achieve very high frequencies [9]. Fig. 5.13 shows the first and third extensional

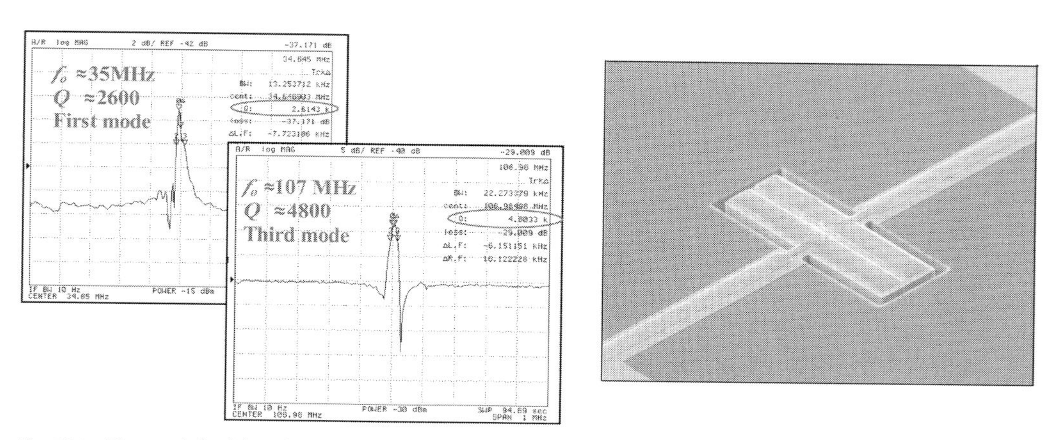

Fig. 5.13 First and third length extensional bulk modes measured at 35 and 107 MHz for a 120×30 μm piezo-on-silicon block

modes for a 120×30 μm block,corresponding to the modes shown in Fig. 5.4. The frequency at the third mode is approximately three times larger than that at the first mode. These modes show relatively high Q at VHF frequencies. The deposited piezo thin film causes a decrease in Q of these resonators compared with that of the SCS capacitive resonators. However, the Q is still in the range of a few thousand, which is high enough for many applications.

5.3.3
Polysilicon Capacitive MEMresonators

A number of polysilicon micromachining processes have been used to implement various types of low-frequency flexural mode and high-frequency bulk mode capacitive resonators.

Early research on low-frequency polysilicon resonators involved comb-drive actuation and capacitive sensing [10] because of the linearity that comb-drive electrostatic actuation provided with displacement. Comb-drive resonators were effective in frequency ranges up to few megahertz. For higher frequency operations, the resonators required high spring stiffness and low mass, and therefore the use of bulky comb-driven resonators were not effective. Simple geometry resonators such as beams and disks provided a better means of producing filters in high-frequency operations. Fig. 5.14 shows an SEM view of a 20 μm diameter capacitive polysilicon disk resonator with a center-post support that has a high order bulk radial mode resonant frequency of 1.14 GHz with a $Q \approx 1500$ in both vacuum and air [11]. This device was fabricated using a self-aligned process that creates a self-aligned center-post and transducer electrodes. Lower frequency radial bulk modes of this device have shown higher quality factors [11].

Thick polysilicon capacitive resonators with silicon electrodes have been realized using the HARPSS process. Thick polysilicon structures with high aspect ratio were created by refilling narrow trenches (2–6 μm wide) with conformal LPCVD polysilicon layers. Fig. 5.15 shows an SEM picture of a 35 μm thick clamped-clamped TR polysilicon resonator fabricated in HARPSS [12]. The drive and sense electrodes for this capacitive device are made of thick single-crystal silicon (SCS) with 0.9 μm inter-electrode gap spacing. Discussion on the fabrication process can be found elsewhere [13].

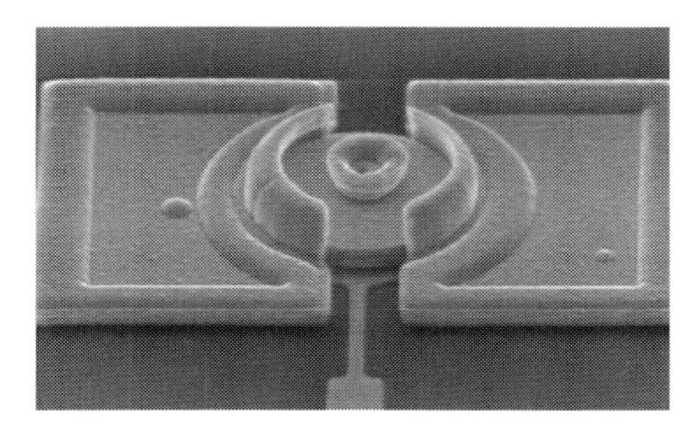

Fig. 5.14 SEM view of a 1.14 GHz self-aligned radial contour mode polysilicon disk resonators. (Courtesy of Professor Clark T. C. Nguyen, University of Michigan)

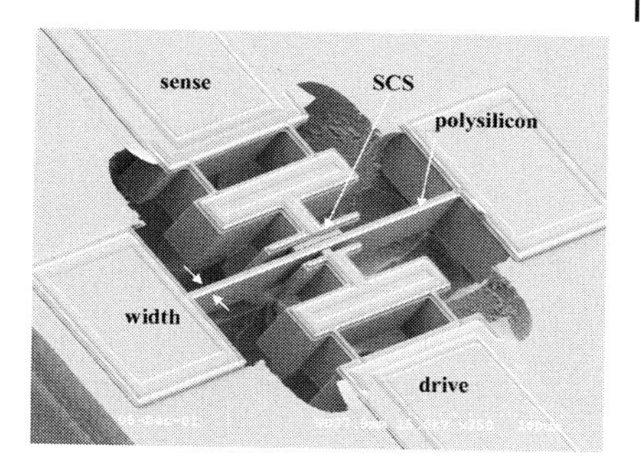

Fig. 5.15 Thick trench-refilled (TR) polysilicon beam resonator fabricated using the HARPSS process (length = 300 μm, width = 5 μm, height = 35 μm)

In the process of refilling the trenches with LPCVD layers, a void is usually created in the middle of the polysilicon structures. This void does not change the resonance frequency of the beam compared with a solid beam of the same dimensions, but it affects the heat transfer across the width of the beam and hence the thermoelastic damping behavior. Because of this, trench-refilled polysilicon resonators can show interesting *Q* characteristics that are not typically observed in solid beams of the same dimensions [12], namely a thickness-dependent *Q* (for a constant frequency) that can even exceed the thermoelastic damping limits and an increase in *Q* with frequency.

5.3.4
Poly-Silicon-Germanium MEMresonators

The feasibility of using thin films of poly-silicon-germanium (poly-$Si_{0.35}Ge_{0.65}$) deposited through LPCVD on top of CMOS-processed wafers for implementation of

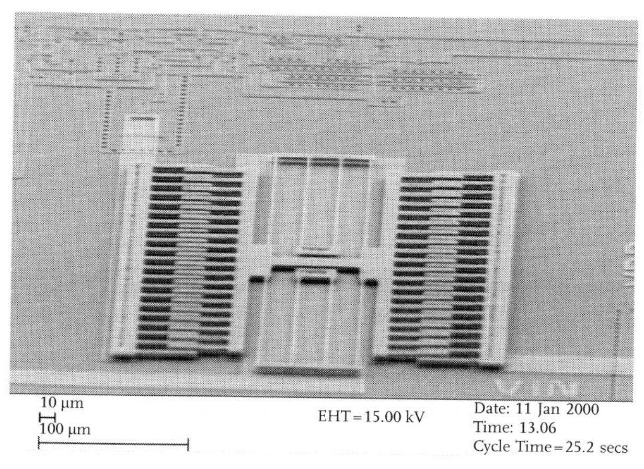

Fig. 5.16 SEM micrograph of a poly-$Si_{0.35}Ge_{0.65}$ resonator post-fabricated on a CMOS wafer. (Courtesy of Professor Roger Howe, UC Berkeley)

capacitive MEMresonators has been demonstrated. Low-frequency (~20 kHz) comb-drive resonators made of p-type poly-$Si_{0.35}Ge_{0.65}$ (2.5 μm thick) were deposited at 450 °C on top of CMOS wafers, using poly-Ge as a sacrificial material [14]. An SEM picture of a fabricated device is shown in Fig. 5.16. The advantage of poly-$Si_{0.35}Ge_{0.65}$ is that it can be deposited at a significantly lower temperature than polysilicon, making it potentially suitable for post-integration on copper/low dielectric constant metallization CMOS. Research is under way to extend the frequency of these resonators into the VHF and UHF range using bulk mode capacitive resonators.

5.3.5
Nanoelectromechanical Resonators

Nanoelectromechanical resonators (NEMresonators) can be considered as MEMresonators scaled to the sub-micron domain. Owing to their very small mass, NEMresonators are capable of reaching very high frequencies. Nanomechanical resonating beam structures with frequencies in the hundreds of megahertz have been demonstrated in a variety of materials such as Si [15], AlN [16], SiC [17] and carbon nanotubes [18]. However, the resonance characteristics of all the devices reported so far were studied using non-integrated large-scale optical, magnetic or transmission electron microscope (TEM) inspection methods [15–19]. The integration of efficient (i.e. low output impedance) nanoelectromechanical transducers with nanomechanical resonators remains a tough challenge. Two approaches can be taken in making NEMresonators: the *top-down approach*, which is based on downscaling physical sizes defined by lithography on a substrate, and the *bottom-up approach*, which is based on chemical assembly at the molecular level to create a particular resonator structure (e.g. carbon nanotubes). The implementation of integrated transducers for NEMresonators can be accomplished more easily using a top-down approach to accelerate the investigation of the scaling limits of various electromechanical transduction methods. Once the scaling limits have been identified, suitable transduction methods can also be utilized for bottom-up approaches.

5.4
Filter Implementations

A MEMresonator is an electromechanical system with a single degree of freedom that is described by the second-order differential equation of motion:

$$M\frac{d^2u}{dt^2} + D\frac{du}{dt} + Ku = f_{\text{transducer}} \tag{5}$$

where u is the displacement of the microresonator, M, D and K are the effective mass, damping and stiffness of the resonating element, respectively, and $f_{\text{transducer}}$ is the force applied to the microresonator by the electromechanical transducers. A single high-Q resonator cannot be effectively used as a narrowband filter by itself, because it does not provide sufficient out-of-band rejection to discriminate between adjacent bands or channels. The ideal characteristic of a passband filter is the so-called 'brick-wall' characteristic (passband gain = 1, out-of-band gain = 0) that sharply rolls off outside of the 3 dB passband of the filter. Such characteristics are best described using high-order transfer functions with multiple poles. If a number of MEMresonators are coupled to one another, a high-order (or multiple degree of freedom) system will result that can be used as a bandpass filter. Owing to the electromechanical nature of the resonators, the coupling can be performed in either the mechanical or electrical domain. We shall discuss various coupling techniques next.

5.4.1 Mechanically Coupled Filters

A mechanical system with n degrees of freedom has n resonant modes. Therefore, if a number of microresonators are connected to each other with compliant coupling elements (ideally massless) as shown in Fig. 5.17, the resulting structure will have multiple resonances.

As an example, consider the mechanical system depicted in Fig. 5.18. Two identical resonators each having a mass of M and a stiffness of K are mechanically coupled to each other by a spring K_C. The first resonance of this two degrees of freedom two-resonator system occurs with the individual resonators vibrating in-phase at their original resonant frequency f_1. For this resonance, the spring K_C is not excited. The second resonance of the system occurs with the resonators vibrating out-of-phase for which the coupling spring is excited, resulting in a frequency of

$$f_2 \cong f_1\left(1+\frac{K_C}{K}\right) \tag{6}$$

assuming that $K >> K_C$. Therefore, the frequency separation Δf between the two resonance modes is determined by the ratio of the coupling stiffness to the effective stiffness of the resonator:

$$\Delta f = f_2 - f_1 \approx f_1 \frac{K_C}{K_{eff}} \tag{7}$$

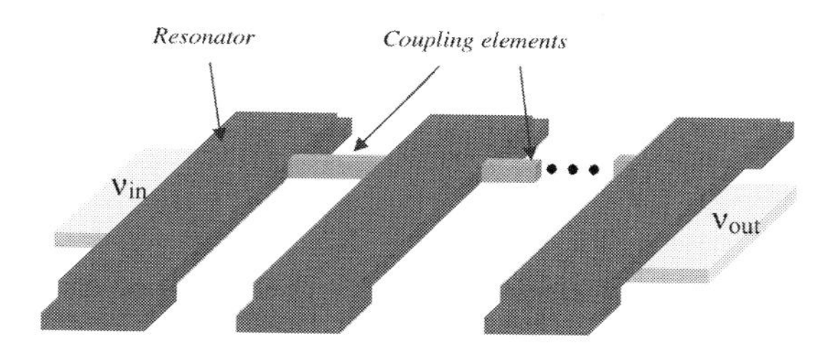

Fig. 5.17 Schematic diagram of a mechanically coupled filter array comprised of beam resonators coupled with compliant coupling elements

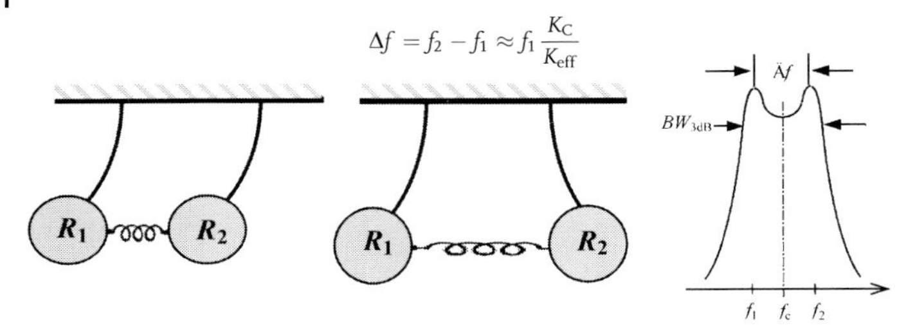

Fig. 5.18 Schematic diagram of a second-order mechanically coupled system, comprised of two identical resonators

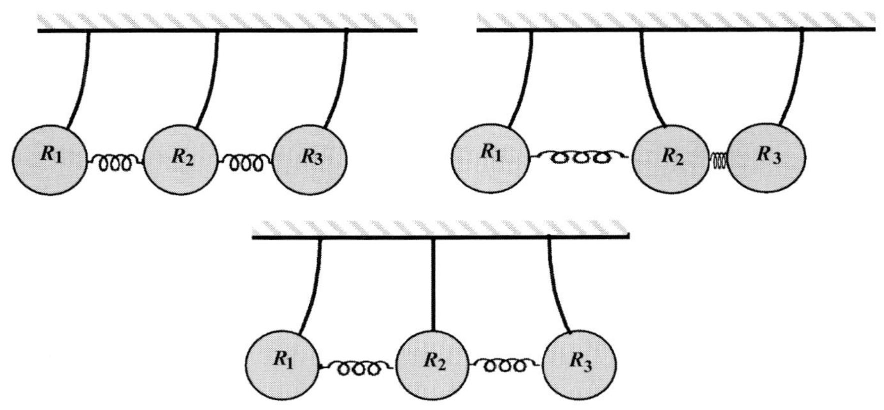

Fig. 5.19 Schematic diagram of a third-order mechanically coupled system, comprised of three identical resonators

The resonance modes for a third-order system are shown in Fig. 5.19.

Several mechanically coupled filters using polysilicon micromachining technology have been reported [20–23]. Wong et al. [22] demonstrated a two-resonator 68 MHz spring-coupled polysilicon capacitive micromechanical filter with a configuration very similar to that in Fig. 5.17. Placement of a discrete coupling element at low-velocity points resulted in less than 1% bandwidth at 68 MHz. However, extension of this approach to higher frequencies and smaller bandwidths remains a great challenge. For high-frequency applications (UHF range), owing to the very small size of the resonator element (< 10 μm), mechanical coupling will require sub-micron size coupling elements (e.g. wires), which are difficult to fabricate using low-cost optical lithography.

5.4.1.1 Through-support-coupled Micromechanical Filters

To reduce further the effective stiffness at the coupling location of mechanically coupled resonators and achieve ultra-small filter bandwidth, the novel concept of through-support coupling has been investigated [24]. In structural analyses,

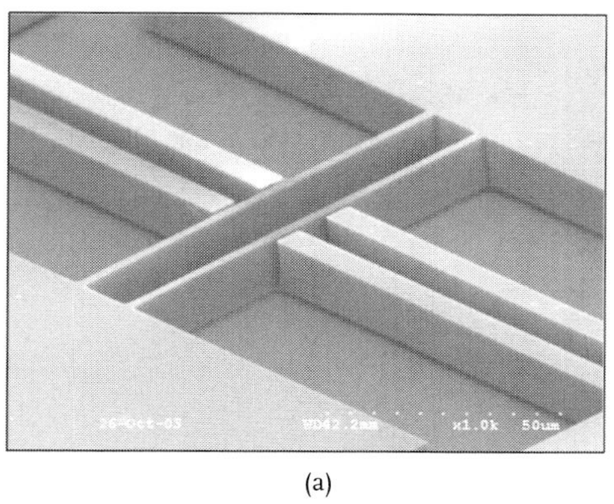

(a)

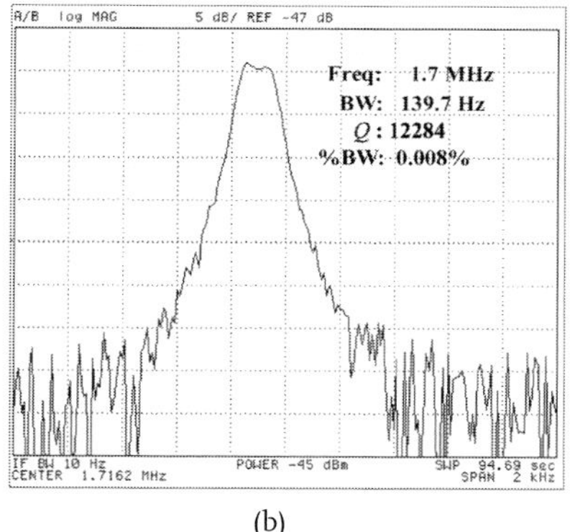

(b)

Fig. 5.20 (a) SEM view of an array of two 100×2 μm clamped-clamped beams with 9 μm spacing. (b) Frequency response of an array of two 100×2 μm beams with a beam spacing of 18 μm

clamped boundaries are commonly assumed to be ideal (i.e. infinitely rigid). However, solids are elastic with a finite stiffness. Any deviation from the equilibrium position will cause displacement at these elastic clamped boundaries. This finite stiffness and small non-zero displacement enables resonators to be coupled through their support. We exploited the elasticity of the support medium to provide coupling between adjacent resonators. Using this integrated coupling technique, there are no additional fabrication requirements for a discrete coupling element; attainment of higher frequency filter arrays is limited only by the ability to fabricate the dimensionally smaller resonators.

Fig. 5.20 shows an SEM picture of an SOI-based second-order through-support-coupled filter and its measured frequency response in vacuum. This filter, comprised of two 100×2 μm clamped-clamped silicon beams that are mechanically coupled through their support, demonstrated a record ultra-small bandwidth of 0.008% at 1.7 MHz [24]. This technique holds great promise for extension into the very high frequency range and further research on application to bulk mode resonators is under way.

5.4.2 Electrically Coupled Filters

Bandpass characteristic can also be realized by coupling the MEMresonators *electrically* to one another using passive or active coupling elements. Electrical coupling techniques can be used to overcome some of the difficulties associated with the mechanical coupling approach and provide greater potential for extension into the UHF frequency range. Both capacitors and active electronic buffers have been used to couple MEMS resonators to each other and provide a high order transfer function.

5.4.2.1 **Capacitive Coupling**

In the capacitive coupling approach [25], as depicted in Fig. 5.21, micromechanical resonators are cascaded with a shunt capacitor to ground between every two adjacent resonators. The interaction of the coupling capacitors and the resonators' equivalent RLC tank circuits results in several resonance modes in the system and consequently a multiple-order bandpass frequency response.

If we consider a two-resonator system, the capacitive coupling of two resonators with identical center frequencies (f_0), quality factors ($Q > 1000$) and motional resistances (R) results in a new pair of conjugate poles at a frequency of

$$f_1 = f_0 \sqrt{\frac{1 + \pi f_0 C_c R Q}{\pi f_0 C_c R Q}} \qquad (8)$$

where C_c is the coupling capacitor. This will introduce a new resonance in addition to the inherent resonance mode of the individual resonators at f_0. Looking at the frequency response of the two-resonator system, the first resonance occurs at the mechanical resonant frequency of the individual resonators. At the first resonance, as shown in Fig. 5.22 a, the two resonators resonate in-phase and the coupling capacitor has no contribution (while C_c is being charged by the first resonator, the other resonator is discharging it). At the second resonance (f_1), the two resonators operate with a 180° phase difference and hence the coupling capacitor comes into the picture (it is being charged and discharged at the same time by both resonators). Owing to its symmetry, the system can be reduced to a half circuit with one resonator and a series capacitor $C_c/2$ to ground. The series capacitor reduces the total capacitance of the RLC tank, causing the second resonance mode to occur. The case will be more complicated for a three-resonator system with two coupling capacitors, as shown in Fig. 5.22 b.

The asymmetry in the frequency response of the third-order filter is due to the fact that the end resonators have only one coupling capacitor attached to them but the one in the middle is terminated with two coupling capacitors at the two ends. This asymmetry can be compensated by slight frequency tuning of the end resonators of the chain with respect to the other resonators, but it can result in an increase in an insertion loss. A better solution to this problem is to use a closed chain of coupled resonators [23] to have complete symmetry for all the resonators.

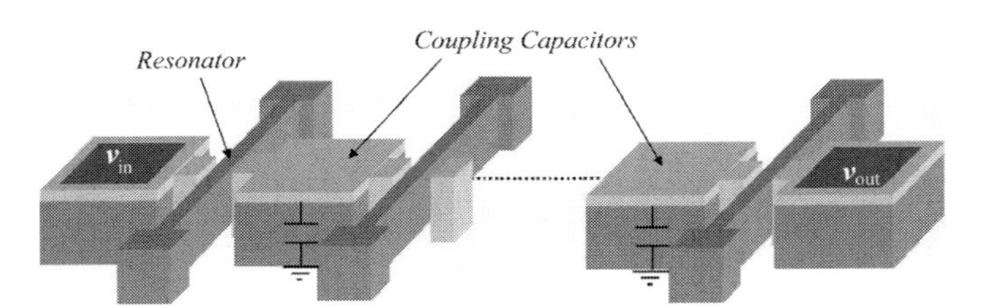

Fig. 5.21 Schematic diagram of a capacitively coupled microelectromechanical filter

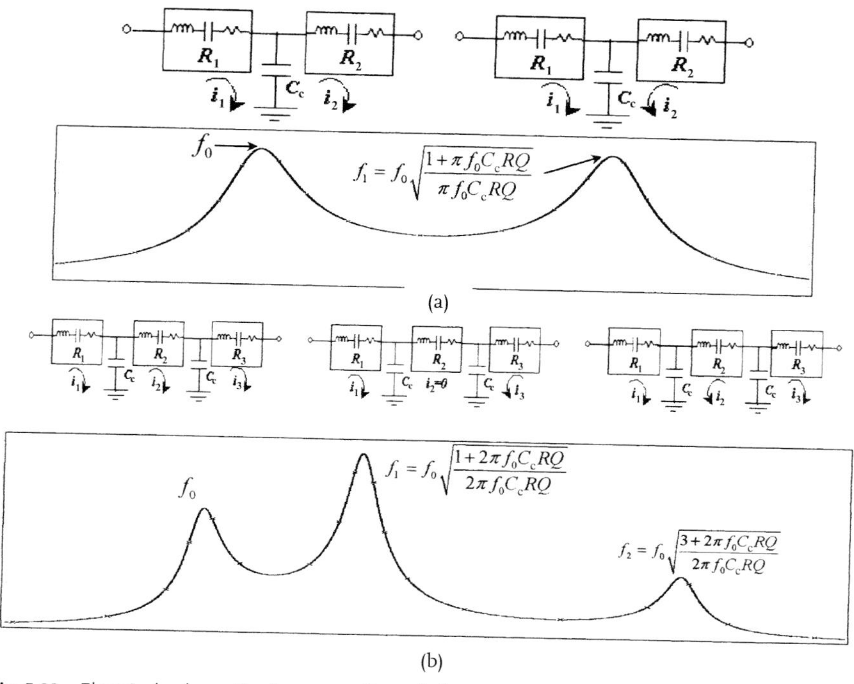

Fig. 5.22 Electrical schematic diagrams of coupled resonators and their frequency response: (a) second-order system; (b) third-order system

Fig. 5.23 illustrates simulation results of capacitively coupled electromechanical filters at 600 kHz with different resonator quality factors, showing the dependence of the insertion loss on the Q of individual resonators. The value of the coupling capacitors can be extracted from the resonators' Q, the desired filter bandwidth and the desired passband ripple. For the specific filter characteristics in Fig. 5.23, coupling capacitors of 0.2 pF are required that can be easily fabricated on-chip.

The insertion loss of capacitively coupled filters (assuming ideal lossless coupling capacitors) is determined by the Q of the individual resonators, the order of the filter and the termination resistors added to flatten the passband:

$$\text{insertion-loss (dB)} = 20 \log\left(\frac{nR_r + 2R_{\text{term}}}{2R_{\text{term}}}\right) \tag{9}$$

where n is the order of the filter, R_r is the equivalent motional resistance of the resonators ($R_r \propto 1/Q$) and R_{term} is the termination resistor. Fig. 5.24 shows the simulation results with non-ideal (lossy) coupling capacitors. Since the coupling capacitors are not involved in the first resonance mode, the finite Q of the coupling capacitors does not have a significant effect on the first resonance peak but its attenuation effect becomes more pronounced in higher resonance modes.

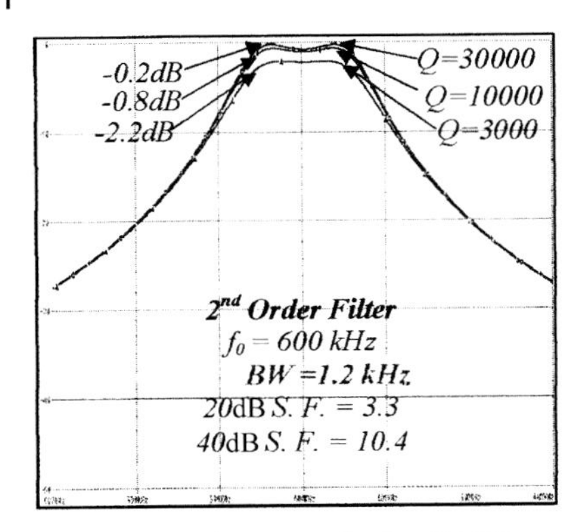

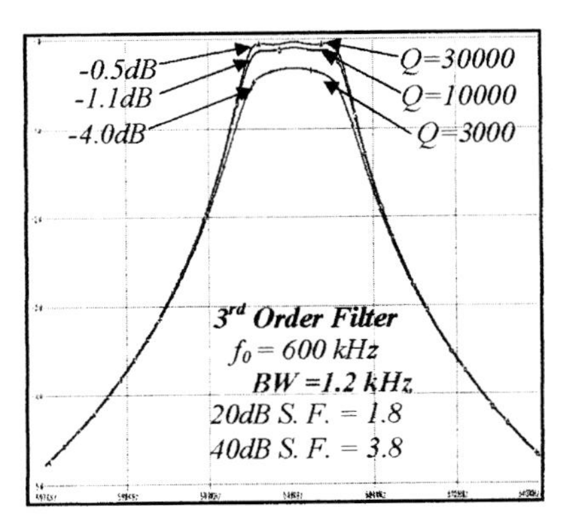

Fig. 5.23 Simulation results for 600 kHz capacitively coupled second- and third-order filters. A higher resonator Q results in a lower insertion loss. Coupling cap: C_c=0.15 pF (second order) and C_c=0.2 pF (third order)

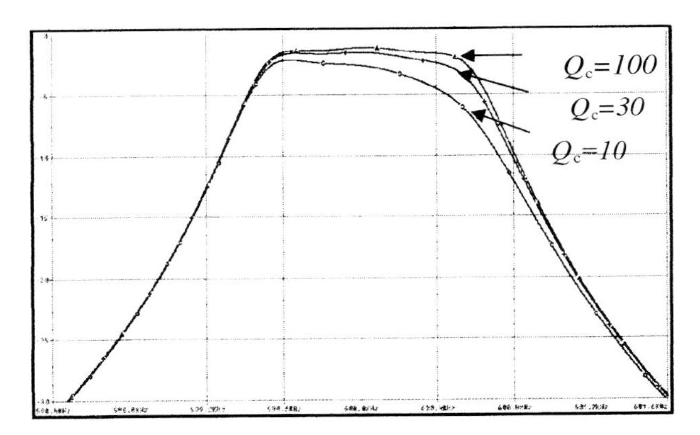

Fig. 5.24 Effect of the finite Q of the coupling capacitors on the frequency response of a third-order capacitively coupled filter

Fig. 5.25 shows the measured bandpass filter response obtained by coupling two 600 kHz SCS HARPSS resonators with a 3 pF shunt capacitor to ground [25]. Research is under way to implement capacitively coupled filters in the VHF and UHF range.

5.4.2.2 **Electrical Cascading**

Another approach used for implementation of high-order MEMS filters is the electrical cascading of MEMresonators using active components [25]. The electrical cascading of resonators with buffers or amplifiers in between (to eliminate the loading effect) results in multiplication of the transfer functions and an overall higher order transfer function with several pairs of conjugate poles.

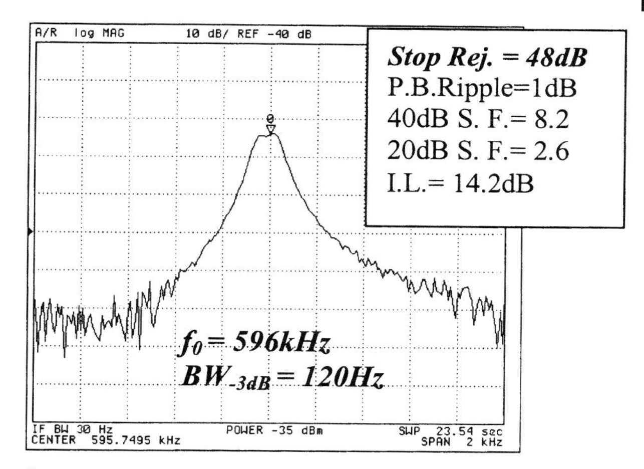

Fig. 5.25 Frequency response of capacitively coupled resonators at 600 kHz

When all the stages have equal center frequencies, cascading will result in order multiplication of poles, which can be interpreted as an overall higher equivalent quality factor. Mathematically, it can be shown that if n identical second-order resonators with individual quality factors Q_i are cascaded, the resultant Q factor of the cascade is given by

$$Q_{total} = \frac{Q_i}{\sqrt{10^{0.3/n} - 1}} \rightarrow Q_{total} \cong 1.2\sqrt{n}Q_i \tag{10}$$

if $n >> 1$.

This concept can be used to increase the equivalent quality factor of MEMS resonators for filtering or frequency synthesis applications, in case their intrinsic Q is not high enough. In addition, according to the following equation, the shape factor (SF) for the cascaded resonators is determined only by the order of the system, independent of the quality factor:

$$SF_{40\text{ dB}} = \sqrt{\frac{10^{4/n} - 1}{10^{0.3/n} - 1}} \rightarrow 1 \quad \text{as } n \text{ becomes large.} \tag{11}$$

Fig. 5.26 a illustrates simulation results of cascaded resonators with different orders showing the overall Q amplification by increasing the order of the system. The comparison between cascaded resonators with different number of stages but identical overall Q (Fig. 5.26 b) confirms that despite having equal quality factors, higher order cascades provide sharper roll-off and better selectivity.

To achieve larger bandwidths without sacrificing the sidewall sharpness in electrically cascaded micromechanical filters, one can take advantage of the frequency tuning characteristics of capacitive resonators. Introducing a slight mismatch be-

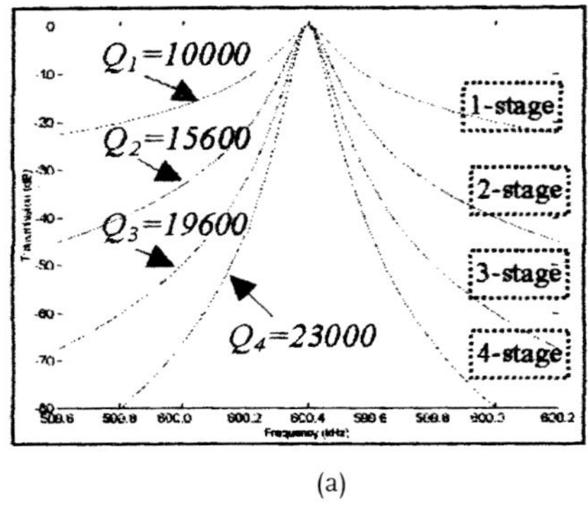

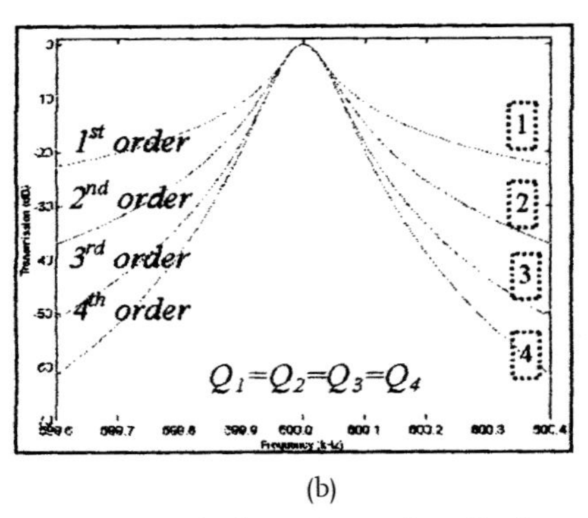

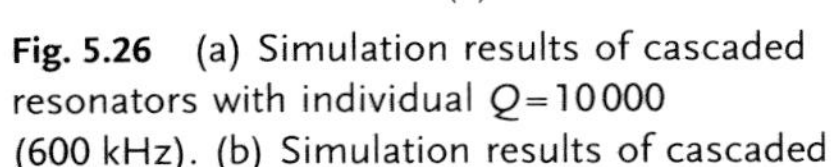

(a) (b)

Fig. 5.26 (a) Simulation results of cascaded resonators with individual $Q=10\,000$ (600 kHz). (b) Simulation results of cascaded resonators with identical overall quality factors and different orders

tween the center frequencies of cascaded resonators results in separation of poles and hence a wider bandwidth. However, center frequencies of cascaded devices should be close enough to avoid extra attenuation of each stage by the other stages (ultra-narrow bandwidth applications).

The frequency response of the active cascade was investigated using a test setup comprised of cascaded HARPSS resonators (600 kHz) with off-chip amplifiers in between. Second- and third-order bandpass filters were achieved using two and three cascaded resonator stages, respectively, as shown in Fig. 5.27 [25]. A pass-band gain can be achieved in this case because of the amplifiers. When the three cascaded resonators were tuned to have exactly identical center frequencies (600 kHz), an overall cascade Q of 19 300 was achieved from resonators with individual Q of 10 000.

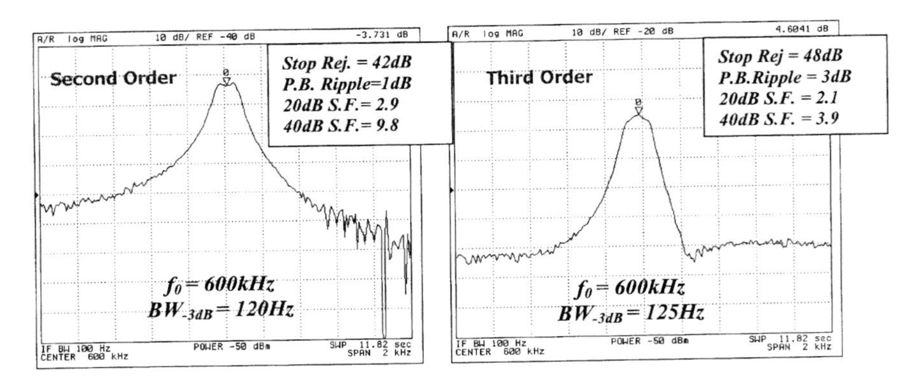

Fig. 5.27 Frequency response of second- and third-order bandpass filters achieved by active electrical cascading of HARPSS resonators

5.4.2.3 Electrostatic Coupling

Similarly to the through-support filters (Section 5.4.1.1) that do not use any discrete mechanical elements to couple the resonators, electrical coupling of MEMresonators can be accomplished without using distinct coupling elements. In the electrostatic coupling approach [26], an electrostatic force between the resonating bodies of two closely spaced microresonators causes coupling and results in a higher order resonant system, without the need for any physical coupling element. Fig. 5.28 shows an SEM picture of an SOI-based second-order electrostatically coupled filter and its measured frequency response in vacuum. The filter is comprised of two 15 μm thick, 5 μm wide, 500 μm long clamped-clamped beam resonators with a 2.8 μm wide coupling gap in between. Filter quality factors as high as 6800 (0.015% BW) with more than one decade of bandwidth tunability were demonstrated in the second-order electrostatically coupled beam filter of Fig. 5.28 with a center frequency of 170 kHz [26].

5.5 Dissipation Mechanisms

The unloaded quality factor of a micromechanical resonator is determined from various energy loss mechanisms and can be expressed by the following:

$$Q = \left(\frac{1}{Q_{\text{viscous}}} + \frac{1}{Q_{\text{TED}}} + \frac{1}{Q_{\text{support}}} + \frac{1}{Q_{\text{surface}}} + \frac{1}{Q_{\text{internal}}}\right)^{-1} \tag{12}$$

The contribution of each loss mechanism may vary depending on the operating conditions, geometry, frequency and mode shape of the microresonator. It should be kept in mind that electrical components/networks connected to a MEMresonator can alter the Q of a resonator (loaded Q).

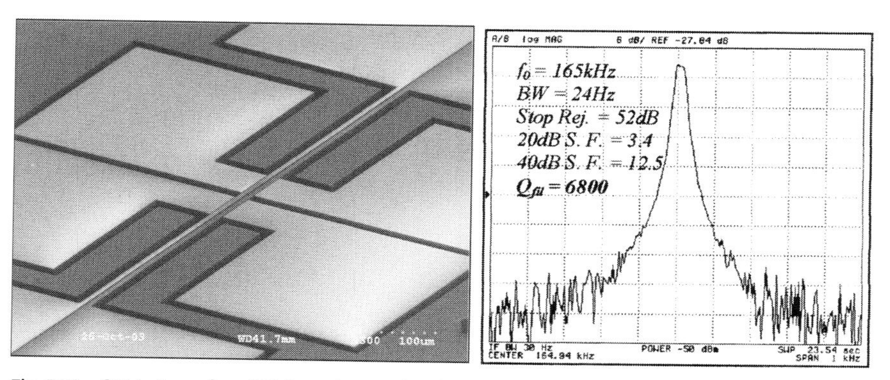

Fig. 5.28 SEM view of an SOI-based second-order clamped-clamped electrostatically coupled beam filter and its frequency response [26]

$Q_{viscous}$. Viscous damping occurs for micromechanical resonators that operate in air or a viscous environment. In order to achieve maximum Q, many MEMresonators are designed to operate in vacuum (<10 mTorr). Fig. 5.29 shows the experimental data for quality factor versus air pressure for two piezo-on-silicon clamped-clamped beams (Fig. 5.12), 100 and 200 μm long, with identical width and thickness of 20 and 4 μm, respectively. The figure illustrates how viscous damping becomes the dominant loss mechanism at higher air pressures. The quality factor tends to drop at higher pressure in higher frequency (higher stiffness) beams [27].

Although the Q of low-frequency (kHz), low-stiffness microresonators can drop by as much as 2–3 orders of magnitude in air compared with that of the vacuum, high-frequency ultra-stiff resonators tend to maintain their high Q even in air. For example, the 150 MHz SCS disk resonator in Fig. 5.10 showed a quality factor of 8200 when operated at atmospheric pressure (a factor of five reduction compared with its Q_{vacuum}), while the 1.14 GHz disk resonator of Fig. 5.14 had the same Q in air and vacuum.

Q_{TED}. Thermoelastic damping is a result of irreversible heat flow across temperature gradients produced by inhomogeneous compression and expansion of a resonating structure. While thermoelastic damping can become the main loss mechanism in high aspect ratio beam resonators, its contribution is negligible in disk resonators owing to the large dimensions of the structures and longer thermal path. According to Zener [28], thermoelastic loss of a solid homogeneous beam in the fundamental flexural mode is approximated by

$$Q^{-1} = \Delta \frac{\omega \tau_{TH}}{1 + (\omega \tau_{TH})^2} \tag{13}$$

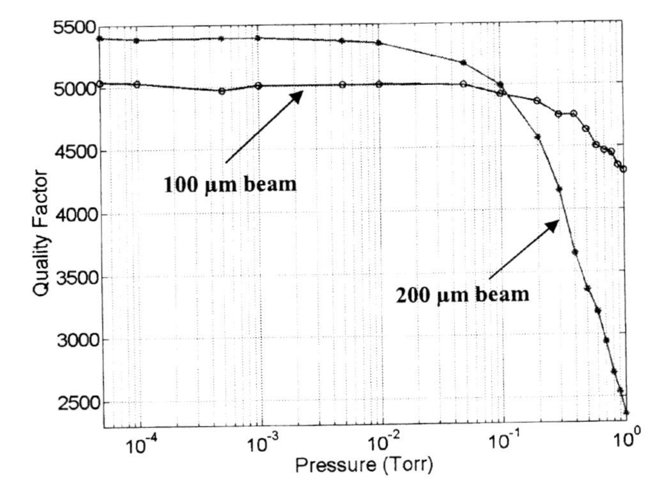

Fig. 5.29 Quality factor versus air pressure for the beam resonators in Fig. 5.12

where

$$\Delta = \frac{E\alpha_{TH}^2 T_0}{\rho C_p}, \quad \tau_{TH} = \frac{\rho C_p d^2}{\kappa \pi^2} \tag{14}$$

Δ is the relaxation strength and τ_{TH} is the thermal relaxation time constant with d as the effective thermal path length. T_0 is the resonator equilibrium temperature, ρ is the density, E is the Young's modulus, C_p is the specific heat capacity, α_{TH} is the linear thermal expansion coefficient, and κ is thermal conductivity of the resonator material. In the case of a flexural solid beam, d is equal to the width of the beam. Therefore, as shown in Fig. 5.30, the thermoelastic loss of a beam has a Lorentzian behavior and the maxima occurs at $\omega\tau_{TH}=1$. Hence the frequency of minimal Q is

$$f_{Q_{min}} = \frac{1}{2\pi\tau_{TH}} = \frac{\kappa\pi}{2\rho C_p d^2} \tag{15}$$

An interesting variation of the original TED characteristic in solid beam resonators applies to the polysilicon trench-refilled (TR) resonators in Fig. 5.15. The void created in TR polysilicon beams causes a discontinuity in thermal transport across the width, altering the thermoelastic behavior of the resonator [12]. Because of this, two thermal paths will be present in a vibrating TR beam, as opposed to only one thermal path in a solid resonator (shown in Fig. 5.31). In the hollow structure, one thermal path is across the thickness of the beam wall (P1) and the other exists around the void (P2). P1 and P2 illustrate the flow of heat from hot to cool regions within the structure. The individual thermal paths have a unique frequency-dependent behavior of their own, which will cause a double-dip TED characteristic as shown in Fig. 5.32. Experimental results of polysilicon TR beams have confirmed the TED of Fig. 5.32 [12]. It should be mentioned that our analy-

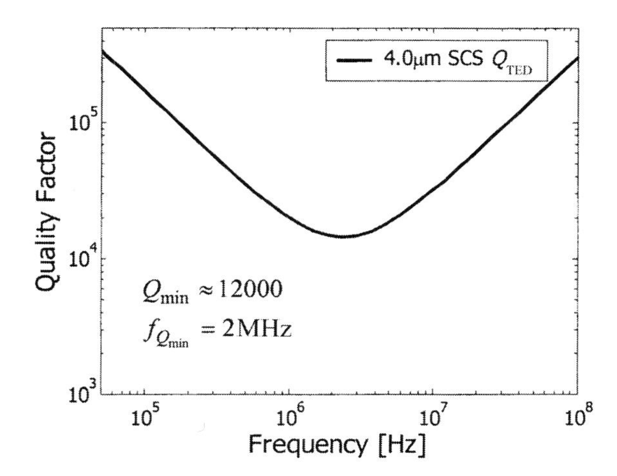

Fig. 5.30 Thermoelastic damping limited quality factor (Q_{TED}) versus frequency in a silicon beam resonator

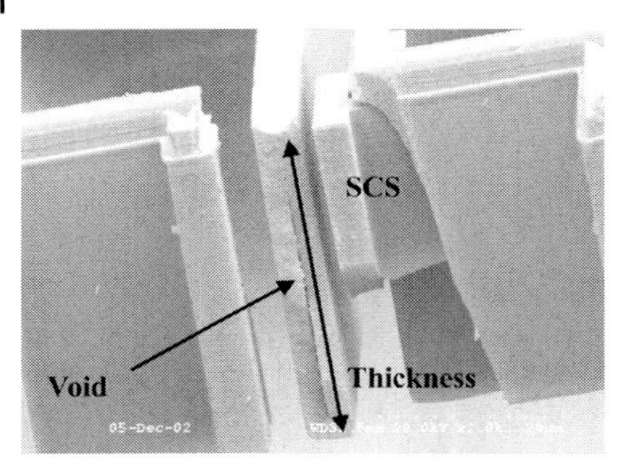

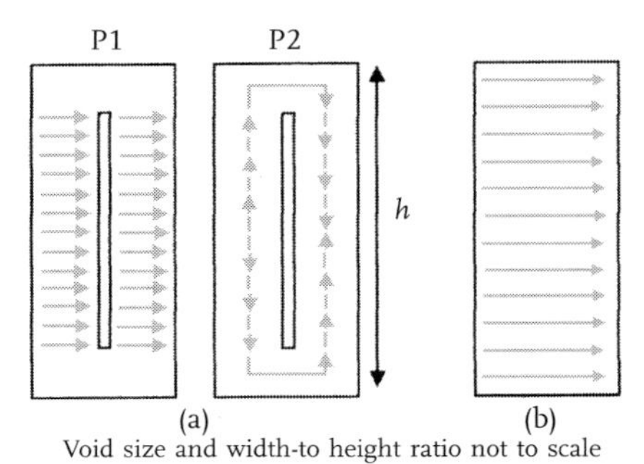

Fig. 5.31 SEM picture of a broken resonator, showing the void in the polysilicon trench-refilled beam. Two thermal paths exist in (a) a hollow beam (TR beam) and one thermal path in (b) a solid homogeneous beam

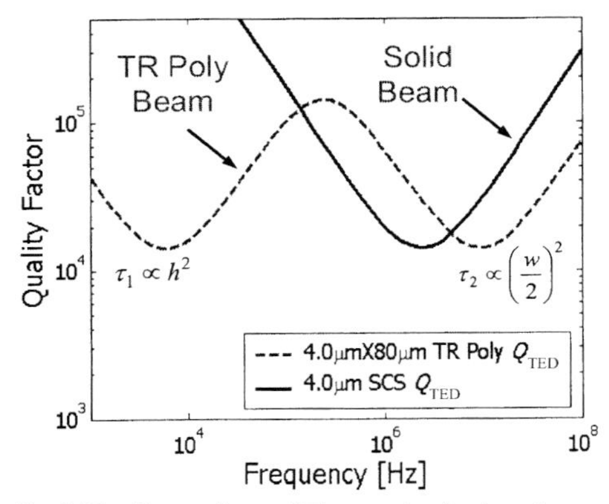

Fig. 5.32 Comparison of Thermoelastic damping limited quality factor (Q_{TED}) in a TR and a solid silicon beam

sis does not include the intracrystalline thermoelastic damping in polysilicon structures [29] that will come into effect at very high frequencies (gigahertz range).

$Q_{support}$. Support loss, also known as clamping loss, is the vibration energy of a resonator dissipated by transmission through its support structure. During its flexural vibration, a beam resonator will exert both vibrating shear force and moment on its clamped ends. Acting as excitation sources, these vibrating shear force and moment will excite elastic waves propagating into the support. Therefore, the support structure absorbs some of the vibration energy of the beam resonator. It can be shown that the support loss of the beam resonator is proportional to the third power of the aspect ratio of the beam [30, 31]:

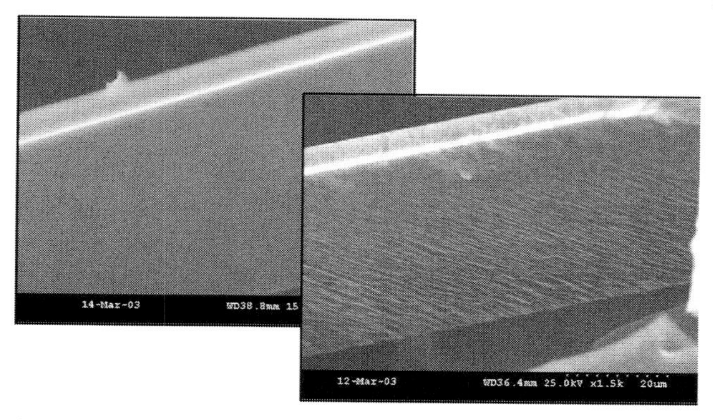

Fig. 5.33 Part-to-part variations in the surface roughness of the fabricated TR polysilicon beams

$$Q_{\text{support}} \propto \left[\frac{L}{b}\right]^3 \tag{16}$$

The support loss is the dominant dissipation mechanism in side-supported bulk-mode disk resonators and can be minimized by optimizing the dimensions of the supporting elements and reducing their numbers.

Q_{surface}. It has been observed that resonators with rough surfaces have lower quality factors than those with smooth surfaces. Fig. 5.33 shows the difference in the surface roughness of two polysilicon beams (trench-refilled) from two different fabrication lots. Surface roughness can also occur on the sidewalls of SCS resonators that are etched using deep or regular RIE processes. Beam microresonators with smooth sidewalls have shown between 20 and 100% higher Q factors. The surface roughness and irregularities of silicon resonators can be improved by the growth and subsequent removal of a thin layer of oxide on the surfaces.

5.6 Acknowledgments

The author acknowledges substantial contributions from his former and current graduate students, especially Yoel No, Aki Hashimura, Siavash Pourkamali, Gavin Ho, Gianluca Piazza, Reza Abdolvand and Shweta Humad, to the development of single-crystal silicon resonators and filters presented in this chapter.

5.7 References

1. C.T.-C. Nguyen, L.P.B. Katehi, G.M. Rebeiz, *Proc. IEEE* **1998**, *86*, pp. 1756–1768.
2. S.Y. No, A. Hashimura, S. Pourkamali, F. Ayazi, in: *Solid-State Sensor and Actuator Workshop, Hilton Head Island, SC*; **2002**, pp. 281–284.
3. G. Piazza, R. Abdolvand, F. Ayazi, in: *Proceedings of the Sixth IEEE International Micro Electro Mechanical Systems Conference (MEMS '03), Kyoto*; **2003**, pp. 149–152.
4. S. Pourkamali, A. Hashimura, R. Abdolvand, G. Ho, A. Erbil, F. Ayazi, *High-Q Single Crystal Silicon HARPSS Capacitive Beam Resonators with Sub-micron Transduction Gaps, IEEE Journal of Microelectromechanical Syststems*, Vol. 12, No. 4, August **2003**, pp. 487–496.
5. S. Pourkamali, F. Ayazi, in: *Digest of the 12th International Conference on Solid State Sensors, Actuators and Microsystems (Transducers '03), Boston, MA*; **2003**, pp. 837–840.
6. A. Hashimura, *Single-crystal Silicon HARPSS Capacitive Beam Resonators*; MSc Thesis, Georgia Institute of Technology, **2002**.
7. D.L. DeVoe, *Sens. Actuators A* **2001**, *88*, 263–272.
8. R. Ruby, et al., *ISSCC Dig. Tech. Pap.* **2001**, *44*, 121–122.
9. S. Humad, et al., presented at the 2003 IEEE International Electron Devices Meeting, Washington, DC, 7–10 December **2003**, pp. 957–960.
10. C.T.C. Nguyen, R.T. Howe, in: *Tech. Dig. IEEE Int. Electron Devices Meeting, Washington DC;* **1993**, pp. 199–202.
11. J. Wang, et al., in: *Digest of the 12th International Conference on Solid State Sensors, Actuators and Microsystems (Transducers '03), Boston, MA*; **2003**, pp. 947–950.
12. R. Abdolvand, G. K. Ho, A. Erbil, F. Ayazi, in: *Digest of the 12th International Conference on Solid State Sensors, Actuators and Microsystems (Transducers '03), Boston, MA;* **2003**, pp. 324–327.
13. F. Ayazi, K. Najafi, *High Aspect-Ratio Combined Poly and Single-Crystal Silicon (HARPSS) MEMS Technology, IEEE Journal of Microelectromechanical Systems*, Vol. 9, September **2000**, pp. 288–294.
14. A. E. Franke, et al., *IEEE J. Microelectromech. Syst.* **2003**, *12*, 160–171.
15. A.N. Cleland, M. L. Roukes, *Appl. Phys. Lett.* **1996**, *69*, 2653.
16. A.N. Cleland, et al., *Appl. Phys. Lett.* **2001**, *79*, 2070–2072.
17. X.M.H. Huang, et al., in: *Digest of the 12th International Conference on Solid State Sensors, Actuators and Microsystems (Transducers '03), Boston, MA*; **2003**, pp. 342–343.
18. P. Poncharal, Z.L. Wang, et al., *Science* **1999**, *283*, 1513.
19. M.L. Roukes, in: *Solid-State Sensor and Actuator Workshop, Hilton Head Island, SC*; **2000**, pp. 367–376.
20. L. Lin, et al., *IEEE J. Microelectromech. Syst.* **1998**, *7*, 286.
21. K. Wang, et al., *IEEE J. Microelectromech. Syst.* **1999**, *8*, 534.
22. A.-C. Wong, et al., in: *Digest of the 10th International Conference on Solid State Sensors, Actuators and Microsystems (Transducers '99), Sendai*; **1999**, pp. 1390–1393.
23. D.S. Greywall, et al.,. *Micromech. Microeng.* **2002**, *12*, 925–938.
24. G.K. Ho, R. Abdolvand, F. Ayazi, in: *Proceedings of the Seventh IEEE International Micro Electro Mechanical Systems Conference (MEMS '04), Maastricht*; **2004**, pp. 769–772.
25. S. Pourkamali, R. Abdolvand, F. Ayazi, in: *Proceedings of the Sixth IEEE International Micro Electro Mechanical Systems Conference (MEMS '03), Kyoto*; **2003**, pp. 702–705.
26. S. Pourkamali, et al., in: *Proceedings of the seventh IEEE International Micro Electro Mechanical Systems Conference (MEMS '04)*, Maastricht; **2004**, pp. 584–587.
27. F.R. Blom, S. Bouwstra, M. Elwenspoek, J.H.J. Fluitman, *J. Vac. Sci. Technol. B* **1992**, *10*, 19–26.
28. C. Zener, *Phys. Rev.* **1937**, *52*, 230–235.
29. V.T. Srikar, S.D. Senturia, *IEEE J. Microelectromech. Syst.* **2002**, *11*, 499–504.
30. Y. Jimbo, K. Itao, *J. Horol. Inst. Jpn.* **1968**, *47*, 1–15 (in Japanese).
31. Z. Hao, A. Erbil, F. Ayazi, *An Analytical Model for Support Loss in Micromachined Beam Resonators with In-plane Flexural Vibrations, Sensors and Actuators A*, Vol. 109, December **2003**, pp. 156–164.

6
MEMS in Mass Storage Systems

T. R. Albrecht, M. Despont, E. Eleftheriou, IBM Zurich Research Laboratory, Rüschlikon, Switzerland, J. U. Bu, LG Electronics Institute of Technology, Seoul, Korea and T. Hirano, Hitachi Global Storage Technologies, San Jose Research Center, San Jose, CA, USA*

Abstract
MEMS devices may play a variety of roles in future mass storage devices. In conventional magnetic and optical storage devices, MEMS components replace or augment critical subassemblies, such as head positioning actuators or optical head elements. The motivation for, design of and performance of these MEMS subassemblies are discussed. A new class of 'MEMS-centric' storage devices may emerge in which MEMS structures completely replace conventional mechanical elements and novel recording schemes derived from scanning probe technology are used. The recording methods used, the MEMS component design and integration, and the control system and architecture of MEMS-centric devices are discussed.

Keywords
MEMS microactuator; MEMS optical recording; near-field optical recording; probe storage; micromechanical storage; high-density recording.

*) Recently moved to: Hitachi Global Storage Technologies, San Jose Research Center, San Jose, CA, USA

Advanced Micro and Nanosystems. Vol. 1
Edited by H. Baltes, O. Brand, G. K. Fedder, C. Hierold, J. Korvink, O. Tabata

ISBN: 3-527-30746-X

6.1 Introduction

Forecasts of future directions in data storage often predict a prominent role for MEMS technology. The continuing evolution of mass storage devices toward higher data density, faster access and smaller form factors is thought to offer excellent opportunities to exploit the general advantages of MEMS. This chapter analyzes specific implementations of MEMS technology in mass storage devices as subcomponents in conventional types of storage devices in addition to 'MEMS-centric' devices, where MEMS technology enables and becomes the heart of an entirely new genre of mass storage devices. In these devices, MEMS technology will not be limited to enhancing a device's mechanical capability – it may also be the key to exploiting new ultra-high density recording schemes that trace their lineage to the nanometer-scale resolution of scanning probe microscopes.

Conventional mass storage devices fall into two categories – those which use mechanical access (magnetic and optical disks, in addition to magnetic tape systems) and those which use electrical access (solid-state technologies such as Flash

EEPROM). Mechanical access devices have the highest areal density and lowest cost per gigabyte and therefore have been predominant in the high-capacity and larger form factor segments of the market, whereas solid-state storage devices have emerged in the past decade as the solution of choice for small, low-power, portable applications for which limited capacities are sufficient. Since MEMS technology could be a route to both superior mechanical performance and higher recording densities, MEMS-centric devices could exhibit some of the best attributes of both solid-state and mechanical access devices: the small form factor, fast access and low power consumption of solid-state devices, together with an areal density and cost per gigabyte more typical of mechanical access devices. Although MEMS devices are unlikely to replace either of these existing classes of storage devices completely, they could emerge as an important new category in the hierarchy of mass storage options.

6.2 MEMS in Conventional Storage Devices: Hard Disk Drives

Spindles and head actuators in today's hard disk drives (HDDs) are built with conventional copper coils, sintered magnets and stamped or laminated iron yokes and cores. Miniaturization per se has not yet driven the industry to MEMS solutions; witness the 1-inch Microdrive, with its conventional spindle motor and rotary actuator [1]. In the near future, however, mechanical performance requirements combined with size and cost concerns appear likely to propel the industry towards a MEMS solution for a key component – a microactuator to work in tandem with the conventional rotary actuator that positions the head over the target track. Although the impending adoption of microactuators in HDDs is often regarded as the first high-volume use of MEMS components in storage devices, a modest broadening of the definition of 'MEMS' reveals that MEMS technology in the form of advanced air-bearing sliders has already been present in HDDs for many years.

6.2.1 Track-following Servo in HDDs

Modern HDDs (see Fig. 6.1) access data by means of two actuators: a spindle motor, which rotates the disk(s), and a rotary voice-coil motor (VCM), which swings the magnetic head(s) radially across the disk(s). Data are recorded in circular tracks that are subdivided into sectors; thus, when a user writes or reads data, the VCM rotates to position the head accurately over the target track and the write or read operation occurs when the correct sector along the target track travels under the head. With the increasing areal density in each new generation of HDDs, the width and pitch of the data tracks become narrower. A typical track pitch of an HDD in 2003 was $\sim$250 nm, a figure that is expected to continue to fall by about 30% per year. The positioning accuracy of the read/write head must be improved accordingly. In an HDD, the head position relative to the target track center is monitored and adjusted of the order of 100 times per revolution by a closed-loop

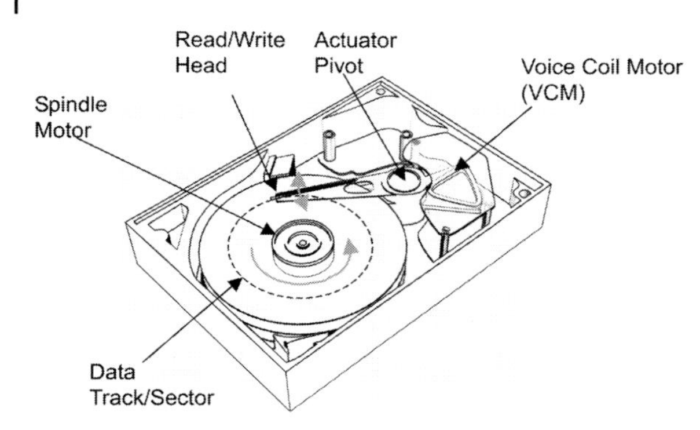

Fig. 6.1 Schematic of a magnetic hard disk drive showing the rotary actuator, suspension and slider

control system (servo system), which drives the rotary actuator. The accuracy of this system is governed by a number of parameters; one of the most important is the servo bandwidth, which roughly means the maximum frequency at which the servo system can respond to and correct a tracking error. In general, the higher the bandwidth, the better is the positioning accuracy. Conventional VCM actuators, however, impose limits on servo bandwidth. If the servo bandwidth is increased to include frequency ranges where the actuator has multiple mechanical resonances, the servo system ceases to function properly. Today's typical VCM actuators (after decades of refinement) have resonances from ~6 kHz and upwards, and this generally limits the achievable servo bandwidth to about 2 kHz.

6.2.2
Dual-stage Actuators

One of the most promising ways to increase servo bandwidth is to use a dual-stage actuator system. The first stage is a traditional VCM that acts as a coarse actuator with low speed and long stroke. The second stage is a smaller actuator which rides on the moving arm of the VCM and acts as a fine actuator with high speed and short stroke. Three types of secondary actuator systems (see Fig. 6.2) have been proposed so far: (1) the moving suspension type, (2) the moving slider type and (3) the moving head type. They are also called generations 1–3 to reflect the increasing complexity and benefit that they offer [2].

In conventional HDDs using single-stage actuators, the read/write elements are located on an air-bearing slider, which floats over the disk surface on a cushion of air several nanometers thick. The slider is attached to the arm of the VCM via a spring-like suspension. In a first-generation (moving suspension) dual-stage system, the secondary actuator is located at the end of the actuator arm and swings the entire suspension and slider as a unit. Since the secondary actuator in this implementation has to move the relatively heavy mass of the suspension (~100 mg), conventional actuators such as macroscopic piezoelectric actuators [3, 4] are used (see Fig. 6.3). This type of secondary actuator is relatively easy to fabricate but the maximum attainable servo bandwidth is limited by the resonant modes of the sus-

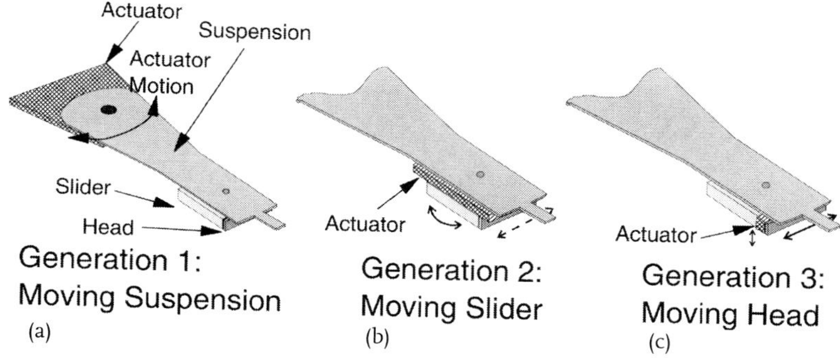

Fig. 6.2 Secondary actuator locations for (a) moving suspension, (b) moving slider and (c) moving head implementations. Reprinted with permission from [2]. © 2003 ASME International

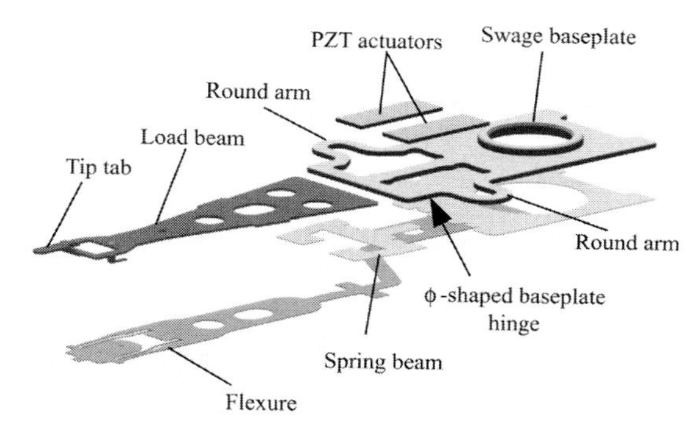

Fig. 6.3 Example of moving suspension secondary actuator [4]. Courtesy of M. Tokuyama et al.

pension. The maximum bandwidth reported so far is $\sim$3 kHz – only a modest improvement over single-stage VCMs.

In generation 2, the actuator is located between the suspension and the slider, moving the slider relative to the suspension. Rotational actuation is generally preferred because the effects of acceleration generated by VCM on the secondary actuator can be greatly reduced. Several MEMS designs have been published [5–8]. Since the actuator and slider are co-located, there are no significant resonance issues; thus, fairly high servo bandwidth can be achieved. An example of a moving slider MEMS actuator will be discussed in the next section.

In generation 3, the actuator is located between slider body and the read/write element, moving the element relative to the slider body. The bandwidth can be very high, but the fabrication of this actuator and its integration with a magnetic head are extremely challenging. Although a MEMS actuator of this type has been fabricated [9–11] (see Fig. 6.4), it has not yet been demonstrated integrated with read/write elements. From a servo bandwidth point of view, there is currently not much impetus to implement generation 3.

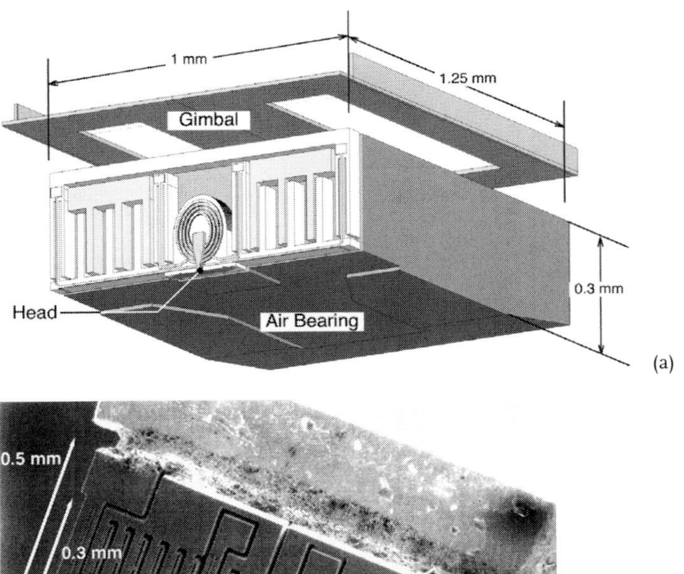

(b)

Fig. 6.4 (a) Conceptual diagram of an electrostatic moving head actuator. (b) SEM image of fabricated actuator (no magnetic head present). Reprinted with permission from [10], © IEEJ 2004

6.2.3 Moving Slider MEMS Microactuators

Of the three dual-stage actuator types, the moving slider type appears to offer an attractive compromise between performance and complexity and is also well suited for implementation with MEMS technology. The design requirements for a microactuator of this type include (1) sufficiently high output force to move a slider (~ 2 mg) over distances on the micrometer scale at high frequency, (2) sufficient Z stiffness to sustain the loading force of the suspension (~ 20 mN) on the slider, (3) good robustness against particulate contamination, (4) a means to provide electrical interconnection between the suspension and the moving slider, and (5) compatibility with slider and suspension assembly methods.

An example [8] of this type of microactuator is shown in Fig. 6.5. In this design, fine adjustment of the head position is performed by rotating the slider;

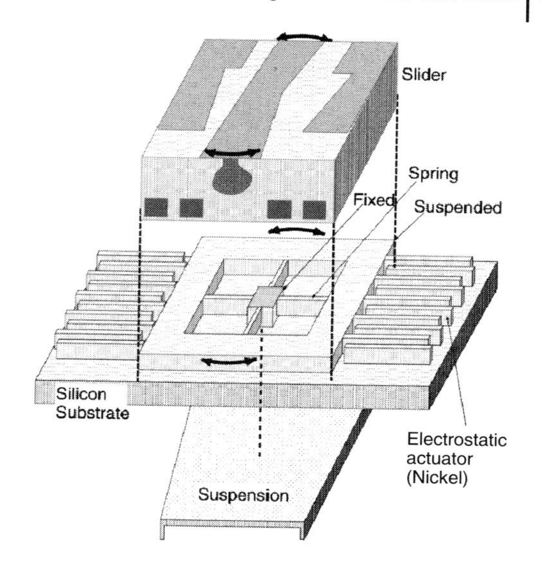

Fig. 6.5 Schematic diagram of MEMS moving slider microactuator. Reprinted with permission from [8]. © 1999 IEEE

since the read/write elements are located in the center of the trailing edge of the slider, rotation of the slider provides the desired cross-track motion of the elements. The rotational pivot is realized by a spoke-like flexure anchored at its center. The slider is attached to the suspended frame of this flexure. An electrostatic comb-type actuator is distributed around this frame to create a torque about the pivot point. Because the slider loading force is applied through the flexure, a relatively large Z stiffness is required; at the same time, a low in-plane stiffness is needed to achieve sufficient stroke. To meet these two requirements, the flexure has a very high aspect ratio (3 μm wide×35 μm high). The flexure and comb-drive elements of this design are fabricated with multiple steps of electroplating of nickel into high aspect ratio polymer masks formed by anisotropic plasma etching. The entire microactuator is covered by a protective cover layer made of electroplated nickel (not shown in Fig. 6.5) to keep particles out.

An SEM image of the completed microactuator [12] is shown in Fig. 6.6. The chip is 2.1 mm wide, 1.7 mm long and 0.16 mm thick. The air gaps of the electrostatic comb are 2 μm wide and 35 μm high. The bonding pads on the slider are connected to bonding pads on the moving platform of the microactuator; flexible conductors route the signals from these pads back to corresponding pads on the fixed substrate of the microactuator, which mate with conductors on the suspension. The stroke of this microactuator is ~1 μm (peak to peak) with a driving voltage of 50 V.

The microactuator is integrated with the slider and suspension as follows. First, the microactuator is attached to the load point of the suspension flexure with adhesive. Then, the suspension leads and microactuator bonding pads are electrically connected by a solder-ball placement/reflow process [13, 14]. Next, the slider is attached with adhesive to the top of the microactuator and the slider's bonding

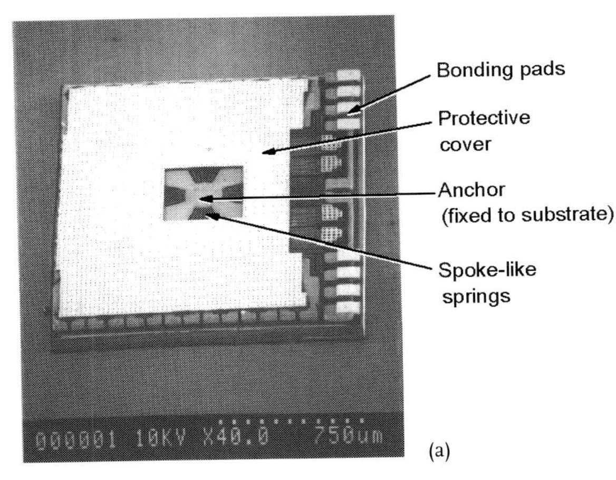

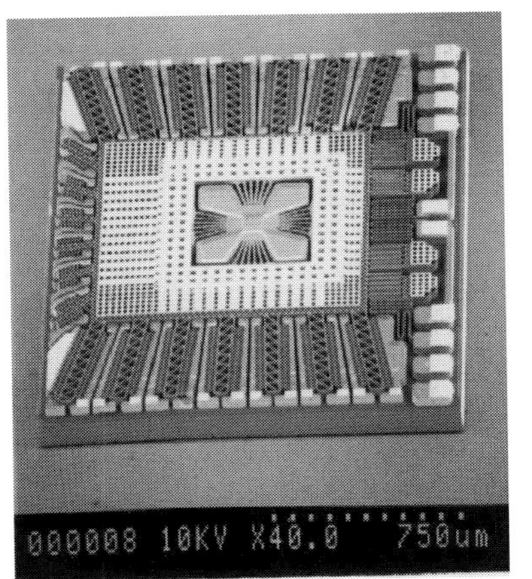

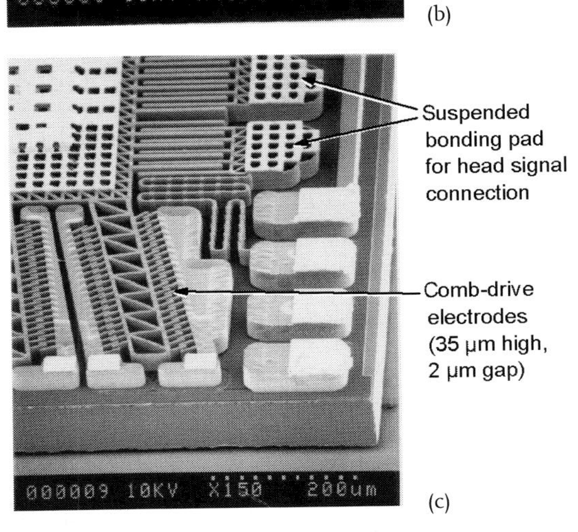

Fig. 6.6 SEM images of (a) the complete microactuator, (b) the microactuator with the protective cover removed, revealing the flexure and comb drive, and (c) an enlarged view of the electrodes. Reprinted with permission from [12]. © 2001 ASME International

pads and the pads on the microactuator's moving platform are connected by a similar solder-ball process. Fig. 6.7a is an SEM image of the complete assembly. Fig. 6.7b and c show the interconnections between the slider, microactuator and and suspension before and after the solder-ball process.

The mechanical frequency response of a conventional VCM and that of a microactuator as described above are compared in Fig. 6.8. Although the microactuator has a main resonance at 3 kHz, this mode has no ill-effect because it is the main spring-mass mode of the system with which the servo system is designed to deal. Above this peak, the response is free of undesired additional resonances out to beyond 25 kHz. The phase delay seen above 10 kHz is an artifact of the measurement

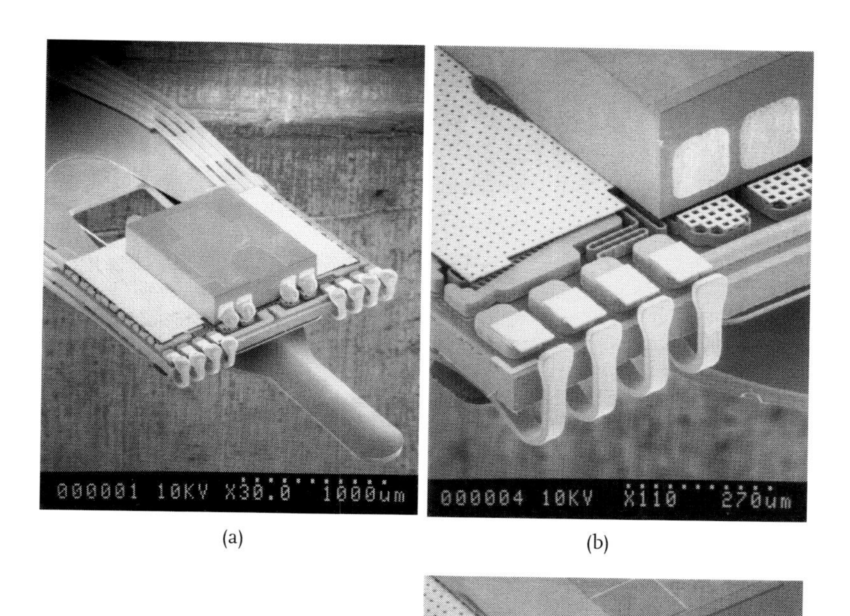

(a) (b)

000002 10KV X110 270um

(c)

Fig. 6.7 SEM images of (a) the complete slider/microactuator/suspension assembly, (b) the interconnects before the solder ball process and (c) after the solder-ball process. Reprinted with permission from [12]. © 2001 ASME International

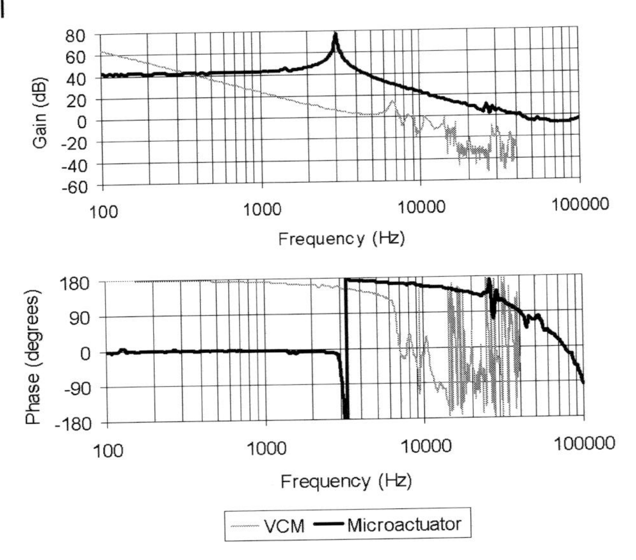

Fig. 6.8 Frequency response of a microactuator and a conventional VCM actuator

system (low-pass filter in laser Doppler vibrometer). The VCM, by comparison, has multiple resonances starting around 6 kHz, with rapid deterioration in phase margin. The increased servo bandwidth made possible by these improvements is expected to constitute a significant improvement in the overall servo control system [15].

6.2.4 The Air-bearing Slider: an Unconventional MEMS Device

In addition to the microactuator itself, its payload – the air-bearing slider (see Figs. 6.5 and 6.7), which carries the magnetic read/write elements – is another MEMS device in its own right. Although rarely discussed as 'MEMS' components, air-bearing sliders are in fact mechanical devices with micrometer- and nanometer-scale dimensions. Although the technology used to fabricate sliders includes some tools and techniques not commonly used for other MEMS devices, the overall approach is consistent with the broad definition of MEMS. The general lack of awareness of sliders as MEMS devices stems from the fact that sliders and their fabrication methods have evolved over many decades from macroscopic millimeter-scale devices fabricated by conventional cutting and grinding tools to modern micrometer-scale, batch-fabricated (wafer-processed) devices made by etching, plating, vacuum deposition and photolithographic patterning [16]. Mechanical grinding and fine-scale lapping are still used, but on a scale of precision never anticipated decades ago. Another reason that sliders are often overlooked as MEMS devices is that the substrate material, which becomes the slider body material, is not silicon, but rather AlTiC ceramic, an uncommon material in the world of MEMS.

A surprising facet of HDD slider technology is that state-of-the-art magnetic read/write elements under development today contain features with dimensions smaller than those used in the most advanced integrated circuits. Therefore, lithography tools and processes used in the semiconductor industry no longer provide the resolution needed by the HDD industry. As a result, HDD magnetic heads are becoming the first very large-volume products to use e-beam lithography [17]. Although e-beam lithography has been considered and rejected in the past for semiconductor device fabrication owing to low throughput and high cost, it turns out that magnetic read/write elements only have one or two critical features that require nanometer-scale lithographic resolution (the magnetoresistive stripe in the read element and the pole of the write head that determines the track width). The density of these features on a wafer is sufficiently low that throughput with e-beam lithography is cost effective.

Although it has often been predicted that magnetic recording would eventually be replaced by a higher density, 'nanotechnology'-based recording scheme, the use of e-beam-defined nanostructures in magnetic heads and the surprisingly high recording densities achieved mean that magnetic recording has itself become a nanotechnology, competing alongside other nanotechnologies for use in future storage devices (see Section 6.4.2.1).

6.3 MEMS in Conventional Storage Devices: Optical Disk Drives

Although the areal density of optical storage devices is lower than that of HDDs, optical disks are very inexpensive and are far more robust than magnetic disks for removable media. As a result, optical disks have become predominant in market segments where removability has strong advantages, such as for audio/video recording, distribution of prerecorded content and in certain backup and archival applications. The highest capacity optical disks to date can store 25 Gbytes and have an areal density of 18 Gbit/in^2, which is achieved through the use of a GaN blue laser diode and a two-element objective lens with high numerical aperture. Although the use of multilayer recording provides a route to higher capacity in optical disks, moving to a significantly higher density for an individual layer would require that the wavelength of the laser diode be decreased further, a challenging task for which no ideal solution has yet been found. An alternative way to increase density is to exploit near-field optics and evanescent energy to achieve a spot size smaller than that attainable with conventional optics [18]. Development of such a system requires advances in the understanding and modeling of surface plasmon effects and also the development of high-efficiency probes using MEMS technology and nanofabrication.

In addition to increasing density, another important direction for the optical storage industry is miniaturization. In this era of digital convergence, massive amounts of information and multimedia content are being handled in portable devices. The decreasing cost of data storage and the increasing capacity of ever smaller storage devices are among the key enablers of this revolution [19]. However, optical storage devices do not yet play a leading role in small form factor portable devices, even though the usual advantages of optical storage – low cost per GByte and removability – are appealing for portable devices also [20]. MEMS technology may be part of the solution to create miniature optical disk drives suitable for these space-critical applications.

6.3.1
MEMS in Near-field Optical Recording Devices

The first demonstration of near-field optical recording used a tapered optical fiber as a probe to create marks smaller than the conventional diffraction limit on magneto-optical media [21]. With this approach, which is based on near-field scanning optical microscopy (NSOM), an optical beam of sub-wavelength diameter is emitted from a tiny metal aperture in the end of the probe, which is scanned in close proximity to the media to write and image domains with dimensions as small as 60 nm in diameter. This optical fiber approach, however, has shortcomings, which include low optical efficiency and low data transfer rate. Thus, fabrication of a high output power light source has been a critical issue in the progress of NSOM-based optical recording. Other promising approaches include solid immersion lens (SIL) schemes [22] and super-resolution near-field structures (super-RENS) [23].

The principle of the SIL is that by focusing light inside a high index of refraction material, where the speed of light is slower, the spot size can be reduced below the minimum achievable spot size in air. As shown in Fig. 6.9 b, an incident collimated laser beam is focused using an objective lens toward the base of a hemispherical SIL. In addition to reducing the wavelength inside the SIL, the spot size is further reduced by the convergence of the refracted rays from higher angles, which increases θ_{max}. The large angle rays, however, may be at angles greater than the critical angle at the base of the SIL. Hence both the reduced wavelength and the large angle rays exist only within the high-index SIL and must be coupled via their evanescent field

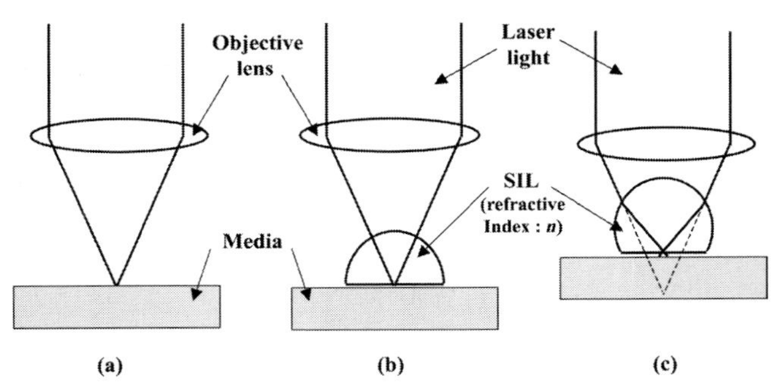

Fig. 6.9 SIL optical systems: (a) conventional optical system; (b) near-field system using a hemispherical SIL; (c) system using a super-hemispherical SIL

to the recording medium located at the base of the SIL. The small spot of light can reach across an air gap to the recording medium as long as the gap width is less than the evanescent decay length. In the case of a *super*hemispherical SIL, a focused spot is obtained at the base of the SIL when the incident rays converge at a point located at a distance nr below the center of the sphere. A preferable gap height between the near-field optics and the recording medium was reported to be no more than $\lambda/10$ in order to achieve sufficient evanescent coupling. Practical implementations of the SIL approach involve mounting the SIL within an HDD-type air-bearing slider, which flies several nanometers above the surface of the medium. One of the challenges for this approach is to control the fluctuations of the head-disk spacing tightly enough to fall within the tolerances of the SIL system.

In probe-type optics, the achievable resolution is basically determined by the size of the aperture fabricated at the end of the probe, which can be as small as 50 nm. However, tapered metal-aperture probes have an intrinsically low coupling efficiency because the optical power decreases drastically as the aperture size decreases and the optical loss in the optical fiber itself is severe in the region where the core diameter is smaller than the wavelength. Without a breakthrough in optical throughput, this limitation will continue to hinder real applications of near-field metal aperture probes in optical storage devices. Simply increasing the input power is problematic, however, because excessive power can damage the probe tip.

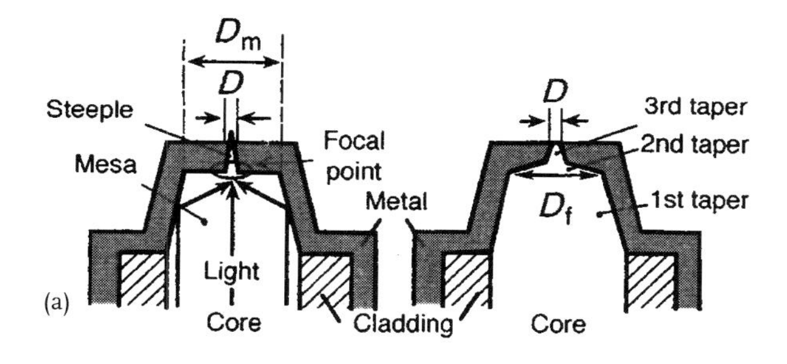

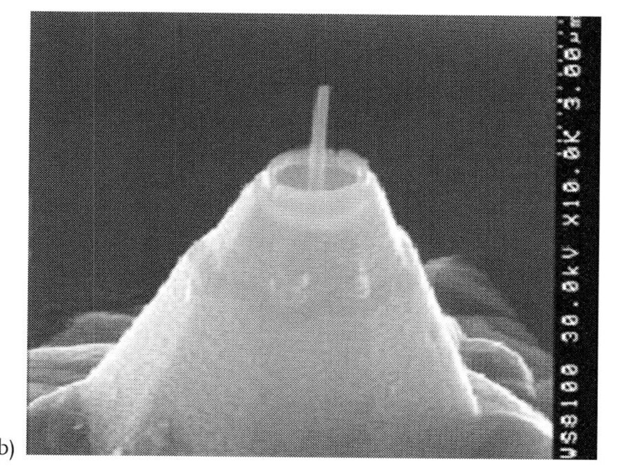

(b)

Fig. 6.10 Optical probe designs which improve the optical efficiency. (a) Steeple-on-mesa probe and triple-tapered aperture probe. Reprinted with permission from [25]. (b) coaxial probe structure. Reprinted with permission from [26]

Part of the solution to this problem may be the use of a localized plasmon resonance to enhance transmission through the small aperture at the probe tip [24]. Another option may be the use of multitaper probes as shown in Fig. 6.10a, which can be fabricated via a multistep etching process [25]. Other solutions using MEMS processing techniques have been applied to increase the optical efficiency of metal aperture probes, including the coaxial probe structure [26] shown in Fig. 6.10b. This type of structure, which generates a strong field at its apex, is potentially suitable for integration with other MEMS structures.

Recording devices based on probe-type optics eliminate the need for a flying air-bearing head. Instead, multiple probes can be used to channel light to localized spots on the medium to create a MEMS-centric (see Section 6.4) near-field optical probe storage device. Fig. 6.11a shows an example of an optical probe design [27], where all of the necessary optical elements, including the light source, waveguide, nanometer-scale aperture and photodetector, have been integrated into an Si microcantilever. The light emitted from the source reacts with the recording medium, where it is scattered and subsequently detected by a photodiode integrated with the Si cantilever. Arrays of optical probes can be integrated in such a manner that each probe has its own electrostatic Z actuator, as shown in Fig. 6.11b. Although the probes shown in Fig. 6.11b are magnetic rather than optical probes, the same type of concept could be applied to arrays of optical probes [28].

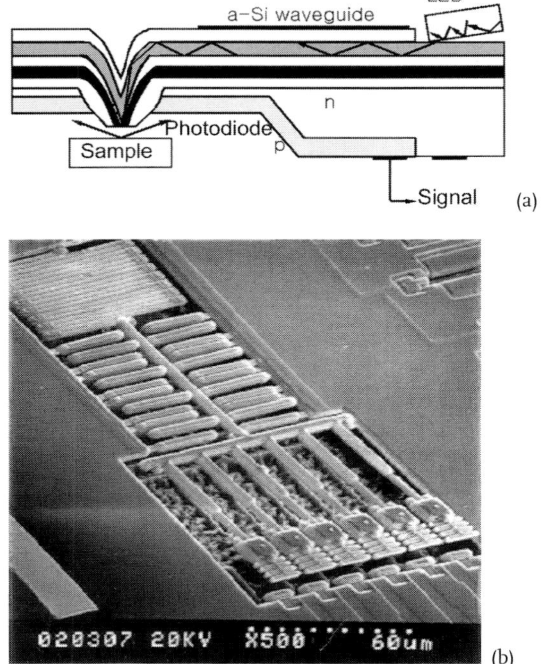

Fig. 6.11 (a) Highly integrated cantilever optical probe structure. Reprinted with permission from [27]. © 2000 The Physical Society of Japan. (b) An array of individually actuated MEMS cantilever probes. Reprinted from [28]

6.3.2
MEMS in Small Form Factor Optical Storage Devices

Although it is still premature to discuss a standard physical format for small form factor (SFF) optical storage devices, several companies are engaged in significant efforts to demonstrate suitable technology. MEMS technology may play a key role in miniaturizing the necessary optical and mechanical components to fit into a suitably small form factor.

There are two types of tracking mechanisms in SFF optical storage devices: linear sled-type and rotary swing-arm-type mechanisms. Although most conventional optical drives have used linear sled-type actuators, rotary swing-arm actuators (similar to those used in HDDs) are gaining favor for SFF drives. Placing several key optical elements on a flying slider is another design cue from HDDs that can help reduce the complexity and size of the overall device. Fig. 6.12 illustrates a swing-arm concept in which MEMS-based micro-optical elements are used for magneto-optical recording [29]. The design and fabrication of this type of SFF MEMS optical pickup are described below.

Fig. 6.13 shows a schematic drawing of a micro-optical pickup head composed of MEMS-based micro-optical elements and SEM images of key components. An integrated planar microcoil produces a magnetic field either upwards or downwards at the optical focal point on the recording layer and changes the direction of the magnetization of the heated material during the writing process. An integrated hemispherical lens surrounded by the microcoil serves to increase the numerical aperture of the pickup optics in conjunction with a macroscopic objective lens. On the bottom surface of the pickup, patterned recesses and plateaus form the topology of a hydrodynamic air bearing, which provides stable aerodynamic levitation over the spinning optical disk at a submicrometer height.

The microlens can be manufactured by a batch process at the wafer level via reactive ion etching (RIE) of the fused-silica substrate followed by thermal reflow of the masking photoresist. The SEM images show the reflowed photoresist and the corresponding spherical shape transferred to the substrate by using a fluorine-based chemistry in an inductively coupled plasma (ICP) etcher. Good uniformity and reproducibility of the spherical shape of the lens require tight control over the photolithography process from run to run. The etched surface of the glass should be sufficiently smooth to avoid degradation of the optical properties of the microlens. A commonly acceptable criterion is that the r.m.s. roughness of the lens should be less than one-twentieth of the wavelength for which the lens will be used.

The air-bearing surface (ABS) is similar to those used in HDDs; however, since the optical components that it carries have more mass than an HDD head, it is designed for greater stiffness and load force than an HDD ABS. The optical focal point is located in the rear pad, which flies closest to the disk surface. The partial hemispherical lens is located on the opposite side of the slider, focusing light through the slider body to the rear ABS foot.

An integrated focusing actuator is used to adjust actively the position of the objective lens to maintain proper focus of the optical beam through the microlens

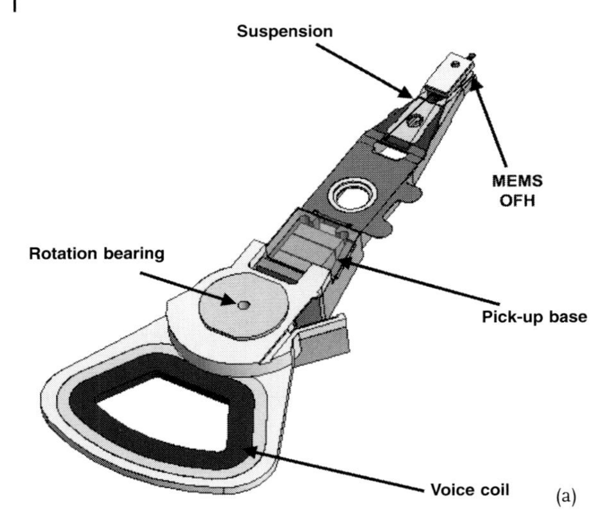

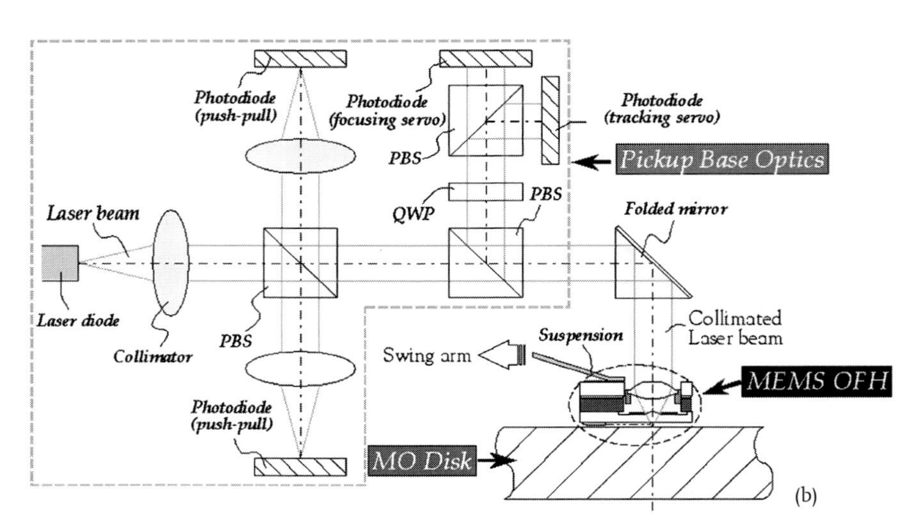

Fig. 6.12 (a) Swing-arm type micro-optical pickup head using MEMS-based micro-optical components integrated into an optical flying head (OFH) in combination with additional components in the pickup base. (b) Functional block diagram of the components used in the swing-arm pickup system

and across the air gap to the recording layer of the disk. The focusing actuator shown in Fig. 6.13 is an electrostatic comb-type vertical actuator [20].

One of the challenges in developing an SFF optical drive using an integrated flying optical head is the complexity of fabricating the integrated head. Another challenge is to make the head-disk interface sufficiently robust against contamination for removable media. Use of a flying optical head counteracts one of the advantages of conventional optical disk systems over removable magnetic disk systems – the absence of a head flying within nanometers of the surface of the medium, which may be damaged by the particles and contamination present in remov-

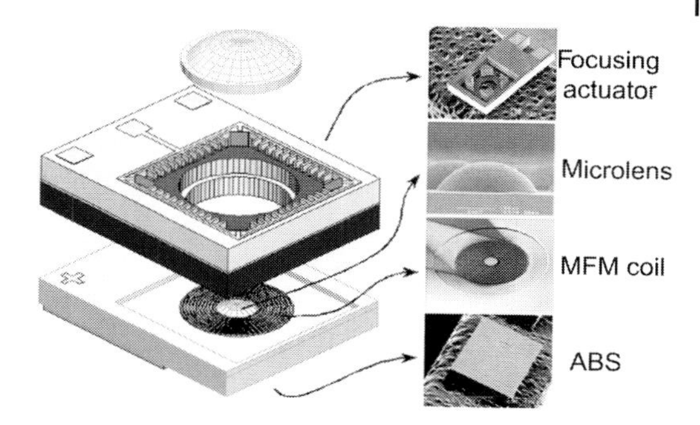

Fig. 6.13 Micro-optical pickup head

able media devices. Since optical heads can fly at greater heights than magnetic heads, adequately robust solutions may be feasible.

6.4 New MEMS-centric Storage Devices

MEMS technology offers promise not only in critical subcomponents of magnetic and optical drives, but also as the basis for an entirely new class of storage devices, in which MEMS technology plays a central role in the overall device architecture and fabrication process. These *MEMS-centric* devices will still be based on mechanical access, but MEMS actuators and mechanical structures will entirely replace the macroscopic motors, actuators and mechanical structures of conventional devices. MEMS-centric devices, by virtue of the unique capabilities of MEMS structures, are thought to have the potential to offer faster mechanical access, low power consumption and good shock robustness – all in very small form factors suitable for the most compact host devices such as mobile phones, handheld computing devices, digital video and still cameras and digital music players.

In addition to replacing conventional mechanical components with their MEMS counterparts, MEMS-centric devices are envisioned to exploit newly developed ultra-high density data recording technologies. Some of these new approaches are enhancements of conventional methods such as magnetic recording or optical phase-change recording, whereas others are new recording schemes derived from scanning probe (SP) technology. From the very earliest days of SP techniques, there have been successful attempts not only to use SPs to image surfaces, but also to modify them for the purpose of data storage [30, 31]. The unprecedented spatial resolution of SP techniques in their native imaging mode also applies when SPs are used to modify surfaces. The ultimate example is the manipulation of individual atoms on well-ordered surfaces [32]. This feat demonstrates not only the ultimate capability for dense data storage, but also reveals a primary shortcoming of many SP-based data recording techniques – a very low data rate. Fortunately, SP techniques are well suited to use batch-fabricated probes, so the prospect of using low-cost integrated arrays of probes to achieve high data rate via parallelism is practical.

6.4.1 Technology Building Blocks and Hardware Architecture of MEMS-centric Storage Devices

MEMS-centric storage devices require a combination of several technological building blocks: a reliable and high-density basic data recording phenomenon, suitable probes to perform data recording and retrieval, integrated MEMS structures which include a probe array and media translation system (scanner), and a suitable control system that includes probe drivers, data channels, media navigation system and host interface.

The basic hardware architectures of mechanical-access storage devices can be classified into two groups: (1) *long-range* and (2) *short-range translation* architectures. Conventional disk and tape drives fall into the long-range translation category. In this architecture, a small read/write transducer or a small array of transducers mechanically accesses a medium surface area far larger than the transducer itself. For example, a tape cartridge may typically have more than 5 m^2 of tape surface area, compared with a head module with millimeter-scale dimensions overall, which includes several micrometer-scale read/write elements. Although the ratio of medium area to head size is smaller, a similar situation exists for disk-based devices. These long-range translation devices typically have a huge data capacity (tens to hundreds of gigabytes) and a low cost per gigabyte owing to a fixed drive cost amortized over a large area of storage medium. Although it is certainly possible to conceive of MEMS-centric devices with long-range translation, most current efforts focus on short-range translation because they are mechanically much simpler. To create, for example, a disk-based MEMS-centric device, there is a need for rotational bearings with tight tolerance, low friction and long life. Good rotational bearings that can be successfully fabricated by MEMS techniques do not yet exist. One futuristic vision of a MEMS-centric storage system with long-range translation uses multiple independent miniature robots, which roam over a large area of storage medium [33].

Short-translation MEMS-centric devices generally use a read/write transducer array roughly as large as the medium itself. For example, a 64×64 array of probe-type transducers, arranged in a $100 \times 100\ m^2$ grid pattern, can access a $6.4 \times 6.4\ mm^2$ area of storage medium with a total motion of 100 μm in X and Y. This type of short-range scanning requirement is well suited to MEMS actuators. An example of such a device is shown in Fig. 6.14.

A consequence of this short-range translation architecture is that the transducer array is generally large (and therefore relatively expensive) and the total fixed cost of the overall device is amortized over only a small area of medium. This means that for a given areal density, a MEMS-centric device is not competitive on a cost per gigabyte basis with a traditional long-range access device unless its total cost is orders of magnitude lower. Conversely, if the total device costs are similar

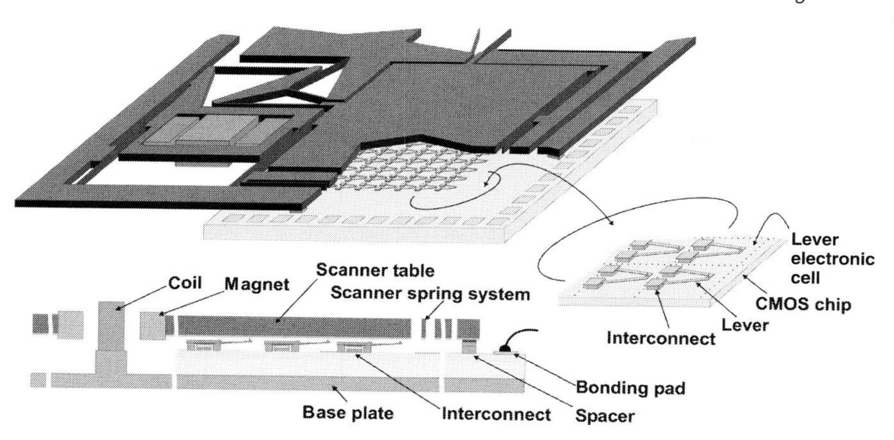

Fig. 6.14 Schematic illustration of a short translation-type MEMS-centric data storage device, showing the layout of the probe array, analog front end electronics chip, microscanner and storage medium. Reprinted with permission from [87]. © 2003 VLDB Endowment

(which is more realistic), a MEMS-centric device cannot compete with a disk or tape device unless its areal density is orders of magnitude higher. Conventional flash EEPROM devices used for relatively low-capacity storage (especially in portable devices) have much higher cost per gigabyte; it is in this market segment that MEMS-centric devices may initially find a competitive advantage.

6.4.2 Basic Data Recording Methods

To create a MEMS storage device of the short-range translation type which could enter the market within a few years with a competitive capacity and cost, an areal density of perhaps 500 Gbit/in^2 or more will probably be necessary. To achieve these densities, SP-based recording methods rely on bit sizes determined by the interaction of a sharp probe tip with the medium. The use of a sharp probe tip sets these techniques apart from existing techniques, such as conventional magnetic recording, where the bit size is determined by lithographically patterned structures in the head, and optical recording, where the optical spot size determines the bit size. Although several SP-based methods show promise to reach the densities needed, none has yet achieved a product-worthy level of long-term reliability in terms of error rate, write/erase cycles and robustness against aging and environmental conditions. One of the challenges is to keep the probe tips sharp and free of contamination, especially for methods that bring the tip into contact with the storage medium. Several of the leading recording methods are discussed below.

6.4.2.1 Magnetic Recording

In the light of the magnitude of the effort required to develop new recording schemes, one option is to adapt a well-proven method such as magnetic recording for use in MEMS-centric devices [28, 34]. Magnetic recording not only supports very high data rates and unlimited write/erase cycles, it also currently achieves the highest areal density ($\sim$100 Gbit/in^2) of any commonly used storage method. Moreover, the magnitude of investment behind it suggests that further advances are likely. However, simple scaling of magnetic grains in magnetic media to accommodate higher densities (which has been successfully applied to increase areal density by more than seven orders of magnitude since the 1950s) is no longer straightforward because of challenges imposed by the superparamagnetic effect [35]. To achieve densities of 500 Gbit/in^2 and beyond, fairly radical changes such as perpendicular recording [36] or patterned media [37, 38] may be needed. Patterned media are often considered too expensive for HDD disks. However, the small area needed in MEMS-centric devices and the rectilinear raster-scan track format they use could allow techniques such as interference lithography [39] (which is better suited to rectilinear than to circular patterns) or soft lithography [40] to be cost-effective.

A variety of probe types can be used for magnetic recording. In one approach, an inductive write element and a read sensor based on the giant magnetoresistive (GMR) effect are integrated into a cantilever probe [28, 41]. Alternatively, a ferromagnetic tip on a cantilever (with or without externally applied field) can be used [34, 41–43]. Another option for writing, although more speculative, is to exploit current-induced magnetic switching with current pulses from tip into certain types of multilayer magnetic media [44].

In the first approach (see Fig. 6.15), a combination of a horizontal-coil monopole write head (incorporating a sharp tip) and a yoke-type GMR read sensor are used in conjunction with a bilayer perpendicular medium (soft magnetic underlayer plus hard magnetic PtCo recording layer). The soft underlayer completes the magnetic circuit, which routes a high flux density through the tip (for localized writing) and a lower flux density (not sufficient to write the medium) across the head-media air gap through the large-area yoke wing. The GMR read sensor straddles a gap in the yoke. Although relatively complex compared with other probe types, this structure can potentially be fabricated by use of conventional deposition and patterning techniques along with several planarization steps using chemical mechanical polishing (CMP). The tip itself must be made of a material with high saturation magnetization, such as iron nitride. The Spindt process (originally developed for field emission cathodes) has been proposed to fabricate the tip [45]. Modeling results [28] suggest that such a system could meet the relevant thermal and SNR requirements to achieve at least 250 Gbit/in^2 and a power consumption of about 0.1 mW for both writing and reading (higher densities will be needed, however). Data rate per probe could in principle approach that of an HDD head (up to 1 Gbit/s), although media translation speeds in MEMS devices are likely to limit this to a lower value.

The second approach (see Fig. 6.16) is based on magnetic force microscopy (MFM), in which a force or force gradient acting between a ferromagnetic tip and

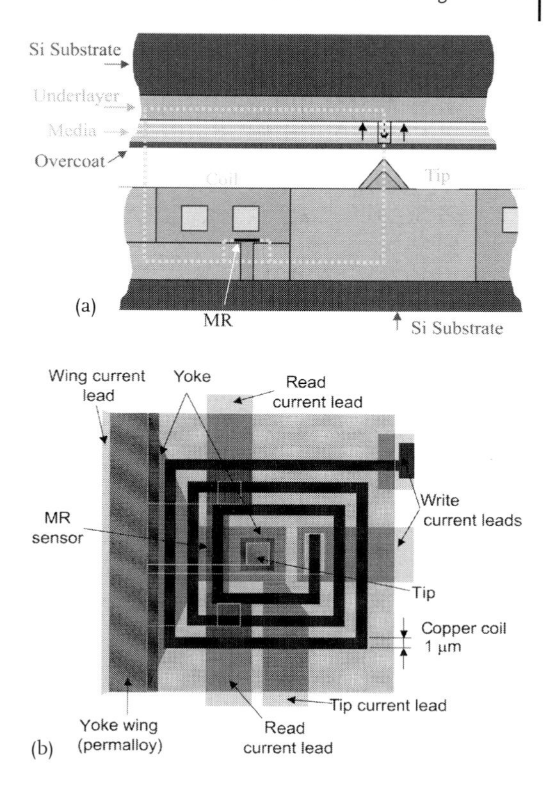

Fig. 6.15 (a) Cross-section and (b) plan view of a probe head for perpendicular magnetic recording, implementing a monopole inductive write and a GMR read sensor. The sharp tip concentrates the flux for localized writing; the flux return path is through the soft magnetic underlayer of the medium, across the air gap and into the large-area yoke wing. During reading, flux originating in the medium passes through the same loop, where it is detected by the GMR sensor that bridges a gap in the yoke. Reprinted with permission from [28]

the magnetic medium is detected via its effect on the deflection or resonant frequency of the flexible microcantilever on which the tip is placed. Writing is accomplished by bringing the magnetic tip close to or into contact with the medium, where the localized field of the tip can reverse the magnetization of the medium [46]. Since data recording involves writing regions of both magnetic polarities, it is necessary either to provide two such tips with opposite polarity or to provide a means such as a switchable external field to reverse the polarity of a single tip. For readback, the tip hovers above the medium at a height such that the field of the tip cannot reverse the magnetization of the medium (but close enough that it still responds to recorded information). Readback sensing requires the ability to detect cantilever deflection at the nanometer scale. Approaches such as capacitance sensing, piezoelectric sensing or piezoresistive sensing would be likely candidates. Simulations [47] predict that such a system can meet SNR requirements at densities of at least 300 Gbit/in^2. Power consumption in the probe itself is significantly less than for inductive writers and GMR sensors. However, a tradeoff is that the mechanical response of the cantilever operating in a noncontact mode would typically limit the data rate to less than 1 Mbit/s.

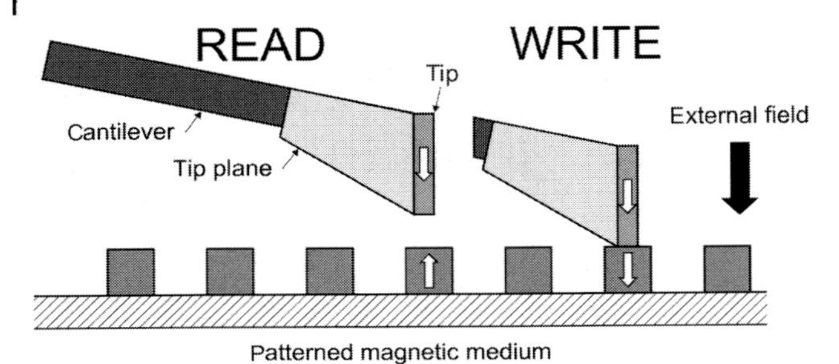

Fig. 6.16 MFM probes with permanent magnetic tips can be used for magnetic recording. In writing, two tips of opposite polarity (only one shown) can flip the magnetization of the medium by bringing the tip into close proximity to or contact with the medium. For readback a single tip is held some distance away from the medium, where the magnetization of the medium causes detectable deflections of the lever without changing the magnetic state of the medium. Image courtesy of L. Abelmann

6.4.2.2 **Thermomechanical Recording**

Although magnetic recording is used widely in conventional disk and tape drives, its usefulness in MEMS storage devices is not yet proven. One recording method which has been shown to achieve very high densities is *thermomechanical* recording [48, 49]. This recording method uses a sharp tip to create nanometer-scale indentations in a thin polymer film. To create a pit, the tip is heated while a momentary indentation force is applied. For a given polymer film, the size of the indentation depends on the magnitude of the force and temperature applied.

A variety of polymer materials, both linear and cross-linked, can be used as thermomechanical media. Examples include polyimide, polysulfone, polystyrene, poly(methyl methacrylate) (PMMA) and SU-8 epoxy-based photoresist. The onset temperature for the formation of indentations has been observed to scale with the bulk glass transition temperature of the material. In addition, the time-temperature superposition principle, which predicts that short-duration, high-temperature pulses are equivalent to long-duration, low-temperature pulses, has been found to hold for thermomechanical writing with pulse lengths of microsecond-scale duration and longer.

Thermomechanical data storage was first demonstrated using an external laser to heat the tip on a single lever in contact with a bulk PMMA medium. Readback was accomplished by using an optical lever to monitor cantilever motion as the tip traversed indentations in contact with the medium [50, 51]. More recently, arrays of cantilevers with integrated resistive heaters and tips have been developed. Readback sensing has been demonstrated using piezoresistive [52], piezoelectric [53, 54] and thermal [48] displacement sensing. Capacitive sensing [55] is also an option.

In the case of levers with piezoelectric readback (see Fig. 6.17), a sputtered ZnO or sol-gel deposited PZT film on a single-crystal Si cantilever directly converts strain due to deflection of the cantilever into electric charge. Piezoelectric sensing offers better power efficiency than either piezoresistive or thermal sensing be-

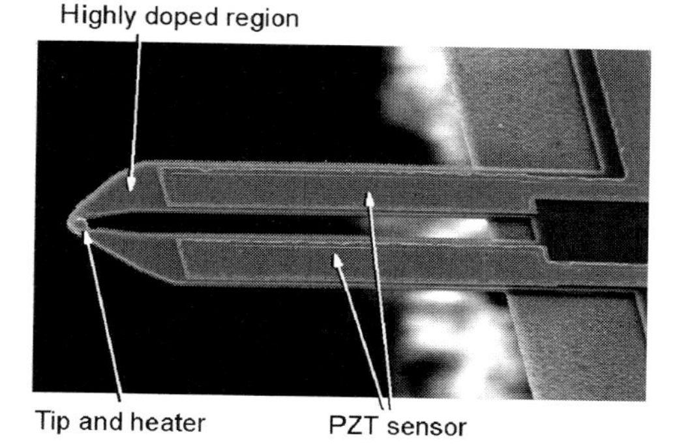

Fig. 6.17 A piezoelectric film of PZT deposited on a cantilever probe can be used for high-speed, low-power read-back in thermomechanical recording. Reprinted with permission from [53]. © 2003 IEEE

cause no steady-state power needs to be applied to the sensor to obtain a response to lever deflection changes. Although speed is limited by the mechanical response of the lever, piezoelectric sensing introduces no additional speed limitations.

In the case of thermal readback, cantilever probe structures are particularly simple because the same type of resistor used to heat the tip can be used for read-back detection. A hot resistor can be used to sense vertical motion of the lever by virtue of the changes in temperature and resistance that occur as the distance (and therefore the thermal conduction through the air gap) between the hot resistor and room-temperature surface of the medium is varied. In a simple U-shaped design, as shown in Fig. 6.18a, a single resistor is heated to a relatively high temperature (e.g. 400 °C) for writing and a lower temperature for reading (e.g. 300 °C). The read temperature is chosen to be insufficiently hot to cause any permanent indentations. For this type of lever, with a spacing of ~0.5 μm between heater and sample, a change in heater resistance on the order of 10^{-5} per nanometer of spacing change occurs. Unlike piezoelectric sensing, thermal sensing is limited not only by the mechanical response of the lever, but also by the thermal time constant of the heater, which is typically around several microseconds and limits the per-tip data rate to around 100 kbit/s.

An improved three-leg lever design for thermomechanical recording is shown in Fig. 6.18b. In this case, separate read and write resistors are used to optimize each for their separate roles. The tip protrudes from the write resistor, while the read resistor is relatively far away. This allows the read resistor to be run hotter (for better sensitivity) without the risk of modifying the medium because the tip remains cool. Furthermore, a large central platform has been added to provide increased capacitive force at low voltage for writing.

Thermomechanical recording is a fully rewritable technique. Data can be either bulk-erased by heating the entire medium to a high temperature to allow indentations to reflow back to a flat surface or it can be erased bit by bit. Bitwise erasing (erasure of short sequences within a larger pre-written area) relies on the observation that if a new indentation is formed too close (i.e. within half a pit diameter) to an existing indentation, the existing indentation is erased. Heat spreading during the formation of the new indentation, coupled with the pressure gradients oc-

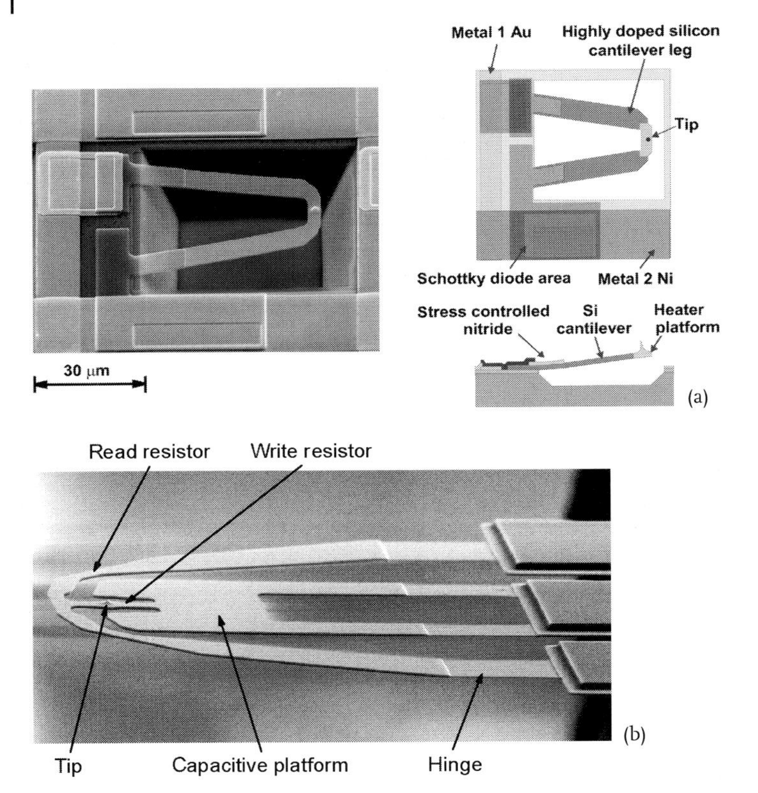

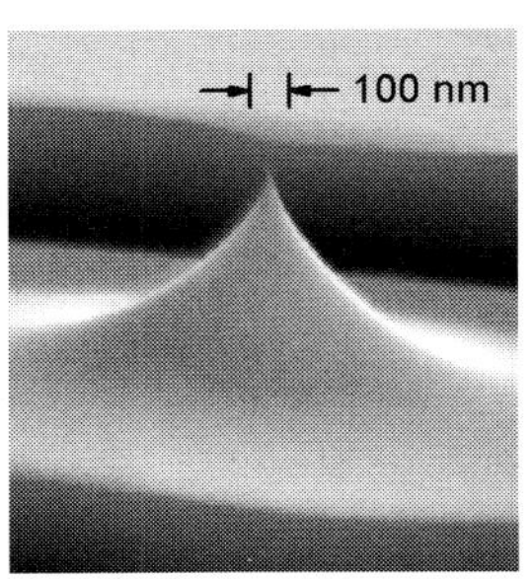

Fig. 6.18 (a) A simple U-shaped, two-terminal cantilever with a single resistor can be used for both writing and reading. Reprinted with permission from [81a]. © 1999 IEEE. (b) A more elaborate three-terminal design incorporates separate read and write resistors as well as a platform for capacitive force. (c) Close-up view of the tip reveals the sub-20 nm radius, which is essential for achieving high density. Reprinted with permission from [56], © 2004 IEEE

curring near the new pit, result in relaxation of the existing pit. Repeating this procedure sequentially down a track of existing bits results in erasure of the track, leaving only one final bit at the end of the sequence. A variation on the bitwise erasure method described above can be used for direct overwriting (writing new data over old without erasing in between). Interrupting an otherwise continuous train of closely spaced write pulses leaves a single pit at each of the interruption points, which are timed properly to write the new data track.

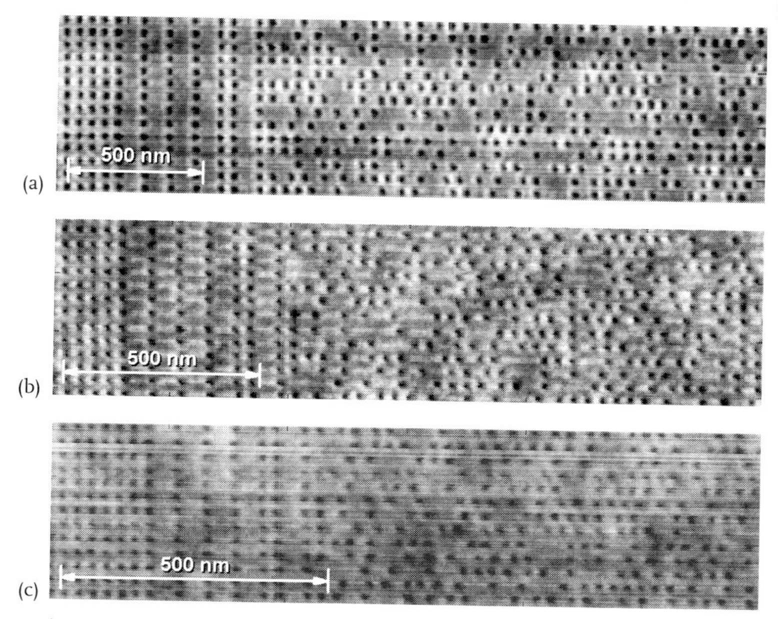

Fig. 6.19 Thermal readback images of recorded fields at different areal densities of (a) 410 Gbit/in^2, (b) 641 Gbit/in^2 and (c) 1.14 Tbit/in^2. In all cases, (1, 7) modulation coding is applied in the on-track direction. Reprinted with permission from [56], © 2004 IEEE

Thermomechanical data recording has been demonstrated to achieve raw bit error rates of 10^{-4} or lower at a density of 641 Gbit/in^2 [56]. In this demonstration, a single lever tester (rather than an array of levers) was used to write data fields (see Fig. 6.19) of at least 400 000 bits with 37 nm minimum indentation spacing along the tracks and 37 nm track pitch. Use of (1, 7) modulation coding, which allows an increase in linear density without the partial erasure that occurs when indentations are placed too closely together, enhances the linear density by a factor of 4/3 (see Section 6.4.5.2). Although thermomechanical and other recording techniques can be used to make marks at densities higher than this, the more rigorous type of demonstration reported [56] is more indicative of the actual data recording capability of a particular recording method at its current state of development.

Thermomechanical recording is not known to have any fundamental density limitations analogous to the superparamagnetic effect encountered in magnetic recording. The ultimate density limit of thermomechanical recording may occur when the indentations approach atomic dimensions. Thus, thermomechanical recording has plenty of potential for future density increases.

6.4.2.3 Phase-change Recording

As early as the 1980s, local modification of phase-change materials with a scanning tunneling microscope (STM) was explored for data storage [57]. Bits are re-

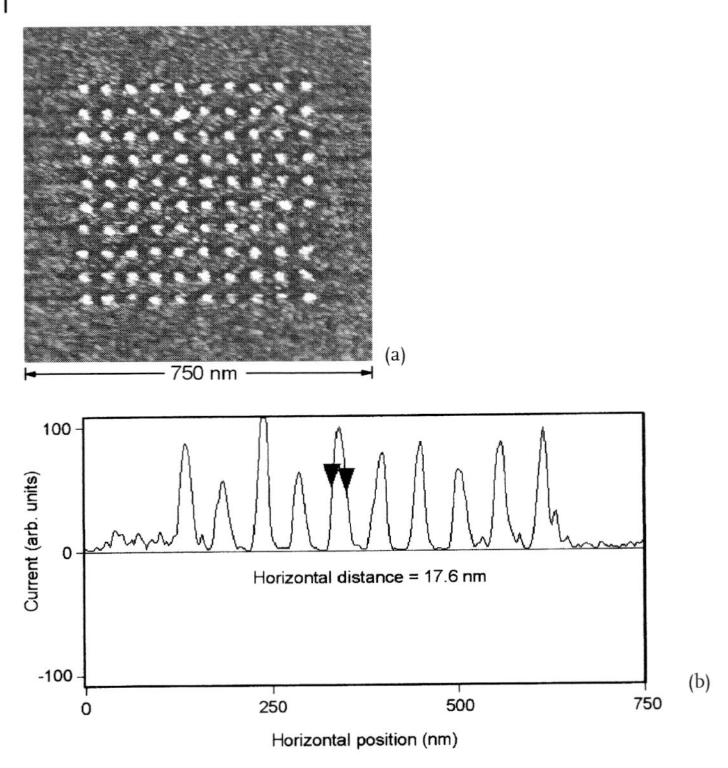

Fig. 6.20 (a) Dots of crystalline material recorded in a 50×50 nm^2 grid on a 10 nm thick film of $Ge_2Sb_2Te_5$ phase change material. (b) Conductivity trace through a row of dots reveals a high contrast and a dot width of approximately 18 nm. Images courtesy of CAE-LETI France

corded as localized regions of amorphous vs. crystalline phase in a manner similar to that used in phase-change optical media. Rather than using a heated tip as in the case of thermomechanical recording, phase-change recording relies on direct heating of the medium via current flowing from the probe into or through the medium. In general, heating above the melting temperature followed by rapid quenching creates a local zone of amorphous material, whereas heating to a lower annealing temperature for a longer duration of time recrystallizes (and therefore erases) such a zone. Readback is accomplished by detecting differences in electrical conductivity between the two phases. One suitable medium is an InSe/GaSe bilayer grown epitaxially on an Si(111) substrate [58]. Other media choices include AgInSbTe and GeSbTe alloys [59] for which resistivity ratios on the order of 10^3 between amorphous and crystalline phases are reported.

Phase-change media may be written and read using an electrically conducting tip in contact with the medium. A capping layer of doped amorphous carbon prevents oxidation of the medium and improves its wear resistance [60]. During writing, a self-focusing of current occurs owing to the high conductivity of crystalline material compared to the amorphous surrounding material. Arrays of dots as small as 10 nm in diameter have been created by this method (see Fig. 6.20). A slight topographic change accompanies the phase change, with the surface of the crystalline material sinking less than 1 nm.

When the tip is used in contact with the medium, tip wear and contamination can compromise resolution, reproducibility, signal-to-noise ratio and lifetime. Use of a noncontact microfabricated field emission probe with integrated extraction and focusing elements has been proposed as a solution [58, 61, 62]. A focused beam from this structure is used to induce phase changes directly. During read-back, a phenomenon referred to as electron beam-induced current (EBIC) is exploited. A potential is applied between a surface layer of the medium and the Si substrate. As the focused beam from the field emission probe passes over different phases of the medium, a larger current flows between the surface and substrate when the beam impinges on crystalline as opposed to amorphous material. A current ratio of 14 is reported between the two phases [58]. Readback bandwidths up to 550 kHz have been reported.

6.4.2.4 Other Recording Schemes

A variety of other recording methods have been proposed and studied experimentally for MEMS storage systems, including near-field optical recording, charge injection in nitride oxide-semiconductor (NOS) media, polarization switching in ferroelectric materials, and state switching in molecular films. Near-field optical recording using probe-type optics was discussed in Section 6.3.1. Charge storage in NOS media was one of the first rewritable methods to show a high level of reproducibility [63]. However, it appears to be difficult to scale to the densities needed in the future. Recording in ferroelectric films such as PZT is also rewritable and has been used to create bits as small as 40 nm [64]; however, retention loss (self-erasure) phenomena continue to raise questions about the usefulness of this method for nonvolatile recording. Recently, it has been proposed that monolayer molecular films of certain types can exhibit a switching behavior at the single molecule scale, offering extremely high density (bits as small as 1 nm in diameter, which would correspond to more than 600 Tbit/in^2). One example using an organic complex has been reported [65], although questions remain concerning the nature of the contrast mechanism [66]. Another material reported to exhibit molecular-scale switching is rotaxane [67], although this appears to be an irreversible (i.e. write once) phenomenon.

6.4.3 Microscanner

To create a functional storage device, mechanical motion is required between the read/write probes and the storage medium. This function is accomplished by a microscanner, which consists of a movable table on which the medium (or, alternatively, the probe array) is located, a spring system with sufficient compliance to accommodate the required amount of motion and an actuation means with sufficient force to move the table the desired distance and with the desired acceleration.

Although the microscanner is a conceptually simple device, its design requirements are stringent, and include the following: (1) independent and orthogonal *X–Y* travel range of 50–200 µm; (2) good repeatability and linearity; (3) a maximum undesired out-of-plane motion of a small fraction of a micrometer; (4) actuation bandwidth sufficient to accommodate a 1–5 ms seek/settle time; (5) small size suitable for flash card form factors; (6) shock and vibration resistance, both operational (typically 5 g) and nonoperational (survive up to 2000 g or more); (7) low power consumption; (8) low driving voltage; and (9) low cost. In some designs, *Z*-axis and tilt correction actuation are additional requirements.

These requirements have led to very different approaches, including electromagnetic scanners [68–70], another with an electrostatic interdigitated comb drive [28], a scanner with a stepper-like electrostatic surface drive [71], an electrostatic clamp and slip design [72, 73], and one with thermal actuation [74]. Notably absent is a piezoelectric design; although piezoelectric actuators are common in SP microscopes, it is difficult to generate the desired stroke in the small space available in a microscanner.

To minimize the driving force and power consumption, the table mass and the *X–Y* spring stiffness should be minimized. On the other hand, high *Z* stiffness is required unless *Z*-axis actuation is also included. To meet these requirements, high aspect ratio (as high as 40:1) cantilever springs are used, which can be formed by deep reactive ion etching (DRIE) of Si [75]. Constructing the entire scanner from single-crystal Si eliminates drifts and deflections due to thermal mismatches; Si springs also have a very long life with minimal aging effects due to material fatigue, provided they are not overstressed. For narrow (~10 µm wide) springs, keeping the scanner table travel range to within less than 10% of the spring length not only prevents overstressing the spring, but also results in a highly linear displacement versus actuation force.

The above assumes that a single actuator is used to create motion of the medium relative to a rigid probe array. In some designs, individual actuation of probes is desired, either to counter the effects of thermal expansion mismatches between the probe array and medium (which would result in tracking errors that vary over the active area of the probe array) or to increase the degrees of freedom – allowing individual probes to follow individually optimized scans. Actuators on individual probes could also be much faster owing to their smaller mass; however, their complexity is increased dramatically compared with single-actuator designs. An example of probes with individual electrostatic *Z* actuators is shown in Fig. 6.11 b.

6.4.3.1 **Electrostatic Interdigitated Comb Drive**

Comb-drive actuators, which have been used in MEMS for years, consist of two partially meshed comb structures, one a fixed stator and the other movable. By applying an electrical potential between the two combs, a force is generated in the direction of increasing the overlap of the combs. Comb actuators have the advantage of easy fabrication – a single mask level and one DRIE step are sufficient to delineate the combs, the table, the springs and the frame. By measuring the capacitance across a pair of combs, the position of the table can be determined for closed-loop position control. Disadvantages of comb drives, however, include the relatively weak force that they generate and the high voltage that they need. To achieve a 50 μm motion with 200 fingers, each 500 μm high and with a 16 μm gap between them, as much as 120 V may have to be applied [28].

6.4.3.2 Electrostatic Surface Drive

This type of scanner, which works much like a stepper motor [71], also consists of two periodic structures with one fixed and the other movable. Unlike a comb drive, however, the structures are planar and face one another. Depending on the spatial phase relationship between the two sets of periodic electrodes and the voltages applied to them, in-plane forces are generated, which move the table to stable positions where the periodic structures are aligned. For continuous motion, each surface has multiple interleaved periodic electrodes, which can be successively activated to create a voltage wave to propel the table continuously in the desired direction. Along with the desired in-plane forces, this type of actuator also exerts large Z-axis forces, which must be taken into account in the design of the flexure springs.

As in the case of other stepping-type actuators, the position of the table is known without an independent position sensor to within a single step unless an external force overcomes the actuation force. A position resolution as small as 5 nm has been reported [76] for an actuator of this type with seven addressable periodic electrodes. Since the intrinsic electrostatic centering force at each step position acts as a restoring force, the overall in-plane stiffness of the scanner when operating can be significantly greater than the stiffness of the flexure springs. Hence the flexure springs can be made relatively soft for better power efficiency, while maintaining a strong holding force to counter external disturbances. Other actuation systems can accomplish the same thing via closed-loop servo, but this requires an independent position sensor and control system.

6.4.3.3 Electrostatic Shuffle Drive

The *shuffle drive* (also known as an *inchworm* actuator) is an electrostatic design based on a clamp and slip motion. It consists of two feet, which can be independently clamped to the substrate via electrostatic force, and a membrane, which can be reversibly collapsed by electrostatic force. The motion cycle is described in Fig. 6.21. Since the moving element is always clamped by at least one foot, it has high in-plane and out-of-plane rigidity for immunity to external disturbances without a need for a stiff flexure and its power consumption penalty. A microactuator based on a similar principle in which the out-of-plane deformation of the membrane has been replaced by in-plane deformable beams has been reported [77].

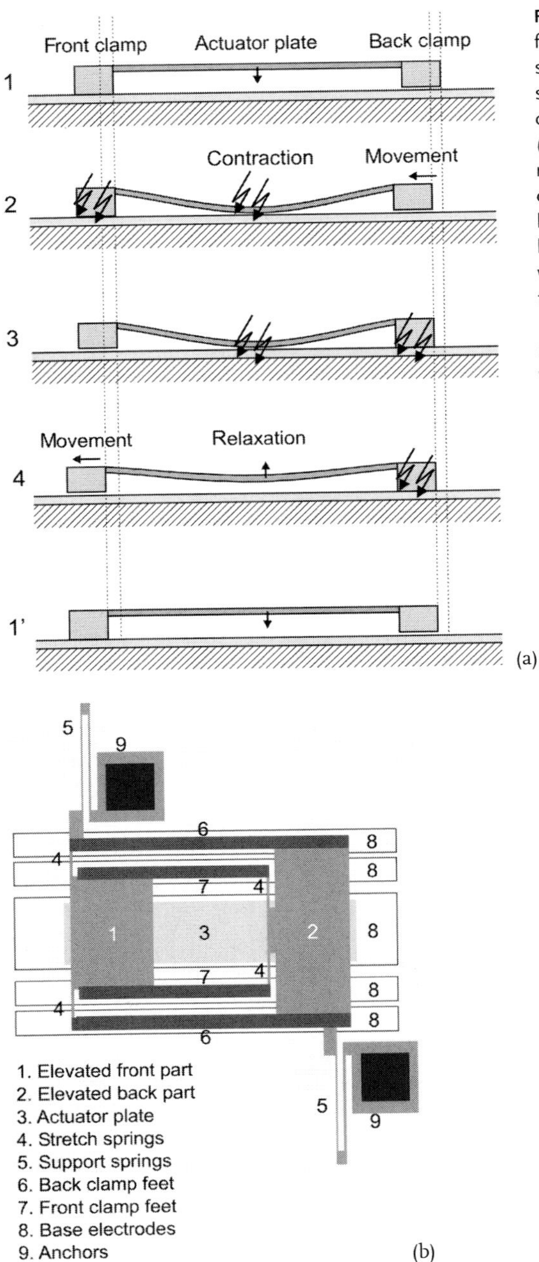

Fig. 6.21 (a) One-dimensional shuffle motor concept based on stick–slip motion. Reprinted with permission from [33]. The motion cycle consists of (1) clamping one foot, (2) deforming the membrane, which makes the second foot slip, (3) clamping the second foot and (4) releasing the voltage on the membrane and unclamping the first foot, which slips forward, ready to start the cycle over from a new position (1′). (b) Layout (top view) of the shuffle motor. Reprinted from [72], with permission from Elsevier Science. © 1998 Elsevier Science

Although for both concepts only one-dimensional devices have been reported, two-dimensional versions are conceivable.

6.4.3.4 Electromagnetic Drive

This means of actuation, which is widely used because of its high force and efficiency in conventional macroscopic actuators, generates a force via the flow of current through a coil of wire in the presence of a magnetic field. Conventional fabrication methods using stampings and wire-wound coils can be used to produce efficient miniature electromagnetic motors and actuators, even for devices as small as 1 mm, such as those used in watches. In MEMS scanners, although planar integrated coils and magnets created by depositing magnetic films would appear to be attractive from a fabrication standpoint, they are difficult to design and fabricate with a sufficient number of coil turns and a sufficient mass of magnetic material to generate large forces efficiently. As a result, hybrid solutions are often preferred, in which MEMS techniques are used to fabricate the passive mechanical components such as the frame, flexures and table and conventional discrete magnets and wire-wound coils serve as the active elements.

The challenge in creating planar coils stems from the fact that in order to be efficient, the coils generally must fill the available volume with multiple layers of turns. Each layer may require a planarization step, lithography, thick-film electroplating, an insulation layer to avoid unwanted shorts between layers and a means of forming a vertical electrical interconnection to the next layer. Scanners using integrated planar coils (along with discrete conventional magnets) have been implemented for scanners with *Z*-axis and tilt actuation in addition to *X*–*Y* motion [68–70]. In these designs, the complexity of the coil arrangement justifies the use of integrated coils. Since integrated coils are lighter than permanent magnets, these scanners rely on moving coils rather than moving magnets. Copper flexures, which are fabricated in a single step with the coils, serve both as flexures and as electrical connections to the moving coils.

An advanced hybrid design is shown in Fig. 6.22. The passive structures are defined by DRIE etching of single-crystal Si. The fixed coils are located between pairs of opposing magnets attached to moving shuttles. An unusual aspect of this design is that it utilizes pivoting reversing levers between the shuttle and table for both axes of motion. By balancing the masses on each side of the reversing lever, a 'mass balanced' design is achieved, in which external shock and vibration have minimal effect on the motion of the table. The forces on the table and shuttle due to external acceleration cancel one another, leaving the scanner largely immune to external influences. Although mass balancing is commonly used in rotary actuators, it is less often applied to linear actuators owing to the complexity of the mechanism needed. In a MEMS scanner, this complex mechanism can be fabricated with high precision and low cost.

When all of these scanner options are compared, no single approach stands out as optimal. Although several designs have been successfully built and tested, full evaluation of their performance requires integration into complete working

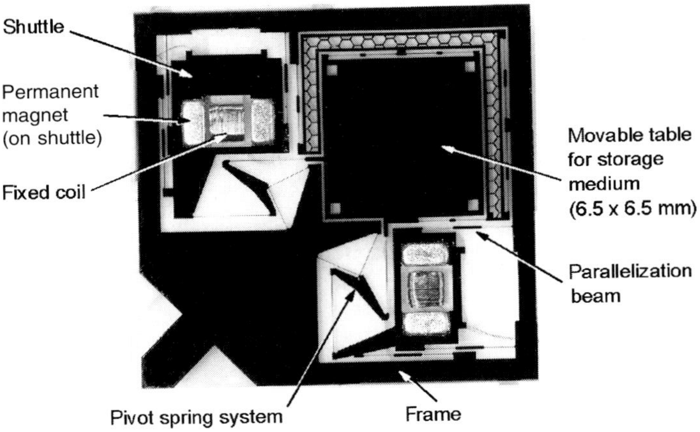

Fig. 6.22 Two-dimensional mass-balanced electromagnetic scanner. Displacement of the actuation shuttle is transferred to the scanner table via a reversing pivot. Since the shuttle and the media table have equal mass and are constrained to move in opposite directions, the scanner is largely immune to external disturbances

MEMS-centric storage devices, a step that will not be completed until all of the other building blocks of the system are ready.

6.4.4
MEMS Integration

Creation of a functional MEMS-centric storage device as shown in Fig. 6.14 involves several integration steps, which include: (1) integration of hundreds or thousands of probes into a probe array, (2) integration of the probe array with its driving electronics, (3) integration of the storage medium with the microscanner, and (4) final integration of the scanner/media subassembly with the probe array/electronics subassembly.

6.4.4.1 Two-dimensional Probe Array

The first integration step is the fabrication of a two-dimensional array of probes. Although probe arrays have been fabricated not only for storage applications but also for parallel imaging and lithography applications [78–81], only in a few cases has successful operation of arrays been reported [54, 69, 82]. The array shown in Fig. 6.23 is a 32×32 array of U-shaped probes for thermomechanical recording. The 1024 probes are placed on a $92 \times 92\ \mu m^2$ array, yielding a $3 \times 3\ mm^2$ overall array size. This array includes passive crosspoint addressing of the levers via series isolation diodes with row and column lines. The cantilevers in this array are formed by surface micromachining and have been released from the Si substrate by a wet anisotropic underetching technique.

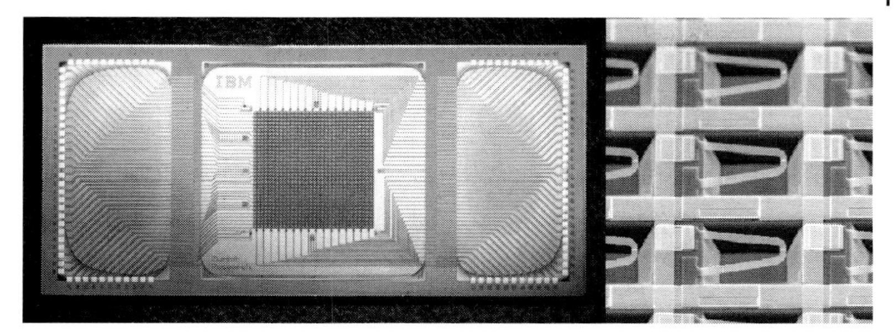

Fig. 6.23 Optical and SEM views of a 32×32 cantilever array chip. Reprinted with permission from [81a]. © 1999 IEEE

Homogeneity of probes over the entire area of the array is essential. For the array shown in Fig. 6.23, a tip height tolerance of 80 nm, a heater resistance standard deviation of 1% and tip apex radii generally below 20 nm have been reported [82]. Since arrays of this type do not have actuators for individual levers, it is also necessary that cantilever flatness (or bending) be controlled for a well-defined Z height of the probes relative to the medium across the entire array. In a thermomechanical recording test with this type of array, more than 80% of the 1024 probes were found to be functional [69].

6.4.4.2 **Cantilever Array/CMOS Integration**

For the passive array shown in Fig. 6.23, time multiplexing can be used to operate the array in a sequential column-by-column fashion. Since the maximum number of probes that can operate simultaneously is limited to one full column, the maximum data rate of the complete device may be limited to an unsatisfactorily low level. To operate more probes in parallel, it is necessary to integrate addressable driving electronics with individual probes or at least with small groups of probes. However, methods of merging of MEMS devices, each having its own unique fabrication process, with conventional CMOS electronics are not obvious.

Among the solutions that have been investigated are post-CMOS MEMS processing [83] and through-wafer interconnects to allow a CMOS chip to connect to the backside of the array chip [79, 80, 84]. Both solutions, however, have shortcomings: the limitations of post-CMOS processing excessively constrain the probe design and through-wafer vias have been found to be unusually challenging and complex to fabricate.

A promising new solution is the *device transfer method* [85], in which MEMS structures originally fabricated on one wafer are physically transferred en masse to the surface of a CMOS wafer, where they are electrically and mechanically bonded in place. This transfer process involves bonding the device side of the MEMS wafer to a temporary glass carrier (wafer) with polyimide, followed by complete removal of the original MEMS substrate via grinding and etching. Bonding pads and polyimide socket-type alignment features are then fabricated on the

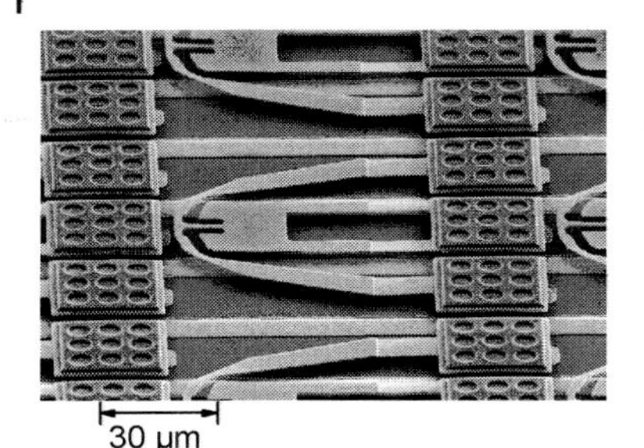

Fig. 6.24 SEM view of a section of a cantilever array transferred to a CMOS-like wiring chip using the device transfer method. Reprinted with permission from [85]. © 2003 IEEE

exposed bottom side of the MEMS structures. The bonding pads are bonded to matching solder-capped studs on the CMOS wafer using heat and pressure. Finally, the glass carrier is released by laser ablation of the polyimide through the transparent glass, leaving the MEMS structures mechanically and electrically bonded in place on the CMOS wafer. This process has been demonstrated for wafer-scale transfer of multiple arrays of 4096 freestanding cantilevers (with a cantilever density of 100 levers/mm^2 and an interconnect density of 300 interconnects/mm^2) to a CMOS-like wafer. Mechanical parameters have been shown to be fairly uniform (see Fig. 6.24), with the slope of the cantilevers controlled to within 0.1°. The electrical resistance of the interconnects is well below 1 Ω and their yield was found to be >99.9%. The cantilevers, which are 70 µm long and 300 nm thick, survive the process unscathed, with even their 20 nm radius tips (which are well protected throughout the process) fully intact.

6.4.4.3 **Medium/Scanner Integration**

Although there is at least one case of a design in which the lever array is placed on the scanner [74], it is generally easier to put the storage medium on the scanner to avoid the need for multiple electrical interconnections to the moving scanner platform. Locating the medium on the scanner, however, also has its challenges. Three approaches can be envisioned: (1) depositing the medium on the substrate before the scanner itself is fabricated, (2) depositing the medium after the scanner has been fabricated, and (3) attaching a 'storage medium chip' to the finished scanner. For a thermomechanical polymer medium, which is most easily deposited by spin coating, the first approach has been chosen [48] because spin coating requires a surface free of significant topography and the finished scanner is much too fragile to survive the spinning process. However, this option requires that the medium be protected from damage through the subsequent scanner fabrication process. Depositing the medium after completion of the scanner would be an option for vacuum-deposited material, such as magnetic or phase-change media, as long as patterning of the medium is not required. For polymer media, spray coating could be an option, as well as physical or chemical vacuum deposition. Mounting an additional chip with the storage medium on the finished scanner presents challenges in maintaining height and tilt tolerances for the plane of the storage medium relative to the scanner frame; systems which have *Z* height and tilt motion available could accommodate this strategy.

6.4.4.4 Final MEMS Integration

Once the two major subassemblies (lever array + CMOS, scanner + medium) have been completed, the final mechanical integration step is to combine them into a single unit. Since this step involves bringing two large mating chips of $\sim 1\ \text{cm}^2$ area together with a gap of no more than a few micrometers between them, excellent flatness and height tolerance control are needed. Cleanliness, particularly with respect to particles, is also essential and, because both subassemblies contain fragile micromechanical elements, cleaning options are limited. For designs which include *Z* height and tilt actuation in the scanner or individual *Z* actuation on the probes [28], some relaxation of assembly tolerances is possible.

6.4.5 Storage Device Control System and Architecture

MEMS-centric storage devices are hybrid approaches that employ technologies used in both semiconductor memories and conventional mechanical access storage devices such as HDDs. As MEMS devices, they have adopted low-cost batch fabrication processes from the semiconductor world. As mechanical access devices, they have adopted the concept of using mechanical positioning to address data on a continuous thin film of storage medium. Given the capabilities and limitations of MEMS design and the specific requirements of SP-based recording techniques, probe-based devices generally abandon the rotating-disk and single head-per-surface paradigm of HDDs in favor of MEMS-based *X*/*Y* actuators which position large arrays of probe tips over the storage medium. The unique mechanical architecture of these devices requires innovation in the architecture of their control systems.

Irrespective of the basic recording method employed and of whether MEMS-centric storage devices are based on a long-range or a short-range translation architecture, they resemble conventional HDDs or optical drives from the system's functional viewpoint. Fig. 6.25 shows a simplified block diagram flowchart of a MEMS-centric storage device. Such a device consists of a suitable transducer array-chip to perform the parallel data recording and retrieval, a suitable medium to record information, the microscanner serving as the medium-translation unit and the integrated control system, which includes write and scanner drivers, data detection and servo subsystems, the microcontroller, the error correction coding (ECC) unit and the host interface. The capabilities of this control system are as important for achieving the density and performance of the device as the MEMS components of the device.

Each probe tip can write data to and read data from a dedicated area of the medium in a rectangular region whose dimensions are equal to the maximum *X*/*Y* actuation distances. This rectangular region is called a storage field. There are as

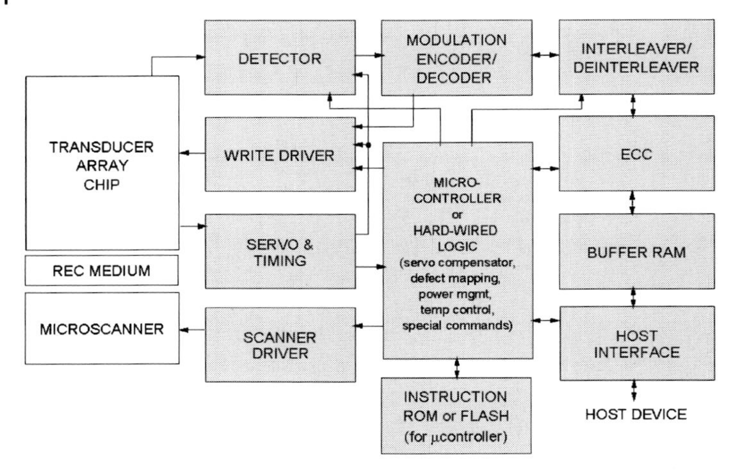

Fig. 6.25 Architecture of MEMS-centric storage system

many storage fields as there are probe tips on the 2D transducer array. In each storage field, the presence (absence) of a mark, for example in the thermomechanical recording approach, or the magnetic polarity of a mark, such as in the magnetic recording approach, corresponds to a logical '1' ('0'). All marks are nominally of equal size. The marks may be placed only at fixed periodic symbol positions along a data track. We refer to the distance between symbol positions as the symbol pitch. The cross-track distance between bit positions, referred to as the track pitch (TP), is also fixed. Unlike most conventional HDDs, multiple probe tips can access the medium in parallel. Efficient parallel operations of large 2D probe arrays can be achieved by a row/column time-multiplexed addressing scheme similar to that implemented in DRAMs. As discussed in Section 6.4.4.2, a row/column multiplexing scheme can be used to allow full parallel write/read operations within one column, or, if address-decoding circuitry is distributed within the array itself, any arbitrary subset of the probes may be operated in parallel. Clearly, the latter solution yields higher data rates, whereas the former leads to a lower implementation complexity of the electronics. In view of power consumption it is unlikely that all probe tips could be active simultaneously; accordingly we currently expect that only a subset will be active at any given time.

6.4.5.1 **Servo for MEMS-centric Storage Devices**

High areal-density MEMS-centric storage devices with very narrow tracks require a closed-loop servo system to write uniform tracks and to read them with a high degree of accuracy to ensure low error rates. In general, the servo system in such a device has two functions. First, it locates the track in which information is to be written or from which information is to be read, starting from an arbitrary initial scanner position. This is achieved by the so-called seek and settle procedure. During seek, the scanner is moved rapidly with the help of positioning sensors so that the read/write probes are at a position close to the beginning of the target track. A smaller further move in the cross-track direction from that position to the center of the target track is performed in the settle mode. As the actuation distances during the seek and settle modes are very small, i.e. on the order of 100 μm, the average data-access time is also expected to be short.

In the case of an electrostatic comb-drive scanner, the scanner itself provides both *X*/*Y* positioning and the capability to sense the scanner's exact position (see Section 6.4.3.1). This may allow closed-loop *X*/*Y* position control with better than 10 nm precision, even in the presence of external shock and vibration [86]. Assuming microscanner displacements of 100 μm in the *X* and *Y* directions, it has been estimated that the maximum seek time starting from the center point of a storage field is ~0.5 ms [86]. The thermomechanically-based approach [48, 49] utilizes two thermal position sensors during the seek and settle modes. These thermal sensors exhibit a closed-loop accuracy of 3–5 nm and experimental results indicate a data access time on the order 4 ms or less [87]. Capacitive or other types of position sensors could also be used.

The second function of the servo system is to maintain the position of the read/write probe in the center of the target track during normal read/write operation. This is achieved by the so-called track-follow procedure. Track following controls the fine positioning of the read/write probe in the cross-track direction and is critical for reliable storage and retrieval of user data. It is typically performed in a feedback loop driven by a medium-derived position-error signal that indicates the deviation of the current position from the track-center line. There are two types of track-follow servo architectures in practical use. In the first type, which is referred to as *embedded servo*, segments of position information are interspersed within a data track. In the second, which is referred to as *dedicated servo*, certain probe tips and corresponding storage fields are dedicated solely to providing position information for the servo system. Embedded servo has been demonstrated in a disk-based (non-MEMS) thermomechanical recording prototype [51]. In this demonstration, special pre-written servo marks at specified locations around the disk were utilized to provide a signal regarding the position of the probe tip relative to the center line of the track. The servo marks covered 15% of the track length. Similarly, the proposed MEMS-centric data-storage system [86] appears to rely on an embedded servo. In this case the servo information is expected to require about 10% of the device capacity.

Alternatively, a MEMS-centric thermomechanical storage device [48, 49] employs dedicated servo to achieve both timing synchronization and servo control by reserving a small number of storage fields exclusively for timing recovery and servo-control purposes, as illustrated in Fig. 6.26. The approach is based on the concept of vertically displaced bursts, arranged in such a way as to produce two signals that guarantee a position error signal (PES) that is uniquely decodable, i.e. each PES value should be mapped to a unique cross-track position. The servo marks for the in-phase signal are labeled *A* and *B* bursts, whereas those for the quadrature signal are labeled *C* and *D* bursts. Each of the four types of bursts is pre-writ-

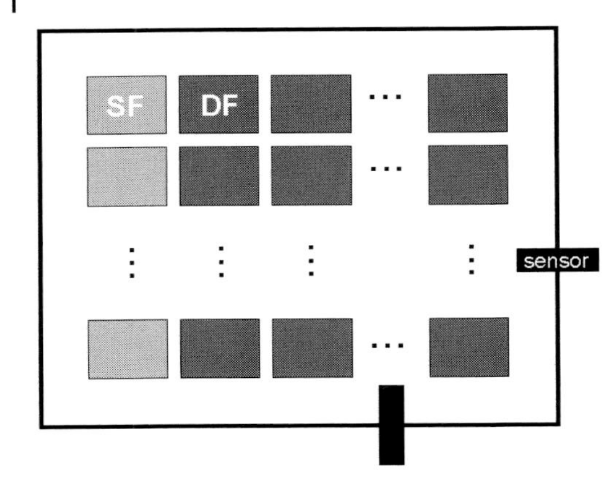

Fig. 6.26 Layout of data and servo and timing fields. The illustration shows an *X/Y* top view of the medium on the scanner. Each rectangle outlines the area accessible by one probe tip. Light-gray boxes (SF) indicate servo/timing fields; dark-gray boxes (DF) indicate data fields. The black boxes (sensor) indicate the location of the thermal or capacitive (position) sensors above the medium. Reprinted with permission from [87]. © 2003 VLDB Endowment

ten in a separate servo field. These four servo fields are identical, except for the position of the servo marks in the cross-track direction. Fig. 6.27 illustrates parts of the four servo fields according to this configuration. The *A*, *B*, *C* and *D* servo fields are placed in the 2D array in such a way that they are always accessed in parallel irrespective of the addressing scheme.

The process of writing servo information on the medium is referred to as *servo writing* and is analogous to servo writing procedures in HDD manufacturing. Regardless of whether sector or dedicated servo is used, servo writing is performed once during the device manufacturing process and remains fixed for the life of the device.

Similarly to obtaining servo information via dedicated servo fields, one can use the same strategy to obtain timing information by adopting dedicated clock fields. The basic concept is to have continuous access to a pilot signal for synchronization purposes. Servo and timing functions can also be combined in the same dedicated fields. For example, the periodic single-frequency fields shown in

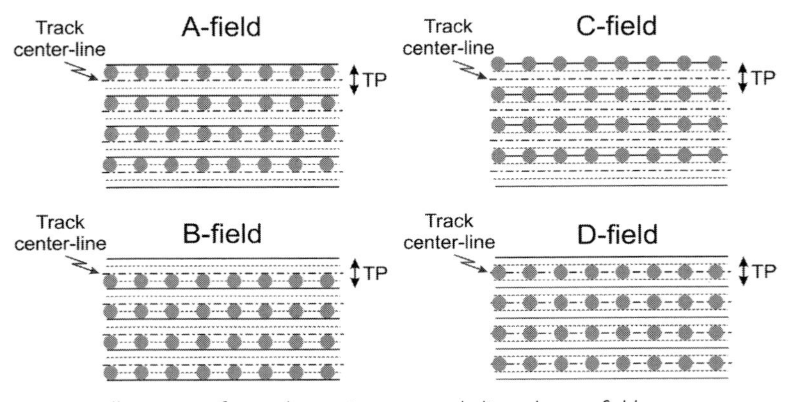

Fig. 6.27 Illustration of servo bursts in separate dedicated servo fields

Fig. 6.27 also lend themselves well to timing synchronization. Owing to the large number of levers in the 2D transducer arrays, this solution appears to be advantageous in terms of overhead compared with the alternative of having timing and servo information embedded in all data fields. It has been estimated that the dedicated servo- and timing-field strategy incurs a very low overhead of <3%. Fig. 6.26 could be regarded as showing the general layout of data and servo/timing fields for the dedicated servo architecture. The two position sensors used during the seek and settle modes of operation are also indicated.

6.4.5.2 MEMS-centric Data-storage Capacities and Data Rates

The extremely high position resolution provided by probe tips with nanometer-scale sharpness provides a potential pathway to the high areal densities that will be needed in the future. The intrinsically nonlinear interactions between closely spaced indentations on the polymer medium or transitions in the magnetic medium may, however, limit the minimum distance between successive indentations/transitions and hence the areal density. For example, as discussed in Section 6.4.2.2, the creation of a new indentation too close to an existing one can result in partial or full erasure of the existing indentation. One way to mitigate this nonlinear interaction, while also increasing the storage capacity of the device, is to apply *(d, k)-constrained* or *modulation* codes [88]. Similar encoding schemes have been successfully used in conventional magnetic and optical storage. The code parameters d and k are non-negative integers with $k>d$, where d and k indicate the minimum and maximum number of '0's between two successive '1's, respectively. For applications in which dedicated clock fields are used, the k constraint does not really play an important role and therefore can in principle be set to infinity, thereby facilitating the code-design process. In a code design where the presence or absence of an indentation (transition) represents a '1' or '0', respectively, the d constraint is instrumental in limiting the interference between successive marks (here a mark represents a polymer indentation or a magnetic transition) while at the same time increasing the effective areal density of the storage device. In particular, the quantity $(d+1)R$, where R denotes the rate of the (d, k) code, is a direct measure of the increase in linear recording density. Whereas the effective density can be increased by increasing d, large values of d result in a lower code rate. Lowering the code rate in turns lowers the user data rate, which renders codes with a high d value less attractive for storage systems that are limited by clock speed. These coding schemes played a crucial role in achieving the ultra-high densities reported for thermomechanical data recording [56]. Specifically, for thermomechanical recording, $d=1$ has been found to be a good compromise. A similar coding scheme may also be beneficial for MEMS-centric devices based on magnetic recording [47, 89].

Tab. 6.1 shows the achievable areal densities and storage capacities for a MEMS-centric storage device with an 80×80 transducer array corresponding to 6400 storage fields, each having an area of $100\times100\ \mu m^2$, resulting in a total storage area of $8.0\times8.0\ mm^2$. In this table, symbol pitch refers to the distance between any

Tab. 6.1 Total accessible medium 0.64 cm^2, (d=1, ∞) code

Symbol pitch (nm)	*Track pitch (nm)*	*Linear density (kbit/in)*	*Track density (ktrack/in)*	*Areal density (Gbit/in²)*	*User capacity (Gbyte)*
12.5	25	1354.7	1016.0	1376.3	14.5
15.0	30	1128.9	846.7	955.8	10.1
17.5	35	967.6	725.7	702.2	7.4
20.0	40	846.7	635.0	537.6	5.7
22.5	45	752.6	564.4	424.8	4.5
25.0	50	677.3	508.0	344.1	3.6
27.5	55	615.8	461.8	284.4	3.0
30.0	60	564.4	423.3	238.9	2.5

two successive logical symbols (corresponding to '1' or '0') written on the medium. Owing to the $d=1$ coding, the symbol pitch is half the minimum distance between two successive indentations, in the case of thermomechanical recording, or between two successive transitions in the case of magnetic recording. For the computation of the storage capacity, an overall efficiency of 85% has been assumed, taking into account the redundancy of the outer error-correction coding as well as the presence of dedicated servo and clock fields. For the reported thermomechanical areal density demonstration [58], in which large data sets were recorded at 641 Gbit/in^2 and read-back with raw error rates lower than 10^{-4}, the symbol and track pitch were 18 and 37 nm, respectively. Therefore, assuming the 80×80 transducer array in Tab. 6.1, the resulting storage capacity in the 0.64 cm^2 area would be 6.8 Gbyte.

Another important characteristic of a storage device is the sustained data rate for storing or retrieving information. MEMS-centric probe storage is inherently slow in recording or reading back information with only a single probe or sensor (particularly when recording techniques relying on a mechanical or thermal response in the probe are used). The total sustained data rate of the device depends on the number of simultaneously active probe tips together with the per-tip data rate. Like conventional magnetic disk drives, a MEMS-centric storage system has to switch tracks when data transfers cross track boundaries. Unlike conventional magnetic disk drives, the rotational speed of which is independent of the arm positioning, the track switching time of a MEMS-centric storage device depends directly on the access velocity. Specifically, the microscanner must turn around whenever the data transfer is such that it crosses a track boundary. Reversing directions requires decelerating, changing direction and then re-accelerating to the target access velocity. Thus, in MEMS-centric probe-storage systems, the sustained data rate depends not only on the actuator access velocity but also on the actuator acceleration and turn-around times [90]. Tab. 6.2 shows the data rate per probe for various symbol-clock periods T. In this analysis a fixed symbol pitch of 20 nm has been assumed and a rate $R=2/3$, ($d=1$, $k=\infty$)-constrained code and a sector overhead factor (including ECC redundancy, etc.) of 0.8797 have been assumed. Finally, depending on the linear velocity, different average retrace and settle times have

Tab. 6.2 Data rate per probe for various symbol-clock periods *T* 9 [symbol pitch 20 nm; track pitch 40 nm; rate –2/3; (d=1, ∞) code]

Symbol clock T (μs)	*Sector over-head factor (ECC, etc.)*	*Linear speed (mm/s)*	*100 μm scan line time (ms)*	*Retrace and settle time (ms)*	*Sustained data rate per lever (kbit/s)*
20	0.8797	1	100	5	27.9
10	0.8797	2	50	4	54.3
5	0.8797	4	25	3	104.7
2	0.8797	10	10	2	244.3
1	0.8797	20	5	1	488.7

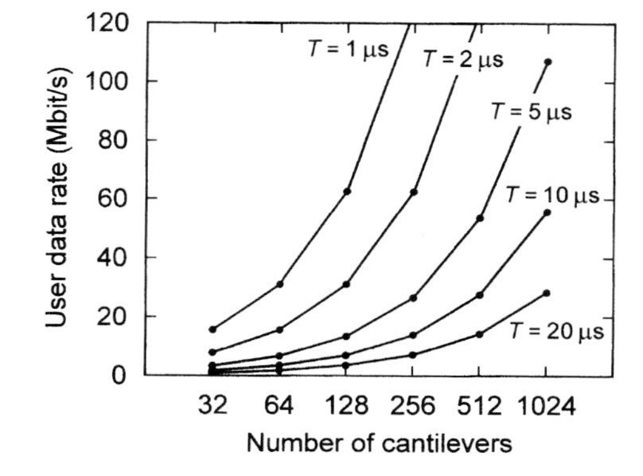

Fig. 6.28 User data rate versus number of active cantilevers for the (d=1, k=∞) coding scheme

been considered. A typical future design point would be the case of a 20 nm symbol pitch and a symbol clock period of T=5 μs, for which a sustained data rate per probe of 104.7 kbit/s can be achieved. Fig. 6.28 shows the user data rate as a function of the total number of cantilevers accessed simultaneously, using the per-cantilever assumptions listed in Tab. 6.2. A future device using a symbol clock period of T=5 μs with an array of 256 active cantilevers would achieve a sustained data rate of 26.8 Mbit/s.

6.5 Conclusions

MEMS technology is expected to make inroads into mass storage devices. In the short run, MEMS will most likely appear as subcomponents that enhance the capacity and performance of conventional magnetic and optical storage devices. Further in the future, a new genre of MEMS-centric devices may appear, which will replace conventional mechanics with MEMS and rely on ultra-high density probe-based recording schemes. Whether these devices succeed will depend on whether they can achieve sufficiently high density and performance to offer advantages over conventional storage options.

A full technical demonstration of a complete MEMS-centric device using probe storage technology has not yet been achieved. Further advances are needed in many areas, including the basic recording methods (to achieve both ultra-high density and good reliability), MEMS design and fabrication, servo and overall device integration.

The first MEMS-centric devices to become viable will likely be short-range translation-type devices because they are technically much simpler to implement. Although such devices are not likely to completely replace conventional mechanical or solid-state storage devices in the short run, they may be competitive for high-capacity storage in portable devices. In the longer term, they are also thought to have attributes that could make MEMS-centric devices attractive in a wide variety of storage systems.

6.6 References

1. T.R. Albrecht, D.W. Albrecht, K. Kuroki, T.C. Reiley, K. Takahashi, *J. Inf. Storage Process. Syst.* **2000**, *2*, 47–51.
2. T. Hirano, H. Yang, S. Pattanaik, M. White, S. Arya, in: *Proc. STLE and ASME Joint Tribology Conf., Ponte Verde Beach, FL*; **2003**.
3. R. Evans, K. Griesbach, W. Messner, *IEEE Trans. Magn.* **1999**, *35*, 977–982.
4. M. Tokuyama, T. Shimizu, H. Masuda, S. Nakamura, M. Hanya, O. Iriuchijima, J. Soga, *IEEE Trans. Magn.* **2001**, *37*, 1884–1886.
5. D. Horsley, et al., *IEEE/ASME Trans. Mechatron.* **1998**, *3*, 175–183.
6. P. Crane, W.A. Bonin, B. Zhang, *US Patent* 6 414 822, **2002**.
7. M. Del Sarto, S. Sassolini, L. Baldo, M. Marchi, M.J. McCaslin, *Proc. SPIE* **2003**, *4797*, 155–164.
8. T. Hirano, L.-S. Fan, T. Semba, W.Y. Lee, J. Hong, S. Pattanaik, P. Webb, W.-H. Juan, S. Chan, et al., *IEEE Trans. Magn.* **1999**, *35*, 3670–3672.
9. H. Toshiyoshi, M. Mita, H. Fujita, *IEEE J. Microelectromech. Syst.* **2002**, *11*, 648–654.
10. T. Yoshino, H. Toshiyoshi, M. Mita, D. Kobayashi, H. Fujita, *IEEJ Trans. SM* **2004**, *124*, 21–27.
11. T. Imamura, M. Katayama, Y, Ikegawa, T. Ohwe, R. Koishi, T. Koshikawa, *IEEE/ASME Trans. Mechatron.* **1998**, *3*, 166–174.
12. T. Hirano, M. White, X.-H. Yang, T. Semba, V. Shum, S. Pattanaik, S. Arya, D. Kercher, L.-S. Fan, in: *Proc. ASME Int. Mech. Eng. Congress and Exposition, New York*; **2001**.
13. T. Hirano, L.-S. Fan, D. Kercher, S. Pattanaik, T.-S. Pan, in: *Proc. ASME Int. Mech. Eng. Congress and Exposition, Orlando, FL*; **2000**, pp. 449–452.
14. S. Pattanaik, *US Patent* 5828031, **1998**.
15. M.T. White, T. Hirano, H. Yang, K. Scott, S. Pattanaik, F.-Y. Huang, *Proc. IEEE Int. Workshop on Adv. Motion Control, Kawasaki, Japan*, **2004**, pp. 299–304.
16. B. Liu, M. Zhang, S. Yu, L. Gonzaga, Y.S. Hor, J. Xu, *IEEE Trans. Magn.* **2003**, *39*, 909–914.
17. R.E. Fontana, J. Katine, M. Rooks, R. Viswanathan, J. Lille, S. MacDonald, E. Kratschmer, C. Tsang, S. Nguyen, N. Robertson, P. Kasiraj, *IEEE Trans. Magn.* **2002**, *38*, 95–100.
18. T.D. Miller, *Proc. IEEE* **2000**, *88*, 1480–1490.
19. S. Kim, J. Lee, J. Park, G. Park, J. Lee, C. Lee, D. Son, J.-Y. Kim, S.-H. Kim, Y. Yee, in: *Technical Digest, Optical Data Storage Conf., Vancouver, BC*; **2003**, pp. 5–7.
20. S. Kim, Y. Yee, J. Choi, H. Kwon, M.-H. Ha, C. Oh, J. Bu, in *Proc. 12th Int. Conf. Solid-State Sensors, Actuators and Microsystems, Boston, MA*; **2003**, pp. 607–610.
21. E. Betzig, J.K. Trautman, R. Wolfe, E.M. Gyorgy, P.L. Finn, *Appl. Phys. Lett.* **1992**, *61*, 142–144.

22 S. M. Mansfield, W. R. Studenmund, G. S. Kino, K. Osato, *Opt. Lett.* **1993**, *18*, 305–307.

23 J. Tominaga, H. Fuji, A. Sato, T. Nakano, N. Atoda, *Jpn. J. Appl. Phys.* **2000**, *39*, 957–961.

24 M. A. Paesler, P. J. Moyer, *Near-Field Optics: Theory, Instrumentation and Application*; New York: Wiley, **1996**, pp. 249–255.

25 T. Yatsui, M. Kourogi, M. Ohtsu, *Appl. Phys. Lett.* **1998**, *73*, 2090–2092.

26 A. Vollkopf, O. Rudow, T. Leinhos, C. Mihalcea, E. Oesterschulze, *J. Microsc.* **1999**, *194*, 344–348.

27 M. Sasaki, K. Tanaka, K. Hane, *Jpn. J. Appl. Phys.* **2000**, *39*, 7150–7153.

28 L. R. Carley, J. A. Bain, G. K. Fedder, D. W. Greve, D. F. Guillou, M. S. C. Lu, T. Mukherjee, S. Santhanam, *J. Appl. Phys.* **2000**, *87*, 6680–6685.

29 J. Bu, Y. Yee, S. Lee, J. Kim, in *Proc. 12th Int. Conf. Solid-State Sensors, Actuators and Microsystems, Boston, MA*; **2003**, pp. 1762–1767.

30 C. F. Quate, *US Patent* 4 575 822, **1986**.

31 J. S. Foster, J. E. Frommer, P. C. Arnett, *Nature* **1988**, *331*, 324–326.

32 D. M. Eigler, E. K. Schweizer, *Nature* **1990**, *344*, 524–526.

33 L. Abelmann, T. Bolhuis, A. M. Hoesum, G. J. M. Krijnen, J. C. Lodder, *IEE Proc. Sci. Meas. Technol.* **2003**, *150*, 218–221.

34 T. Onoue, M. H. Siekman, L. Abelman, J. C. Lodder, *J. Magn. Magn. Mater.* **2004** (in press).

35 D. Weller, A. Moser, *IEEE Trans. Magn.* **1999**, *35*, 4423–4439.

36 J. H. Judy, *J. Magn. Magn. Mater.* **2001**, *235*, 235–240.

37 S. Y. Chou, *Proc. IEEE* **1997**, *85*, 652–671.

38 M. Albrecht, S. Ganesan, C. T. Rettner, A. Moser, M. E. Best, R. L. White, B. D. Terris, *IEEE Trans. Magn.* **2003**, *39*, 2323–2325.

39 C. A. Ross, et al., *J. Vac. Sci. Technol. B* **1999**, *17*, 3168–3176.

40 M. Bremer, L. Conrad, L. Funk, *J. Dispers. Sci. Technol.* **2003**, *24*, 291–304.

41 M. H. Kryder, S. H. Charap, *US Patent* 6 011 644, **2000**.

42 O. Watanuki, Y. Sonobe, S. Tsuji, F. Sai, *IEEE Trans. Magn.* **1991**, *27*, 5289–5291.

43 T. Ohkubo, J. Kishigami, K. Yanagisawa, R. Kaneko, *IEEE Trans. Magn.* **1993**, *29*, 4086–4088.

44 E. B. Myers, D. C. Ralph, J. A. Katine, R. N. Louie, R. A. Buhrman, *Science* **1999**, *285*, 867–870.

45 C. A. Spindt, *J. Appl. Phys.* **1968**, *39*, 3504.

46 S. Manalis, K. Babcock, J. Massie, V. Elings, M. Dugas, *Appl. Phys. Lett.* **1995**, *66*, 2585–2587.

47 R. T. El-Sayed, L. R. Carley, *IEEE Trans. Magn.* **2003**, *39*, 3566–3574.

48 P. Vettiger, G. Gross, M. Despont, U. Drechsler, U. Dürig, B. Gotsmann, W. Häberle, M. A. Lantz, H. E. Rothuizen, R. Stutz, G. K. Binnig, *IEEE Trans. Nanotech.* **2002**, *1*, 39–55.

49 E. Eleftheriou, T. Antonakopoulos, G. K. binnig, G. Cherubini, M. Despont, A. Dholakia, U. Dürig, M. A. Lantz, H. Pozidis, H. E. Rothuizen, P. Vettiger, *IEEE Trans. Magn.* **2003**, *39*, 938–945.

50 H. J. Mamin, D. Rugar, *Appl. Phys. Lett.* **1992**, *61*, 1003–1005.

51 H. J. Mamin, B. D. Terris, L. S. Fan, S. Hoen, R. C. Barrett, D. Rugar, *IBM J. Res. Dev.* **1995**, *39*, 681–699.

52 M. Lutwyche, C. Andreoli, G. Binnig, J. Brugger, U. Drechsler, W. Häberle, H. Rohrer, H. Rothuizen, P. Vettiger, in *Proc. 11th IEEE Int. Workshop on Micro Electro Mechanical Systems*; Piscataway, NJ: IEEE, **1998**, pp. 8–11.

53 C. S. Lee, W. H. Jin, H. J. Nam, S. M. Cho, Y. S. Kim, J. U. Bu, in *Proc. 16th Ann. Int. Conf. Micro Electro Mechanical Systems*; Piscataway, NJ: IEEE, **2003**, pp. 28–32.

54 S. H. Lee, S. S. Lee, J. U. Jeon, K. Ro, in *Proc. 16th Ann. Int. Conf. Micro Electro Mechanical Systems*; Piscataway, NJ: IEEE, **2003**, pp. 72–75.

55 N. Blanc, J. Brugger, N. F. de Rooij, U. Dürig, *J. Vac. Sci. Technol. B* **1996**, *14*, 901–905.

56 H. Pozidis, W. Häberle, D. Wiesmann, U. Drechsler, M. Despont, T. R. Albrecht, E. Eleftheriou, *IEEE Trans. Magn.* **2004** (in press).

57 J. S. Foster, K. A. Rubin, D. Rugar, *US Patent* 4 916 688, **1990**.

58 C. C. Yang, presented at the *INSIC Annual Meeting Symposium on Alternative Storage Technologies, Monterey, CA*, **2003**.

59 D. Saluel, J. Daval, B. Bechevet, C. Germain, B. Valon, *J. Magn. Magn. Mater.* **1999**, *193*, 488–491.

60 S. Gidon (CAE-LETI, Grenoble, France) personal communication.

61 G. Gibson, T. I. Kamins, M. S. Keshner, S. L. Neberhuis, C.M. Perlov, C. C. Yang, *US Patent* 5 556 596, **1996**.

62 S. Neberhuis, *J. Magn. Magn. Mater.* **2002**, *249*, 447–451.

63 R. C. Barrett, C. F. Quate, *J. Appl. Phys.* **1991**, *70*, 2725–2733.

64 H. Shin, S. Hong, J. Moon, J. U. Jeon, *Ultramicroscopy* **2002**, *91*, 103–110.

65 H. J. Gao, K. Sohlberg, Z. Q. Xue, H. Y. Chen, S. M. Hou, L. P. Ma, X. W. Fang, S. J.

PANG, S.J. PENNYCOOK, *Phys. Rev. Lett.* **2000**, *84*, 1780–1783.

66 Y. ZHAO, A. FEIN, C.A. PETERSON, D. SARID, *Phys. Rev. Lett.* **2001**, *87*, 1797.

67 C.P. COLLIER, E.W. WONG, M. BELOHRADSKY, F.M. RAYMO, J.F. STODDART, P.J. KUEKES, R.S. WILLIAMS, J.R. HEATH, *Science* **1999**, *285*, 391–394.

68 H. ROTHUIZEN, M. DESPONT, U. DRECHSLER, G. GENOLET, W. HÄBERLE, M. LUTWYCHE, R. STUTZ, P. VETTIGER, in: *Technical Digest, 15th IEEE Int. Conf. on Micro Electro Mechanical Systems 'MEMS 2002', Las Vegas, NV*; Piscataway, NJ: IEEE, **2002**, pp. 582–585.

69 M. LUTWYCHE, U. DRECHSLER, W. HÄBERLE, H. ROTHUIZEN, R. WIDMER, P. VETTIGER, J. THAYSEN, in: *Magnetic Materials, Processes and Devices: Applications to Storage and Micromechanical Systems (MEMS)*, L. ROMANTKIV, S. KRONGELD, C.H. AHN (eds.), Pennington, NJ: Electrochemical Society, **1999**, Vol. 98–20, pp. 423–433.

70 J.-J. CHOI, H. PARK, K.Y. KIM, J.U. JEON, *J. Semicond. Technol. Sci.* **2001**, *1*, 84–93.

71 S. HOEN, P. MERCHANT, G. KOKE, J. WILLIAMS, in: *Proc. Int. Conf. Solid-State Sensors and Actuators, Chicago, IL*; **1997**, pp. 41–44.

72 N. TAS, J. WISSINK, L. SANDER, T. LAMMERINK, M. ELWENSPOEK, *Sens. Actuators A* **1998**, *70*, 171–178.

73 E. SARAJLIC, E. BERENSCHOT, G. KRIJNEN, M. ELWENSPOEK, *Microelectron. Eng.* **2003**, *67*, 430–437.

74 T. RUST, presented at the *INSIC Annual Meeting Symposium on Alternative Storage Technologies, Monterey, CA*, **2003**.

75 F. LAERMER, A. SCHILP, K. FUNK, M. OFFENBERG, in: *Proc. IEEE 12th Int. Conf. Micro Electro Mechanical Systems, Orlando, FL*; Piscataway, NJ: IEEE **1999**, pp. 211–216.

76 S. HOEN, Q. BAI, J.A. HARLEY, D.A. HORSLEY, F. MATTA, T. VERHOEVEN, J. WILLIAMS, K.R. WILLIAMS, in: *Technical Digest, Transducers'03*; **2003**, pp. 344–347.

77 E. SARAJLIC, E. BERENSCHOT, G. KRIJNEN, M. ELWENSPOEK, **2003**, personal communication.

78 D. SAYA, in: *Proc. 14th Int. Conf. Micro Electro Mechanical Systems, Interlaken*; Piscataway, NJ: IEEE, **2001**, pp. 131–134.

79 D.W. LEE, T. ONO, T. ABE, M. ESASHI, in *Proc. 14th Int. Conf. Micro Electro Mechanical Systems, Interlaken*; Piscataway, NJ: IEEE **2001**, pp. 204–207.

80 E.M. CHOW, V. CHANDRASEKARAN, A. PARTRIDGE, T. NISHIDA, M. SHEPLAK, C.F. QUATE, T.W. KENNY, *J. Microelec. Mech. Syst.* **2002**, *11*, 631–640.

81 (a) M. DESPONT, J. BRUGGER, U. DRECHSLER, U. DÜRIG, W. HÄBERLE, M. LUTWYCHE, H. ROTHUIZEN, R. STUTZ, R. WIDMER, H. ROHRER, G. BINNIG, P. VETTIGER, in: *Technical Digest, Twelfth IEEE Int. Conf. on Micro Electro Mechanical Systems (MEMS '99), Orlando, FL*; Piscataway, NJ: IEEE, **1999**, pp. 564–569; (b) M. DESPONT, J. BRUGGER, U. DRECHSLER, U. DÜRIG, W. HÄBERLE, M. LUTWYCHE, H. ROTHUIZEN, R. STUTZ, R. WIDMER, G. BINNIG, H. ROHRER, P. VETTIGER, *Sens. Actuators A* **2000**, *80*, 100–107.

82 G.K. BINNIG, G. CHERUBINI, M. DESPONT, U. DÜRIG, E. ELEFTHERIOU, H. POZIDIS, P. VETTIGER, in: *Handbook of Nanotechnology*, B. BHUSHAN (ed.); Heidelberg: Springer, **2004**, pp. 921–950.

83 D. LANGE, in: *Technical Digest, Twelfth IEEE Int. Conf. on Micro Electro Mechanical Systems (MEMS '99), Orlando, FL*; Piscataway, NJ: IEEE, **1999**, pp. 447–452.

84 M. DESPONT, U. DRECHSLER, R. YU, M. GEISSLER, E. DELAMARCHE, H.B. BOGGE, P. VETTIGER, in: *IBM Research Report*, RZ 3491 (May 2003).

85 M. DESPONT, U. DRECHSLER, R. YU, H.B. POGGE, P. VETTIGER, in *Transducers '03 – Digest of Technical Papers*; Piscataway, NJ: IEEE, **2003**, Vol. 2, pp. 1907–1910.

86 L.R. CARLEY, G. GANGER, D.F. GUILLOU, D. NAGLE, *IEEE Trans. Magn.* **2001**, *37*, 657–662.

87 E. ELEFTHERIOU, P. BÄCHTOLD, G. CHERUBINI, A. DHOLAKIA, C. HAGLEITNER, T. LOELIGER, A. PANTAZI, H. POZIDIS, T.R. ALBRECHT, G.K. BINNIG, M. DESPONT, U. DRECHSLER, U. DÜRIG, B. GOTSMANN, D. JUBIN, W. HÄBERLE, M.A. LANTZ, H. ROTHUIZEN, R. STUTZ, P. VETTIGER, D. WIESMANN, in: *Proc. 29th Int. Conf. on Very Large Data Bases, Berlin, Germany*, J.C. FREYTAG, P.C. LOCKEMANN, S. ABITEBOUL, M.J. CAREY, P.G. SELINGER, A. HEUER (eds.); San Francisco: Morgan Kaufman, **2003**, pp 3–7.

88 K.A.S. IMMINK, *Coding Techniques for Digital Recorders*; Hemel Hempstead: Prentice Hall International (UK), **1991**.

89 L.R. CARLEY, G.R. GANGER, D. NAGLE, *Commun. ACM* **2000**, *43*, 73–80.

90 J.L. GRIFFIN, S.W. SCHLOSSER, G.R. GANGER, D.F. NAGLE, in: *Proc. ACM Sigmetrics 2000, Santa Clara, CA, USA*; **2000**, pp. 56–65.

7
Scanning Micro- and Nanoprobes for Electrochemical Imaging

C. Kranz, A. Kueng, B. Mizaikoff, School of Chemistry and Biochemistry, Georgia Institute of Technology, Atlanta, GA, USA

Abstract
Scanning probe techniques have a remarkable impact on the characterization of in situ processes occurring at solid/liquid and liquid/liquid interfaces. Electrochemical processes can be studied at molecular levels and microscopic changes of the surface structure can be correlated with dynamic electrochemical processes. Initially, electrochemcial scanning tunneling microscopy and electrochemical atomic force microscopy were the predominant scanning probe techniques in electrochemical surface sciences. However, within the last decade scanning electrochemical microscopy and scanning ion conductance microscopy have gained significance, especially for the investigation of complex redox processes in biology and life sciences. Combined scanning probe techniques provide new insight into increasingly complex systems providing simultaneous complementary information along with increasing chemical specificity. Fundamental principles of individual and combined scanning probe techniques relevant to electrochemical investigations and selected applications are addressed in this chapter, with particular focus on novel strategies for increasing the lateral resolution and the information content of in situ scanning probes.

Keywords
scanning probe microscopy; combined scanning probe techniques; scanning electrochemical microscopy; integrated SECM/AFM probes; nanoelectrodes; microfabrication.

Advanced Micro and Nanosystems. Vol. 1
Edited by H. Baltes, O. Brand, G.K. Fedder, C. Hierold, J. Korvink, O. Tabata

ISBN: 3-527-30746-X

7.1 Introduction to Scanning Probe Techniques

7.1.1 The Electrochemical Interface

The combination of surface sciences, surface measurement technologies and electrochemistry has achieved remarkable insight into interfacial electrochemical processes over recent decades. Electrochemical processes play a major role in a variety of fields ranging from biochemical transport mechanisms to material sciences, including fuel cell technology, catalysis, electroanalytical sensors and industrial and environmental processes. Historically, scanning probe-based investigations of electron transfer processes have studied electrode/electrolyte interfaces. However, there is an increasing demand for the investigation of biochemically and biologically relevant redox processes in complex matrices in real time with high lateral resolution.

In conventional electrochemical experiments, electron transfer occurs at the solid/liquid boundary between an electrode surface and an electroactive molecule. At this interface, known as the electric double layer (or Helmholtz layer), some of the most important solution-phase physical phenomena such as ionic solvation and electron transfer occur, along with surface-relevant processes such as chemisorption and catalysis. Initially, these processes were investigated with traditional voltammetric methods based on current and voltage measurements at a macroscopic level [1, 2]. However, understanding fundamental mechanistic and dynamic phenomena in electrochemical surface science at a microscopic level requires surface-sensitive and structure-specific analytical techniques elucidating the involved processes at an atomic or molecular level.

Classically, the electric double layer is defined as a plate capacitor with molecular dimensions as shown schematically in Fig. 7.1. The first description of the metal/solution interface goes back to Helmholtz in 1874. In principle, this model describes a capacitor with a solid metal with an excess surface charge as one plate and several layers of solvated ions as a second plate. The solution layer closest to the metal electrode is called the Stern layer or inner Helmholtz plane (IHP) with a distance d_1 (see Fig. 7.1) and consists of solvent molecules and specifically adsorbed ions or molecules. The distance d_2 is called the outer Helmholtz plane (OHP), which is defined by solvatized ions non-specifically attracted by electrostatic forces over few monolayers. Owing to thermal agitation in the solution, these nonspecifically adsorbed ions further extend from the OHP into the bulk solution, a region called the diffusion layer. ϕ_M, ϕ_i, and ϕ_d are the potentials inside

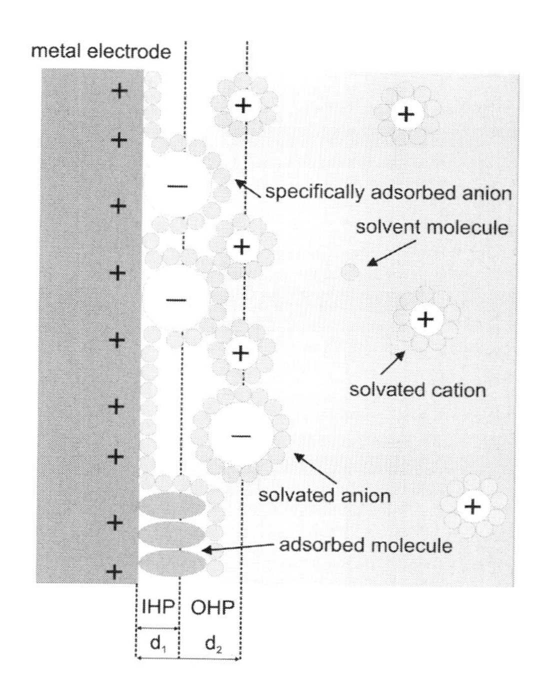

Fig. 7.1 Simplified model of the electrode/electrolyte interface and the double-layer region. The location of the electrical centers of the specifically adsorbed ions defines the inner Helmholtz plane (IHP). Solvated ions can approach the metal electrode surface only to a certain distance. The location of centers of these solvated ions defines the outer Helmholtz plane (OHP)

the metal, the electrolyte and the outer Helmholtz plane, respectively. A linear potential drop across the electrochemical interface is assumed for the region of non-specifically adsorbed ions, as described by the capacitor model. Specific adsorption at the electrode surface leads to a steeper potential gradient close to the metal surface [3].

The strong dependence of the charge-transfer or capacitive charging event on surface structures and the detailed arrangement of atoms and molecules at this interface is central to electrochemical reactivity, which encompasses processes occurring within the inner Helmholtz plane at a distance of few atomic diameters from the electrode surface. Hence, for fundamental investigations of electron-transfer processes, a structurally well-defined and clean electrode surface is a prerequisite. Therefore, for many years, mercury electrodes were used for studying electrode processes at the electrochemical double layer [4, 5].

The key development for the investigation and detailed understanding of surface and interface processes was the introduction of surface-sensitive measurement techniques providing information on an atomic or molecular level, along with ultra-high vacuum (UHV)-based techniques applied for the preparation and characterization of structurally well-defined electrode surfaces in the 1960s [6]. These mainly electron-based physical analysis techniques provide detailed insight into surfaces processes, but they are not applicable for in situ electrochemical investigations. Hence combined investigations (ex situ approach) involving UHV surface analytical probing before and after electrochemical experiments are used. After initial electrode preparation and characterization with surface-sensitive UHV techniques, the interface was studied again after electrochemical treatment [7]. Systematic studies on the relationship between surface structure and electrochemical behavior were performed with a combined UHV chamber and electrochemical cell using a vacuum-enclosed sample transfer system [8–10]. However, the fundamental limitation of all ex situ techniques to establish a relationship between the surface prior to and after the emersion from solution into vacuum conditions remained along with the uncertainty of affecting the electrode surface due to the experimental procedure.

In situ electrochemical information became accessible on a routine basis by flame annealing techniques introduced by Clavilier et al. [11, 12]. Well-ordered monocrystalline metal electrodes can be prepared in a laboratory environment using a Bunsen burner followed by cooling in water and/or in an inert gas stream followed by transfer into an electrochemical cell. Optical methods such as spectroscopic and diffraction techniques have been widely used to study the solid/electrolyte interface. However, information obtained by these techniques usually averages over a macroscopically large sample region. Since defects and impurities at the electrode surface can significantly influence the electrochemical reactions that occur, in situ characterization of the solid/liquid interface with high spatial resolution plays a major role in understanding electrochemical surface processes.

This chapter is focused exclusively on scanning probe techniques with relevance to electrochemically governed surface processes under ambient conditions. The application of scanning probe microscopy for electrochemical application is demonstrated with selected examples. New developments regarding combined scanning probe techniques for electrochemical imaging are discussed in detail.

7.1.2 Scanning Probe Techniques

The major breakthrough for in situ investigation of interfacial electrochemical processes was the introduction of scanning probe techniques. This family of surface-sensitive analytical tools permits in situ investigations of electrode surfaces with atomic-scale resolution in real time. Binnig et al. [13, 14] laid the fundamentals for a whole family of scanning probe techniques, which was honored with the Nobel Prize in physics in 1986. The basic principle of scanning probe microscopic (SPM) techniques is based on the specific tip–sample interaction when the tip is scanned in close proximity across the sample surface. Common to most SPM instrumentation is the probe positioning unit consisting of an x, y, z positioning device usually based on piezoelectric elements [15] and a feedback control system to keep the probe at a constant distance from the sample surface. The major difference between individual SPM techniques is usually determined by the regulation of the probe distance, which depends on the fundamental physical aspects of the selected tip–sample interaction principle. An overview of SPM techniques and a schematic representation of the underlying near-field interaction commonly applied in electrochemical surface sciences are given in Tab. 7.1.

7.1.2.1 Scanning Tunneling Microscopy (STM)

STM provides spatially resolved surface images on the atomic level by recording the electron tunneling current between a sharp probe tip and a conductive sample surface. By applying a voltage between the tip and the sample, a tunneling current is generated, which depends exponentially on the distance between the sample surface and tip. Usually, the STM tip is scanned across the sample surface keeping the tunnel current and, hence, the distance between tip and surface constant ('constant current mode'). Initially, STM was predominantly developed for surface investigations under UHV conditions. However, early studies demonstrated that this technique could also operate in electrolyte solutions [16–18]. Sonnenfeld and Hansma showed the investigation of HOPG as one of the first in situ applications of STM in solution [16]. Modifications of STM instrumentation were reported to control independently the electrode potential of the substrate and the tip with respect to a reference electrode [19, 20]. Electrochemical STM (ECSTM) [21–24] allows the in situ investigation of electrochemical reactions under potentiostatic conditions using a four-electrode configuration (Fig. 7.2). However, the total current measured in an ECSTM experiment is a convolution of the tunneling current and the Faraday current, which is generated by the electrochemical reactions at the substrate/electrolyte and/or tip/electrolyte interface. In order to minimize the contribution of the Faraday current, the electroactive area of the STM probe has to be reduced, which is described in Section 7.1.3.1. Furthermore, the

Tab. 7.1 Overview of scanning probe techniques and principles of tip–sample interaction

Scanning probe technique	***Physical interaction***	***Schematic***
Scanning tunneling microscopy	Local electron density; resolution: atomic	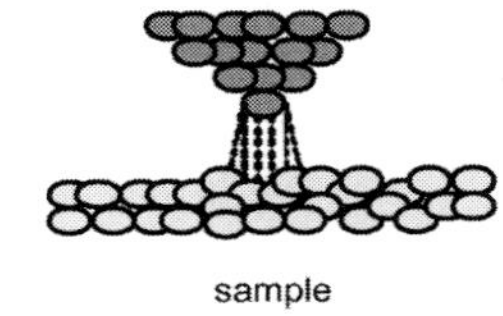
Atomic force microscopy	Interaction forces: Coulomb forces (repulsive); van der Waals forces (attractive); fluid surface tension (attractive); electrostatic forces (attractive); resolution: molecular to atomic	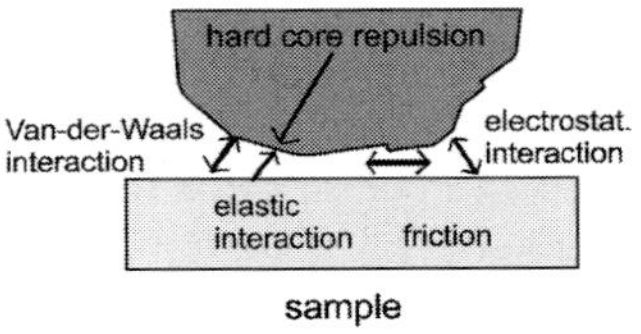
Near-field scanning optical microscopy	Optical near-field interaction of light with matter; resolution: ~ 30 nm	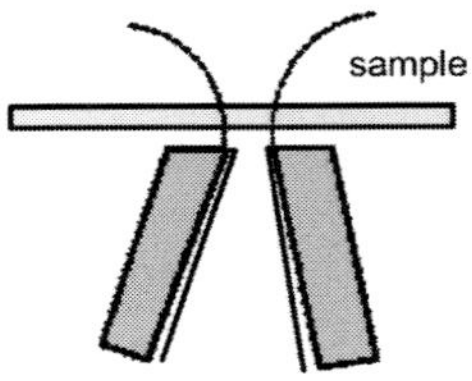
Scanning ion conductance microscopy	Change of ion conductance between electrode inside and outside the micropipette in close proximity to sample surface; resolution: ~ 50 nm	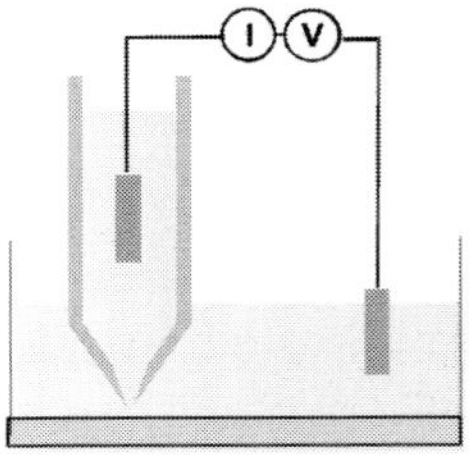
Scanning electrochemical microscopy	Changes in Faraday current at UME in close proximity to sample surface; resolution: μm to nm range, strongly dependent on tip size	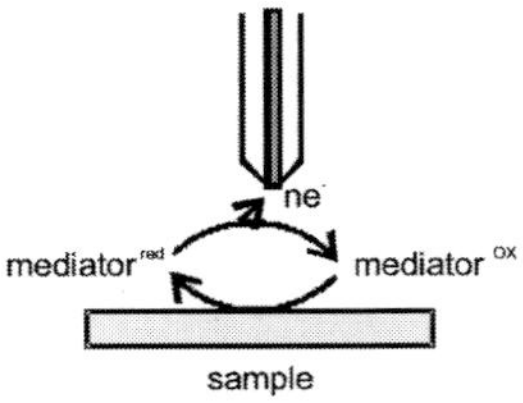

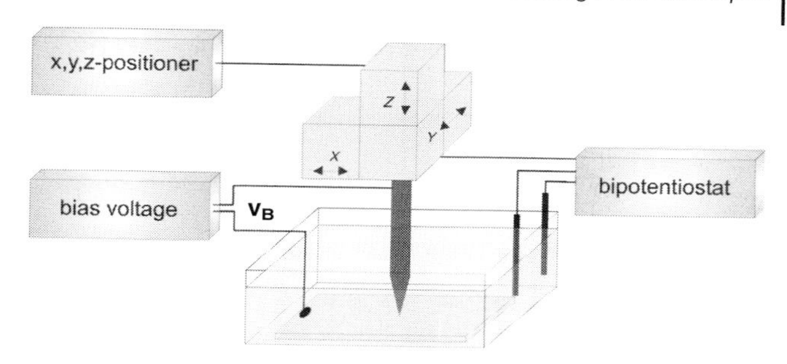

Fig. 7.2 Schematic of the principle components of ECSTM instrumentation

potentials applied to the substrate and the STM tip have to be controlled separately. Using a bipotentiostat allows independent adjustment of the tip potential E_T and the substrate potential E_S relative to a reference electrode. A tip potential can be selected where in principle no electrochemical reactions occur. Hence the Faradaic contribution to the current signal should be zero. The tunnel bias voltage U is defined then as $U = E_S - E_T$ with the substrate potential determined by the electrochemistry and the tip potential set to a value that allows minimization of the Faradaic contribution. Mechanistic studies on electron tunneling through water layers have been performed by a number of research groups and found to be either close to or smaller than the vacuum value [20, 25]. Lindsay and co-workers [26, 27] performed systematic studies of the tunneling barrier in STM as a function of bias voltage and electrochemical potential in the presence and absence of organic adsorbates. The lateral resolution achievable with STM is superior to that with other SPM techniques, allowing atomistic insight into electrode surface processes.

ECSTM has been applied to many different areas in electrochemistry. Selected examples highlighting the versatility of ECSTM are discussed in Section 7.2.1.

Despite its inherent utility, STM has certain limitations. A common drawback characteristic to a variety of SPM techniques is the limited availability of surface information with high chemical specificity, which is of particular importance during biologically or biomedically relevant investigations of complex systems. Furthermore, the benefit of high lateral resolution is usually traded against information obtained only on a very small fraction of the total sample surface. As a result, the correlation between laterally resolved information obtained by STM and macroscopic information obtained by bulk electrochemical techniques might be difficult or not representative for the overall reactivity or surface process. Furthermore, STM measurements require conducting substrates, which excludes application to many biologically relevant redox processes. However, these limitations cannot derogate the impact of STM on fundamental investigations of electrochemically active interfaces.

7.1.2.2 Atomic Force Microscopy (AFM)

AFM was invented in 1986 [28] and has rapidly developed into a routine tool for surface scientists in vacuum and ambient environments. Similarly to STM, a probe is scanned across the sample surface while detecting variations in physical and chemical forces at molecular and atomic levels between a sharp tip, which is attached to a cantilever arm, and the sample surface. The tip is scanned across a surface utilizing feedback mechanisms, enabling the piezoelectric x, y, z positioning system to maintain the tip at a constant force (to obtain height information) or at constant height (to obtain force information). In noncontact mode (at distances >10 Å between the tip and the sample surface), Van der Waals forces, electrostatic forces, magnetic forces or capillary forces produce images of topography, whereas in contact mode, ionic repulsion forces take the leading role. The proportional relationship between force and displacement of the cantilever is described by Hooke's law. The cantilever deflection is monitored and a feedback mechanism actuates the piezoelectric positioning elements maintaining a constant cantilever deflection and, hence, a constant force between tip and sample. Most instrumental AFM designs are nowadays based on optical detection using a segmented photodiode to measure the deflection of a laser beam reflected from the back of the cantilever [29, 30], although, capacitance sensing [31] and detection by optical interferometry [32, 33] can be found in the literature. The absence of current during imaging and the inert nature of the AFM probe facilitate the application of AFM for in situ electrochemical investigations. However, imaging and localizing electrochemical reactivity in electrochemical AFM (ECAFM) is mainly based on a change in sample topography induced by electrochemical processes at the substrate. The basic principles of in situ AFM will be omitted here as they have been reviewed in numerous publications during the last two decades [34].

The most commonly applied modes of AFM operation for in situ investigations [35] are shown schematically in Fig. 7.3. In contact mode (or DC mode) AFM [28, 36], the tip apex – ideally single atoms – is in direct physical contact with the sample surface. This contact area is characterized by short-range repulsion forces due to overlapping orbitals of tip and sample atoms. Since the range and contact area of the interaction are at Ångstrom levels, atomic resolution can be achieved. Long-range forces such as magnetic force, electrical force, Van der Waals forces and capillary forces occur in addition and increase the force at the contact area. The usual data representation comprises force–distance curves (see Fig. 7.4), which describe the tip–sample interaction in close proximity. At larger distances no interaction forces are recorded (1). Upon approaching the sample surface, probe/sample interaction resumes, but with forces much lower than the restoring force of the cantilever (2). At point (3) the cantilever abruptly engages with the sample surface due to attractive Van der Waals interactions, which are greater than the restoring force of the cantilever. Decreasing the distance further increases the cantilever deflection (4). If the distance is increased again, a similar cantilever deflection profile is traced as the tip and sample remain in contact (5). At point (6) the cantilever restoring force exceeds the adhesive forces resulting from the tip–sample con-

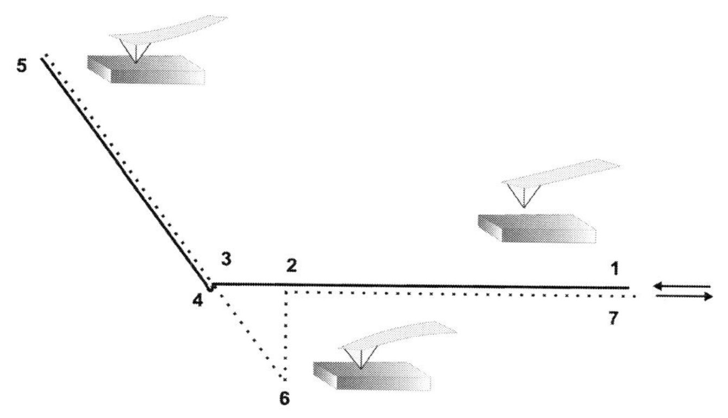

Fig. 7.3 Schematic representation of an idealized force-distance curve. (1) The tip is moved towards the sample without interaction; (2) probe and sample resume interaction; however, the interaction strength is less than the restoring force of the cantilever; (3) the cantilever 'jumps into contact' since the interaction force now exceeds the restoring force of the cantilever; (4) the distance is further decreased and the cantilever deflection increases; (5) the distance is decreased and a similar cantilever deflection trace appears as the tip and sample remain in contact; (6) the cantilever force exceeds the adhesive forces at the tip–sample contact and the contact between tip and sample is interrupted ('snaps off'); (7) the cantilever retracts to a distance without force interaction with the sample surface

tact and the cantilever disengages from the surface. The cantilever is removed to a distance where no force interaction with the sample occurs (7).

Contact mode imaging is less suitable for the investigation of soft samples such as biomolecules and polymers, since direct contact and resulting forces occurring at the tip/sample interface may damage the investigated species. Dynamic operation modes can be classified into two basic methods: amplitude-modulation (AM) operation [37] and frequency-modulation (FM) operation [38]. In AM operation the cantilever is driven at a certain oscillation frequency. As the frequency modulation mode is mainly used in vacuum a detailed discussion is omitted here, focusing on SPM operation under ambient conditions. The AM mode permits application under ambient conditions involving repulsive tip–sample interactions, while 'tapping' on to the sample surface. The 'Tapping Mode' (trademark of Digital Instruments) was introduced for imaging in air [39] and liquids [40] with a reduced tip–sample contact time resulting in minimized frictional forces, avoiding damage of the sample surface [41, 42]. AFM operates in many more modes depending on the specific force interaction (e.g. double-layer forces, electric forces, magnetic forces, Van der Waals forces). Excellent reviews on different operation modes in AFM can be found in the literature [29, 43–45].

For in situ investigations of electrochemical processes, AFM has to be equipped with a liquid cell providing the opportunity to add a reference and a counter electrode. Two different arrangements frequently used for electrochemical experiments are shown in Fig. 7.5. The electrochemical cell is located on top of the

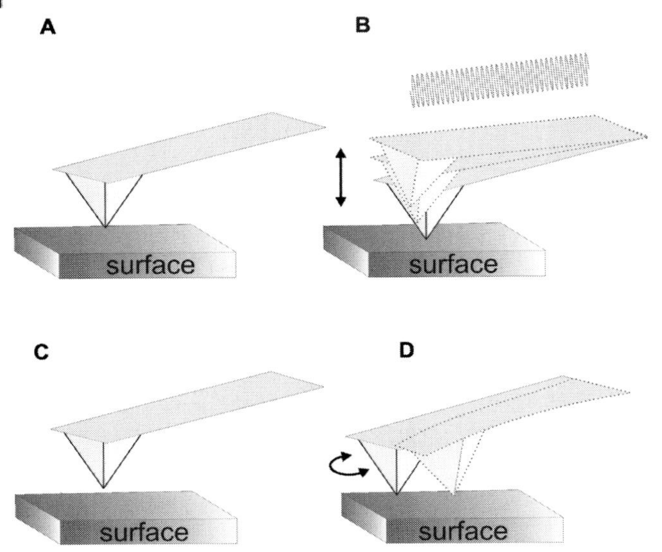

Fig. 7.4 Operation modes in scanning force microscopy: (A) classical contact mode for imaging topography; (B) AC mode (also known as intermittent mode, dynamic mode, 'tapping' mode) for imaging of soft samples; (C) DC noncontact mode for detection of long-range interactions; (D) DC mode for measuring friction forces and elasticity

three-axis piezoelectric positioner. Usually, a glass cell is pressed via an O-ring on to the sample. An alternative approach, providing more flexibility for electrochemical investigations, is based on an inverted design with the cantilever attached to the bottom of the piezoelectric positioner mounted on a free-standing sample stage, which incorporates the electrochemical cell. In contrast to ECSTM, ECAFM is based on an insulating tip, which is primarily used as an imaging tool for topographical changes induced by electrochemical processes occurring at conducting substrates. Owing to the insulating nature of regular AFM tips, no electrochemical processes occur directly at the SPM probe. As an example, structured deposition involving in situ ECAFM is mainly based on hindered mass transfer caused by the physical presence of the AFM tip in close proximity to the sample surface as described in Section 7.2.1. Recently, electrochemically induced dip pen lithographic processes for the deposition of nanometer-sized metal structures have been published [46, 47]. Electrochemical reactions involving redox processes directly at the AFM tip can be achieved by metallization of the cantilever, which is mainly applied for studying dissolution or deposition processes. However, modification of the AFM tip with a metal layer frequently leads to a significant increase in the tip curvature, which results in reduced topographical resolution.

The lateral resolution obtained with AFM is usually lower than that with STM owing to the larger tip–sample contact area. In STM, each imaging point is char-

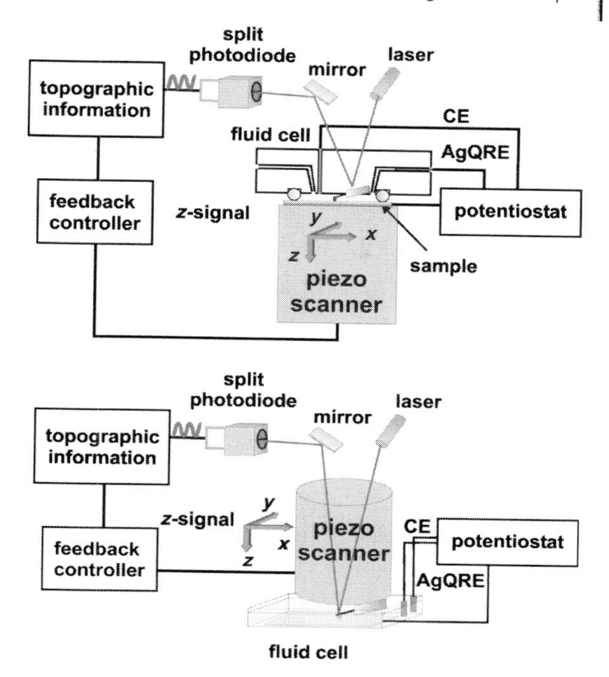

Fig. 7.5 Schematic of ECAFM set-up. Top: the sample stage is mounted on the piezo-positioner. Bottom: the cantilever is mounted on the piezo-positioner. The feasibility of conducting electrochemical experiments is enhanced by improved access to the electrochemical cell

acterized by the dependence of the current, which flows between ideally an atomic apex of the STM tip and the sample surface. The resolution attainable in AFM imaging is generally determined by the increased probe tip area and is affected by a variety of tip and sample properties, including the tip sample microcontact, the dependence of the contact area on the force, the size and shape of the tip terminus and the elasticity of the sample. The routinely achievable resolution is usually in the range of few nanometers and requires sharp tips, low loads and relatively rigid sample films. Nevertheless, atomic resolution has been demonstrated with AFM [48, 49].

7.1.2.3 Near-field Scanning Optical Microscopy (NSOM)

NSOM is based on the near-field interaction of light with solid matter and potentially useful for in situ investigations of electrochemical processes. This noninvasive optical technique was predicted by Synge more than 80 years ago [50, 51] and realized in practice by Pohl et al. [52] and Betzig et al. [53] at the beginning of the 1980s. Synge envisaged an intriguing concept to overcome the limitations in spatial resolution for conventional optical microscopy determined by the diffraction limit of $\lambda/2$ (Abbé criterion). In brief, a laser beam is guided through a tiny aperture (e.g. optical fiber) to the sample surface while the aperture is scanned across the sample. Light transmitted by aperture and sample is collected by a classical optical microscope or by a focusing mirror and usually detected with a photomultiplier. The detected light flux varies with the dielectric and topographical properties of the sample surface. The spatial resolution in NSOM is therefore limited only by the size of the aperture and the proximity of the aperture to the sample surface and not by the wavelength of light, as is the case in conventional microscopy. As shown schematically in Fig. 7.6, two main concepts of near-field optical arrangements have been developed: the 'aperture approach' and the 'apertureless approach'. In the aperture approach, a thin optical fiber coated with a metal film at the sidewalls guides radiation through a sub-micrometer to nanometer sized aperture at the apex of the fiber to the sample (transmission mode) [54] or collects light, which is reflected from an illuminated surface through this aperture (reflection mode) [55]. In contrast, light is scattered at a sharp metallized tip in the apertureless approach [56]. As a result, these two concepts require different designs of the NSOM probe. The main motivation for introducing the apertureless approach resulted from efforts to improve the lateral resolution obtainable in NSOM based on the aperture approach by decreasing the dimensions of the aperture, which is limited by low radiation output efficiency. In the apertureless approach, light is

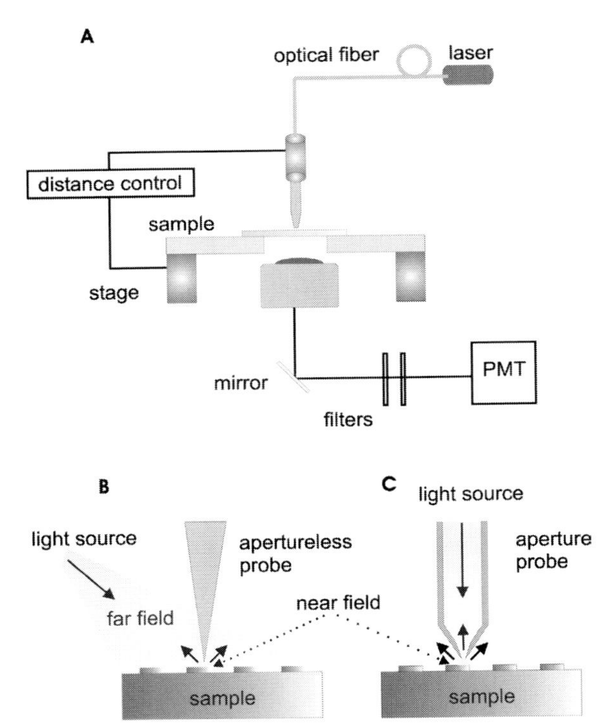

Fig. 7.6 Schematic of aperture and apertureless approach in NSOM. (A) Principle components of NSOM instrumentation (aperture approach). (B) Aperture probe: radiation is guided through a metallized fiber probe, which has a tiny non-metal coated aperture at the apex of the fiber. (C) Apertureless approach based on far-field sample irradiation and near-field scattering in presence of a sharp metallized tip

evanescently generated and a sharp metallic tip positioned in the near-field regime of the sample surface interferes with the evanescent field. Scanning the tip in the near-field regime above the surface allows detection of the probe–sample interaction in the x,y-plane and provides a local map of the near-field intensity. In this case, the resolution is determined by the curvature of the apertureless near-field tip. Both methods appear of equal utility for NSOM experiments as documented by extensive reviews on instrumental development, operation principles and applications [57, 58].

Compared with AFM imaging, NSOM operation in liquids is still not a routine optical imaging procedure. Some of the difficulties of optical imaging in fluids arise from the tip–sample distance control. Distance control in NSOM is usually based on a shear-force detection scheme. The fiber tip dithers at its resonance frequency. In close proximity (5–15 nm) to the sample surface, the dithering amplitude is damped. Optical detection [59, 60] and nonoptical shear force detection [61, 62] are applied for implementing a feedback loop to keep the tip–sample distance constant. Since all optical contrast methods known from far-field optical microscopy can be applied in NSOM, this technique has great potential for imaging electrochemically relevant processes or spectroelectrochemical investigations. Nevertheless, imaging in liquids is still demanding in NSOM and to date the best reported resolution of NSOM operated in liquids is ~60 nm using noncontact AFM for distance control [63]. First attempts towards hybrid technology combining NSOM and SECM (scanning electrochemical microscopy) were reported for electrochemical investigations [64, 65].

7.1.2.4 Scanning Ion Conductance Microscopy (SICM)

SICM is a noninvasive in situ technique that has great potential for studying biochemically relevant electrochemical processes owing to its capability of mapping ion fluxes [66]. In SICM, an electrolyte-filled micropipette housing an electrode is scanned over the surface of a sample immersed in an electrolytic solution (see scheme in Fig. 7.7). A bias voltage is applied to the electrode inside the micropi-

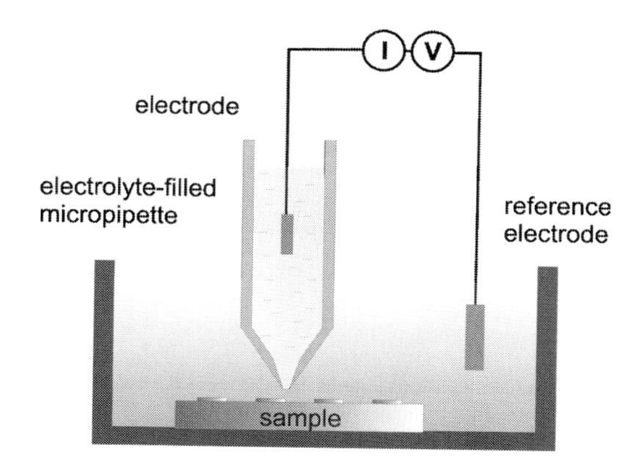

Fig. 7.7 Schematic of the basic principle of SICM. The micropipette is scanned across the sample surface by keeping the ion current constant

pette as ions migrate through the aperture to the electrode outside the micropipette. The pipette–sample distance is kept constant using a feedback loop by controlling the ion current developing across the pipette aperture. Since the current is dependent on the distance between the SICM micropipette and the sample surface, local ion currents can be mapped as the micropipette is scanned at constant height across the sample surface. Improvements in distance control [67, 68] and combination with patch-clamp techniques [69, 70] allow high-resolution imaging of living cells, ion channels and sub-cellular structures. The spatial resolution achievable using SICM is dependent on the size of the tip aperture, which typically ranges from 50 nm to 1.5 μm.

7.1.2.5 **Scanning Electrochemical Microscopy (SECM)**

SECM has gained increasing importance among the scanning probe techniques for the investigation of in situ processes occurring at solid/liquid and liquid/liquid interfaces. Measurements providing information on local (electro)chemical reactivity are based on the disturbance of Faraday currents at a scanned microelectrode depending on the surface chemistry. In SECM the probe is typically a disk ultramicroelectrode (UME) with diameters ranging from 1 to 25 μm. Although, different electrode geometries such as conical electrodes and ring electrodes and electrodes with dimensions <1 μm (see Section 7.1.3.4) have been applied in SECM experiments [71–75]. The unique behavior of microelectrodes including reduced double-layer charging effects, well-defined steady-state currents and a reduced iR drop [76] allows their application as scanning probes. The first experiment investigating local concentrations of redox species at a macroscopic electrode using a microelectrode was reported in 1986 by Engstrom et al. [77]. Simultaneously, Bard and co-workers published results on ECSTM experiments involving Faraday currents at large sample–tip distances [78], followed by a quantitative description of the Faraday current at a microelectrode as a function of the tip–sample distance [79, 80]. Several modes of operation have been developed for SECM. The most prominent imaging modes are the 'feedback mode' [80] and 'generator/collector mode' [81]. Signal generation in SECM is based on surface-induced changes of a Faraday current due to hemispherical diffusion of redox-active species in solution to a biased UME, which is scanned in close proximity across the sample surface [82]. In feedback mode an artificial redox species is added to the solution. The concentration of the redox mediator is influenced by the surface properties of the sample if the tip and sample are in close vicinity. Fig. 7.8 shows a schematic representation of the principle of the feedback mode (A), generation collection mode (B) and direct mode (C). If the disk UME is distant from the surface, the steady-state current $i_{T\infty}$ is given by the following equation (Fig. 7.9 A):

$$i_{T\infty} = 4nFcDa \tag{1}$$

where F is the Faraday constant, n is the number of electrons transferred during the electrochemical reaction at the tip, D is the diffusion coefficient of the redox

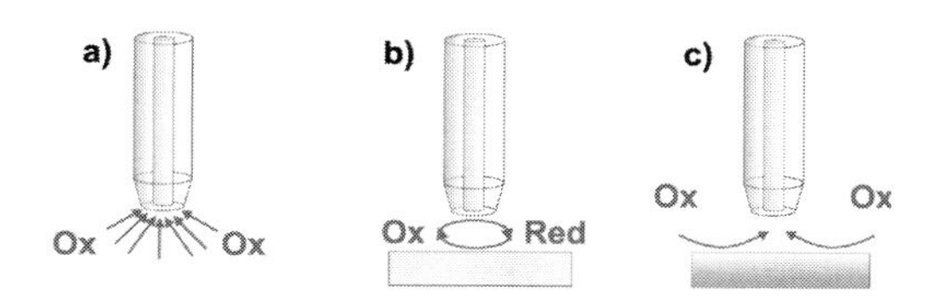

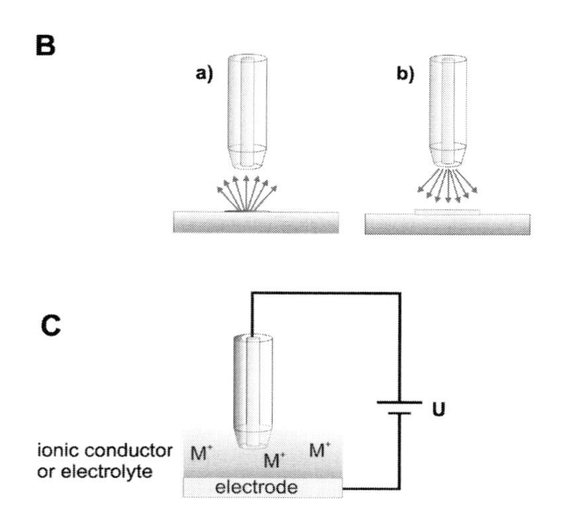

Fig. 7.8 Schematic representation of the basic operation modes of SECM. (A) Principle of feedback mode. (B) Generator/collector mode: (a) substrate generation/tip collection and (b) tip generation/substrate collection. (C) Direct mode

mediator, c is the bulk concentration of the redox species and a is the tip radius. When the tip is brought close (i.e. within a few tip radii) to a sample surface, depending on the nature of the sample, current–distance curves (approach curves) can be obtained (Fig. 7.9). When the electrode is in close proximity (i.e. within a few electrode radii) to a conductive substrate, the mediator oxidized or reduced at the UME diffuses to the substrate where it can be re-oxidized or re-reduced, respectively, leading to an enhanced concentration of redox species within the gap between the UME and the sample surface. Hence an increased current is recorded at the UME ($i_T > i_{T\infty}$). This effect is called 'positive feedback' (Fig. 7.9 B). If the UME is approaching an insulating sample, hemispherical diffusion to the UME is blocked by the sample surface and the current at the UME decreases ($i_T < i_{T\infty}$), called 'negative feedback' (Fig. 7.9 C).

In generator collector mode, the UME detects electroactive species, which are generated at the sample surface. Alternatively, detectable species are generated at the UME and then detected at the sample surface (Fig. 7.8 B) [81]. In both modes, the distance between the UME and sample surface, the size of the UME and the sample properties influence the Faraday current significantly. Surface modifications using deposition or etching processes are mainly performed in the direct

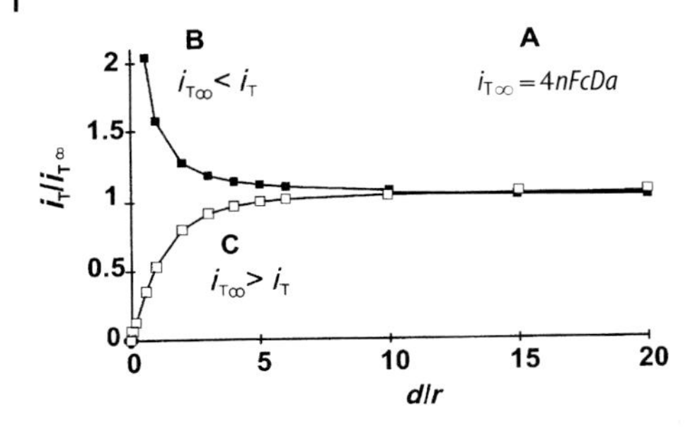

Fig. 7.9 Normalized current–distance curves. (A) UME in bulk solution; (B) UME is approached to a conducting surface and the redox mediator is converted at the UME and recycled at the sample surface (positive feedback); (C) UME is approached to an insulating surface and hemispherical diffusion towards the UME is hindered (negative feedback)

mode of SECM [83]. In direct mode, the UME and a macroscopic substrate electrode form an electrochemical cell (Fig. 7.8 C). The electrodes are immersed in electrolyte solution or an ionic conductor. By applying an electrical field, surface processes can be induced and the modification is restricted to the sample surface directly underneath the UME. Although this technique could be also applied for imaging, it is used almost exclusively for surface modification. Excellent reviews on different modes applied in SECM have been published in recent years [84–91]. Electroinactive species can be detected by using potentiometric electrodes such as ion-selective microelectrodes [92–95]. The optimum working distance (several radii of the active disk electrode) is determined by recording the Faraday current at the UME while the tip is approaching the surface (z-approach curves). The distance is then evaluated by comparing the Faradaic response with theoretically calculated curves [96], assuming an ideal geometry of the UME.

A valuable asset of SECM compared with other SPM techniques is that electrochemical data obtained in SECM experiments can be theoretically characterized and quantitative information on mass transport can be derived. Mass transport characteristics of SECM have been theoretically described for a variety of experimental conditions and different geometries with electrodes under diffusion-limited conditions [85, 97–102]. For quantification of the data obtained, the geometry of the UME and the absolute distance d between the electrochemical probe and the sample surface have to be known parameters, which can be derived by recording approach curves. Usually, the steady-state current at the UME in feedback mode is recorded as a function of the distance to the sample surface.

Usually, imaging is achieved by scanning the UME in a fixed height in the x,y-plane across the sample surface ('constant-height imaging'). Consequently, the currently achievable lateral resolution obtained with SECM in constant-height mode is still not comparable to the lateral resolution provided by AFM and STM imaging, owing to the 'fixed-height' problem. The working distance is correlated with the UME size and decreases with reduced electrode size. Additionally, in constant-height imaging the electrochemical response at the UME may be convoluted by topographic surface features and changing reactivity of the sample surface [103]. Improving the lateral resolution in SECM requires (i) separation of current information from topographic information and (ii) reduction of the UME size. Recently, several papers have been published describing fabrication of nanoelectrodes based on electrochemical etching [104–108] or based on laser-based micropipette pulling of metal microwires [96, 109]. However, accurate control to a few nanometers distance between the nanoelectrode and the sample surface cannot be achieved in conventional constant-height imaging. Tip crashes due to surface roughness, topographical features and tilt of the sample may destroy the sample and the tip.

Within the last decade, several groups have focused on approaches for alternative tip positioning to overcome the described limitations [103, 110–121].

The first successful separation of tip positioning from the electrochemical response was published by Ludwig et al. [113]. In principle, tip position control based on optical shear forces has been derived from near-field scanning optical microscopy and was adapted to SECM by laterally dithering a fiber-shaped microelectrode at its resonant frequency. In close proximity to the sample surface, the vibration amplitude of the microelectrode is attenuated owing to hydrodynamic effects, which is recorded using a laser beam focused on the fiber-shaped electrode and detected via a split photodiode. Based on lock-in amplification, a feedback loop can be integrated into the tip positioning software routine of the SECM keeping the distance between the tip and the sample surface constant. An alternative to optical shear force detection was developed by Karrai and Grober based on a tuning fork as vibration amplitude detector [122]. Furthermore, variations of the nonoptical shear force detection are frequently used in SECM experiments [117, 119, 121]. Recently, a highly sensitive, nonoptical shear force-based distance control mechanism has been developed for SECM tips to simplify constant-distance SECM [120]. The key component of the SECM setup with nonoptical, shear force-based constant-distance control is an integrated, piezoelectric shear force detection system consisting of two piezoelectric plates. One plate acts as stimulator and the other detects the resonance frequency. Phase-sensitive amplification of its alternating voltage is used to monitor the amplitude of the tip oscillation via a lock-in amplifier. Again, computer-controlled closed-loop feedback is used, which constantly regulates the tip-to-sample spacing. Fig. 7.10 provides a schematic overview of the different distance control schemes used in SECM. Given the delicacy of accurate positioning, the application of nanoelectrodes as scanning probes in SECM experiments is still limited, albeit their fabrication is well established. Hence the majority of routine SECM experiments are performed with electrode sizes ranging from 1 to 25 μm.

7.1.2.6 Combined SPM Techniques

Based on the limitations of individual scanning probe microscopies, the combination of complementary techniques and in particular the combination of SECM with high-resolution imaging provided by NSOM, ECSTM and AFM is gaining substantial interest. With respect to the instrumental limitations of electrochemical imaging, several advantages derive from combined SPM techniques: (i) enhanced information on electrochemical processes at the solid/liquid interface, (ii) simultaneously obtained complementary parameters and (iii) facilitated positioning of nanometer-sized probes.

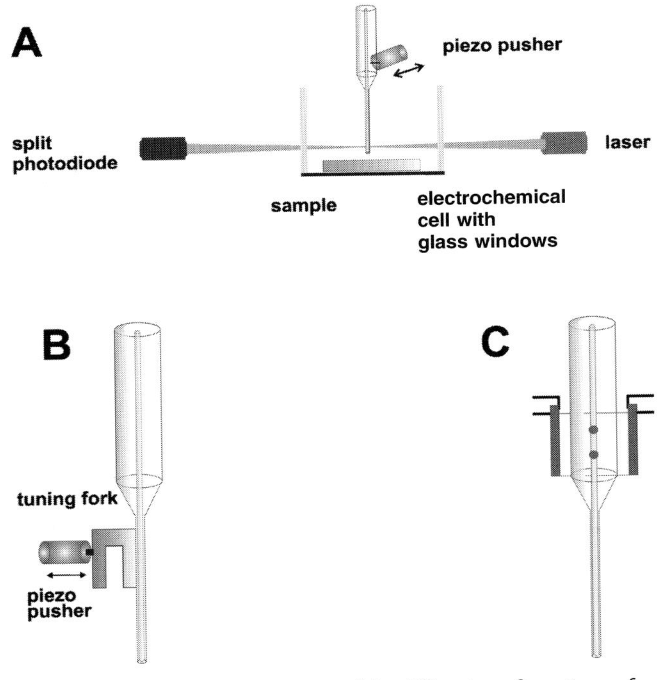

Fig. 7.10 Schematic representation of the different configurations of SECM tips for current independent positioning. (A) Optical shear force detection; (B) and (C) nonoptical shear force distance control based on tuning fork (B) and piezo elements (C)

Initially, applications of combined scanning probe techniques were mainly utilized to investigate corrosion processes at oxide-passivated metal electrodes. Williams et al. reported a combination of SECM and ECSTM for the investigation of pitting corrosion of stainless steel [123]. A commercial AFM was adapted to perform combined ECSTM–SECM experiments. For this purpose, a Pt–Ir wire was etched and insulated similarly to the fabrication of ECSTM probes. After positioning of the probe using the tunneling current of STM, the tip was retracted to a designated distance suitable for SECM imaging. The lateral resolution achievable with this approach was in the micrometer range. Treutler and Wittstock reported combined tips with an electroactive area down to 4 nm by improving the tip fabrication [124]. Hence, using the positioning principle of ECSTM (usually based on modified commercial AFM/STM instrumentation) and retracting the combined electrode to the sensitive working distance for the SECM measurement, the lateral resolution could be significantly improved. However, both approaches are based on a sequential acquisition of STM and SECM data. After recording the ECSTM image, the tip is retracted and the electrochemical image is recorded. Hence data evaluation of time-sensitive surface processes or materials with temporally changing surface properties may be difficult.

The first attempts at combining NSOM and SECM were described by Smyrl and co-workers [64]. Although no combined measurements were presented, a modified optical fiber was reported with the potential to serve as a combined NSOM–SECM probe. These combined probes were mainly used for SECM imaging with independent height control [117]. A complementary optical and electrochemical image of an interdigitated electrode array was published recently [65]. However, the lateral resolution of the optical image obtained with this device was in the far-field regime.

The combination of SICM and NSOM has high potential for the investigation of interfacial processes. The NSOM probe is modified to house the electrode immersed in electrolyte solution, which permits simultaneous recording of optical images during ion concentration mapping. By coupling laser light into the aluminum-coated NSOM–SICM pipette and controlling the distance keeping the ion current constant, simultaneous optical and topographical imaging can be achieved [125].

The motivation to overcome the limitations of constant-height imaging in SECM has recently pioneered a series of exciting developments. Due to the high spatial resolution and versatility of AFM, its combination with SECM is of particular interest for electrochemical imaging at the nanoscale. The powerful benefits of merging AFM technology with SECM are based on the direct correlation of structural information with chemical surface activity at excellent lateral resolution [126]. In principle, two different approaches have been devised [127, 128] with both concepts based on the development of a combined scanning probe comprising AFM and SECM functionality. The first approach of combined SECM–AFM imaging in liquid was realized by fabricating bent Pt wire nanoelectrodes shaped as AFM cantilevers merging both functionalities into a single tip [127, 129]. The second approach, developed in our research group, is based on integrating a submicro- or nanoelectrode into an AFM tip by three-dimensional microfabrication. This technology retains the curvature of the AFM tip for high-resolution imaging while the electrode is recessed from the apex of the tip for simultaneous electrochemical imaging [128, 130, 131]. Both approaches for combined AFM–SECM imaging are applicable to any standard AFM instrument equipped with a liquid AFM cell and an additional channel to read in the electrochemical data.

Comparing both concepts, the lateral resolution for electrochemical imaging is so far superior utilizing cantilever-shaped bent wire electrodes. However, simultaneous imaging is usually limited to contact mode and robust, nonconducting samples owing to the contact of the cantilever-shaped nanoelectrode and the sample surface. Conducting samples require noncontact imaging, avoiding contact of the electroactive area located at the very end of the tip. In the bent wire approach, this requires scanning the sample surface twice: first, topographical information is recorded in contact mode operation without electrochemistry involved; second, the sample surface is scanned again using the recorded topographical information to keep the tip at a defined distance from the surface (lift mode) to perform electrochemical imaging [129].

In the microfabrication approach, the smallest electrode integrated into a standard AFM tip is in the diameter region of ~ 150 nm [132]. However, there are several distinct advantages to this technology. Due to the design of the integrated AFM–SECM tip with a recessed electroactive area, there is no restriction with respect to the nature of the sample, i.e. whether the surface is conducting or insulating. As the AFM functionality of the integrated tip is based on insulating tip materials (e.g. silicone nitride), topographic imaging can be performed in AFM contact mode or dynamic mode entirely independently of the properties of the investigated surface. Consequently, topographical and electrochemical information can be obtained simultaneously on any sample and dynamic operation modes (e.g. tapping mode) of AFM for investigation of soft samples can be applied for combined simultaneous SECM–AFM measurements [132].

7.1.3 Fabrication of Scanning Probes for Electrochemical Application

The lateral resolution in scanning probe microscopy is mainly determined by the size and shape of the scanning probe tip. Hence fabrication processes based on reproducible, highly accurate procedures for manufacturing SPM probes are of great importance. In the following section, a survey of procedures for SPM tip preparation suitable for in situ electrochemical investigations is discussed.

7.1.3.1 ECSTM Probes

Imaging artifacts due to improper probe geometries have always been a concern in imaging at atomic resolution in topographic STM. Ideally, tunneling takes place between a single atom terminating the STM tip and the sample surface. However, in practice more than one atom may have the same distance to the sample surface, leading to 'multiple tips' and thus imaging artifacts [133]. For rough sample surfaces, an imperfect tip geometry consisting of several 'mini tips' or an improper macroscopic shape such as a relatively shallow angle of the tip cone can also lead to imaging artifacts.

Since the tip may be in direct contact with the sample surface during the imaging procedure, robust tip materials are preferable. After iridium and molybdenum, tungsten is the most commonly used tip material. Several strategies of tip preparation were applied and reviewed in great detail by Melmed [134]. Frequently, iridium–platinum alloys are used in ECSTM, since they provide sufficient stiffness and inertness to oxidation. These tips can be fabricated by simple cutting; however, this procedure can lead to multiple 'mini tips' resulting in tip artifacts during imaging. Controlled electrochemical etching of metal wires is widely used to produce atomically sharp tips. Immersing a metallic wire vertically into an electrolyte solution and applying a potential between the wire and a counter-electrode initiates the electrochemical etching process [135–143]. After a certain etching period, the lower part of the wire simply drops off and a sharp tip is formed at the breaking junction. Depending on the material, KOH or NaOH solutions are used for etching tungsten [136, 141] and $CaCl_2$–HCl [138] or KCN–KOH [144] solutions are preferred for etching Pt–Ir. In the case of oxidizable tip materials, the oxide layer has to be removed prior to application. Ion milling can be applied, which additionally helps to decrease further the curvature of the tips. Focused ion beam (FIB) milling, as discussed in Section 7.1.3.5, leads to tip radii as small as 4 nm and cone angles $<10^\circ$ [145]. More detailed descriptions of the influence of etching parameters on the tip geometry can be found elsewhere [139–142]. In ECSTM, the tunneling current is superimposed by the Faraday current resulting from electrochemical processes at the substrate–electrolyte and tip–electrolyte interfaces. Hence the electroactive tip surface in contact with the electrolyte solution has to be coated with an electrically insulating material except the very end of the STM tip. Several coating materials including glass [16], electrophoretic paint [146], paraffin [144] Apiezon wax [104] and chemically inert thermoplasts [20] have been applied. Fig. 7.11 shows an image of an etched Pt–Ir tip insulated with Apiezon wax. By reducing the electroactive area to a size of $\sim 10^{-8}$–10^{-7} cm^2, the contribution of the Faraday current is <50 pA, which does not interfere with tunneling currents, which are typically in the range 1–10 nA for STM imaging.

A

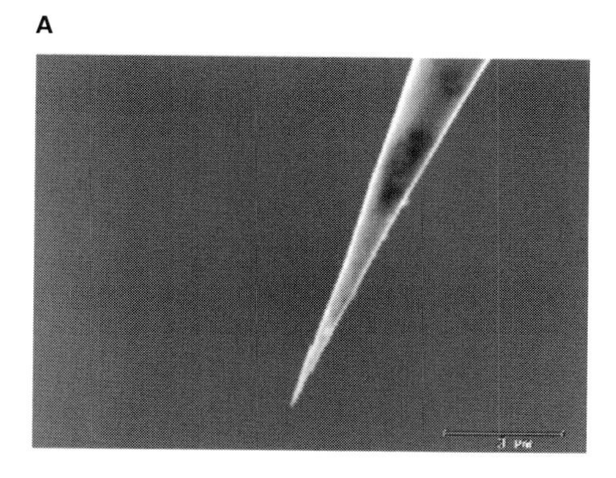

B

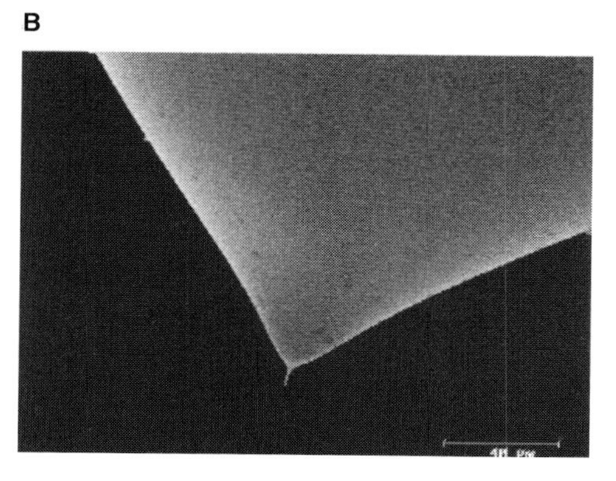

Fig. 7.11 SEM image of etched STM tips: (A) etched tungsten tip; (B) etched tungsten tip insulated with Apiezon wax. The etched fiber is pulled twice through molten Apiezon wax. The apex of the tip is not insulated. Reprinted with permission from [141]

7.1.3.2 Metallized AFM Tips

ECAFM investigations of electrochemical processes occurring at solid/electrolyte interfaces are usually performed with unmodified AFM cantilevers. By applying a potential to a macroscopic electrode surface in electrolyte solution, topographical changes in microscopic domains induced by electrochemical surfaces processes are imaged in situ with AFM. However, several approaches have been published in which the electrochemical reactions – mainly dissolution or deposition processes – are actively triggered by applying a potential to a metallized AFM tip [147–149]. Metallized AFM cantilevers can be easily produced by sputtering or evaporating a thin metal layer on to commercially available AFM cantilevers [150]. However, sputtering an additional layer on to the AFM cantilever increases the curvature of the AFM tip, which usually results in decreased lateral resolution. Furthermore, the electroactive area covers the entire AFM tip and cantilever. Hence lateral contrast on electrochemical information is attainable if only a small part of the cantilever is immersed in the electrolyte solution. Alternatively, manual deposition procedures can be applied to insulate the cantilever mount, resulting in an electroactive area of the size of the cantilever [147].

7.1.3.3 Probes for NSOM and SICM

Two different probe designs dominate in NSOM: (i) fiber-shaped probes with a defined aperture and (ii) apertureless probes. The resolution and quality of the images obtained in aperture NSOM are strongly dependent on the quality of the aperture delivering radiation to the sample surface. Betzig et al. successful introduced a design for reproducible fabrication of tapered NSOM fibers, based on heating and pulling an optical glass fiber under controlled conditions [60]. Alternatively, glass fibers can be etched by penetrating hydrofluoric acid through a protective organic layer [151]. Although the output of high-quality tips from an etching process is lower compared with pulled fibers, the optical throughput of etched fibers is increased owing to larger cone angles [152, 153]. Following the fiber processing, an opaque metal layer (mainly aluminum) with a thickness of 50–100 nm is coated on the fiber core except at the apex of the tip. Thus, a small aperture is created at the fiber tip, avoiding radiation loss of guided light along the fiber. The quality of the metal coating is crucial for the performance of the NSOM tip. Typical apertures produced by fiber pulling range around 50 nm with a cone angle of $\sim 20^\circ$. More recently, focused ion beam milling has been used to create highly defined apertures at the apex of the optical fibers [154, 155]. FIB milling allows precise cutting off of the very end of the fiber-optic NSOM probe, producing smooth, unstructured aperture regions eliminating problems with, e.g., material grain structures.

Combined AFM–NSOM probes have been fabricated by modifying the AFM cantilever with optically transparent tips instead of the usual Si or Si_3N_4 materials [156]. The lateral resolution achievable with aperture probes is limited to the tens of nanometers range by the achievable light output of the aperture. Hence increasing optical resolution requires alternative concepts for NSOM probes. In principle, an apertureless NSOM probe is a sharpened metal tip. Hence many probe designs or even lasing nanoparticles can be envisaged to serve as apertureless NSOM probes. Alternatively, apertureless probes can be sharpened by FIB technology, increasing the spatial resolution of the optical information. More detailed information on the state-of-the-art in NSOM technology was provided by Dunn [57].

In SICM, the lateral resolution is dependent on the dimensions of the orifice at the end of the micropipette, which is similar to the aperture of fiber-shaped NSOM probes. Hence similar fabrication strategies based on heating and pulling utilizing commercially available pipette pullers have been applied [125]. Micromachined SICM probes based on conventional photolithographic processes have also been described [157]. Improvements to the SICM probe orifice can also be achieved by FIB technology in combination with microfabricated SICM probes, in particular since reproducible fabrication of nanometer-sized orifices will dramatically increase the lateral resolution attainable.

7.1.3.4 Fabrication of Micro- and Nanoelectrodes

Microelectrodes have been used for a long time for neurophysiological investigations. However, design modifications are required for utilizing microelectrodes as components of imaging tools. The conventional design of microelectrodes used in SECM experiments comprises a disk-shaped electrode with robust, geometrically defined glass insulation. However, different electrode geometries have been used for SECM imaging (see Fig. 7.12). Disk electrodes are fabricated by coaxial placement of a noble metal wire with diameters of 5–50 μm inside a glass capillary sealed at one end. Capillaries are usually made of borosilicate or soda-lime glass for gold wire electrodes. The wire is melted under vacuum into the glass tube and the sealed end is ground with sandpaper until the cross-section of the wire is exposed. Successive polishing with a series of diamond pastes results in a corrugation-free surface of the disk-shaped electrode. Electrical connection of the microwire is usually achieved by attaching to a copper wire with silver epoxy paint to the top of the capillary. For a more detailed description, several reviews on fabrication of microelectrodes are recommended [158, 159]. Disk microelectrodes with diameters of 1–5 μm are usually fabricated from Wollaston wire following the same procedure. However, prior to sealing off the capillary, the silver coating has to be removed with nitric acid. For imaging experiments the insulating glass shielding is conically reduced to a ratio of the glass shielding to the UME radius (*RG* value) of ~ 10. Geometrically reducing the glass insulation avoids contact of the shielding and substrate in case of a deviation in the axial alignment between the sample surface and electrode, while still providing sufficient contrast for imaging of insulating substrates. Chemical selectivity may be increased by introducing amperometric UMEs functionalized with appropriate recognition chemistries at the electrode surface, which is particularly relevant for the investigation of biological/biochemical redox processes [160].

Ring-shaped electrodes can be fabricated by deposition of a metal layer on a cylindrical support material with micrometer-sized diameter. Microfabrication tech-

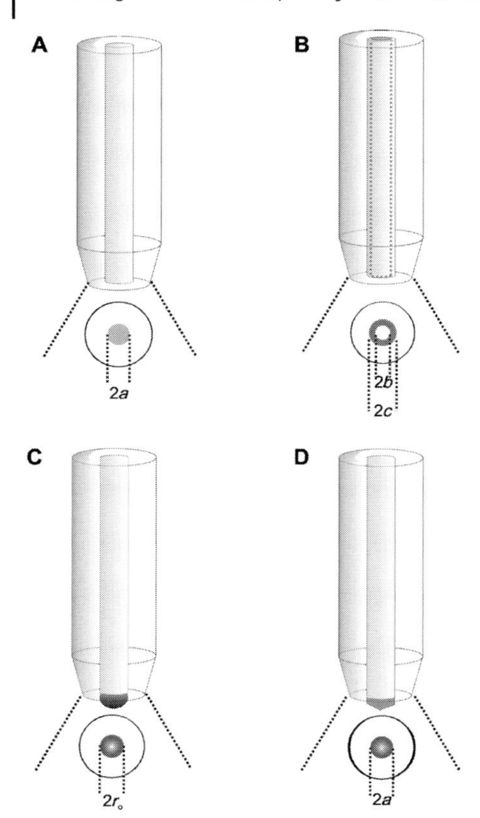

Fig. 7.12 Microelectrode geometries used in SECM experiments: (A) disk microelectrode; a=radius of the disk. (B) Ring microelectrode; b=inner radius of the ring, c=outer radius of the ring. (C) Hemispherical microelectrode; r_0=radius of the hemisphere. (D) Finite conical microelectrode; a=radius of the disk

niques such as sputtering or vapor deposition are usually applied to deposit the electrode material, providing ring electrodes with a layer thickness of 10 nm–5 µm. Metal coatings can be insulated from the ambient solution by epoxy resins. Carbon ring electrodes have been fabricated by pyrolyzing methane within a pulled quartz capillary. After a carbon ring is formed, the capillary is filled with epoxy and beveled to expose the carbon ring electrode [161]. Combining NSOM with SECM can be realized by coating of an NSOM fiber probe with a metal layer and successive insulation towards to solution providing an exposed microring electrode around the aperture of the optical fiber tip. Shi et al. demonstrated this concept by coating an optical fiber with gold, providing a SECM microelectrode [162]. A combined probe suitable for simultaneous optical and electrochemical measurements was published recently by Bard and co-workers [71].

The fabrication of nanometer-sized electrodes (also called 'nanodes') was reported at the beginning of the 1990s [74, 163]. However, application as an electrochemical imaging tool was realized only 10 years later. The fabrication of nanoelectrodes is mainly based on etching processes similar to the fabrication of STM tips [108, 144] or by drawing an embedded metal wire with laser pipette pulling techniques [96, 109]. Electrodes with diameters of ~4 nm have been reported [109]. Several strategies and materials have been published for insulation of these probes in order to expose a well-defined electroactive area suitable for electrochemical imaging [107, 164, 165]. Bent nanoelectrodes shaped like AFM cantilevers have been fabricated by etching and successive insulation with electrophoretic paint providing an electroactive area at the very end of the tip [127]. A problem still not solved with these approaches is the mediocre reproducibility of the fabrication process and the lack of control for defined electrode geometries [126]. The dimensions of the nanoelectrode can be derived experimentally by measuring the diffusion-controlled steady-state current. The electrode shape is frequently derived by comparing experimentally obtained approach curves with theoretically calculated curves. However, for quantification of data obtained in nanoscale electrochemical imaging experiments, a precise knowledge of the distance between the scanning tip and the sample surface and known and well-defined electrode geometries are prerequisites.

7.1.3.5 Microfabricated Combined Scanning Probe Tips

One of the major advantages of AFM among all the scanning probe techniques, in addition to its versatility and broad applicability, is the well-established fabrication of AFM cantilevers [166]. Based on microfabrication technology, batch processing of silicon and silicon nitride cantilevers can be routinely achieved at comparatively low cost and with high reproducibility.

Several strategies for combined scanning probes for AFM–NSOM [156, 167] and combined STM–potentiometric sensing probes [168] based on microfabrication techniques have been published in recent years. Our group introduced a successful concept for combining SECM and AFM based on integrating a microelectrode into an AFM tip based on microfabrication technology [128, 130, 169, 170]. The design in Fig. 7.13 permits simultaneous recording of topographical and electrochemical properties of the sample surface while keeping the distance between the integrated electrode and the surface constant. Predominantly thin-film chemical vapor deposition (CVD) and sputter techniques along with focused ion beam (FIB) micromachining originating from microelectronics are used, ensuring reproducible fabrication of nanoelectrodes integrated into conventional AFM tips. This development enables an electrode to be integrated at a precisely defined and deliberately selected distance from the apex of a scanning probe tip. Conventional silicon nitride cantilevers are sputtered with a metal electrode layer (mainly gold and platinum) at a thickness of ~100 nm. The cantilever is then insulated based on plasma-enhanced chemical vapor deposition (PECVD) for silicon nitride or sandwiched layers of silicon nitride and silicon oxide or by using chemical vapor polymerization (CVP) for parylene C. Owing to its high uniformity and biocompatibility, parylene C is the preferred choice for many applications in the biomedical field [170]. In order to expose an electroactive area recessed from the apex of the tip, a fabrication step allowing 3D structuring has to be applied. FIB technology is an attractive tool for various maskless structuring processes and a well-established

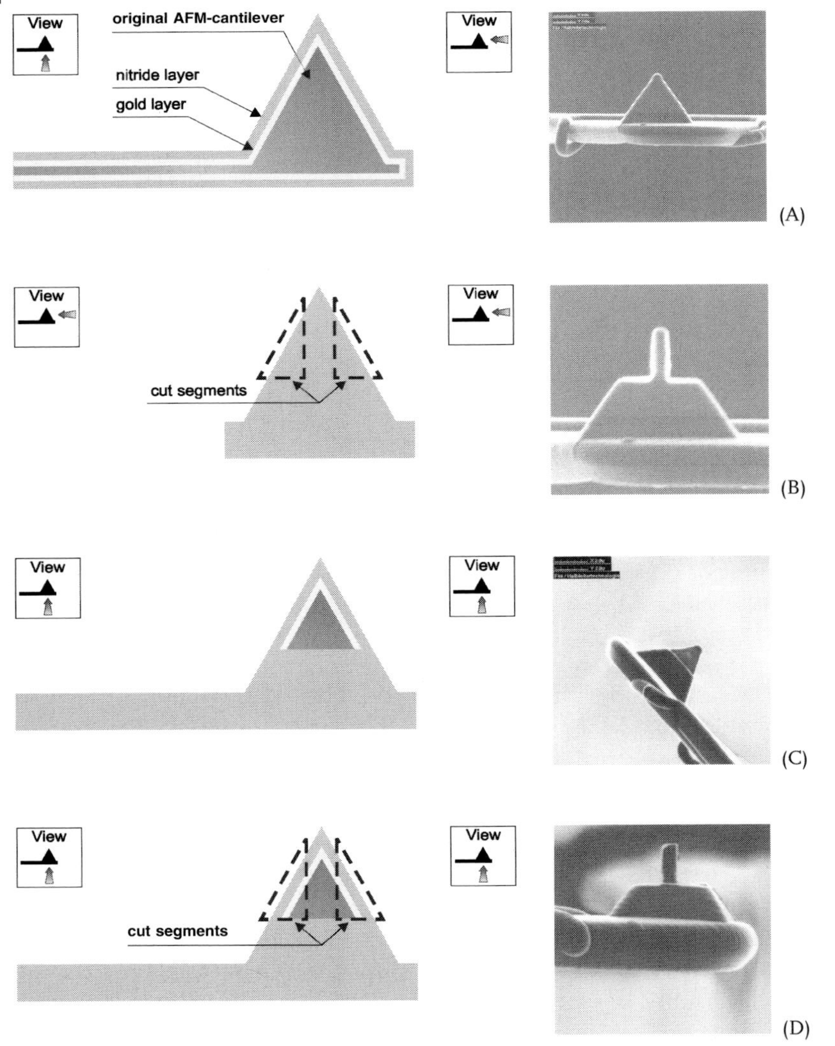

Fig. 7.13 Design of an integrated SECM–AFM probe. Processing steps of a modified AFM tip using FIB technology showing in the left column a schematic view of the processing step and in the right column the corresponding FIB images. (A) AFM cantilever after coating with the gold layer and the silicon nitride insulation. (B) Diametrically opposed FIB cuttings along the dotted lines. (C) Side view after step (B). (D) Repetition of the diametrically opposed FIB cuttings of step 2 along the dotted lines after turning the cantilever by 90° creating a free-standing square pillar. (E) Re-modeling of the non-conductive AFM tip by FIB cuttings along the dotted lines on all four sides of the square pillar. (F) FIB image of the final *integrated frame microelectrode* (edge length: 2.2 μm) after 'single-pass milling' along the dotted lines for removal of re-deposited material from the electroactive surface. Reprinted with permission from [128]

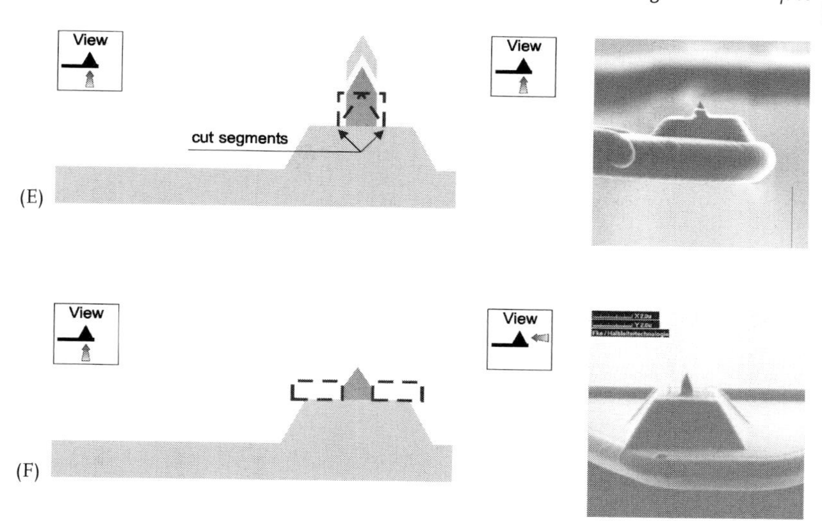

Fig. 7.13 (continued)

technique in the semiconductor industry [171]. Recently, FIB milling has been increasingly applied for scanning probe manufacturing, e.g. ultra-sharp AFM tips with needle-like probes, providing high aspect ratios for the investigation of narrow grooves [172]. Several FIB milling steps are performed to expose reproducibly an electrode recessed from the apex of the AFM tip (see Fig. 7.13). Finally, the original AFM tip is reshaped, correlating the tip height to the dimensions of the integrated electrode. Hence the optimum working distance between electrode and sample surface and high quality for AFM imaging are ensured. Several different electrode geometries have successfully been integrated into AFM probes, including frame-, ring- and disk-shaped electrode structures (see Fig. 7.14).

The resolution of FIB systems promises integration of nanoelectrodes in the near future, although electrode dimensions accessible with etching and pulling techniques have not yet been reached based on the illustrated microfabrication technology. So far, disk-shaped integrated nanoelectrodes with diameters as small as 150 nm have been fabricated [132]. A distinct advantage of this technology is the integrity of the basic features of the AFM cantilever. The quality of topographical images is determined by the precision of FIB milling, which is suitable for the fabrication of ultra-sharp AFM tips [172]. Furthermore, the bifunctional probe can be used in all operational modes provided by AFM, e.g. dynamic mode for imaging of soft sample surfaces [173]. It should be noted that the concept of integrated electrodes can be extended to a variety of different electrode materials, e.g. pH-sensitive oxides and to the integration of multiple electrodes.

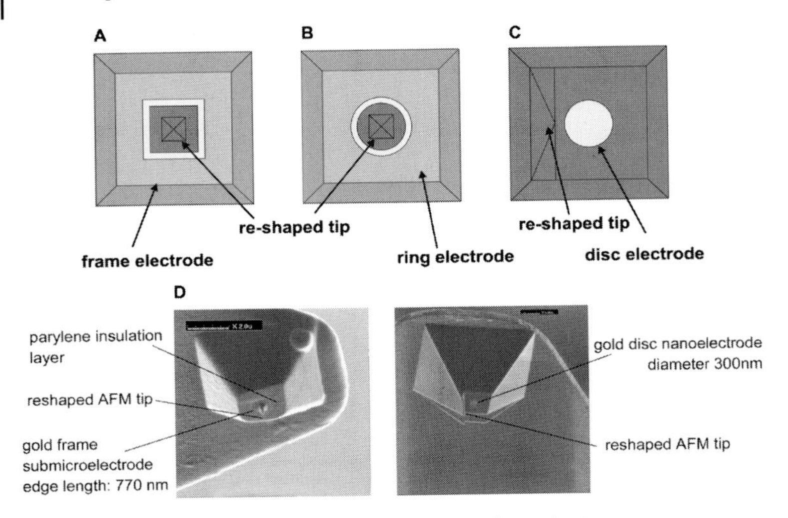

Fig. 7.14 Attainable geometries of micro- and nanoelectrodes integrated into AFM tips: (A) frame electrode; (B) ring electrode; (C) disk electrode. (D) FIB image of (left) an integrated frame sub-microelectrode with an edge length of 770 nm and a tip height of 700 nm and (right) an integrated disk nanoelectrode with a diameter of 300 nm. Reprinted with permission from [132]

7.1.3.6 Carbon Nanotubes

Very recently, carbon nanotubes have been identified as having promising potential as electrochemical scanning probe tips. The electrochemical behavior of carbon nanotubes and their application as nanometer-sized electrochemical sensors in analytical and biological applications have already been demonstrated [174]. Their unique features combining a high aspect ratio and sub-nanometer radius of curvature with chemical stability and mechanical robustness have identified carbon nanotubes as an interesting alternative to conventional scanning probe sensors. They provide a unique combination of high resolution and high aspect ratio imaging owing to their exceptionally large Young's modulus [175]. In addition, carbon nanotubes bend elastically instead of fracturing under large forces, which makes them highly robust for AFM probe applications. The first approaches using carbon nanotubes as SPM probes have been published recently [176–178]. Early attempts at using carbon nanotubes as conductive probes used the concept of adhesive pick-up of carbon nanotubes with an AFM probe [179]. The assembly of a carbon nanotube attached to the AFM tip was then modified with a conducting metal layer. The resulting nanowire has a high aspect ratio, providing a beneficial geometry for both electric and magnetic force microscopy and an ideal geometry for imaging small, deep pore structures. However, for fabrication of electrochemical scanning probe tips similar to the SECM probes, the metallized carbon nanotube has to be insulated, which increases the total diameter of the wire and demands further microstructuring for developing a combined electrochemical scanning probe tip. Merging the benefits of carbon nanotube probes for SPM along with sophisticated microstructuring may lead to combined electrochemical scanning probes tips with unique imaging properties in the near future.

7.2 Applications

The introduction of scanning probe microscopy had a substantial impact on the fundamental understanding of processes occurring at the solid/liquid interface. Investigations of the initial states of corrosion and heterogeneous electron transfer reactions with high lateral resolution and in real time provided by in situ SPM revolutionized electrochemical surface science. Furthermore, a wide variety of biologically and biomedically relevant reactions occur at surfaces or interfaces and are frequently related to structural changes of the surface. SPM techniques have successfully been applied to correlate structural changes with physical or chemical properties. Ongoing progress in probing electrochemical surface features with scanning probe microscopy and selected fundamental applications will be discussed in the following section.

7.2.1 Material Characterization and Modification

Prior to the development of scanning probe microscopy, structural information about electrode surfaces had to be inferred from the results of spectroscopic or macroscopic electrochemical measurements. The development of in situ electrochemical SPM had a strong impact on many areas of interfacial science. SPM techniques have gradually matured into powerful tools for imaging electrode surfaces in an electrochemical environment with atomic resolution, contributing to an in-depth understanding of a wide variety of electrochemical processes.

Applications of electrochemical STM in the field of probing solid/liquid interfaces include investigating electron transfer processes [180, 181] and in situ imaging of gold and platinum electrode surfaces [182–184]. Furthermore, ECSTM has been applied for fabricating nanostructures and for studying thin-film depositions [185–187]. ECSTM studies imaging potential-dependent phase transitions (surface reconstruction) of multiple single-crystal electrodes including gold [188] and platinum [189] have been described. Fig. 7.15 A shows an in situ STM image in 0.1 M H_2SO_4 electrolyte solution of a freshly prepared flame annealed Au(100) surface demonstrating hexagonal-close-packed surface atoms. When the electrode potential is scanned in positive direction, reconstruction is lifted at about +0.36 V vs. SCE due to specific adsorption of sulfate ions. The (hex) $\rightarrow$ (1×1) transition causes a pronounced current peak visible in the cyclic voltammogram at this potential. The unreconstructed surface, which has a fourfold symmetry, is covered with small mono-atomic gold islands (Fig. 7.15 B) formed due to the excess of surface atoms in the more densely packed (hex)-structure. Scanning the potential

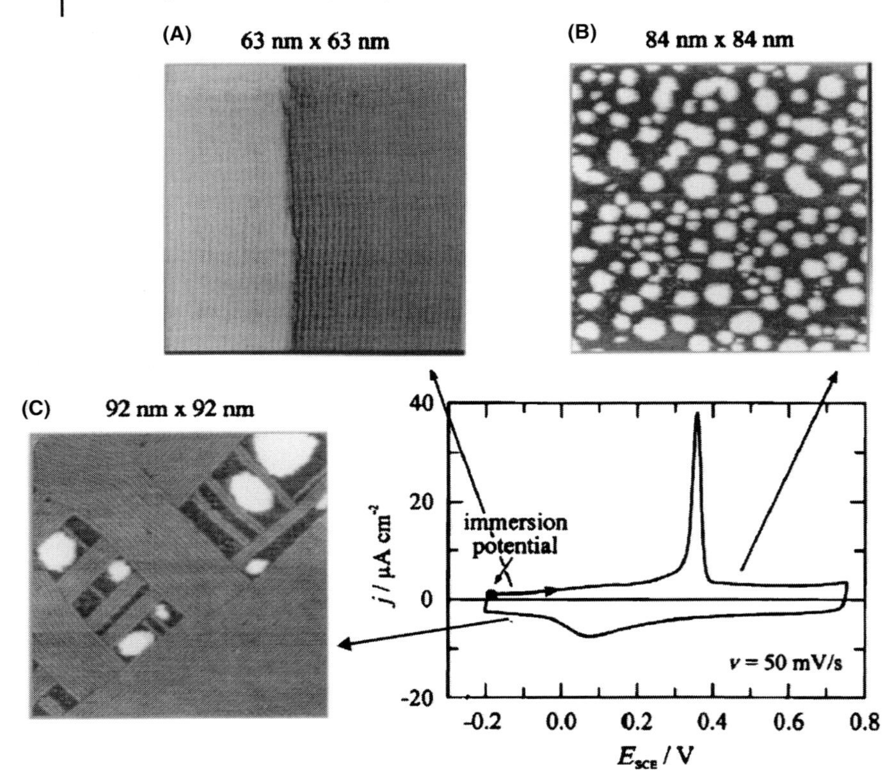

Fig. 7.15 Sequence of ECSTM images representing surface reconstruction of Au(100) in 0.1 M H_2SO_4. (A) The freshly prepared reconstructed surface at –0.2 V vs. SCE, (B) the unreconstructed surface immediately after lifting of the (hex) structure at +0.36 V vs. SCE and (C) the surface after potential-induced reconstruction. The corresponding cyclic voltammogram shows the current peak at +0.36 V vs. SCE representing the (hex) → (1×1) transition. Reprinted with permission from [190]

back from positive values (e.g. +0.75 V vs. SCE) to the starting point at –0.2 V vs. SCE leads to a slow reconstruction of the (hex) surface illustrated in Fig. 7.15 C.

Electrochemical SPM has become an indispensable tool for gaining mechanistic insights into electrode reactions at single-crystal electrodes. Excellent reviews describing structure studies of metal electrodes by in situ STM [25, 190–192] and electrochemical applications of in situ scanning probe microscopy [193–195] have been published. Specific topics which are addressed during ECSTM studies include the characterization of the structure of bare metal surfaces, monolayers and multilayers of metals formed on top and of semiconductor electrodes and adsorbates. Especially the occurrence of adsorbates and adsorbate ordering is a central issue in electrochemistry. Electrochemical AFM has been used to monitor electrodeposition processes of metals [196–198], electrodeposition and characterization of conducting polymers [199–201] and thin-film deposition on highly oriented pyrolytic graphite (HOPG) [202, 203]. For the characterization of adsorbate layers, different operating modes of AFM have been applied to perform adhesion and elasticity measurements of monolayers in electrochemical environments. Recently, dip-pen nanolithography induced by electrochemical reactions has been reported [46].

The SECM configuration of positioning a UME in close proximity to a sample surface has been widely used for surface modification. Microstructured deposition of metals [204, 205] and conducting polymers [112, 206, 207] and etching processes of semiconductors [208, 209] have been performed with the direct and feedback mode of SECM. An attractive unique feature of SECM is the possibility of inducing surface reactions and characterizing the results or simultaneously monitoring the reaction rate. A comprehensive review on surface modification with SECM has been published [83]. Microscopic analysis and reaction rate imaging of composite materials and modified electrodes have been performed with SECM [87]. Recently, Bard and co-workers published several contributions on SECM investigations concerning the electrical properties of self-assembled monolayers at a molecular level, including charge transfer through these films [210, 211].

7.2.1.2 Dissolution and Crystallization Processes

Dissolution and crystallization processes occurring at solid/liquid interfaces are of key importance in a wide variety of chemical reactions [212] and have been studied extensively using a variety of electrochemical SPM techniques. Focus on in situ investigations of these processes is centered on exploring the initial stages of dissolution and crystallization, similar to studying deposition processes. SPM studies on corrosion aim at elucidating the initiating steps and the structural evolution of the corroding material. In situ studies of corrosion and dissolution have been conducted at various surfaces including Cu [213–219], steel [220, 221], Ag [222] and other materials [223–227]. Excellent reviews on corrosion have been published [25, 190, 192, 193]. Dissolution processes are usually potentiostatically controlled. A potential is applied to the conducting sample and topographical changes at the electrode surface are monitored at the nanometer level in real time as a function of the applied potential or current. An example for the observation of anodic dissolution of Cu using in situ AFM is depicted in Fig. 7.16. Fig. 7.16 A shows a cyclic voltammogram recorded at a highly oriented graphite electrode for the oxidation–reduction behavior of Cu in a $CuSO_4$–HCl electrolyte solution. In Fig. 7.16 B, a sequence (images a–f) of in situ AFM images at different stages of the Cu deposition are shown. All electrochemical experiments were conducted with a Cu reference and counter-electrode. The condition of the initial freshly cleaved highly oriented graphite surface prior to electrochemical experiments (at 0 V) is shown in Fig. 7.16 B(a). During a first anodic cycle (0 V to +400 mV and back) the graphite surface remained unchanged. At a cathodic potential of –80 mV, small crystals nucleated and grew, as shown in Fig. 7.16 B(b). Fig. 7.16 B(c) corresponds to the graphite surface at the end of the Cu deposition cycle (–40 mV) when the current became anodic. Between –40 and 40 mV, the majority of small Cu crystals had dissolved, but at 40 mV significant growth of the remaining crystals was observed

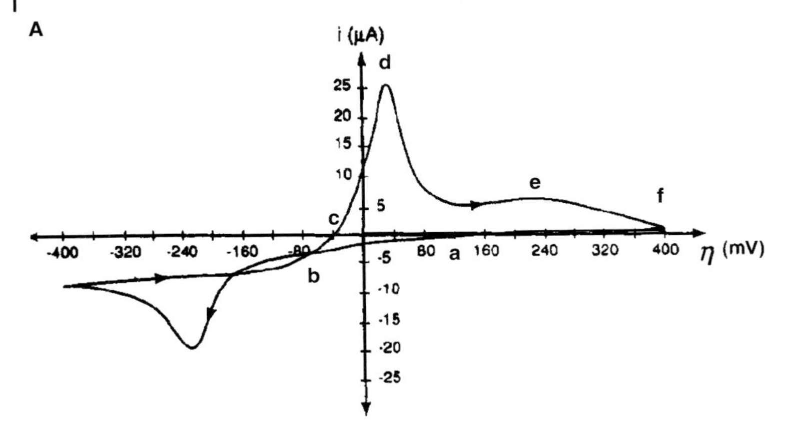
A
i (μA)
d
25
20
15
10
5
c
e
f
-400
-320
-240
-160
-80
80
a
160
240
320
400
η (mV)
b
-5
-10
-15
-20
-25

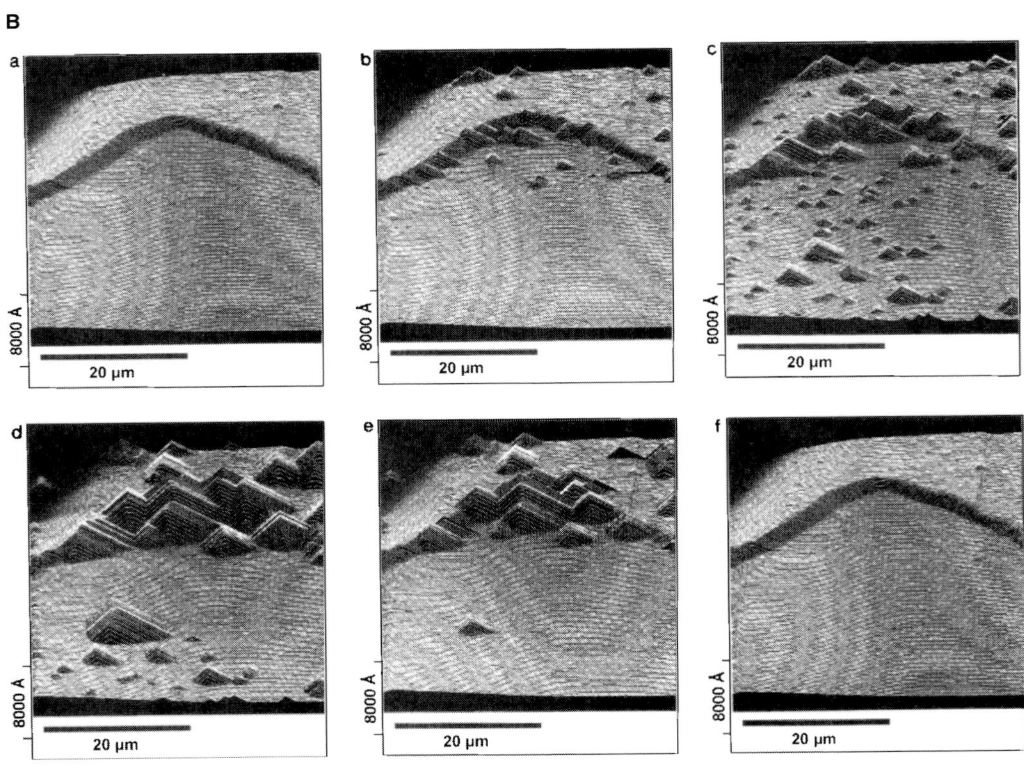
B
a
b
c
d
e
f
8000 Å
20 μm

(Fig. 7.16 B(d)). These crystals with a unit cell more than three times larger than Cu were identified as CuCl using Auger spectroscopic and X-ray diffraction (XRD) analysis. At the current peak of 240 mV, crystal dissolution started (Fig. 7.16 B(e)) and was completed revealing the surface in its original starting condition at 400 mV as shown in Fig. 7.16 B(f).

Although electrochemical STM and AFM permit in situ imaging of electrochemically induced processes with high resolution down to the atomic level, only topographical information on induced surface changes can be obtained. Hence there is a lack of chemical specificity on the involved redox processes compensated by the application of element-specific ex situ surface techniques, such as Auger spectroscopy and XRD. SECM is a promising SPM technique adding chemical information to the investigation of such interface processes. Moreover, electrochemical processes can be locally induced independent of the nature of the substrate, in contrast to STM, which is limited to conducting samples. However, conventional SECM provides mediocre lateral resolution in comparison with STM and AFM.

Hansma et al. introduced the promising technique of SICM, providing chemical information on ionic processes [66]. This in situ technique based on measuring ionic fluxes is suitable for high-resolution imaging at conducting and insulating surfaces. Simultaneous imaging of topography and local ion currents has been shown using a combined ion-conductance and shear-force microscope [67] and AC mode SICM [68]. Recently, first approaches have been published using SICM for processing microcircuits and for localized electrochemical deposition [228, 229].

SECM is widely used to characterize and investigate processes at solid/liquid interfaces [230]. The concept of positioning a probe close to a phase boundary provides the opportunity to measure transfer rates of processes across interfaces in a wide range of applications. Furthermore, SECM combined with a thin-layer liquid cell configuration allows probing of electrochemical reactions in a small volume. Thus, the analysis of heterogeneous transfer processes and laterally resolved investigations primarily of electron transfer represent an application uniquely accessible with SECM [231]. Consequently, SECM is ideally suited for corrosion studies: (i) there is no limitation based on the conductivity of the sample as SECM works for both insulating and conducting surfaces, (ii) structural surface effects can be correlated with localized electrochemical activity, (iii) the UME can be used either to initiate the reaction and/or to detect electroactive corrosion products at the

Fig. 7.16 In situ AFM imaging of electrochemical Cu deposition and dissolution. (A) Oxidation–reduction behavior of Cu on a highly oriented graphite working electrode. Electrolyte solution, 4×10^{-4} mol/L $CuSO_4$ and 2×10^{-3} mol/L HCl; scan rate, 2 mV/s; initial potential, 0 V. (B) EC–AFM images of (a) freshly cleaved graphite working electrode surface at the beginning of the CV scan (0 V), (b) the beginning of the Cu deposition on the graphite step defect (–80 mV) and (c) Cu crystals deposited at the end of the cathodic deposition cycle (0 V). At a potential of 40 mV, the AFM image (d) shows the growth of larger crystals at sites previously occupied by Cu; further increase of the potential to 120 mV leads to dissolution of the crystals (e) until, at the end of the CV scanning cycle (400 mV) (f), the surface was free of crystalline deposits. Reprinted with permission from [214]

UME and (iv) the current measured at the UME and/or in the case of a conducting sample at the substrate provides quantitative information on the processes involved. Corrosion investigations have been performed at several metal surfaces including steel [232–234], Al [235], Ni [236], Ti [237, 238] and semiconductors [239]. Furthermore, SECM is a powerful tool for investigating the dissolution characteristics of electrically insulating materials, in particular ionic single-crystal surfaces [230, 240]. While SECM studies provide valuable quantitative information on local dissolution rates, complementary topographical information is usually derived from microscopic studies owing to the limited lateral resolution obtainable in conventional SECM [241]. A significant advance in the measurement of dissolution kinetics was accomplished by the introduction of a complementary AFM technique, in which a biased Pt-coated AFM cantilever is used to measure the topography of a dissolving crystal surface while simultaneously inducing electrochemical dissolution under conditions closely mimicking those for kinetic SECM measurements. This technique has been applied to study the dissolution of KBr single crystals in acetonitrile solution [147], for in situ observation of the dissolution of potassium ferrocyanide trihydrate [148] and recently for the observation of surface processes involved in the dissolution of calcite in aqueous solution [149]. One major drawback of this technique is the size of the electroactive area of the scanning probe resulting from coating the entire AFM cantilever with the electrode material. Hence the lateral resolution of the electrochemical information obtained is limited. In addition, the conductivity of the AFM tip limits the application to insulating surfaces in AFM contact mode.

Corrosion processes of surface-coated metals have been intensively studied by scanning Kelvin probes (SKPs) [242]. However, to date the lateral resolution of the SKP has been limited by the diameter of the probe (typically 20–100 μm) and a comparatively large working distance (~10 μm). Owing to the inherent limitations of this technique, further detailed discussion is omitted in this chapter. However, the combination of AFM and SKP (SKPFM) is noteworthy, since the lateral resolution is dramatically improved based on the resolution of a metal-coated AFM cantilever and the AFM distance control. Simultaneous information on surface topography and surface potentials developing during electrochemical reactions can be obtained [243] providing laterally resolved potential information [244].

7.2.1.3 Combined SPM Techniques for Material Characterization and Modification

Recently, Macpherson and co-workers published results on dissolution experiments revealing significant improvement of the lateral resolution by combined SECM–AFM measurements [127]. Bent Pt-wire nanoelectrodes shaped as AFM cantilevers have been used to induce dissolution electrochemically from a localized region of a nonconducting solid substrate. This technique is limited to insulating substrates in AFM contact mode; however, it provides excellent spatial resolution due to the nanometer-sized electrode. Since the electroactive area is located at the very end of the tip, conducting samples have to be investigated using lift-mode imaging [129]. Similarly to combined ECSTM–SECM, the sample surface has to be scanned twice. First, the topographical information is recorded in AFM contact mode operation followed by a second scan in SECM mode using the recorded topographical information to keep the tip at a defined distance to the surface tracing the contours of the sample.

Simultaneous SECM–AFM imaging of electrode surfaces can be achieved by integration of submicro- and nanoelectrodes into AFM tips using microfabrication techniques [128]. Since the distance between the integrated recessed electrode and the surface remains constant, simultaneous imaging of the topography and electrochemical properties of the sample surface independent of the nature of the sample can be achieved. Due to the well defined geometry of the bifunctional probe, quantification of the current signal and theoretical modeling can be performed [245].

Fig. 7.17 shows simultaneously obtained height (B) and current (C) images of disk-shaped platinum structures embedded in an insulating silicon nitride surface recorded in AFM contact and SECM feedback mode using 0.05 mol/L $[Fe(CN)_6]^{4-}$

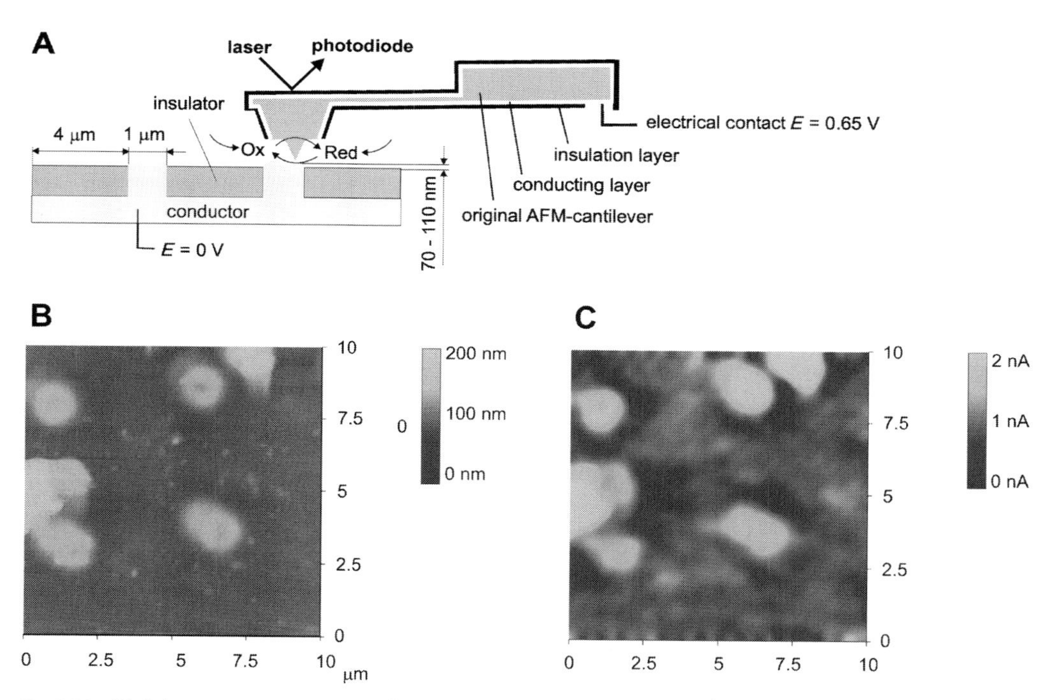

Fig. 7.17 (A) Schematic representation of simultaneous AFM–SECM imaging of disk-shaped platinum structures embedded into an insulating silicon nitride surface. (B) Topography recorded in AFM contact mode and (C) simultaneously obtained current image recorded in SECM feedback mode. Redox mediator, 0.05 mol/L $[Fe(CN)_6]^{4-}$ in 0.5 mol/L KCl electrolyte solution. The tip was held at a potential of 0.65 V and the sample at 0 V vs AgQRE to ensure diffusion-controlled processes. Electrode edge length, 770 nm; tip height, 406 nm; scan rate, 2 Hz

as redox mediator in 0.5 mol/L KCl electrolyte solution. Biasing the integrated tip at a constant potential (0.65 V) oxidizes ferrocyanide to ferricyanide in solution. The sample was held at 0 V to ensure diffusion-controlled processes. According to the theory of the feedback mode, the diffusion-limited Faradaic current measured at the electrode is influenced by the properties of the sample surface. Accordingly, scanning the AFM–SECM tip across the insulating silicon nitride layer results in blocked diffusion to the tip integrated electrode and a decreased current in contrast to the steady-state current in bulk solution (negative feedback). Due to the re-reduction of the redox mediator at the biased platinum spots, the tip current is increased compared with the steady-state current in the bulk solution (positive feedback) above the conducting features. The periodicity of the disk-shaped conducting surface features is evident in the electrochemical image and corresponds well with the topographical information.

7.2.2 Biological Applications

Molecular monolayers, including layers of functional biological macromolecules, have been investigated successfully by probing single-molecule electronic properties directly in aqueous solutions using electrochemical in situ STM. In addition to high structural resolution, in situ STM offers the perspective to obtain single-molecule spectroscopic data. This opportunity emerges most clearly when adsorbate molecules have accessible redox levels and the tunneling current decomposes into successive single-molecule interfacial electron-transfer steps. Excellent reviews on this topic have been published recently [246–248]. Proteins adsorbed in monolayers on neat and modified Au(111) electrodes have been studied. Comprehensive surface characterization including resolving structural features at single-molecule levels combined with information on the electrochemical potential windows for functional monolayer stability of these proteins monolayers could be achieved by in situ SPM. Investigations include blue copper protein *Pseudomonas aeruginosa* azurin [249–253], *Saccharomyces cerevisiae* cytochrome *c* (single-haem protein) [254], poplar plastocyanin mutants [255] and a synthetic 4-α-helix bundle carboprotein [256]. ECSTM investigations require attaching the target molecule to a conductive layer, limiting the applicability of this technique. Imaging of DNA adsorbed on mica was achieved with STM [257], but the imaging mechanism was controversially discussed and may in fact be based on an SECM measurement owing to a monolayer of water and dissolved ions at the mica substrate [258, 259]. Despite excellent resolution, in situ STM is not applicable for investigations of biological samples such as entire cells.

In contrast, AFM has developed into a routine tool for biological investigations and certainly is among the most commonly applied scanning probe techniques in the biological field. The investigation of biological interfaces is mainly based on structural changes, which can be detected by conventional AFM techniques. Chemical functionalization of AFM tips extends the information space beyond topographical data provided by conventional force microscopy. Force spectroscopy and molecular recognition force microscopy (MRFM) yield information on molecular interactions characterizing, e.g., ligand–receptor binding and dissociation processes [260]. However, to the best of our knowledge, these techniques have not yet been applied to redox-active processes and are considered beyond the scope of this chapter.

SICM has proven to be an excellent tool for investigations on living cells. In contrast to other SPM techniques, SICM operates in buffered solutions and is highly suitable for noninvasive imaging of soft biological samples. The first successful imaging of living cell surfaces using SICM was accomplished by Korchev et al. in the mid-1990s [69, 261]. Living cardiac myocytes have been imaged with a lateral resolution of 50 nm [261]. SICM has also been applied for assessing the volume of living cells, which allows quantitative characterization of dynamic changes in cell volume while retaining the cell functionality with high resolution (2.5×10^{-20} L) [262]. More recently, a novel combined technique based on SICM and patch-clamp [70] has been used for mapping of single active ion channels in intact cell plasma membranes for the first time [263]. Local application of potassium ions is enabled via a micropipette and detection of the resulting ion flow via a channel using the patch-clamp technique. Based on this configuration, the distribution of ATP-regulated K^+ channels (K_{ATP} channels) in cardiac myocytes was imaged (Fig. 7.18 B). A photograph and a schematic view of the experimental set-up showing the SICM micropipette for imaging the topography and the patch-clamp micropipette for simultaneously monitoring the electrical response is shown in Fig. 7.18 A. A similar method using a combined SICM–patch-clamp technique was applied to image ion channels of epithelial microvilli, cardiomyocyte T-tubules and sperm cells [70]. SICM utilizing a feedback-controlled fast responding distance control permitted imaging of contracting cardiac myocytes [264]. By combining this method with confocal laser microscopy, simultaneous measurements on the nanometric motion of the cell and local calcium ion concentrations just underneath the cell membrane have been demonstrated [264]. Recently, SICM was successfully applied for imaging the life cycle of microvilli [265], which previously was a challenging imaging task with AFM owing to substantial mechanical interaction between the specimen and the AFM probe.

SECM is a powerful technique for obtaining electrochemical information on a wide range of biological species. In contrast to AFM, information on specific electroactive species generated at or involved in biologically relevant surface processes can be obtained. SECM has been applied successfully for monitoring the respiratory activity of PC12 cells [266] and HeLa cells [267, 268] at a single cell level and for the detection of neurotransmitter release [118]. Furthermore, electron-transfer rates between metallic electrode surfaces and enzymes [269] and of tip-generated species occurring at nonconductive surfaces containing a redox-active enzyme [270] have been investigated in feedback mode SECM. The utility of SECM for imaging molecular transport across membranes has recently been reviewed [271]. SECM in generator/collector mode has been applied to probe ion transfer at bilayer lipid membranes [272]. Imaging of enzyme activity has been performed in SECM feedback [273–275] and generator/collector modes [276, 277]. A combination of localized desorption induced by SECM followed by chemical derivatization

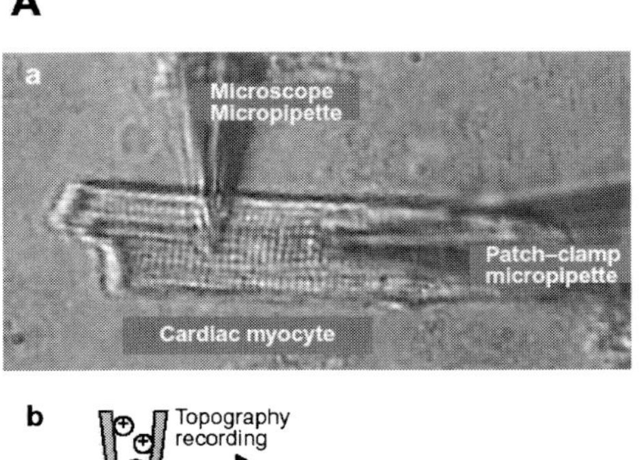

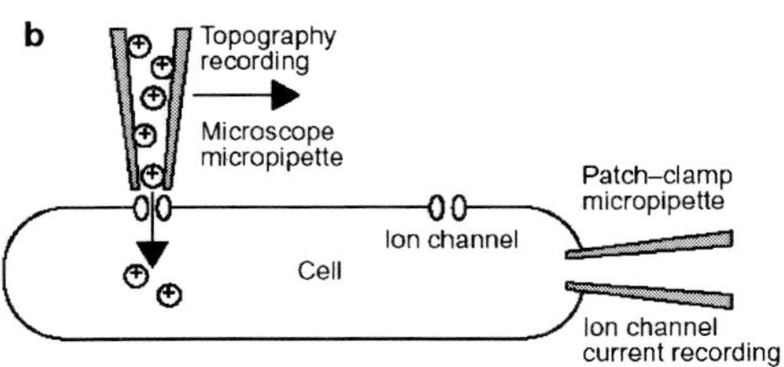

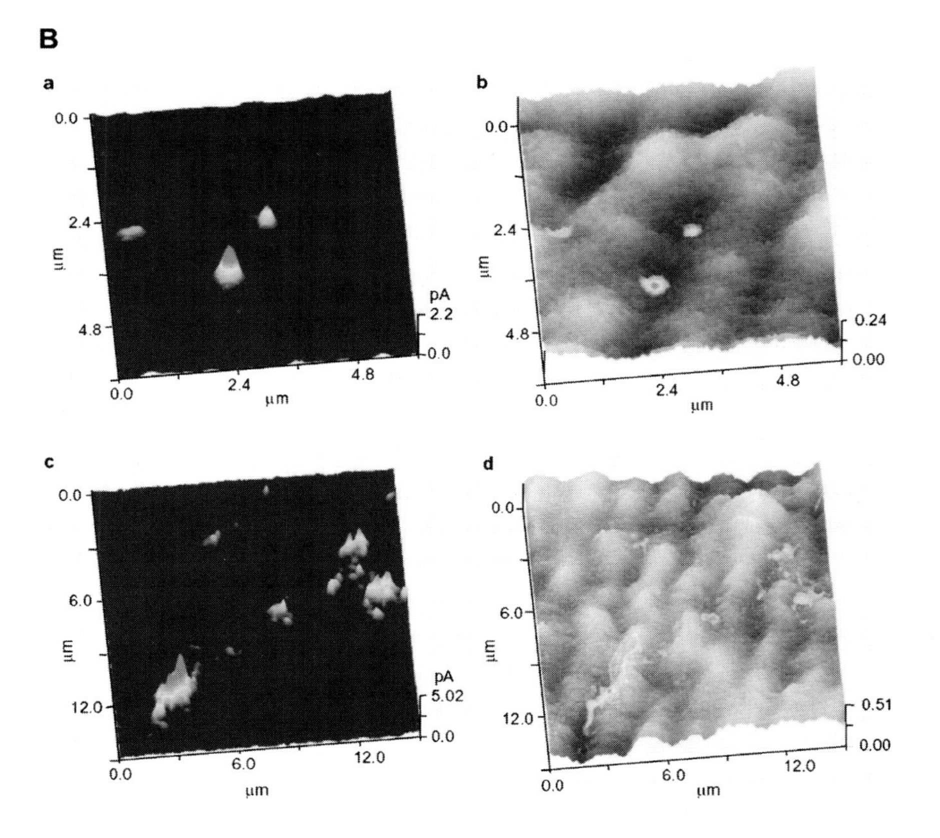

Fig. 7.18 Functional localization of single ion channels in intact cell membranes. (A) Photograph of the experimental arrangement showing the SICM and the patch-clamp pipette and the cardiac myocyte (a) and corresponding scheme of the sensing mechanism using a combined SICM–whole-cell voltage-clamp technique (b). (B) Functional imaging of K_{ATP} channel distribution of the cardiomyocyte sarcolemma. Maps of K^+ current (a, c) and surface topography superimposed on the K^+ current map (b, d). Reprinted with permission from [262]

was used to create and image enzymatically active spots on alkanethiolate-covered gold electrodes [278]. SECM offers several mechanisms for local surface modification and characterization of biochemically active microscopic regions and for the analysis of interactions in patterned multi-enzyme layers [279].

Surface modification of electrodes with bioreceptors such as enzymes or application of potentiometric electrodes in SECM is gaining substantial interest owing to the specificity added to information gained during investigations of complex biological systems. However, utilizing the current signal for positioning of SECM probes is only applicable for amperometric electrodes. Consequently, nonamperometric tips, such as biosensors, require alternative tip positioning strategies for imaging SECM experiments. Utilizing optical microscopy for tip positioning yields only qualitative results [280]. A few approaches have been reported for the application of potentiometric UMEs [281–284] and enzyme UMEs [160], despite the fact that potentiometric probes are of great interest for detecting electroinactive species [285] or for detecting heavy metals [286]. Horrocks et al. [281] used antimony SECM electrodes operating also in amperometric mode for positioning of the tip prior to pH mapping. Wei et al. [282] introduced positioning techniques based on DC measurements of the solution resistance using Ag/AgCl micropipette electrodes or potentiometric measurements of steady-state concentration profiles with ion-selective microelectrodes providing the tip–sample distance. Furthermore, potentiometric UMEs with a dual-electrode configuration have been reported. One channel is configured as an ion-selective electrode and the second channel as an amperometric electrode for distance control during the approach of the UME. A hydrogen peroxide microbiosensor was positioned by applying a high-frequency alternating potential to the tip and measuring the solution resistance between the tip and the auxiliary electrode [119]. Although good agreement of the experimentally obtained data with theory has been achieved, this approach is limited to micrometer-sized sensors, since the relative contribution of the solution resistance to the impedance decreases with the dimensions of the sensor. Shear force-based constant-distance SECM has also been applied for positioning of microbiosensors [103].

Nevertheless, a major drawback of conventional SECM is signal-dependent positioning of microelectrodes resulting in insufficient lateral resolution and convolution of the electrochemical response and topographical information. In addition, imaging of enzyme activity in the generator/collector mode does not provide topographical information when the substrate of the enzymatic reaction is absent. While SECM imaging of enzyme activity using the shear force feedback mode to control the substrate–probe distance has been described [103, 287], to date few results have been published using nanoelectrodes in SECM experiments.

7.2.2.1 Combined SPM Techniques for Biological Applications

Combined SPM techniques enable complex processes such as biological activity at the surface of living cells to be addressed. As an example, cell activity has been imaged with a combination of NSOM and SICM based on fluorescence detection of the optically active Ca-fluo-3 complex [288].

As many biological processes induce structural changes along with chemical signals, the combination of AFM with SECM appears an attractive choice for imaging electrochemical processes at biologically active interfaces. Recently, integrated AFM–SECM probes have been used to image enzyme activity with high resolution [132]. In Fig. 7.19, simultaneous imaging of immobilized peroxidase activity in generator/collector mode SECM and contact mode AFM is demonstrated using bifunctional SECM–AFM probes [289]. Enzyme spots were created by chemisorption of a functionalized thiol monolayer (cystaminium chloride) on micropatterned disk-shaped gold structures embedded in a silicon nitride layer. Subsequently, peroxidase was covalently attached to the gold spots by adding a mixture of glutaraldehyde and peroxidase, forming a surface grafted cross-linked protein gel at the aminated surface of the self-assembled monolayer [278]. Electrochemical imaging of peroxidase activity was obtained in the generation/collection mode [276]. The substrate of the enzymatic conversion (H_2O_2) is reduced to water with hydroxymethylferrocene (FMA), a metal–organic electron donor (containing Fe^{2+}), which is added to the solution. Hence the oxidized species ferrocinium methylhydroxide (FMA^+, containing Fe^{3+}) is continuously generated during the

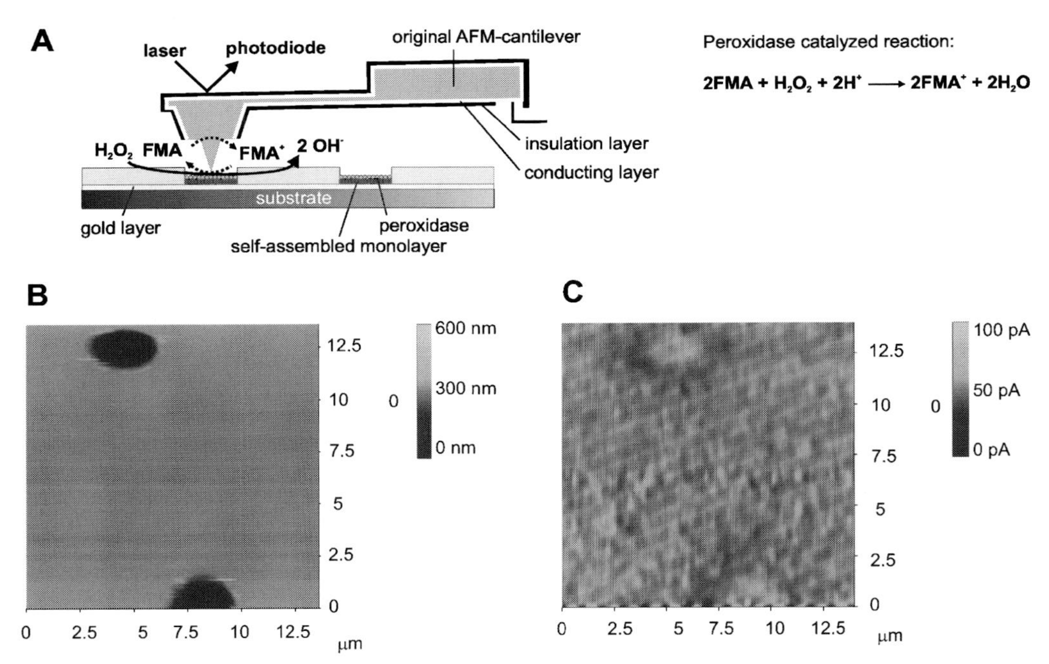

Fig. 7.19 (A) Schematic representation of simultaneous AFM–SECM imaging of peroxidase activity in contact mode generator/collector mode AFM and SECM. (B) Top view of the height and (C) simultaneously recorded current in the presence of the enzymatic substrate H_2O_2 (0.5 mmol/L). Images were recorded in 2 mmol/L FMA, 0.1 mol/L KCl in phosphate buffer (0.1 mol/L, pH 7.0). The tip was held at a potential of 0.05 V vs. Ag/AgCl. Electrode edge length, 860 nm; tip height, 410 nm; scan rate, 1 Hz [294]

enzymatic reaction (Fig. 7.19 A). FMA^+ diffuses to the integrated electrode and can be detected amperometrically at $E=0.05$ V vs. Ag/AgCl, a potential where no other electroactive species present in solution can be converted. Fig. 7.19 B and C show the topography and the electrochemical response, respectively, corresponding to the enzymatic activity of the peroxidase in the presence of H_2O_2.

Recent developments of AFM–SECM probes in our research group were focused on imaging electrochemical activity at soft biological samples during tapping mode AFM operation. Although several approaches of combined AFM–SECM using bent cantilever-shaped nanoelectrodes have been published [148, 290], to date none of these reported measurements have involved the tapping mode AFM owing to the spring constant of these hand-made tips [126].

The application of microfabricated AFM–SECM tips for dynamic ('tapping') mode AFM operation has recently been demonstrated [173]. Despite the cantilever modification during tip fabrication, the resonance frequency of the unmodified silicon nitride cantilever is not significantly altered, which was confirmed by comparing frequency spectra of unmodified and modified cantilevers. Electrochemical images recorded in tapping mode revealed current responses and lateral resolution comparable to the electrochemical images obtained in contact mode.

Recently, AFM tip integrated electrodes have been applied for imaging enzyme activity in tapping mode operation for the first time [132]. The enzymatic activity of glucose oxidase (GOD) entrapped in a soft polymer matrix deposited inside the pores of a periodic micropattern has successfully been imaged in tapping mode AFM and generation/collection mode SECM.

7.2.3
Imaging with Tip-integrated Sensors

Integrated SECM–AFM probes provide the opportunity to integrate and position micro- and nanobiosensors at a well-defined and deliberately selected distance above the sample surface. Standard surface modification processes for biosensors can be used to immobilize biological recognition elements at the surface of the integrated electrode. Utilizing electrochemical techniques, the site of immobilization is confined to the electrode surface, providing highly localized and reproducible integration of a biosensing interface [291, 292]. With the immobilization of enzymes via self-assembled thiol monolayers with reactive headgroups [293], conversion of electroactive and -inactive species generated at the sample surface is enabled. Successful imaging with a tip-integrated biosensor based on horseradish peroxidase has recently been demonstrated [294]. The AFM–SECM technique is by no means limited to the integration of biosensors and can be extended to potentiometric sensors, such as pH-microsensors or amalgam electrodes. Many biological processes including enzymatic reactions, biocorrosion and metabolic processes of microorganisms involve or are driven by pH changes. Hence simultaneous AFM imaging and mapping of local pH modulations at the nanoscale are of substantial interest. For example, antimony and iridium/iridium oxide electrodes have been reported for electrochemical pH measurements of biological systems [295–297]. As thin films of these materials can be sputtered with conventional microfabrication techniques, the fabrication steps described above for the integration of gold and platinum electrodes can be modified and optimized to integrate pH-sensitive electrochemical sensors into AFM tips. The potential of tip integrated electrodes and biosensors for biological application and in addition to force imaging is discussed in detail elsewhere [298].

7.3 Conclusions

In situ scanning probe techniques have been applied successfully to a variety of studies on interfacial processes at solid/liquid and liquid/liquid interfaces. Recent decades were characterized by substantial advances in instrumental development, theoretical models and fundamental studies rendering in situ SPM technique key technologies in electrochemical surface science. Depending on the tip–sample interaction, a variety of processes can be studied ranging from topographical changes accompanying electrochemical processes at the interface to the determination and quantification of electron transfer rates. The beginnings of SPM in electrochemical surface science were focused on investigating the surface structures of single-crystal electrodes in ambient environments and adsorbate formation processes. Current trends of SPM application in the life sciences clearly address increasingly complex interfaces and the redox surface processes involved, representing the next 'grand challenge' in this field. Distinct correlation of physical properties with (bio)activity and reactivity is among the key questions to be answered by suitable in situ scanning probe techniques. Recently, innovative solutions for combined and integrated scanning probe techniques have been achieved, providing the exciting opportunity to obtain simultaneously information on complementary parameters with high spatial and temporal resolution at bio-related interfaces.

7.4 Multifunctional Scanning Probes – Quo Vadis?

Surface electrochemistry in combination with scanning probe techniques is a highly interdisciplinary field reaching out into many scientific and engineering disciplines. In addition to fundamental studies and consideration of unsolved problems related to the structure of the electrical double layer, a variety of novel applications for multifunctional SPM techniques in bioelectrochemistry, biocorrosion, fuel cell research, electrochemical nanotechnology, sensor technology and material sciences can be envisaged. Especially the multi-parametric investigation of dynamic processes correlated in time and space provides an innovative and important role for scanning probe techniques in these specified areas, demanding smart combination and integration of existing techniques. Progress in nanotechnology may have a substantial impact on scanning probe microscopy, providing the next-generation technology for improved scanning probes, such as the application of carbon nanotubes as nanoelectrodes or scanning SPM probes. Coupled with strategies for chemical modification of these materials, this will be an important step towards enhancing lateral resolution and providing novel mechanisms of imaging contrast. However, increasing complexity of the probe demands advanced fabrication strategies providing a sound technological basis for continuous advancement. Examples highlighted in this chapter, including combined AFM–SICM, AFM–NSOM and AFM–SECM, clearly demonstrate that ongoing progress requires synchronized efforts in the scientific and engineering aspects of this dynamic area. Without doubt, microfabrication will continue to have a major impact on the reliable and reproducible fabrication of scanning probe tips. Novel probe designs such as multifunctional scanning probes integrating AFM–SECM–NSOM functionality or multiple electrodes into a single tip cannot be envisaged without corresponding progress in micro- and nanofabrication technology. Continuing this vision leads to the integration of nanoelectrodes into arrays of cantilevers as the ultimate tool for electrochemical surface analysis. Although there is still a long way to go towards this goal, the first successful steps have already demonstrated that yesterday's visions maybe turned into today's technology.

7.5 Acknowledgments

The authors acknowledge support for their research on multifunctional scanning probes from the National Science Foundation (project 0216368, Biocomplexity/IDEA), the National Institute of Health (project EB000508) and the Fonds zur Förderung der wissenschaftlichen Forschung, Austria (projects P14122-CHE and J2230). A. Lugstein and E. Bertagnolli of the Institute of Solid State Electronics, Vienna University of Technology, are gratefully acknowledged for their collaboration and contributions to the development of integrated SECM–AFM tips.

7.6 References

1 P. Parsons, in *Modern Aspects of Electrochemistry*, J. O'M. Bockris (ed.); New York: Academic Press, **1954**.
2 P. Delahay, *Double Layer and Electrode Kinetics*; New York: Wiley-Interscience, **1965**.
3 J. O'M. Bockris, A. K. N. Reddy, *Modern Electrochemistry*, 2nd edn.; New York: Plenum Press, **1998**.
4 D. C. Grahame, *Chem. Rev.* **1947**, *41*, 441–501.
5 D. M. Mohilner, in: *Electroanalytical Chemistry*, A. J. Bard (ed.); New York: Marcel Dekker, **1966**, Vol. 1, pp. 241 ff.
6 G. A. Somorjai, *Introduction to Surface Chemistry and Catalysis*; New York: Wiley, **1994**.
7 M. P. Soriaga, *Surf. Sci.* **1992**, *39*, 325–443.
8 A. T. Hubbard, *Crit. Rev. Anal. Chem.* **1973**, *3*, 201–242.

9 W. E. O'Grady, M. Y. C. Woo, P. L. Hagans, E. Yeager, *J. Vac. Sci. Technol.* **1977**, *14*, 365–368.
10 F. T. Wagner, P. N. Ross, *J. Electroanal. Chem.* **1983**, *150*, 141–164.
11 J. Clavilier, R. Faure, G. Guinet, R. Durand, *J. Electroanal. Chem.* **1980**, *107*, 205–209.
12 J. Clavilier, R. Faure, G. Guinet, R. Durand, *J. Electroanal. Chem.* **1980**, *107*, 211–216.
13 G. Binnig, H. Rohrer, C. Gerber, E. Weibel, *Appl. Phys. Lett.* **1982**, *40*, 178–180.
14 G. Binnig, H. Rohrer, C. Gerber, E. Weibel, *Phys. Rev. Lett.* **1982**, *49*, 57–61.
15 G. Binnig, H. Rohrer, *Rev. Mod. Phys.* **1987**, *59*, 615–625.
16 R. Sonnenfeld, P. K. Hansma, *Science*, **1986**, *232*, 211–213.
17 H.-L. Liu, F.-R. F. Fan, C. W. Lin, A. J. Bard, *J. Am. Chem. Soc.* **1986**, *108*, 3838–3839.
18 R. Sonnenfeld, B. C. Schardt, *Phys. Lett.* **1986**, *49*, 1172–1174.
19 P. Lustenberger, H. Rohrer, R. Christoph, H. Siegenthaler, *J. Electroanal. Chem.* **1988**, *243*, 225–235.
20 J. Wiechers, T. Twomey, D. M. Kolb, R. J. Behm, *J. Electroanal. Chem.* **1988**, *248*, 451–460.
21 S. Morita, I. Otsuka, T. Okada, H. Yokoyama, T. Iwasaki, N. Mikoshiba, *Jpn. J. Appl. Phys.* **1987**, *26*, L1853–L1855.
22 O. Lev, F.-R. F. Fan, A. J. Bard, *J. Electrochem. Soc.* **1988**, *135*, 783–784.
23 Z. W. Tian, X. D. Zhuo, J. Q. Mu, J. H. Ye, Z. D. Fen, B. W. Mao, C. L. Bai, C. D. Dai, *Ultramicroscopy* **1992**, *42–44*, 460–463.
24 N. Batina, D. M. Kolb, R. J. Nichols, *Langmuir* **1992**, *8*, 2572–2576.
25 N. J. Tao, X. Z. Li, H. X. He, *J. Electroanal. Chem.* **2000**, *492*, 81–93.
26 J. Pan, T. W. Jing, S. M. Lindsay, *J. Phys. Chem.* **1994**, *98*, 4205–4208.
27 A. Vaught, T. W. Jing, S. M. Lindsay, *Chem. Phys. Lett.* **1995**, *236*, 306–310.
28 G. Binnig, C. F. Quate, C. Gerber, *Phys. Rev. Lett.* **1986**, *56*, 930–933.
29 D. Sarid, *Scanning Force Microscopy, Oxford Series in Optical and Imaging Sciences*; New York: University Press, **1991**, Vol. 2.
30 G. Meyer, N. M. Amer, *Appl. Phys. Lett.* **1988**, *53*, 2400–2401.
31 D. W. Abraham, C. Williams, J. Slinkman, H. K. Wickramasinghe, *J. Vac. Sci. Technol. A* **1991**, *9*, 703–706.
32 R. Erlandsson, G. M. McClelland, C. M. Mate, S. Chiang, *J. Vac. Sci. Technol. A* **1988**, *6*, 266–270.
33 D. Rugar, H. J. Mamin, R. Erlandsson, J. E. Stern, B. D. Terris, *Rev. Sci. Instrum.* **1988**, *59*, 2337–2340.
34 J.-B. Green, C. A. McDermott, M. T. McDermott, M. D. Porter, in: *Imaging of Surfaces and Interfaces*, J. Lipkowski, P. N. Ross (eds.); New York: Wiley-VCH, **1999**, Chapter 6, and references cited therein.
35 G. Friedbacher, T. Prohaska, M. Grasserbauer, *Mikrochim. Acta* **1994**, *113*, 179–202.
36 U. Duerig, O. Zueger, A. Stadler, *J. Appl. Phys.* **1992**, *72*, 1778–1798.
37 Y. Martin, C. C. Williams, H. K. Wickramasinghe, *J. Appl. Phys.* **1987**, *61*, 4723–4729.
38 T. R. Albrecht, P. Grutter, H. K. Horne, D. Rugar, *J. Appl. Phys.* **1991**, *69*, 668–673.
39 Q. Zong, D. Innis, K. Kjoller, V. B. Elings, *Surf. Sci. Lett.* **1993**, *290*, L688–L692.
40 P. K. Hansma, J. P. Cleveland, M. Radmacher, D. A. Walters, P. E. Hillner, M. Bezanilla, M. Fritz, D. Vie, H. G. Hansma, C. B. Prater, J. Massie, L. Fukunaga, J. Gurley, V. Ellings, *Appl. Phys. Lett.* **1994**, *64*, 1738–1740.
41 C. A. J. Putman, K. O. van der Werf, B. G. De Grooth, N. F. Van Hulst, J. Greve, *Appl. Phys. Lett.* **1994**, *64*, 254–256.
42 Ch. Le Grimellec, M.-C. Giocondi, R. Pujol, E. Lesniewska, *Single Mol.* **2000**, *1*, 105–107.
43 D. Sarid, V. Ellings, *J. Vac. Sci. Technol. B* **1991**, *9*, 431–337.
44 R. Wiesendanger, *Scanning Probe Microscopy and Spectroscopy, Methods and Applications*; Cambridge: Cambridge University Press, **1994**.

45 H. Takano, J.R. Kenseth, S.-S Wong, J.C. O'Brien, M.D. Porter, *Chem. Rev.* **1999**, *99*, 2845–2890.

46 Y. Li, B.W. Maynor, J. Liu, *J. Am. Chem. Soc.* **2001**, *123*, 2105–2106.

47 B.W. Maynor, Y. Li, J. Liu, *Langmuir* **2001**, *17*, 2575–2578.

48 F. Ohnesorge, G. Binnig, *Science* **1993**, *260*, 1451–1456.

49 F.J. Giessibl, *Science* **1995**, *267*, 68–71.

50 E.H. Synge, *Philos. Mag.* **1928**, *6*, 356–362.

51 E.H. Synge, *Philos. Mag.* **1931**, *11*, 65–80.

52 D.W. Pohl, W. Denk, M. Lanz, *Appl. Phys. Lett.* **1984**, *44*, 651–653.

53 E. Betzig, A. Lewis, A. Harootunian, M. Isaacson, E. Kratschmer, *Biophys. J.* **1986**, *49*, 269–279.

54 D.W. Pohl, L. Novotny, *J. Vac. Sci. Technol. B* **1991**, *9*, 1441–1446.

55 Ch. Paulson, A.B. Ellis, L. McCaughan, B. Hawkins, J. Sun, T.F. Kuech, *Appl. Phys. Lett.* **2000**, *7*, 1943–1945.

56 F. Zenhausern, M.P. O'Boyle, H.K. Wickramasinghe, *Appl. Phys. Lett.* **1994**, *65*, 1623–1625.

57 R.C. Dunn, *Chem. Rev.* **1999**, *99*, 2891–2927.

58 S. Kirstein, *Curr. Opin. Colloid Interface Sci.* **1999**, *4*, 256–264.

59 E. Betzig, P.L. Finn, J.S. Weiner, *Appl. Phys. Lett.* **1992**, *60*, 2484–2486.

60 E. Betzig, J.K. Trautman, T.D. Harris, J.S. Weiner, R.L. Kostelak, *Science* **1991**, *251*, 1468–1470.

61 R. Brunner, O. Hering, O. Marti, O. Hollrichter, *Appl. Phys. Lett.* **1997**, *71*, 3628–3630.

62 O. Hollrichter, R. Brunner, O. Marti, *Ultramicroscopy* **1998**, *71*, 143–147.

63 T.H. Keller, T. Rayment, D. Klenerman, *Biophys. J.* **1998**, *74*, 2076–2079.

64 P.J. James, L.F. Garfias-Mesias, W.H. Smyrl, *J. Electrochem. Soc.* **1998**, *145*, 2011–2016.

65 Y. Lee, Z. Ding, A.J. Bard, *Anal. Chem.* **2002**, *74*, 3634–3643.

66 P.K. Hansma, B. Drake, O. Marti, S.A.C. Gould, C.B. Prater, *Science* **1989**, *243*, 641–643.

67 H. Nitz, J. Kamp, H. Fuchs, *Probe Microsc.* **1998**, *1*, 187–201.

68 D. Pastre, H. Iwamoto, J. Liu, G. Szabo, Z. Shao, *Ultramicroscopy* **2001**, *90*, 13–19.

69 Y. E. Korchev, M. Milovanovic, C.L. Brashford, D.C. Bennett, E.V. Sviderskaya, I. Vodyanoy, M.J. Lab, *J. Microsc.* **1997**, *188*, 17–23.

70 J. Gorelik, Y. Gu, H.A. Spohr, A.I. Shevchuk, M.J. Lab, S.E. Harding, C.R.W. Edwards, M. Whitaker, G.W.J. Moss, D.C.H. Benton, D. Sanchez, A. Darszon, I. Vodyanoy, D. Klenerman, Y. E. Korchev, *Biophys. J.* **2002**, *83*, 3296–3303.

71 Y. Lee, S. Amemiya, A.J. Bard, *Anal. Chem.* **2001**, *73*, 2261–2267.

72 P. Liljeroth, C. Johans, C.J. Slevin, B.M. Quinn, K. Kontturi, *Electrochem. Commun.* **2002**, *4*, 67–71.

73 P. Liljeroth, C. Johans, C.J. Slevin, B.M. Quinn, K. Kontturi, *Anal. Chem.* **2002**, *74*, 1972–1978.

74 M.V. Mirkin, F.-R.F. Fan, A.J. Bard, *J. Electroanal. Chem.* **1992**, *328*, 47–62.

75 C.G. Zoski, M.V. Mirkin, *Anal. Chem.* **2002**, *74*, 1986–1992.

76 R.M. Wightman, S. Pons, D.R. Rolison, P.P. Schmidt, *Ultramicroelectrodes*; Morganton, NC: Datatech Systems, **1987**.

77 R.C. Engstrom, M. Weber, D.J. Wunder, R. Burgess, S. Winquist, *Anal. Chem.* **1986**, *58*, 844–848.

78 H.Y. Liu, F.-R.F. Fan, C.W. Lin, A.J. Bard, *J. Am. Chem. Soc.* **1986**, *108*, 3838–3839.

79 A.J. Bard, F.-R.F. Fan, J. Kwak, O. Lev, *Anal. Chem.* **1989**, *61*, 132–138.

80 J. Kwak, A.J. Bard, *Anal. Chem.* **1989**, *61*, 1221–1227.

81 C.M. Lee, J.Y. Kwak, F.C. Anson, *Anal. Chem.* **1991**, *63*, 1501–1504.

82 A.J. Bard, F.-R.F. Fan, D.T. Pierce, P.R. Unwin, D. O. Wipf, F. Zhou, *Science* **1991**, *254*, 68–74.

83 D. Mandler, in: *Scanning Electrochemical Microscopy*, A.J. Bard, M.V. Mirkin (eds.); New York: Marcel Dekker, **2001**, pp. 593–602.

84 A.J. Bard, F.-R.F. Fan, M.V. Mirkin, in: *Electroanalytical Chemistry*, A.J. Bard (ed.); New York: Marcel Dekker, **1994**, Vol. 18, 243–298.

85 A. J. Bard, F.-R. F. Fan, M. V. Mirkin, in: *Physical Electrochemistry: Principles, Methods and Applications*, I. Rubenstein (ed.); New York: Marcel Dekker, **1995**, p. 209.
86 M. V. Mirkin, *Anal. Chem.* **1996**, *5*, A177–A182.
87 M. V. Mirkin, *Mikrochim. Acta* **1999**, *130*, 127–153.
88 A. L. Barker, M. Gonsalves, J. V. Macpherson, C. J. Slevin, P. R. Unwin, *Anal. Chim. Acta* **1999**, *385*, 223–240.
89 M. V. Mirkin, B. R. Horrocks, *Anal. Chim. Acta* **2000**, *406*, 119–146.
90 A. J. Bard, in: *Scanning Electrochemical Microscopy*, A. J. Bard, M. V. Mirkin (eds.); New York: Marcel Dekker, **2001**, pp. 1–16.
91 G. Wittstock, *Top. Appl. Phys.* **2003**, *85*, 335–364.
92 B. J. Horrocks, M. V. Mirkin, D. T. Pierce, A. J. Bard, G. Nagy, K. Toth, *Anal. Chem.* **1993**, *65*, 1213–1224.
93 C. Wei, A. J. Bard, G. Nagy, K. Toth, *Anal. Chem.* **1995**, *67*, 1346–1356.
94 B. R. Horrocks, M. V. Mirkin, *J. Chem. Soc., Faraday Trans.* **1998**, *94*, 1115–1118.
95 B. M. Quinn, P. Liljeroth, K. Kontturi, *J. Am. Chem. Soc.* **2002**, *124*, 12915–12921.
96 Y. Shao, M. V. Mirkin, G. Fish, S. Kokotov, D. Palanker, A. Lewis, *Anal. Chem.* **1997**, *69*, 1627–1634.
97 J. L. Amphlett, G. J. Denuault, *J. Phys. Chem. B* **1998**, *102*, 9946–9951.
98 Q. Fulian, A. C. Fisher, G. J. Denuault, *J. Phys. Chem. B* **1999**, *103*, 4387–4392.
99 Q. Fulian, A. C. Fisher, G. J. Denuault, *J. Phys. Chem. B* **1999**, *103*, 4393–4398.
100 A. L. Barker, J. V. Macpherson, C. J. Slevin, P. R. Unwin, *J. Phys. Chem. B* **1998**, *102*, 1586–1598.
101 P. Liljeroth, C. Johans, C. J. Slevin, B. M. Quinn, K. Kontturi, *Anal. Chem.* **2002**, *74*, 1972–1978.
102 O. Sklyar, G. Wittstock, *J. Phys. Chem. B* **2002**, *106*, 7499–7508.
103 A. Hengstenberg, C. Kranz, W. Schuhmann, *Chem. Eur. J.* **2000**, *6*, 1547–1554.
104 L. A. Nagahara, T. Thundat, S. M. Lindsay, *Rev. Sci. Instrum.* **1989**, *60*, 3128–3130.
105 R. M. Penner, M. J. Heben, T. L. Longin, N. S. Lewis, *Science* **1990**, *250*, 1118–1121.
106 S. Chen, A. Kucernak, *Electrochem. Commun.* **2002**, *4*, 80–85.
107 P. Sun, Z. Zhang, J. Guo, Y. Shao, *Anal. Chem.* **2001**, *73*, 5346–5351.
108 C. J. Slevin, N. J. Gray, J. V. Macpherson, M. A. Webb, P. R. Unwin, *Electrochem. Commun.* **1999**, *1*, 282–288.
109 B. B. Katemann, W. Schuhmann, *Electroanalysis* **2002**, *14*, 22–28.
110 D. O. Wipf, A. J. Bard, *Anal. Chem.* **1992**, *64*, 1362–1367.
111 D. O. Wipf, A. J. Bard, D. E. Tallman, *Anal. Chem.* **1993**, *65*, 1373–1377.
112 K. Borgwarth, D. G. Ebling, J. Heinze, *Ber. Bunsenges. Phys. Chem.* **1994**, *98*, 1317–1321.
113 M. Ludwig, C. Kranz, W. Schuhmann, H. E. Gaub, *Rev. Sci. Instrum.* **1995**, *66*, 2857–2860.
114 C. Kranz, M. Ludwig, H. E. Gaub, W. Schuhmann, *Adv. Mater.* **1996**, *8*, 634–637.
115 P. J. James, L. F. Garfias-Mesias, P. J. Moyer, W. H. Smyrl, *J. Electrochem. Soc.* **1998**, *145*, L64–L66.
116 A. Hengstenberg, I. D. Dietzel, W. Schuhmann, *Bioforum* **1999**, *22*, 595–599.
117 M. Büchler, S. C. Kelley, W. H. Smyrl, *Electrochem. Solid State Lett.* **2000**, *3*, 35–38.
118 A. Hengstenberg, A. Bloechl, D. Dietzel, W. Schuhmann, W. *Angew. Chem., Int. Ed. Engl.* **2001**, *40*, 905–908.
119 M. A. Alpuche-Aviles, D. O. Wipf, *Anal. Chem.* **2001**, *73*, 4873–4881.
120 B. B. Katemann, A. Schulte, W. Schuhmann, *Chem. Eur. J.* **2003**, *9*, 2025–2033.
121 D. Oyamatsu, Y. Hirano, N. Kanaya, Y. Mase, M. Nishizawa, T. Matsue, *Bioelectrochemistry* **2003**, *60*, 115–121.
122 K. Karrai, R. D. Grober, *Appl. Phys. Lett.* **1995**, *66*, 1842–1844.
123 D. E. Williams, T. F. Mohiuddin, Y. Y. Zhu, *J. Electrochem. Soc.* **1998**, *145*, 2664–2672.
124 T. H. Treutler, G. Wittstock, *Electrochim. Acta* **2003**, *48*, 2923–2932.
125 Y. E. Korchev, M. Raval, M. J. Lab, J. Gorelik, C. R. W. Edwards, T. Rayment,

D. Klenerman, *Biophys. J.* **2000**, *78*, 2675–2679.

126 C. E. Gardner, J. V. Macpherson, *Anal. Chem.* **2002**, *74*, 576A–584A.

127 J. V. Macpherson, P. R. Unwin, *Anal. Chem.* **2000**, *72*, 276–285.

128 C. Kranz, G. Friedbacher, B. Mizaikoff, A. Lugstein, J. Smoliner, E. Bertagnolli, *Anal. Chem.* **2001**, *73*, 2491–2500.

129 J. V. Macpherson, P. R. Unwin, *Anal. Chem.* **2001**, *73*, 550–557.

130 A. Lugstein, E. Bertagnolli, C. Kranz, B. Mizaikoff, *Surf. Interface Anal.* **2002**, *33*, 146–150.

131 A. Lugstein, E. Bertagnolli, C. Kranz, A. Kueng, B. Mizaikoff, *Appl. Phys. Lett.* **2002**, *81*, 349–351.

132 A. Kueng, C. Kranz, B. Mizaikoff, A. Lugstein, E. Bertagnolli, *Angew. Chem., Int. Ed. Engl.* **2003**, *42*, 3237–3240.

133 S. I. Park, J. Nogami, C. F. Quate, *Phys. Rev. B* **1987**, *36*, 2863–2866.

134 A. J. Melmed, *J. Vac. Sci. Technol. B* **1991**, *9*, 601–608.

135 P. J. Bryant, H. S. Kim, Y. C. Zheng, R. Yang, *Rev. Sci. Instrum.* **1987**, *58*, 1115.

136 J. P. Ibe, P. P. Bey, S. L. Brandow, R. A. Brizzolara, N. Burnham, *J. Vac. Sci. Technol. A* **1990**, *8*, 3570–3576.

137 R. Fainchtein, P. R. Zarriello, *Ultramicroscopy* **1992**, *42–44*, 1533–1537.

138 M. Fotino, *Rev. Sci. Instrum.* **1993**, *64*, 159–167.

139 M. Klein, G. Schwitzgebel, *Rev. Sci. Instrum.* **1997**, *68*, 3099–3103.

140 S. Kerfriden, A. H. Nahle, S. A. Campell, C. F. Walsh, J. R. Smith, *Electrochim. Acta* **1998**, *43*, 1939–1944.

141 R. Kazinczi, E. Szocs, E. Kalman, P. Nagy, *Appl. Phys. A* **1998**, *66*, S535–S538.

142 I. Ekvall, E. Wahlstrom, D. Claesson, H. Olin, E. Olsson, *Meas. Sci. Technol.* **1999**, *10*, 11–18.

143 A. H. Sorensen, U. Hvid, M. W. Mortensen, K. A. Morch, *Rev. Sci. Instrum.* **1999**, *70*, 3059–3067.

144 B. Zhang, E. Wang, *Electrochim. Acta* **1994**, *39*, 103–106.

145 M. J. Vasile, D. A. Grigg, J. E. Griffith, E. A. Fitzgerald, P. E. Russell, *Rev. Sci. Instrum.* **1991**, *62*, 2167–2171.

146 C. E. Bach, R. J. Nichols, W. Beckmann, H. Meyer, A. Schulte, J. O. Besenhard, P. D. Jannakoudakis, *J. Electrochem. Soc.* **1993**, *140*, 1281–1284.

147 J. V. Macpherson, P. R. Unwin, A. C. Hillier, A. J. Bard, *J. Am. Chem. Soc.* **1996**, *118*, 6445–6452.

148 C. E. Jones, J. V. Macpherson, P. R. Unwin, *J. Phys. Chem. B* **2000**, *104*, 2351–2359.

149 C. E. Jones, P. R. Unwin, J. V. Macpherson, *Chemphyschem* **2003**, *4*, 139–146.

150 S. J. O'Shea, R. M. Atta, M. E. Welland, *Rev. Sci. Instrum.* **1995**, *65*, 2508–2512.

151 P. Hoffmann, B. Dutoit, R. P. Salathe, *Ultramicroscopy* **1995**, *61*, 165–170.

152 T. Yatsui, M. Kourogi, M. Ohtsu, *Appl. Phys. Lett.* **1998**, *73*, 2090–2092.

153 D. Zeisel, S. Nettesheim, B. Dutoit, R. Zenobi, *Appl. Phys. Lett* **1996**, *68*, 2491–2492.

154 M. Muranishi, K. Sato, S. Hosaka, A. Kikukawa, T. Shintani, K. Ito, *Jpn. J. Appl. Phys., Part 2* **1997**, *36*, L942–L944.

155 J. A. Veerman, A. M. Otter, L. Kuipers, N. F. van Hulst, *Appl. Phys. Lett.* **1998**, *72*, 3115–3117.

156 J. F. Krogmeier, R. C. Dunn, *Appl. Phys. Lett.* **2001**, *79*, 4494–4496.

157 C. B. Prater, P. K. Hansma, M. Tortonese, C. F. Quate, *Rev. Sci. Instrum.* **1991**, *62*, 2634–2638.

158 R. M. Wightman, D. O. Wipf, in: *Electroanalytical Chemistry*, A. J. Bard (ed.); New York: Marcel Dekker, **1989**, Vol. 15, pp. 267–353.

159 F.-R. F. Fan, C. Demaille, in: *Scanning Electrochemical Microscopy*, A. J. Bard, M. V. Mirkin (eds.); New York: Marcel Dekker, **2001**, pp. 75–110.

160 B. R. Horrocks, D. Schmidtke, A. Heller, A. J. Bard, *Anal. Chem.* **1993**, *65*, 3605–3614.

161 Y. T. Kim, M. D. Scarnulis, A. G. Ewing, *Anal. Chem.* **1986**, *58*, 1782–1786.

162 G. Shi, L. F. Garfias-Mesias, W. H. Smyrl, *J. Electrochem. Soc.* **1998**, *145*, 2011–2016.

163 R. M. Penner, M. J. Heben, T. L. Longin, N. S. Lewis, *Science* **1990**, *250*, 1118–1121.

164 K. Potje-Kamloth, J. Janata, M. Josowicz, *Ber. Bunsenges. Phys. Chem.* **1989**, *93*, 1480–1485.

165 A. Schulte, R. H. Chow, *Anal. Chem.* **1996**, *68*, 3054–3058.

166 T. Albrecht, S. Akamine, T. E. Carver, C. F. Quate, *J. Vac. Sci. Technol. A* **1990**, *8*, 3386–3396.

167 T. Dziomba, H. U. Danzebrink, C. Lehrer, L. Frey, T. Sulzbach, O. Ohlsson, *Microscopy* **2001**, *202*, 22–27.

168 E. Amman, C. Beuret, P. F. Indermuhle, R. Kotz, N. F. de Rooij, H. Siegenthaler, *Electrochim. Acta* **2001**, *47*, 327–334.

169 C. Kranz, B. Mizaikoff, A. Lugstein, E. Bertagnolli, in: *Environmental Electrochemistry Analysis Trace Element Biogeochemistry*, M. Taillefert, T. F. Rozan (eds.); Washington, DC: American Chemical Society, **2002**, pp. 320–336.

170 E. L. H. Heintz, C. Kranz, B. Mizaikoff, H.-S. Noh, P. Hesketh, A. Lugstein, E. Bertagnolli, in: *Proc. IEEE Nanotechnology Conference*, **2001**.

171 J. Melngailis, *J. Vac. Sci. Technol. B* **1987**, *5*, 469–495.

172 H. Ximen, P. E. Russell, *Ultramicroscopy* **1992**, *42–44*, 1526–1532.

173 A. Kueng, C. Kranz, A. Lugstein, E. Bertagnolli, B. Mizaikoff, *Appl. Phys. Lett.* **2002**, *82*, 1592–1595.

174 Q. Zhao, Z. Gan, Q. Zhuang, *Electroanalysis* **2002**, *14*, 1609–1613.

175 E. W. Wong, P. E. Sheehan, C. M. Lieber, *Science* **1997**, *277*, 1971–1975.

176 S. S. Wong, E. Joselevich, A. T. Woolley, C. L. Cheung, C. M. Lieber, *Nature* **1998**, *394*, 52–55.

177 J. H. Hafner, C. L. Cheung, T. H. Oosterkamp, C. M. Lieber, *J. Phys. Chem. B* **2001**, *105*, 743–746.

178 K. Moloni, M. R. Buss, R. P. Andres, *Ultramicroscopy* **1999**, *80*, 237–246.

179 N. R. Wilson, J. V. Macpherson, *Nano Lett.* **2003**, *3*, 1365–1369.

180 J. Zhang, Q. Chi, S. Dong, E. Wang, *Bioelectrochem. Bioenerg.* **1996**, *39*, 267–274.

181 Q.-M. Xu, B. Zhang, L.-J. Wan, C. Wang, C.-L. Bai, D.-B. Zhu, *Surf. Sci.* **2002**, *517*, 52–58.

182 I. Otsuka, T. Iwasaki, *J. Microsc.* **1988**, *152*, 289–297.

183 N. Breuer, A. M. Funtikov, U. Stimmig, R. Vogel, *Surf. Sci.* **1995**, *335*, 145–154.

184 S. Dieluweit, M. Giesen, *J. Electroanal. Chem.* **2002**, *524–525*, 194–200.

185 M. Inaba, Z. Siroma, Z. Ogumi, T. Abe, Y. Mizutani, M. Asano, *Chem. Lett.* **1995**, *8*, 661–662.

186 M. Inabe, Y. Kawatate, A. Funabiki, T. Abe, Z. Ogumi, *Proc. Electrochem. Soc.* **1997**, *97*, 103–180.

187 O. Younes, L. Zhu, Y. Rosenberg, Y. Shacham-Diamand, E. Gileadi, *Langmuir* **2001**, *17*, 8270–8275.

188 A. S. Dakkouri, *Solid State Ionics* **1997**, *94*, 99–114.

189 L. A. Kibler, A. Cuesta, M. Kleinert, D. M. Kolb, *J. Electroanal. Chem.* **2000**, *484*, 73–82.

190 D. M. Kolb, *Electrochim. Acta* **2000**, *45*, 2387–2402.

191 D. M. Kolb, *Angew. Chem., Int. Ed. Engl.* **2001**, *40*, 1162–1181.

192 D. M. Kolb, *Surf. Sci.* **2002**, *500*, 722–740.

193 A. A. Gewirth, H. Siegenthaler, *Nanoscale Probes of the Solid/Liquid Interface* (eds) Nato ASI Series, Applied Sciences; Kluwer, Dordrecht, Vol. 288, **1993**.

194 A. A. Gewirth, B. K. Niece, *Chem. Rev.* **1997**, *97*, 1129–1162.

195 W. J. Lorenz, W. Plieth, *Electrochemical Nanotechnology*, Weinheim: Wiley-VCH, **1998**.

196 M. Ge, A. A. Gewirth, *Surf. Sci.* **1995**, *324*, 140–148.

197 K. Tamura, T. Kondo, K. Uosaki, *J. Electrochem. Soc.* **2000**, *147*, 3356–3360.

198 M. E. Hyde, R. Jacobs, R. G. Compton, *J. Phys. Chem. B* **2002**, *106*, 11075–11080.

199 J. Li, E. Wang, M. Green, P. E. West, *Synth. Met.* **1995**, *74*, 127–131.

200 R. Nyffenegger, E. Ammann, H. Siegenthaler, R. Kotz, O. Haas, *Electrochim. Acta* **1995**, *40*, 1411–1415.

201 M. F. Suarez, R. G. Compton, *J. Electroanal. Chem.* **1999**, *462*, 211–221.

202 G. E. Engelmann, J. C. Ziegler, D. M. Kolb, *Surf. Sci.* **1998**, *401*, L420–L424.

203 D. Alliata, R. Kötz, P. Novak, H. Siegenthaler, *Electrochem. Commun.* **2000**, *2*, 436–440.

204 O.E. Huesser, D.H. Craston, A.J. Bard, *J. Electrochem. Soc.* **1989**, *136*, 3222–3229.
205 D. Mandler, A.J. Bard, *J. Electrochem. Soc.* **1990**, *137*, 1079–1086.
206 C. Kranz, M. Ludwig, H.E. Gaub, W. Schuhmann, *Adv. Mater.* **1995**, *7*, 38–40.
207 J. Zhou, D.O. Wipf, *J. Electrochem. Soc.* **1997**, *144*, 1202–1207.
208 D. Mandler, A.J. Bard, *J. Electrochem. Soc.* **1989**, *136*, 3143–3144.
209 D. Mandler, A.J. Bard, *J. Electrochem. Soc.* **1990**, *137*, 2468–2472.
210 F.-R.F. Fan, J. Yang, S.M. Dirk, D.W. Price, D.V. Kosynkin, J.M. Tour, A.J. Bard, *J. Am. Chem. Soc.* **2001**, *123*, 2454–2455.
211 F.-R.F. Fan, J. Yang, L. Cai, D.W. Price, S.M. Dirk, D.V. Kosynkin, Y. Yao, A.M. Rawlett, J.M. Tour, A.J. Bard, *J. Am. Chem. Soc.* **2002**, *124*, 5550–5560.
212 P.R. Unwin, J.V. Macpherson, *Chem. Soc. Rev.* **1995**, *24*, 109–120.
213 T.P. Moffat, F.R.F. Fan, A.J. Bard, *J. Electrochem. Soc.* **1991**, *138*, 3224–3235.
214 J.Y. Josefowicz, L. Xie, G.C. Farrington, *J. Phys. Chem.* **1993**, *97*, 11995–11998.
215 D.W. Suggs, A.J. Bard, *J. Am. Chem. Soc.* **1994**, *116*, 10725–10733.
216 M.R. Vogt, A. Lachenwitzer, O.M. Magnussen, R.J. Behm, *Surf. Sci.* **1998**, *399*, 49–69.
217 E. Szocs, G. Vastag, A. Shaban, G. Konczos, E. Kalman, *J. Appl. Electrochem.* **1999**, *29*, 1339–1345.
218 P. Broekmann, M. Wilms, K. Wandelt, *Surf. Rev. Lett.* **1999**, *6*, 907–916.
219 O. Matsuoka, S.S. Ono, H. Nozoye, S. Yamamoto, *Surf. Sci.* **2003**, *545*, 8–18.
220 F.R.F. Fan, A.J. Bard, *J. Electrochem. Soc.* **1989**, *136*, 166–170.
221 A. Miyasaka, H. Ogawa, *Corros. Sci.* **1990**, *31*, 99–104.
222 S.G. Garcia, D.R. Alinas, C.E. Mayer, W.J. Lorenz, G. Staikov, *Electrochim. Acta* **2003**, *48*, 1279–1285.
223 K. Sashikata, Y. Matsui, K. Itaya, M.P. Soriaga, *J. Phys. Chem.* **1996**, *100*, 20027–20034.
224 D. Zuili, V. Maurice, P. Marcus, *J. Phys. Chem. B* **1999**, *103*, 7896–7905.
225 J. Bearinger, C.A. Orme, J.L. Gilbert, *Surf. Sci.* **2001**, *491*, 370–387.
226 V. Maurice, L.H. Klein, P. Marcus, *Surf. Int. Anal.* **2002**, *34*, 139–143.
227 J. Römer, M. Plaschke, G. Beuchle, J.I. Kim, *J. Nucl. Mater.* **2003**, *203*, 80–86.
228 H. Yhang, L. Wu, F. Huang, *J. Vac. Sci. Technol. B* **1999**, *17*, 269–272.
229 A.D. Müller, F. Müller, M. Hietschold, *Thin Solid Films* **2000**, *266*, 32–36.
230 J.V. Macpherson, P.R. Unwin, in: *Scanning Electrochemical Microscopy*, A.J. Bard, M.V. Mirkin (eds.); New York: Marcel Dekker, **2001**, Chapter 12.
231 K. Borgwarth, J. Heinze, in: *Scanning Electrochemical Microscopy*, A.J. Bard, M.V. Mirkin (eds.); New York: Marcel Dekker, **2001**, Chapter 6.
232 D.O. Wipf, *Colloids Surf. A* **1994**, *93*, 251–261.
233 C.H. Paik, H.S. White, R.C. Alkire, *J. Electrochem. Soc.* **2000**, *147*, 4120–4124.
234 T.E. Lister, P.J. Pinhero, *Electrochim. Acta* **2003**, *48*, 2371–2378.
235 I. Serebrennikova, S. Lee, H.S. White, *Faraday Discuss.* **2002**, *121*, 199–210.
236 C.H. Paik, R.C. Alkire, *J. Electrochem. Soc.* **2001**, *148*, B276–B281.
237 S.B. Basame, H. S White, *J. Phys. Chem.* **1995**, *99*, 16430–16435.
238 L.F. Grafias-Mesias, M. Alodan, P.I. James, W.H. Smyrl, *J. Electrochem. Soc.* **1998**, *145*, 2005–2010.
239 B.R. Horrocks, M.V. Mirkin, A.J. Bard, *J. Phys. Chem.* **1994**, *98*, 9106–9114.
240 J.V. Macpherson, P.R. Unwin, *J. Phys. Chem.* **1995**, *99*, 14824–14831.
241 J.V. Macpherson, P.R. Unwin, *J. Phys. Chem.* **1995**, *99*, 3338–3351.
242 J.H.W. De Wit, *Electrochim. Acta* **2001**, *46*, 3641–3650.
243 M. Boehmisch, F. Burmeister, A. Rettenberger, J. Zimmermann, J. Boneberg, P. Leiderer, *J. Phys. Chem. B* **1997**, *101*, 10162–10165.
244 M. Rohwerder, E. Hornung, M. Stratmann, *Electrochim. Acta* **2003**, *48*, 1235–1243.
245 O. Sklyar, A. Kueng, C. Kranz, B. Mizaikoff, A. Lugstein, E. Bertagnolli, G. Wittstock, submitted to *Anal. Chem.*
246 A.G. Hansen, H. Wackerbarth, J.U. Nielsen, J. Zhang, A.M. Kuznetsov, J.

ULSTRUP, *Russ. J. Electrochem.* **2003**, *39*, 108–117.
247 J. ZHANG, M. GRUBB, A.G. HANSEN, A.M. KUZNETSOV, A. BOISEN, H. WACKERBARTH, J. ULSTRUP, *J. Phys.: Condens. Matter* **2003**, *15*, S1873–S1890.
248 E.P. FRIIS, J.E.T. ANDERSEN, P. MOLLER, J. ULSTRUP, *J. Electroanal. Chem.* **1997**, *431*, 35–38.
249 Q. CHI, J. ZHANG, E.P. FRIIS, J.E.T. ANDERSEN, J. ULSTRUP, *Electrochem. Commun.* **1999**, *1*, 91–96.
250 E.P. FRIIS, J. E.T. ANDERSEN, Y.I. KHARKATS, A.M. KUZNETSOV, R.J. NICHOLS, J.-D. ZHANG, *Proc. Natl. Acad. Sci. USA* **1999**, *96*, 1379–1384.
251 P. FACCI, D. ALLIATA, S. CANNISTRARO, *Ultramicroscopy* **2001**, *89*, 291–298.
252 Q. CHI, J. ZHANG, J.U. NIELSEN, E.P. FRIEES, I. CHORKENDORFF, G.W. CANTERS, J.E.T. ANDERSEN, J. ULSTRUP, *J. Am. Chem. Soc.* **2000**, *122*, 4047–4055.
253 A.G. HANSEN, A. BOISEN, J.U. NIELSEN, H. WACKERBARTH, I. CHORCKENDORFF, J.E.T. ANDERSEN, J. ZHANG, J. ULSTRUP, *Langmuir* **2002**, *19*, 3419–3427.
254 L. ADOLFINI, B. BONANNI, G.W. CANTERS, M. PH. VERBEET, S. CANNISTRARO, *Surf. Sci.* **2003**, *530*, 181–194.
255 J. BRASK, H. WACKERBARTH, K.J. JENSEN, J. ZHANG, I. CHORKENDORFF, J. ULSTRUP, *J. Am. Chem. Soc.* **2003**, *125*, 94–104.
256 R. GUCKENBERGER, M. HEIM, G. CEVC, H.F. KNAPP, W. WIEGRAEBE, A. HILLEBRAND, *Science* **1995**, *266*, 1538–1540.
257 F.-R.F. FAN, A.J. BARD, *Science* **1995**, *270*, 1849–1851.
258 R. GUCKENBERGER, M. HEIM, *Science* **1995**, *270*, 1851–1852.
259 D.P. ALLISON, P. HINTERDORFER, W. HAN, *Curr. Opin. Biotechnol.* **2002**, *13*, 47–51.
260 Y.E. KORCHEV, C.L. BASHFORD, M. MILOVANOVIC, I. VODYANOY, M.J. LAB, *Biophys. J.* **1997**, *73*, 653–658.
261 Y.E. KORCHEV, J. GORELIK, M.J. LAB, E.V. SVIDERSKAYA, C.L. JOHNSTON, C.R. COOMBES, I. VODYANOY, C.R. EDWARDS, *Biophys. J.* **2000**, *78*, 451–457.
262 Y.E. KORCHEV, Y.A. NEGULYAEV, C.R. EDWARDS, I. VODYANOY, M.J. LAB, *Nat. Cell Biol.* **2000**, *2*, 616–619.
263 A.I. SHEVCHUK, J. GORELIK, S.E. HARDING, M.J. LAB, D. KLENERMAN, Y.E. KORCHEV, *Biophys. J.* **2001**, *81*, 1759–1764.
264 J. GORELIK, A.I. SHEVCHUK, G.I. FROLENKOV, I.A. DIAKONOV, M.J. LAB, C.J. KROS, G.P. RICHARDSON, I. VODYANOY, C.R.W. EDWARDS, D. KLENERMAN, Y.E. KORCHEV, *Proc. Natl. Acad. Sci. USA* **2003**, *100*, 5819–5822.
265 Y. TAKII, K. TAKOH, M. NISHIZAWA, T. MATSUE, *Electrochim. Acta* **2003**, *48*, 3381–3385.
266 T. KAYA, Y. TORISAWA, D. OYAMATSU, M. NISHIZAWA, T. MATSUE, *Biosens. Bioelectron.* **2003**, *18*, 1379–1383.
267 M. NISHIZAWA, K. TAKOH, T. MATSUE, *Langmuir* **2002**, *18*, 3645–3649.
268 C. KRANZ, T. LOTZBEYER, H.-L. SCHMIDT, W. SCHUHMANN, *Biosens. Bioelectron.* **1997**, *12*, 257–266.
269 D.T. PIERCE, P.R. UNWIN, A.J. BARD, *Anal. Chem.* **1992**, *64*, 1795–1804.
270 B.D. BATH, H.S. WHITE, E.R. SCOTT, in: *Scanning Electrochemical Microscopy*, A.J. BARD, M.V. MIRKIN (eds.); New York: Marcel Dekker, **2001**, Chapter 9.
271 S. AMEMIYA, A.J. BARD, *Anal. Chem.* **2000**, *72*, 4940–4948.
272 D.T. PIERCE, A.J. BARD, *Anal Chem.* **1993**, *65*, 3598–3604.
273 C. KRANZ, G. WITTSTOCK, H. WOHLSCHLAGER, W. SCHUHMANN, *Electrochim. Acta* **1997**, *42*, 3105–3111.
274 G. WITTSTOCK, T. WILHELM, *Electroanalysis* **2001**, *13*, 669–675.
275 H. SHIKU, Y. HARA, T. MATSUE, I. UCHIDA, T. YAMAUCHI, *J. Electroanal. Chem.* **1997**, *438*, 187–190.
276 G. WITTSTOCK, *Fresenius' J. Anal. Chem.* **2001**, *370*, 303–315.
277 G. WITTSTOCK, W. SCHUHMANN, *Anal. Chem.* **1997**, *69*, 5059–5066.
278 T. WILHELM, G. WITTSTOCK, *Angew. Chem., Int. Ed. Engl.* **2003**, *42*, 2248–2250.
279 G. DENUAULT, M.H.T. FRANK, L.M. PETER, *Faraday Discuss.* **1992**, *94*, 23–35.
280 B.R. HORROCKS, M.V. MIRKIN, D.T. PIERCE, A.J. BARD, G. NAGY, K. TOTH, *Anal. Chem.* **1993**, *65*, 1213–1224.
281 C. WEI, A.J. BARD, G. NAGY, K. TOTH, *Anal. Chem.* **1995**, *67*, 1346–1356.

282 B. R. Horrocks, M. V. Mirkin, *J. Chem. Soc., Faraday Trans.* **1998**, *94*, 1115–1118.

283 B. M. Quinn, P. Liljeroth, K. Kontturi, *J. Am. Chem. Soc.* **2002**, *124*, 12915–12921.

284 M. H. Troise-Frank, G. Denuault, L. M. Peter, *Faraday Discuss.* **1992**, *94*, 23–35.

285 S. Daniele, C. Bragato, I. M. Ciani, A. Baldo, *Electroanalysis* **2003**, *15*, 621–628.

286 D. Oyamatsu, Y. Hirano, N. Kanaya, Y. Mase, M. Nishizawa, T. Matsue, *Bioelectrochemistry* **2003**, *60*, 115–121.

287 A. Bruckbauer, L. Ying, A. M. Rothery, Y. E. Korchev, D. Klenerman, *Anal. Chem.* **2002**, *74*, 2612–2616.

288 C. Kranz, A. Kueng, A. Lugstein, E. Bertagnolli, B. Mizaikoff, *Ultramicroscopy* **2003**, in press.

289 J. V. Macpherson, C. E. Jones, A. L. Barker, P. R. Unwin, *Anal. Chem.* **2002**, *74*, 1841–1848.

290 W. Schuhmann, *Mikrochim. Acta* **1995**, *121*, 1–29.

291 W. Schuhmann, *Rev. Mol. Biotechnol.* **2002**, *82*, 425–441.

292 T. Wilhelm, G. Wittstock, R. Szargan, *Fresenius' J. Anal. Chem.* **1999**, *365*, 163–167.

293 C. Kranz, A. Kueng, A. Lugstein, E. Bertagnolli, B. Mizaikoff, unpublished results.

294 Y. Matsumura, K. Kajino, M. Fujimoto, *Membr. Biochem.* **1980**, *3*, 99–129.

295 C. Giaume, R. T. Kado, *Biochim. Biophys. Acta* **1983**, *762*, 337–343.

296 S. Glab, A. Hulanicki, G. Edwall, F. Ingmann, *Crit. Rev. Anal. Chem.* **1989**, *21*, 29–47.

297 A. Kueng, C. Kranz, B. Mizaikoff, *Sens. Lett.* **2003**, *1*, 2–15.

298 A. Kueng, C. Kranz, B. Mizaikoff, unpublished results.

8
Nanofluidic Modeling and Simulation

M. Geier, A. Greiner, D. Kauzlaric, J. G. Korvink, Institute for Microsystem Technology, Albert Ludwig University Freiburg, Germany

Abstract
Micro and nano-fluidic systems are of increasing importance, especially in the life sciences. This leads to a growing demand for appropriate models capable of representing fluidic micro devices with sufficient accuracy. Mesoscopic in nature, the devices feature nano scale dynamics that are not captured by continuum models of Computational Fluid Dynamics (CFD) on both length and time scales, which are intractable for molecular simulations. Solving the continuum fluid dynamic equations is typically an iterative process, and therefore coupled models between conventional continuum CFD and microscopic or even atomistic methods are computationally demanding. In this article we discuss two time-explicit CFD schemes: Digital Lattice Boltzmann Automata (DLB), attempting to close the mesoscopic gap from the side of the continuum equation (that is, top down), and Dissipative Particle Dynamics (DPD), a coarse-grained molecular dynamics model (that is, bottom up).

Keywords
micro-fluidics; nano-fluidics; computational fluid dynamics; lattice Boltzmann; Boltzmann transport equation; dissipative particle dynamics.

Advanced Micro and Nanosystems. Vol. 1
Edited by H. Baltes, O. Brand, G. K. Fedder, C. Hierold, J. Korvink, O. Tabata

ISBN: 3-527-30746-X

8.1 Introduction

In the past decade, microtechnology has become an increasingly important key technology for a variety of industrial applications. These applications range from simple miniaturized mechanical structures all the way to complex micromechanical systems operating in various physical domains at the same time. The fluidic domain has gained increasing importance because of its very promising adaptability in the biomedical sector, where dispensing of the smallest amounts of liquids is needed, sophisticated biochemical reactions are employed in reaction chambers containing picoliter agents or even less and immobilization of specific molecules out of the liquid phase is to be performed. Even in material process technology, where liquid materials with structures down to several tens of nanometers are important ingredients, modeling and simulation is needed, especially in molding processes when the length scale of the mold approaches the same order of magnitude as microstructures in the material.

A natural development is that increasing attention is being paid to the modeling of fluid transport in confined micrometer-scale geometries (see, for example, [1]). Although the description of fluid flow within the framework of a conventional continuum approach yields good results for a wide range of scales down to several tens of nanometers, it does not guarantee a proper description at increasingly smaller scales. Physical phenomena, which can usually be neglected on the macro scale, become dominant as the length scale drops below the micrometer range. This does not necessarily mean that the structures confining the fluid are of sub-micrometer scale, but rather that their significant features are. Think of a particle-filled feedstock for micropowder injection molding that has a particle size distribution with diameters in the range from 50 nm up to 1 μm. This implies that modeling and simulation must be capable of spanning the length scale, which in the worst case may range over several orders of magnitude.

Modeling and simulation of the behavior of materials in various applications of modern microsystems is a challenging task. High costs or restricted availability of samples for analysis require modern microfluidic systems to operate in the picoliter regime and below. In production environments, nevertheless, the systems have to be optimized with respect to speed and throughput. Special care must be taken in the design of the liquid handling system. Simulation tools emerging from a theoretical modeling, that ranges from quantum physics of atomic-scale phenomena up to continuum descriptions of macroscopic behavior [2], and going even further to system behavior, are the ingredients of still missing engineering tools that support design processes in micro- and nanosystem technology. The use of advanced computer-aided design (CAD) tools promises reduction in the extent of physical testing necessary to prototype a device. Activities to incorporate various physical models at different length scales (see, for example, [3]) exist.

Systems which are mesoscopic in nature, that is, span the nanometer regime up to the micrometer scales and for which microscopic effects stemming from the nanoscale become important, cannot be treated with classical molecular dynamics (MD) since the required computational power is prohibitively large. Therefore, there is an urgent need for the development of new models and their implementation with adequate numerical models into simulation tools that are capable of properly describing fluid flow in the range from 10 nm up to 1 μm.

Simulation of complex processes in a common framework with a reasonable computational effort – as it is essential for its application in a design process – might be realized by superimposing particle dynamics on a continuum computational fluid dynamics simulation. Typically, the particle dynamics simulation will be computationally expensive but suitable for the description of microscopic features. The continuum CFD solvers available are very fast but not suitable for microscopic dimensions. In coupling both methods, the iterative procedure for fluid solvers will, owing to the nonlinear nature of the Navier-Stokes equations, force the costly particle dynamics calculation to be repeated in every iterative loop. To overcome this drawback, it is desirable to couple the microscale simulation, featuring non-iterative and explicit time marching, to an equivalent model capable of acting at a much coarser length scale and thus eventually leading to a feasible simulation process for engineering CAD applications in multi-scale micro- and nanoelectromechanical systems (MEMS and NEMS) [4]. Note that reducing the Navier-Stokes solution to an explicit time step method is undesirable [5]. The challenge is to develop a tool capable of bridging the gap between the microworld and the macroscopic behavior of the system.

Two promising candidates to fulfil the above-mentioned requirements are the lattice gas (LG) methods, with their special variations found in the lattice Boltzmann (LB) methods, to be applied to the large-scale part of the simulation and dissipative particle dynamics (DPD), a particle dynamic simulation technique for mesoscopic simulations above the molecular length scale. In the following we will give an overview of the basic ideas underlying these two methods with respect to their capabilities in simulating fluid dynamics at the respective length scales applicable in MEMS and NEMS. In the following section we shall also provide more detailed descriptions of the respective methods.

8.2 The Abstract LG Approach and the Physically Motivated DPD Model

Lattice gas methods and LB models are methods for the simulation of fluid flows. They are distinct from MD methods or direct discretizations of partial differential equations, as for example the Navier-Stokes equations. The basic idea of LG and LB methods is the assignment of quantities to quasi-particles moving on a regular lattice in arbitrary dimensions. The equation of motion for the respective quantities is reproduced by transporting the particles along the links between lattice sites, that is, executing a so-called propagation step. After the propagation the quantities transported with the quasi-particles are exchanged between the particles occupying a lattice site according to a finite set of rules applied to the finite set of possible states for each lattice site.

The transported quantity is the momentum that a particle carries. Depending on whether this momentum is a Boolean, an integer or a real quantity, we end up with different categories of the lattice methods, that is, the simple lattice gas, the integer lattice gas (ILG) or the lattice Boltzmann description. An overview of these models is given in [6]. Note that LG models, albeit providing fast and efficient ways for solving partial differential equations, show some fundamentally unphysical behavior, such as the Galilean invariance problem, when it comes to the simulation of macroscopic fluid flow obeying the Navier-Stokes equation. This problem is solved in the LB methods applied to fluid flow problems [7].

The advantages of using LB methods for specific applications, such as the dynamics of multicomponent flow or multiphase flow, are evidenced in [8]. LB methods are easily coupled downscale with atomistic or microscopic particle dynamic methods and also 'upwards' to macroscopic grid methods. This powerful scale-spanning property is best illustrated by the various areas to which the method has been applied, such as flow around macroscopic objects, surface gravity waves [9], oscillatory channel flow [10], finite volume schemes [11], applications to flow in porous media [12] and thermodynamic multiphase flows [13] and the method can be applied also to free surface flow [14].

Our goal is a correct modeling of flow problems in micro- and nanosystems. We have to deal with two main challenges. First, continuum models have their limitations in representing correct hydrodynamic behavior on microscopic scales when the discrete microscopic nature of the fluid becomes relevant. Second, microscopic models often deliver too much information and, from the viewpoint of computational cost, we often cannot afford them. On the molecular level, which means at a microscopic length scale of the order of inter-atomic distances in fluids or solids, MD [15, 16] represents an attractive simulation technique for the study of hydrodynamic phenomena. However, at length scales from 10 to 1000 nm a coarse graining of the molecular model, due to rising computational costs, is desirable.

In other words, either the available modeling tools are too coarse-grained or they are too 'microscopic'. Bridging the gap between these two extremes is currently a wide field of research. Along with this challenge, the question arises of how the bridge can be mounted to its shores, that is, adequate techniques of coupling between different models are also of great importance.

A very promising technique to match the microscopic description to coarser models is DPD, first introduced by Hoogerbrugge and Koelman [17]. In essence, it is a treatment of the dynamics of quasi-particles representing small sets of the liquid's molecules by stochastic differential equations similar to a Langevin approach. It combines features from both the MD and LG methods. Since its introduction, this method has been applied to the simulation of a wide variety of phenomena, especially in material science. Thorough investigations have been carried out in order to understand the capabilities of DPD when applied to CFD problems [18–25]. Dissipative particle dynamics represents in our view an attractive technique, which could provide some answers. It provides the advantage of Galilean invariance and isotropy over simple lattice gas automata (LGA) [26] and the conservation of momentum over the Langevin approach. Covering already coarser scales as MD can do, a drawback of DPD is still its small time-scale and its short length scales compared with the LG methods mentioned above.

We would like to call DPD a mesoscopic simulation method. In the task of 'bridging the gap' between atomistic and mesoscopic simulation [27], DPD will be one of the choices. Its application range for different simulation tasks is briefly listed: mesoscopic dynamics of colloids [28], binary fluids and the matching of macroscopic properties with DPD [29], domain growth and phase separation in immiscible fluids [30] and the simulation of rheological properties [31], which is a very important characteristic of this method, augmenting its capability by predicting material properties, so necessary for the CAD process in MEMS.

Recently, DPD has attracted increasing interest from different research groups. Several attempts at its improvement with respect to different aspects have been made. Its algorithmic optimization was the focus, making it a method appropriate for application in the engineering field [32–34]. The inclusion of energy conservation in the particle–particle interaction for the set of stochastic differential equations describing a DPD model have been derived [35], which is an important extension for heat flow applications. Recently, phase change models have been included, built on the energy-conserving DPD models [36] for a solid–liquid phase transition. Modeling of the liquid–vapor co-existence within the DPD approach was for a long time believed to be impossible, unless one was ready to sacrifice the advantages of the method coming from the use of larger time steps than is possible in MD [27]. Recently, this dilemma has been solved by introducing a so-called multi-particle DPD [37]. For static problems the multi-particle DPD model was shown to reproduce correct phase co-existence, for instance for a pending droplet. For the dynamic problems the model has not yet been tested.

At present, no attempts have been made to exploit a combination of LG/LB models and DPD for the simulation in CFD. Especially for MEMS applications, where an increasing need for modeling and simulation of fluid dynamics can be observed, this configuration seems to be a promising candidate.

The overall challenge that arises with the two models mentioned above is to match macroscopic material properties and to provide results for microscopic phenomena on the smallest length scales occurring in the problem. This is where the model parameters have to be tuned in order to obtain reliable results for all length scales of the given simulation.

8.3 Digital Lattice Boltzmann Automata

In this section, an explicit fluid solver is derived based on a digital approximation of the Boltzmann transport equation (BTE). We start by discussing the state-of-the-art approach to simulate fluid dynamic systems by solving the Navier-Stokes equation and the disadvantages of implicit methods in brief. In the following we take a look at a general properties of conservation laws, allowing us to reformulate them in a digital framework, thereby removing the need for real-valued field variables. In this framework, a finite order approximation of the BTE is derived that is capable of recovering the Navier-Stokes equation. The resulting new model is explicit by definition, free from numerical errors and linear in computational complexity.

The equation governing the time evolution of the single particle probability of many particle systems is the BTE. The huge information content imposed by tracing the trajectories of all particles under consideration makes it computationally intractable for any practical problem size. A radical simplification of the BTE is found by a statistical moment expansion of the particle distribution function. Its zero-order moment is then identified as the mass balance law, the first-order moment is the momentum balance law, as it is found in the Navier-Stokes equation, and the second-order moment is the energy balance law.

The first three moments are sufficient to describe compressible flow in the continuum limit. In the incompressible continuum limit only the Navier-Stokes equation is considered. By being a radical simplification of the underlying physics, the Navier-Stokes equation still imposes some challenges for practical problems. As it shows a second-order nonlinearity in velocity and velocity gradient, it is neither very tractable analytically nor easily dealt with numerically. It might, however, be solved with the help of an appropriate discretization scheme and by applying iterative solution procedures. Such implicit methods are currently applied successfully in computational fluid dynamics. A general advantage of iterative methods is that simply increasing the iteration depth for a single time step can reduce numerical errors. Hence they show much better stability properties than explicit schemes. However, convergence might be slow and cannot be guaranteed in general. Once we consider problems where different physical domains interact with each other, implicit methods might lead to a prohibitive increase in computational cost by including the external simulation into its iterative loop and forcing its recomputation. This would be especially cumbersome if the time-scales for the problems do not match, as in couplings between molecular dynamics and fluid solvers. For these kind of problems, explicit schemes are required.

Partial differential equations for fields, such as the Navier-Stokes equations, are typically solved by numerical methods. After applying one of the various discretization techniques, one ends up with a system of algebraic equations, which in turn are to be solved by suitable matrix inversion techniques. The inversion operation is a process requiring numbers to be given as real values. This is numerically impossible on finite memory digital computers since any real-valued number represents an infinite amount of information. Replacing real numbers by floating-point numbers makes the model tractable but it violates the balance laws solved for by introducing numerical errors. A numerical scheme is said to be unstable if the errors increase with time. Easy ways to force numerical stability are to increase the depth of iteration and the number of digits of the floating-point data type. Both can be increased to arbitrary order, resulting in a fashion of brute force stabilization by excessive computational effort. For safety reasons only, much more work is put into the solution process than necessary. Our aim is to reduce work to a minimum.

Stability would obviously be achieved without excessive computational effort if the numerical errors were identical to zero after every applied operation. This would require the arithmetic on the conserved field variables to be exact. On a digital computer, only integer addition, subtraction and multiplication on a finite range can be done without information loss. That is obviously not enough to solve differential equations, owing to the necessary matrix inversion step. However, neglecting for the moment the way in which differential equations are solved numerically, we can argue that addition is a sufficient operation for the communication among discrete field variables representing a conserved quantity. This is a trivial conclusion from the π-theorem stating that the right-hand side of an equation needs to have the same physical dimension as the left-hand side. In order to benefit from this fact, we have to reformulate our model in such a way that makes equation solving obsolete. Owing to the equivalence of all Turing complete languages, it must be possible to describe all facts captured in the terms of differential equations by other means and without equations. For the lattice Boltzmann method we chose the language of cellular automata.

8.4
Cellular Automata Fluid Model

A cellular automaton is a regular arrangement of cells or sites. Every site is in a state taken from a finite set. A finite neighborhood is defined for each site. All sites become updated in an instant owing to a set of rules taking the current state of the cell and its neighbors as input variables. Cellular automata are Turing complete and can hence describe any describable problem. Doing physical simulations on cellular automata is reducing physical laws to logic, thereby making them very easy to solve on a digital computer. Solving the full Boltzmann transport equation is still intractable, but the cellular automaton can easily be used to model a discrete particle distribution identical with the Maxwell-Boltzmann distribution up to a certain order. Any model that re-samples the moments of this distribution to high enough order satisfies the Navier-Stokes equation implicitly.

The lattice Boltzmann automaton is defined as an arrangement of identical isotropic sites on a regular lattice. Every site is connected to a finite set of neighboring sites by links. The links are occupied by particles where a particle is defined as a unification of mass, momentum and energy. It is a computational object, which must not be mistaken for a molecule or something similar. Particles move in a universal streaming step along their links to the respective neighboring sites where they occupy the same link as before. Subsequently, the particle distribution on every site becomes rearranged in a mass-, momentum- and energy- (if taken into account) conserving way. Hydrodynamic behavior emerges when these two steps are repeated for a long time. Depending on the data type of the link occupation and the update rules, we distinguish several models. We call the automaton a lattice gas if only one particle per link is allowed [26]. A floating-point representation is called lattice Boltzmann automaton and can be distinguished further according to the update algorithm. Integer-based automata can be defined similarly to either the lattice gas principle or the lattice Boltzmann automata. They are either called integer lattice gas or digital lattice Boltzmann automata. Integer lattice gases should be seen as automata making a random walk in their state space [38–40] whereas digital lattice Boltzmann automata compute a deterministic approximation of the Maxwell-Boltzmann distribution function in its digital state space. Owing to the small number of local degrees of freedom, this approximation degenerates to a distribution of particles among the finite number of lattice links, retaining only the most important property of the Maxwell-Boltzmann distribution, namely its invariance under a spatial translation hereafter called Galilean invariance.

The underlying lattice is identical for all models. It has to be sufficiently isotropic. In what follows we consider a three-dimensional model for incompressible flows conserving only mass and momentum. It has been shown [41] that no lattice exists in three dimensions satisfying the isotropy condition. It is possible, however, to define such a lattice in four dimensions, the so-called face-centered hypercube (FCHC) [42], which is the set of 24 links obtained from all permutations of $(\pm 1, \pm 1, 0, 0)$. A three-dimensional lattice can then be obtained by projection. The site is a cube with links pointing to the middle of all edges and two links pointing to the center of each face of the cube. A two-dimensional model can be obtained by another projection. Even though the lattice now looks three-dimensional it is very important not to forget where it came from, especially once we take energy into account. Since the distance from the center of the cube to the center of its face is shorter than to the middle of its edge, one might mistakenly think that the particles move at different speeds. In fact, they have a velocity component in the fourth direction that has to be taken into account.

In this model, all particles have the same energy. This might even be assumed for particles that are at rest (alternatively, we could simply admit that energy is not a conserved quantity in this model, as it is not in the incompressible Navier-Stokes equation). Since we have two links pointing to every face of the cube, we might unify them to single links, leaving us with six links pointing to the cubes faces, 12 links pointing to its edges and one resting link in the center. We call this lattice the D3Q19 lattice, since it has 19 speeds in three dimensions (see Fig. 8.1).

The state of a single site can be represented by the vector

$$\vec{s} = (r, nw, w, sw, s, se, e, ne, n, f, wf, sf, ef, nf, b, wb, sb, eb, nb)^T \tag{1}$$

where r denotes the occupation of the resting link. The other links are named intuitively according to the direction in which they point: n denotes north, w west, s

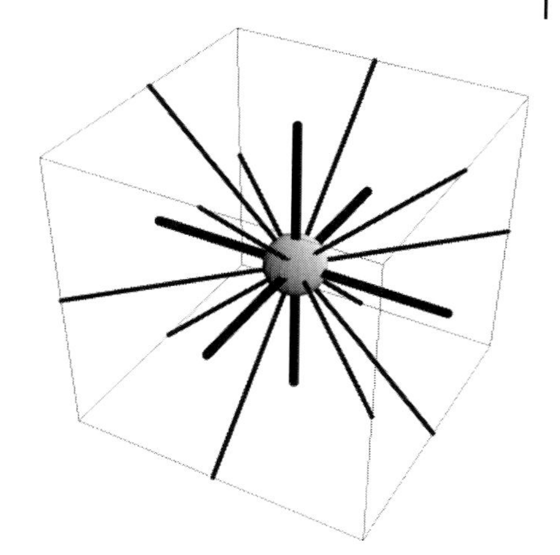

Fig. 8.1 The three-dimensional projection of the FCHC lattice is called the D3Q19 lattice. Particles are either at rest or move along one of the 18 links to a neighboring site. Links pointing to the face of the cube are double links

south, e east, f front, b back and combinations thereof denote links to the respective edges. Four quantities of the state vector should be conserved: the total amount of mass and the momentum in three directions. Those quantities can be obtained from the state vector by multiplication with certain mask vector operators:

$$\left.\begin{array}{l} \rho = \vec{s}\cdot(1,1,1,1,1,1,1,1,1,1,1,1,1,1,1,1,1,1,1) \\ \pi_x = \vec{s}\cdot(0,-1,-1,-1,0,1,1,1,0,0,-1,0,1,0,0,-1,0,1,0) \\ \pi_y = \vec{s}\cdot(0,1,0,-1,-1,-1,0,1,1,0,0,-1,0,1,0,0,-1,0,1) \\ \pi_z = \vec{s}\cdot(0,0,0,0,0,0,0,0,0,1,1,1,1,1,-1,-1,-1,-1,-1) \end{array}\right\} \qquad (2)$$

where ρ is the total amount of mass on the site and π_x, π_y and π_z are the momentum in east, north and front direction, respectively. We can define 15 vectors perpendicular to these mask vectors:

$$\begin{pmatrix} \vec{c}_1^T \\ \vec{c}_2^T \\ \vec{c}_3^T \\ \vec{c}_4^T \\ \vec{c}_5^T \\ \vec{c}_6^T \\ \vec{c}_7^T \\ \vec{c}_8^T \\ \vec{b}_{yx}^T \\ \vec{b}_{yz}^T \\ \vec{b}_{xy}^T \\ \vec{b}_{xz}^T \\ \vec{b}_{zx}^T \\ \vec{b}_{zy}^T \end{pmatrix} = \begin{bmatrix} 2 & -1 & 0 & 0 & 0 & -1 & 0 & 0 & 0 & 0 & 0 & 0 & 0 & 0 & 0 & 0 & 0 & 0 & 0 \\ 2 & 0 & -1 & 0 & 0 & 0 & -1 & 0 & 0 & 0 & 0 & 0 & 0 & 0 & 0 & 0 & 0 & 0 & 0 \\ 2 & 0 & 0 & -1 & 0 & 0 & 0 & -1 & 0 & 0 & 0 & 0 & 0 & 0 & 0 & 0 & 0 & 0 & 0 \\ 2 & 0 & 0 & 0 & -1 & 0 & 0 & 0 & -1 & 0 & 0 & 0 & 0 & 0 & 0 & 0 & 0 & 0 & 0 \\ 2 & 0 & 0 & 0 & 0 & 0 & 0 & 0 & 0 & -1 & 0 & 0 & 0 & 0 & -1 & 0 & 0 & 0 & 0 \\ 2 & 0 & 0 & 0 & 0 & 0 & 0 & 0 & 0 & 0 & -1 & 0 & 0 & 0 & 0 & 0 & 0 & -1 & 0 \\ 2 & 0 & 0 & 0 & 0 & 0 & 0 & 0 & 0 & 0 & 0 & -1 & 0 & 0 & 0 & 0 & 0 & 0 & -1 \\ 2 & 0 & 0 & 0 & 0 & 0 & 0 & 0 & 0 & 0 & 0 & 0 & -1 & 0 & 0 & -1 & 0 & 0 & 0 \\ 2 & 0 & 0 & 0 & 0 & 0 & 0 & 0 & 0 & 0 & 0 & 0 & 0 & -1 & 0 & 0 & -1 & 0 & 0 \\ 0 & 1 & 0 & -1 & 2 & -1 & 0 & 1 & -2 & 0 & 0 & 0 & 0 & 0 & 0 & 0 & 0 & 0 & 0 \\ 0 & 0 & 0 & 0 & 2 & 0 & 0 & 0 & -2 & 0 & 0 & -1 & 0 & 1 & 0 & 0 & -1 & 0 & 1 \\ 0 & -1 & 2 & -1 & 0 & 1 & -2 & 1 & 0 & 0 & 0 & 0 & 0 & 0 & 0 & 0 & 0 & 0 & 0 \\ 0 & 0 & 2 & 0 & 0 & 0 & -2 & 0 & 0 & 0 & -1 & 0 & 1 & 0 & 0 & -1 & 0 & 1 & 0 \\ 0 & 0 & 0 & 0 & 0 & 0 & 0 & 0 & 0 & -2 & 1 & 0 & 1 & 0 & 2 & -1 & 0 & -1 & 0 \\ 0 & 0 & 0 & 0 & 0 & 0 & 0 & 0 & 0 & -2 & 0 & 1 & 0 & 1 & 2 & 0 & -1 & 0 & -1 \end{bmatrix} \qquad (3)$$

The vectors are grouped in two classes according to the hydrodynamic moments that they affect. It is now obvious how we can prevent all numerical errors from reaching the conserved quantities. Everything we are allowed to do is adding integer multiples of the 15 mask vectors perpendicular to mask vectors defining the conserved quantities and thereby we need to avoid under- and overflows. Exactly how often we want to add the different vectors to the state vectors has to be determined from the Galilean invariance condition.

8.5 Scattering Rules for Galilean Invariance

A system in a given frame of reference moving at constant velocity with respect to other frames is Galilean invariant when its behavior is independent of the frame of reference chosen for its description. Ultimately, Nature itself is not Galilean invariant, but it is a reasonable assumption for most practical problems in fluid dynamics to be Galilean invariant. Operating with floating-point numbers or not, Galilean invariance can never be reached exactly in a model stored in a digital computer. In lattice Boltzmann models, the error might become very large owing to the small number of speeds with which we allow the particles to travel. With this finite set of degrees of freedom, only a finite order of Galilean invariance can be obtained. The order of Galilean invariance is a figure of merit of the method. Here we aim only at second-order Galilean invariance in order to be compatible with the incompressible Navier-Stokes equation. By adding more speeds and obtaining higher orders of Galilean invariance, the model might finally reach the quality and the computational complexity of molecular dynamics.

The condition for Galilean invariance is most easily achieved to a given order by forcing the central moments of a single site distribution function with respect to velocity to be independent of flow speed. Central moments κ_n can be computed from the moments μ_n as follows:

$$\left.\begin{aligned} \kappa_0 &= \mu_0 \\ \kappa_1 &= \mu_1 \\ \kappa_2 &= \mu_2 - \mu_1^2 \end{aligned}\right\} \qquad (4)$$

where the μ_n are given by

$$\left.\begin{aligned} \mu_0 &= \int P(x)dx \\ \mu_{m>0} &= \frac{1}{\mu_0}\int x^m P(x)dx \end{aligned}\right\} \qquad (5)$$

for the distribution $P(x)$ with respect to x. Here we are interested in the particle distribution with respect to the fluid speed on a single site. The zeroth order and first order moments are correct by definition since they represent the conservation laws. We need to determine them to remove their impact on higher moments:

$$\left.\begin{aligned} \kappa_0 &= \rho \\ \kappa_1^x &= v_x = \frac{\pi_x}{\rho} \\ \kappa_1^y &= v_y = \frac{\pi_y}{\rho} \\ \kappa_1^z &= v_z = \frac{\pi_z}{\rho} \end{aligned}\right\} \tag{6}$$

For the second-order central moments we look for equilibrium values independent of the conserved quantities. To determine them, we assume the case where the fluid is at rest. All links of the FCHC must then have the same number of particles residing on them. The number of particles in the resting channel is taken to be arbitrary and kept as an adjustable parameter r_0 for the time being. We remember that the links pointing to the faces of the cube are double links with twice the equilibrium occupation of the links pointing to the edges. Since for a fluid at rest $\mu_1 = 0$, we find

$$\left.\begin{aligned} \kappa_2^{xx} &= \kappa_2^{yy} = \kappa_2^{zz} = \frac{\rho - r_0}{\rho} \\ \kappa_2^{xy} &= \kappa_2^{xz} = \kappa_2^{yz} = 0 \end{aligned}\right\} \tag{7}$$

We want to obtain weights ξ_k^{eq} so that $\vec{s}^{\,\mathrm{eq}} = \vec{s} + \sum_k \xi_k^{\mathrm{eq}} \vec{c}_k + \sum_j \varsigma_j^{\mathrm{eq}} \vec{b}_j$, where $\vec{s}^{\,\mathrm{eq}}$ is the equilibrium state vector. For second-order moments only the first sum is of interest. Their weights can now be determined when we set $r_0 = \rho/3$:

$$\left.\begin{aligned} \xi_1^{\mathrm{eq}} &= \rho\left(\frac{-1}{12}\left(\frac{1}{3} + v_x^2 + v_y^2\right) + \frac{v_x v_y}{4}\right) + \frac{nw + se}{2} \\ \xi_2^{\mathrm{eq}} &= \frac{-\rho}{6}\left(\frac{1}{3} + v_x^2 - v_y^2 - v_z^2\right) + \frac{w + e}{2} \\ \xi_3^{\mathrm{eq}} &= \rho\left(\frac{-1}{12}\left(\frac{1}{3} + v_x^2 + v_y^2\right) + \frac{v_x v_y}{4}\right) + \frac{sw + ne}{2} \\ \xi_4^{\mathrm{eq}} &= \frac{-\rho}{6}\left(\frac{1}{3} - v_x^2 + v_y^2 - v_z^2\right) + \frac{s + n}{2} \\ \xi_5^{\mathrm{eq}} &= \frac{-\rho}{6}\left(\frac{1}{3} - v_x^2 - v_y^2 + v_z^2\right) + \frac{f + b}{2} \\ \xi_6^{\mathrm{eq}} &= \rho\left(\frac{-1}{12}\left(\frac{1}{3} + v_x^2 + v_z^2\right) + \frac{v_x v_z}{4}\right) + \frac{wf + eb}{2} \\ \xi_7^{\mathrm{eq}} &= \rho\left(\frac{-1}{12}\left(\frac{1}{3} + v_y^2 + v_z^2\right) + \frac{v_z v_y}{4}\right) + \frac{sf + nb}{2} \\ \xi_8^{\mathrm{eq}} &= \rho\left(\frac{-1}{12}\left(\frac{1}{3} + v_x^2 + v_z^2\right) + \frac{v_x v_z}{4}\right) + \frac{ef + wb}{2} \\ \xi_3^{\mathrm{eq}} &= \rho\left(\frac{-1}{12}\left(\frac{1}{3} + v_z^2 + v_y^2\right) + \frac{v_z v_y}{4}\right) + \frac{nf + sb}{2} \end{aligned}\right\} \tag{8}$$

This choice, however, is not a unique one since the degrees of freedom outnumber the conditions. A proper choice for ς_j^{eq} is much less obvious. They influence the third hydrodynamic moment typically associated with the heat conductivity. Since temperature is not defined in our model, we might conclude that their choices are arbitrary. Any attempt to use them in order to reach third-order Galilean invariance must fail since the number of degrees of freedom offered by them is insufficient. If noise reduction is our major concern we might choose partly empirically

$$\begin{aligned}
\varsigma_{yx}^{eq} &= \frac{ew + se - ne - nw}{4} \\
\varsigma_{yz}^{eq} &= \frac{sf + sb - nf - nb}{4} \\
\varsigma_{xy}^{eq} &= \frac{nw + sw - se - ne}{4} \\
\varsigma_{xz}^{eq} &= \frac{wf + wb - ef - eb}{4} \\
\varsigma_{zx}^{eq} &= \frac{eb + wb - ef - wf}{4} \\
\varsigma_{zy}^{eq} &= \frac{nb + sb - sf - nf}{4}
\end{aligned} \tag{9}$$

We then need to control the scattering process in a way that assures an increase in entropy. We do this with a simple extrapolation scheme and obtain the output state vector:

$$\vec{s}^{\,out} = \vec{s}^{\,in} + \sum_k \lfloor \omega_1 \xi_k^{eq} \rfloor \vec{c}_k + \sum_j \lfloor \omega_2 \varsigma_j^{eq} \rfloor \vec{b}_j \tag{10}$$

8.6
Model Properties and Extensions

The system approaches its equilibrium state if $0 < \omega < 2$. The relaxation parameter ω_1 determines the kinematic viscosity ν of the modeled fluid like [43]

$$\nu = \frac{1}{3}\left(\frac{1}{\omega_1} - \frac{1}{2}\right)$$

The relaxation parameter ω_2 is arbitrary, but it is a good choice to set it to unity since this reduces noise in the solution. The operation is round-off error free in the conserved quantities if the weights for the vectors to be added are chopped to integers. Numerical errors exist, in fact, but they are kept out of sensitive data.

The use of integers adds a new numerical concern to the calculation negligible in floating-point arithmetic. The range of numbers that we are allowed to use is limited. We require the occupation numbers to be in the range $0 \ldots \Lambda$, where Λ is the largest available integer depending on the data type in use. We can now define a 19-sided convex polytope in 15 dimensions to which all allowed operations are restricted:

$$0 \leq \vec{s}^{\text{in}} + \sum_k \lfloor \omega_1 \xi_k^{\text{eq}} \rfloor \vec{c}_k + \sum_j \lfloor \omega_2 \varsigma_j^{\text{eq}} \rfloor \vec{b}_j \leq \vec{I} \Lambda \tag{11}$$

where $\vec{I}$ is the identity vector. Since this polytope is guaranteed to be convex, it is possible to stay inside by adding only one vector at a time and restricting the respective weight to a value not causing over- or underflow. If it is not possible to apply the weights obtained from the Galilean invariance condition, the remainder is saved to a register until all the other vectors have been added. Then one attempts to add the remainders until they are zero or no further improvement can be found.

The authors have successfully implemented this algorithm. It has, however, three undesired features: checking for over- and underflow for every weight individually and starting the process over again after the first iteration is computationally expensive. It is not guaranteed that the computed weights can be applied at all. Galilean invariance and the desired target viscosity hence cannot be guaranteed. Finally, and worst, if the computed weights cannot be applied, if they lie outside the polytope, then there is no guarantee for entropy to increase any more. The gain in entropy is a necessary condition for stability. It would therefore be better if we rejected the condition where over- or underflow occurred. The reason for this kind of error can be traced back to insufficient grid resolution. It is theoretically possible to refine the grid adaptively by splitting one cube into eight identical but smaller cubes. The occupation numbers are then spread among the smaller cubes that need to be updated twice as often as the larger cubes. The kinematic viscosity of the fluid must stay constant. It is given in lattice spacing squared per time steps. The relaxation parameter has to be recomputed for the finer grid as

$$\omega_{1N} = \frac{1}{2^N \left(\dfrac{1}{\omega_1} - \dfrac{1}{2} \right) + \dfrac{1}{2}} \tag{12}$$

where N is the level of refinement. It is easily seen that ω_{1N} will finally approach zero for large N. Since the incoming state vector is always valid, a relaxation parameter of zero must also be valid. Over- and underflow conditions can hence be eliminated by adaptive grid refinement. The remaining question is whether sufficient grid refinement is possible. The largest allowed integer decreases with the level of refinement as $\Lambda_N = \Lambda / 8^N$, resulting in possible refinement factors of 2, 5, 10 and 21 for unsigned integer data types of 8-, 16-, 32- and 64-bit precision, respectively. A refinement factor of 10 means that one large site is split into $\sim 10^9$ smaller sites. Arbitrary levels of refinement could be reached by splitting up individual particles, but that would jeopardize the total correctness of the algorithm.

The option for adaptive grid refinement not only increases stability, it also simplifies the algorithm, since specific knowledge about which operation caused the error is no longer necessary. Individual checks after every operation and the repetition of the circle become obsolete. Checking in the end whether over- or underflow has occurred somewhere is sufficient to decide if grid refinement is necessary or not.

The presented scheme satisfies the incompressible Navier-Stokes equation up to a slight viscosity anomaly, namely that the viscosity decreases at high speeds. The reason for this artifact is that the viscosity, as it may be seen as the diffusion coefficient of momentum, is influenced by higher moments of the particle distribution function. Adding speeds and obtaining third-order Galilean invariance would allow us to define energy and in addition remove this artifact. Even more speeds would then have to be added to remove all artifacts from the heat conductivity. However, adding more and more speeds makes the model impractical and puts its elegance and efficiency in question. The energy behavior obtained would be that of a perfect gas anyway, since the particles have no internal degrees of freedom. It might therefore be better to concentrate on athermal lattice Boltzmann automata and simulate the temperature by other means. The error in viscosity can be ignored if the fluid speed does not exceed a certain value. It becomes visible for speeds exceeding 10% of the speed of sound. In most applications of interest, speeds are in general much lower. However, the lattice Boltzmann method is most often applied to nonacoustic problems depending only on the Reynolds number of the system. The computational efficiency of a Reynolds number simulation can in principle be improved by the application of higher speeds as compared with denser grids. That is why it is so common to run lattice Boltzmann simulations at high Mach numbers.

8.7
Examples and Applications

The digital lattice Boltzmann automata represent a very elegant and simple model. Bounce-back sites, returning incoming particles to where they came from, can model solid boundaries. Source and force terms can be added to the scattering algorithm, allowing us to code inflows, outflows and any kind of arbitrary body forces depending on time, coordinates, speed, mass or simply the wish of the user. The simplicity to set up boundary conditions allows us to model any geometry, however complex, without increasing the computational cost. The phenomenon of vortex shedding, which is not easily captured by continuum methods, without introducing additional modeling of surface roughness emerges naturally in lattice Boltzmann simulations (see Fig. 8.2).

The input might in fact be as inconvenient as a CT scan of a piece of porous rock. Sites can switch from being a fluid site to being a boundary site in an instant, which is useful for modeling moving bodies in the fluid. By achieving all this, the digital lattice Boltzmann automaton is stable by design and linear in computational complexity and memory occupation with respect to the number of

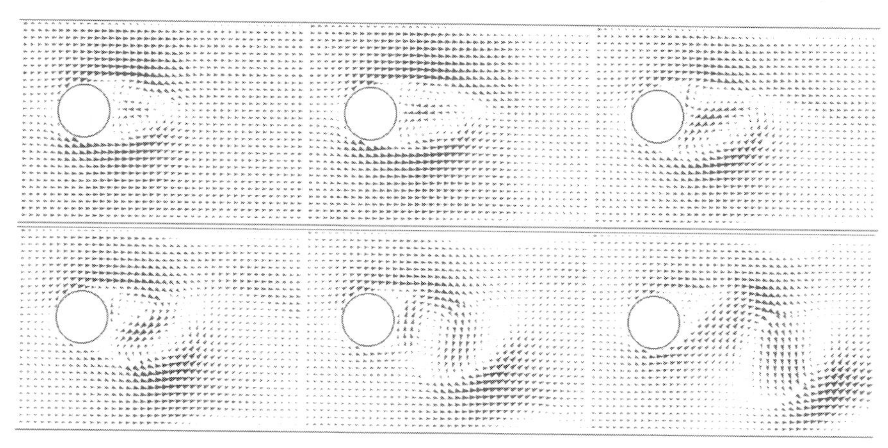

Fig. 8.2 The onset of vortex shedding behind a circular cylinder at a Reynolds number of ~50. The problem is two-dimensional and was simulated with the two-dimensional projection of the FCHC lattice, the D2Q9 lattice

sites. Stability is achieved without excessive computational effort in floating-point precision. The commercial digital lattice Boltzmann simulator [40] uses eight-bit integers with a range of only 0–255 for a 49-speed third-order Galilean invariant model with energy closure [44]. Implemented with floating-point numbers, the same model would require at least 64-bit data types for stability reasons. This means that the unconditionally stable method needs only one-eighth of the memory required for the potentially unstable floating version of the same model.

8.8 Modeling Nano- and Microflows with Dissipative Particle Dynamics

In the following sections we will review DPD and its potential fields of application. First we will present the model and the closing section is devoted to the application of DPD to concrete problems.

Let us first have a look at the basics of the DPD method as introduced by Hoogerbrugge and Koelman [17] and later slightly corrected by Español and Warren [45]. After discussing some extensions to the DPD model, we will see that the method allows for inclusion of energy conservation and thus becomes thermodynamically more consistent.

As in LG models, the continuum is discretized into a set of point particles. DPD particles do not represent single atoms or molecules but, rather, larger clusters of molecules or atoms. In LG models the particles move on a lattice, whereas in DPD the positions and momenta of the particles are defined in a continuous space. Both methods update at discrete time steps.

The updates are computed by applying Newton's second law. For a particle of mass m_i this yields

$$\dot{\mathbf{r}}_i = \mathbf{v}_i = \frac{\mathbf{p}_i}{m_i}, \quad \dot{\mathbf{p}}_i = \mathbf{F}_i^{\text{ext}} + \sum_{j,j \neq i} \mathbf{f}_{ij} \tag{13}$$

with $\mathbf{r}_i$ and $\mathbf{p}_i$ being the position and momentum vector of particle i, respectively. $\mathbf{F}_i^{\text{ext}}$ is an external force field acting on each particle. $\mathbf{f}_{ij}$ is a pair force between two particles i and j. In DPD this pair force is defined as

$$\mathbf{f}_{ij} = \mathbf{F}_{ij}^{\text{D}} + \mathbf{F}_{ij}^{\text{R}} + \mathbf{F}_{ij}^{\text{C}} \tag{14}$$

$\mathbf{F}_{ij}^{\text{D}}$ is a dissipative force and $\mathbf{F}_{ij}^{\text{R}}$ represents a random force. By setting both forces to zero, we are left with the conservative force $\mathbf{F}_{ij}^{\text{C}}$, which shows us that, at least concerning the forces, MD can be seen as a subset of DPD.

For $\mathbf{F}_{ij}^{\text{D}}$ and $\mathbf{F}_{ij}^{\text{R}}$ suitable forms, which fulfil Galilean invariance, are

$$\mathbf{F}_{ij}^{\text{D}} = -\gamma \omega_{\text{D}}(r_{ij})(\mathbf{e}_{ij} \cdot \mathbf{v}_{ij})\mathbf{e}_{ij} \tag{15}$$

and

$$\mathbf{F}_{ij}^{\text{R}} = \sigma \zeta_{ij} \omega_{\text{R}}(r_{ij})\mathbf{e}_{ij} \tag{16}$$

$\mathbf{e}_{ij} = \mathbf{r}_{ij}/r_{ij}$ is the unit vector pointing from particle j to particle i and $\mathbf{v}_{ij} = \mathbf{v}_i - \mathbf{v}_j$ is the relative velocity of the particles. ζ_{ij} is a random number with the properties $\langle \zeta_{ij}(t) \rangle = 0$ and $\langle \zeta_{ij}(t)\zeta_{kl}(t') \rangle = (\delta_{ik}\delta_{jl} + \delta_{il}\delta_{jk})\delta(t - t')$. γ and σ can be interpreted as a friction coefficient and a noise amplitude, respectively. $\omega_{\text{D}}(r_{ij})$ and $\omega_{\text{R}}(r_{ij})$ are weight functions determining the range of the forces and their strength in dependence of the inter-particle distance r_{ij}. For $r_{ij} \geq r_c$ these functions vanish identically, with r_c being the cut-off distance.

For a well-defined equilibrium temperature, detailed balance requires the fluctuation-dissipation theorem for DPD to be fulfilled [45], which is

$$\omega_{\text{D}}(r_{ij}) = \omega_{\text{R}}^2(r_{ij}) \quad \text{and} \quad \frac{2k_{\text{B}}T}{m}\gamma = \sigma^2 \tag{17}$$

m is the mass of a particle, k_{B} is the Boltzmann constant and T stands for the equilibrium temperature. Usually, the weight functions are chosen to be

$$\omega_{\text{D}}(r_{ij}) = \omega_{\text{R}}^2(r_{ij}) = \begin{cases} \left(1 - \frac{r_{ij}}{r_c}\right)^2, & r_{ij} < r_c \\ 0, & r_{ij} \geq r_c \end{cases} \tag{18}$$

A convenient choice for the conservative force is [27]

$$\mathbf{F}_{ij}^{\text{C}} = a\omega_{\text{R}}(r_{ij})\mathbf{e}_{ij} \tag{19}$$

with a representing an amplitude. This force has a purely repulsive nature and is less steep than its pendant in MD. This allows for larger time steps in the simulation. Additionally, the particles may penetrate each other.

The DPD model has two great advantages. The first is its great flexibility in fitting the model parameters to the specific needs of the considered simulation. Below we will see how especially the conservative force $\mathbf{F}_{ij}^{C}$ can be adjusted in order to model, for instance, colloidal suspensions, polymeric solutions or two-phase flow.

The second advantage is the simplicity of the model and its straightforward implementation. However, one could also see this as a drawback of DPD, that is, one could say the model is too simple. At least partially this is true.

One of the major disadvantages of the standard DPD model presented above is its inability to conserve energy. Only with energy conservation temperature gradients and heat flow can be modeled correctly. Avalos and Mackie [35] and Español [46] found independently equivalent solutions on how to introduce energy conservation into DPD. This extension to DPD will be reviewed below.

Additional drawbacks include the following:

1. The physical time and length scales of the DPD model are undefined.
2. The transport coefficients cannot be specified directly, but kinetic theory has to be applied in order to relate the coefficients with the DPD parameters [47].
3. In general it is not clear how to define the shape and amplitude of the conservative force in order to prescribe a certain hydrodynamic behavior.

In order to overcome these deficiencies, two physically similar but algorithmically different approaches were made to establish a thermodynamically more consistent fluid particle model, namely the voronoi fluid particle model [48] and smoothed dissipative particle dynamics (DPD) [49]. The two models will be briefly described below.

8.9 Energy-conserving DPD

Using Español's notation [46], an additional degree of freedom ε_i is introduced for each particle, representing its internal energy, and, additionally, an entropy $s_i = s(\varepsilon_i)$ which is needed for the definition of a 'temperature' $T_i = (\partial s_i/\partial \varepsilon_i)^{-1}$ for each particle.

The additional equation of motion for ε_i is

$$\dot{\varepsilon}_i = \sum_{j,j\neq i}\left[\frac{m}{2}\mathbf{v}_{ij}\cdot\mathbf{F}_{ij}^{D} + \dot{q}_{ij}^{D} - \frac{m}{2}\sigma_{ij}^{2}\omega_{R}^{2}(r_{ij})\right] + \frac{1}{\sqrt{\Delta t}}\sum_{j,j\neq i}\left[\frac{m}{2}\mathbf{v}_{ij}\cdot\mathbf{F}_{ij}^{R} + \dot{q}_{ij}^{R}\right] \tag{20}$$

In [35], $\dot{q}_{ij}^{D}$ is called 'mesoscopic heat flow' and is computed as follows:

$$\dot{q}_{ij}^{D} = \kappa_{ij}\left(\frac{1}{T_i} - \frac{1}{T_j}\right)\omega_{D}^{\varepsilon}(r_{ij}) \tag{21}$$

Additionally, there is a 'random heat flux' [46] $\dot{q}_{ij}^{R}$ with

$$\dot{q}_{ij}^{\mathrm{R}} = a_{ij}\omega_{\mathrm{R}}^{\varepsilon}(r_{ij})\zeta_{ij}^{\varepsilon} \tag{22}$$

κ_{ij} can be interpreted as the thermal conductivity between two particles. It depends on the particle energies of individual particle pairs and it is assumed that $\kappa_{ij} = \kappa_{ji}$. a_{ij} is a noise amplitude. Also, σ and γ from Equations (15) and (16) now transform in principle to coefficients σ_{ij} and γ_{ij} for individual pairs. $\omega_{\mathrm{D}}^{\varepsilon}(r_{ij})$ and $\omega_{\mathrm{R}}^{\varepsilon}(r_{ij})$ are additional weight functions needed for the corresponding changes of the particle energies. ζ_{ij}^{ε} is a second random number with the same characteristics as ζ_{ij} in Equation (16). For simplicity, it is assumed here that all particles possess the same mass m. Otherwise, for the particle interactions, a geometric mean would have to be calculated.

Again, the coefficients are not independent, but the following relations hold:

$$\omega_{\mathrm{R}}^{\varepsilon^2} = \omega_{\mathrm{D}}^{\varepsilon} \quad \text{and} \quad a_{ij}^2 = 2\kappa_{ij} \tag{23}$$

Additionally, the first part of Equation (17) still holds and the second part is replaced by

$$\gamma_{ij} = \frac{m}{2k_{\mathrm{B}}T_{ij}}\sigma_{ij}^2 \tag{24}$$

with

$$T_{ij}^{-1} = \frac{1}{2}\left(\frac{1}{T_i} + \frac{1}{T_j}\right) \tag{25}$$

representing a mean inverse temperature of two particles i and j.

8.10
State of the Art in DPD

In the framework of the so-called GENERIC formalism [50, 51], two thermodynamically consistent models were established, the Voronoi fluid particle model [48] and the smoothed dissipative particle dynamics (SDPD) model [49]. Since both models can be formulated in the GENERIC form, they are guaranteed to conserve energy and to fulfil the second law of thermodynamics, that is, the entropy of the system is a strictly increasing function of time.

While SDPD particles are true point particles, a Voronoi fluid particle is represented by a Voronoi cell in a Voronoi lattice. Each Voronoi cell is associated with a point which is allowed to move freely and the cell defines the volume, which is closer to this point than to any other.

SDPD can be seen as a smoothed particle hydrodynamics (SPH) model (see, for example, [52]), which includes thermal fluctuations. As in SPH, a bell-shaped weight function $W(r)$ is associated with the particles, which helps both in defining the density of the particles and in approximating second spatial derivatives at the particle locations.

In both models, the particles possess a position vector, a momentum vector, a mass and an entropy. In SDPD, the mass is assumed to be constant whereas it is a degree of freedom for Voronoi fluid particles. Therefore, there exists also an equation of motion for the mass. In both cases, the equations of motion represent discretized versions of the Lagrangian equations of hydrodynamics.

When comparing the computational efficiency of both models, two factors are important. First, the construction of the moving Voronoi grid represents an additional effort. However, second, already in 2D, the number of interacting neighbors in SDPD is roughly an order of magnitude larger than for the Voronoi fluid particles, which have six neighbors on average. The relative importance of these two factors is not yet clear.

Concerning the representation of boundaries, the Voronoi fluid particle model seems to have certain advantages over SDPD. Depending on the boundary conditions, values either for the hydrodynamic fields or for the fluxes can be directly prescribed. This lets one also expect that coupling to a coarser continuum description or to a microscopic molecular dynamics model should be implementable in a straightforward way. Therefore, the Voronoi fluid particle model could also serve as a bridge between continuum models and molecular dynamics, which makes it particularly attractive.

8.11 Fields of Application for DPD

In the following, an overview will be given of the areas where DPD was already applied. We will close the presentation with our own results for pressure-driven flow.

Most of the approaches to model complex fluids with DPD modify the conservative force. One example is the local freezing of groups of DPD particles in order to form rigid bodies [53]. In this case, a constraint force fixes the relative position of the particles. Shear thinning curves were obtained for volume fractions up to 30%. Above this level, the results became slightly inconsistent. One of the reasons is probably the fact that the rigid bodies can interpenetrate to a certain extent owing to the nature of the conservative force [see Equation (19)].

By introducing spring forces between the particles, polymer chains can be modeled. The DPD particles represent monomers, that is, not necessarily single atoms. If these chains are combined with freely moving fluid particles, polymer solutions can be simulated. In this way, the solvent quality was determined as a function of the solvent–solvent and solvent–monomer interactions [54]. Additionally, the scaling of the relaxation time or of the radius of gyration with the chain length coincides well with the Rouse-Zimm models [55].

If we introduce two different types of particles together with conservative forces, which produce stronger repulsion between unequal particles than between particles of the same type, we can investigate phase separation. Domain growth was studied with this technique and transitions were found from diffusive growth to the hydrodynamic regime [56].

Up to now, energy-conserving DPD has been rarely applied. It has been used for modeling heat conduction in a solid [57], and it was shown that the model behaves correctly in equilibrium and that Fourier's law is reproduced in nonequilibrium situations.

Another application of energy-conserving DPD is the simulation of phase changes. For this purpose, the model was extended by introducing an enthalpy of fusion, which leads to a stepwise-defined equation of state with an expression for temperatures lower than the melting temperature, equal to the melting temperature and above the melting temperature [36].

Energy-conserving DPD introduces an energy ε_i as an internal variable of the particles. The strategy of introducing internal variables can be generalized according to Español [58]. Examples were given for the simulation of polymer solutions with particles possessing an internal elongation vector or the simulation of chemically reacting mixtures by introducing an internal fraction variable representing the relative amount of two species in a DPD particle.

In our own work, we are especially interested in the question of which information and modeling capabilities the DPD method can deliver that are missing in continuum approaches. Our application field is the simulation of micro powder injection moulding (micro-PIM). The above question is of interest to us since in this area, continuum simulators simply produce wrong results when compared with experimental outcomes [59]. We will present some simple simulations of Poiseuille flow after a short discussion on the type of boundary conditions we have used.

For the study of the bulk behavior of a DPD fluid, periodic boundary conditions (PBC) are usually applied. That is, a particle leaving the simulation box at one side re-enters the domain at the opposite end. Realistic simulations require the introduction of walls.

For Couette flow, the introduction of explicit walls can be circumvented using the Lees-Edwards technique [32], which modifies the periodic boundary conditions by adding velocity components to the particles passing through the boundary. For non-moving walls this is not possible. The usual technique to model walls directly is the introduction of additional particles which remain at fixed positions. If the interaction potential between the wall particles and the fluid particles diverges for decreasing interparticle distance, this by itself is already a sufficient measure.

However, usually in DPD, forces such as that defined in Equation (19) are used, which allow particles to interpenetrate. In this case, an additional reflection of particles hitting the wall has to be introduced. At least three reflection mechanisms exist:

1. Specular reflection keeps the velocity components parallel to the wall unchanged and inverts the velocity component perpendicular to the wall.
2. 'Bounce back' reflection inverts all velocity components.

3. Maxwellian reflection thermalizes the particles to a wall temperature and reflects them back randomly according to a Maxwellian distribution which is centered at the velocity of the wall.

An additional possibility is the derivation of effective forces exerted by the wall on the fluid particles, by calculating the continuum limit of the discrete particles forming the wall [60].

We have applied the Maxwellian reflection of the particles to Poiseuille flow, but without the introduction of discrete frozen particles or effective forces by the wall on the particles. The idea behind this solution is an additional coarse graining of the interaction between fluid particles and the walls. For a wall in the y-direction, this behavior can be realised by randomly drawing the velocity component v_y from a Rayleigh distribution:

$$\Phi(v_y) = \frac{m}{k_B T_w} v_y \exp\left(-\frac{m v_y^2}{2 k_B T_w}\right) \tag{26}$$

and v_x and v_z from a Maxwell distribution:

$$\Phi(v_i) = (2\pi m k_B T_w)^{-\frac{1}{2}} \exp\left(-\frac{m v_i^2}{2 k_B T_w}\right) \tag{27}$$

with zero mean, that is, the walls stay at rest. The index i is a placeholder for x or z and T_w is the wall temperature.

With a slight modification, this mechanism can also be used for boundary conditions at walls in Couette flow. This was done, for example, in [61].

For energy-conserving DPD, Willemsen et al. [36] introduced an interesting new implementation of a constant-temperature boundary, which produces better results than previous approaches for the temperature of the particles close to the wall.

The simulation of Poiseuille flow, that is, pressure-driven flow between two infinitely extended flat walls, shows us some advantages of DPD over continuum approaches. The infinite extension of the walls parallel and perpendicular to the flow direction is realized by introducing PBCs in the corresponding directions, here x and z. In the y-direction, the walls are realized by the described Maxwellian reflection. In all simulations, the wall temperature was set to $T_w = T_{\text{fluid}} = 1$ in reduced units.

Initially, the particles possess a zero mean velocity in all Cartesian directions. The initial configuration of the particles is a simple cubic lattice with lattice constant $l_a = \rho^{-1/3}$ determined from the particle density ρ. This is valid if we define the mass of all particles to be $m = 1$. It can be observed that the symmetric configuration vanishes very rapidly during a simulation and at least does not affect the results presented here.

In order to generate a pressure-driven flow, a constant force f_x is exerted on each particle. Here, the assumption from continuum theory which states that a constant force field leads to a linear pressure gradient is adopted. Since all particles possess a unique mass $m = 1$, the force accelerates each particle equally with $g_x = f_x/m$.

Fig. 8.3 shows that the boundaries influence the flow as expected: the average flow velocity saturates exponentially to a final value.

In conventional continuum fluid dynamics, nonlinear behavior of the fluid must be explicitly built into the applied model, for instance by introducing corrections for the viscosity near walls (see, for example, [62]).

Fig. 8.4 shows the average velocity of the DPD fluid as a function of the force driving the particles. As can be seen, the fluid behaves nonlinearly above $g_x \approx 0.8$. This happens implicitly due to the influence of the boundary conditions, since no explicit model for non-linearity was built in.

The shear viscosity η can be estimated from the velocity profiles using the analytical expression for the velocity profile in the direction of flow for Poiseuille flow [63]:

$$v_x(y) = \frac{\rho g_x l_y^2}{2\eta}(y - y^2) \tag{28}$$

The cross-stream coordinate y is normalized such that $0 \leq y \leq 1$, with $y = 0$ and $y = 1$ representing the two walls. We observe a constant shear viscosity for $g_x \leq 0.06$ and an increase for larger forces, that is, shear thickening. We emphasize that Equation (28) assumes laminar flow, a constant viscosity and incompressibility.

That incompressibility is not prescribed in DPD can be seen in Fig. 8.5. It shows plots of the relative density $\rho_{\text{rel}} = \rho/\langle\rho\rangle$ over the cross-stream coordinate y

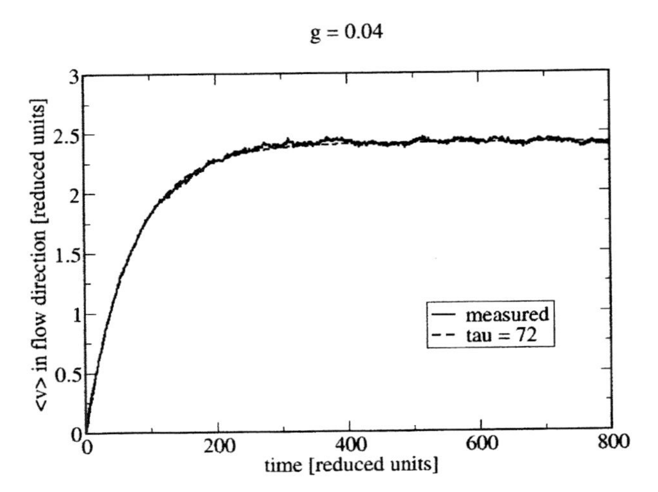

Fig. 8.3 Saturation of flow velocity

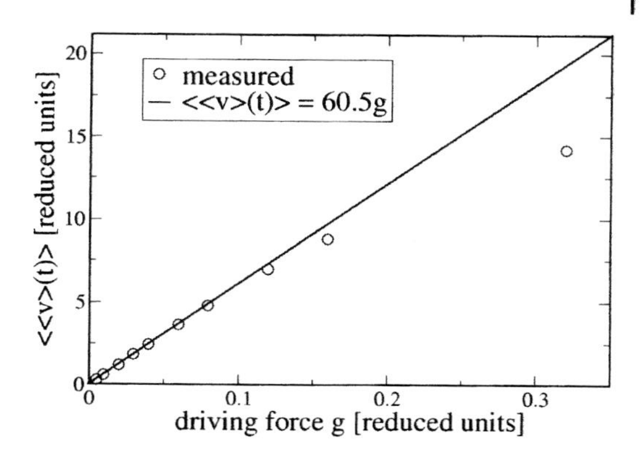

Fig. 8.4 Nonlinear behavior of flow velocity

for $g_x = 0.005$ and $g_x = 0.32$. We can see that the density seems to be constant over a wide range of the channel.

At the walls, there are large fluctuations and Equation (28) is definitely not valid any longer. This can explain why, for all simulation runs, the fitted polynomials deviate more the closer we approach the walls. The consequence is that viscosity is not constant over the whole domain. The computed viscosity should therefore represent the viscosity in the center of the channel.

In Fig. 8.6 we show the same density profiles, but zoomed in on the ordinate to densities $0.85 < \rho_{\text{rel}} < 1$. Here, we can see that the density for $g_x = 0.005$ is really constant far enough from the walls and only suffers from small fluctuations due to the finite number of particles participating in the simulation.

However, the density for $g_x = 0.32$ has a maximum of $\rho_{\text{rel}}(y = 0.5) \approx 0.99$ in the center and decreases to $\rho_{\text{rel}}(y \approx 0.15) = \rho_{\text{rel}}(y \approx 0.85) \approx 0.93$ before the large fluctuations near the wall set in. Additionally, we can see that at $g_x = 0.08$, the density profile starts to change its shape from a flat profile at $g_x = 0.005$ to a 'crooked' profile at $g_x = 0.32$. This coincides perfectly with an increase in the computed shear viscosity η.

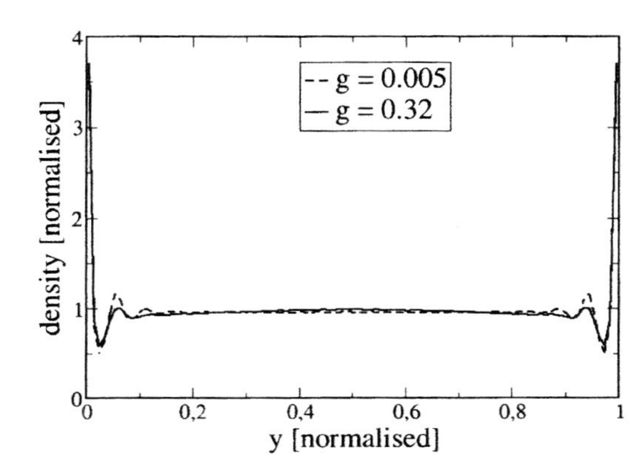

Fig. 8.5 Cross-stream density profile

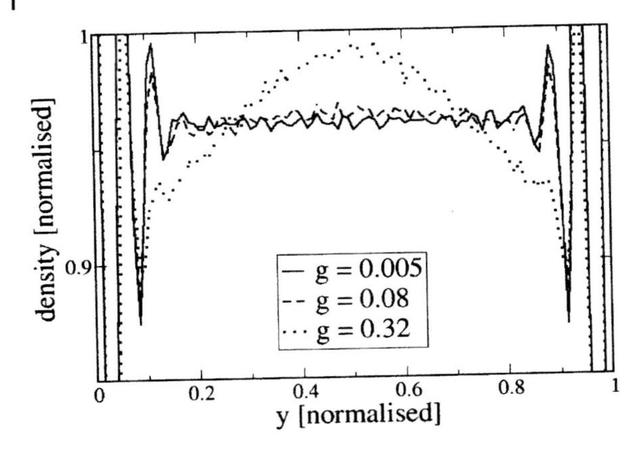

Fig. 8.6 Cross-stream density. Zoom into the range $0.85 < \rho_{\text{rel}} < 1$

Of course, the density results could be wrong because of a wrong model or a wrong implementation of the model. Fig. 8.7 shows that this is not the case. The maximum relative density (which was always observed directly at the walls) is plotted against the particle resolution, that is, the number of particles representing the channel width. The plot shows us that $\rho_{\text{rel}} = 1$ is reached in the continuum limit. Therefore, when a decrease of the dimensions of our system increases the importance of the discrete nature of the fluid, DPD delivers information concerning density fluctuations, which continuum approaches cannot deliver.

One observable effect in micro-PIM is frictional heating. If too much heat is generated, this can cause damage to the fabricated microcomponent. For example, in the Hele-Shaw continuum model [5], this effect is taken into account by introducing a correction term $\eta\dot{\gamma}$ in the energy balance, with $\dot{\gamma}$ being the shear rate. Fig. 8.8 shows measured temperature profiles for the DPD fluid during Poiseuille

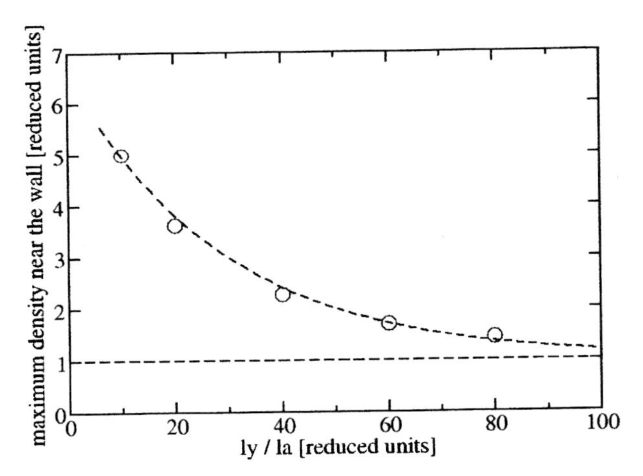

Fig. 8.7 Relative density at the walls as a function of the gap width

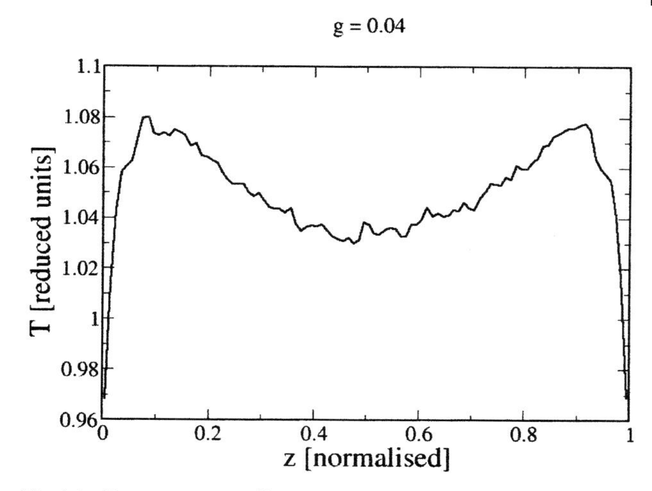

Fig. 8.8 Temperature profile showing frictional heating

flow. Even though no explicit model for frictional heating was built in, the fluid shows an increase in temperature in the vicinity of the highest shear rates.

The currently existing applications of standard DPD show the great potential of the method. Additionally, our results have shown that DPD takes microrheological effects into account, which cannot be represented in continuum approaches, for example the density fluctuations near solid boundaries.

The newest models are even more promising. For example, smoothed dissipative particle dynamics and the Voronoi fluid particle model are candidates with enormous potential, not only because of their thermodynamic consistency but also because of new possibilities arising from them, such as how to couple different modeling domains. Coupling will definitely be one of the central issues for our work in the near future.

8.12 Conclusions

We have reported on two methods suitable for the simulation of micro- and nanofluidic behavior that we believe to be able to bridge the gap between purely microscopic approaches such as molecular dynamics and macroscopic continuum descriptions. This leads us to the conclusion that it is worthwhile to exploit a combination of LG/LB models and DPD for the simulation of CFD. Especially for MEMS applications, where an increasing need for modeling and simulation of fluid dynamics can be observed, this configuration seems to be a promising candidate.

The lattice Boltzmann model presented here is not derived from the Navier-Stokes equation, even though it claims to capture the same physical facts. The question of the extent to which one of the two descriptions might be the more physically consistent might arise. For empirical reasons, the Navier-Stokes equation should be seen as the gold standard in fluid dynamics simulation. Analytic solutions to the Navier-Stokes equation exist and are called 'exact' solutions in the literature. However, we have to keep in mind that an exact solution of the Navier-Stokes equation is still only an approximation of the underlying physics governed by the Boltzmann transport equation. The Boltzmann transport equation by itself is a radical simplification of the physics defined by the Schrödinger equation. We might recognize this as a hierarchy of simplifications always introducing new errors. It is impossible for the Boltzmann equation to capture more physical facts than does the Schrödinger equation. Likewise, there cannot be more truth in the Navier-Stokes equation than in the Boltzmann equation. From the Boltzmann equation we might also derive the digital lattice Boltzmann automaton. Its difference from the classical approach of deriving yet another differential equation comes into play only if we consider problems that cannot be solved analytically and only those are of practical interest for simulation. The digital lattice Boltzmann automaton is an algorithm ready for implementation on a digital computer and without the need for any further simplification. The same is not true for any differential equation being solved numerically. Here the hierarchy of modeling is yet to find its end. Infinitely many differentially small distances and differential time steps are intractable and have to be replaced by finite differences. Real-valued numbers contain an infinite amount of information. They have to be modeled by floating point numbers. At the bottom of the simplification hierarchy, the Navier-Stokes models have few advantages over the digital lattice Boltzmann automaton, which always yields, so to speak, an exact solution by simply applying its rules. The actual quality of this solution depends on the order of Galilean invariance, which is rather low in the presented model. However, since it is not based on the continuum assumption, the lattice Boltzmann automaton can in fact be seen to capture facts ignored by the Navier-Stokes equation. In micro- and nanodimensions the discrete character of real-world fluids becomes obvious through fluctuation effects such as Brownian motion. If the physics of fluids were really only governed by the Navier-Stokes equation, micro- and nanoparticles would drop to the ground and Brownian motion would not exist. The digital lattice Boltzmann automaton shows fluctuations, which can be influenced by the precision of the integer data type and by ω_2. The possibility of modeling fluctuations provides us with a valuable method to interpolate between continuum and particle description of fluids. On the one hand we are able to simulate high Reynolds number flow, and on the other we can study the random walk of single molecules of a dilute substance resolved in a liquid. The potential of the method to study fluctuations is yet to be investigated.

Apart from all its benefits, there are also some general shortcomings of the digital lattice Boltzmann philosophy:

- Arbitrary orders of Galilean invariance can only be reached with great effort.
- The number of speeds that we must add to improve our model increases exponentially with the order of Galilean invariance.

- Particles are not allowed to have internal degrees of freedom.
- For very high resolution or very high order of Galilean invariance, a particle-based scheme such as the DPD method is much more efficient.
- Unless many speeds are added, the dynamics of the model is restricted to subsonic speeds.
- The model is inherently transient.
- Only one steady-state solution shows up regardless of how many are possible.
- To obtain a steady-state solution, the automaton has to be run until no further changes occur. This might take a very long time and the digital lattice Boltzmann method is therefore not recommended for purely steady-state problems.

The ability of the lattice Boltzmann automaton to handle flow in arbitrary complex geometries, such as taken from a CT scan, is of great value for the modeling of real-world setups. By using integer data types instead of floating-point arithmetic, the memory footprint could be significantly reduced, in addition to the general desiderata of unconditional stability. However, the strictly statistical nature of the method captures no specific microscopic interactions among particles or between particles and boundaries. Once the dimensions of the system decrease in such a way that no bulk behavior is present any longer, the rising importance of the discrete nature of the fluid makes a more detailed model such as DPD necessary.

DPD captures information concerning density fluctuations, which continuum approaches cannot deliver. Local density behavior is extremely important for the modeling of the equation of state of the liquid. This, in turn, is extremely important for the description of the co-existence of solid, liquid and gas phases. Modeling of the local interacting behavior in terms of density functionals for the respective interacting energies is a new development in the field of DPD.

It has been shown in the literature, and our results also emphasize, that DPD takes microrheological effects into account without having to model them explicitly as in Navier-Stokes approaches. Moreover, some of the effects cannot even be represented in continuum approaches, for example density fluctuations near solid boundaries. The striking benefit of using more fundamental particle-based methods such as DPD shows up when effects such as frictional heating creep into the results without any explicit modeling.

However, in this discussion we should not forget about the multi-scale nature of micro- and nanosystems. Any of these systems, as special as their applications might be, will always have to be represented for the designer in an environment that allows for the simulation of the functionality of the system in its entity. Therefore, it will be necessary in the future to couple the micro- and nanoworld to a continuum model and even to a global balance description at a macro-model level. This requires new approaches for model order reduction directly from the mesoscopic level represented by the lattice Boltzmann method and the DPD approach.

8.13 Acknowledgments

The authors are grateful for partial support by the German research foundation DFG within the project SFB499 and by the University of Trento, Italy, within the PAT/CNR research project 'Ricerca e Formazione in Microsistemi'.

8.14 References

1 *Technical Proceedings of the 2003 Nanotechnology Conference and Trade Show, Nanotech 2003, San Francisco, 23–27 February 2003*; **2003**, Vols. 1–3.
2 R.M. Nieminen, *J. Phys.: Condens. Matter* **2000**, *14*, 2859–2876.
3 M.G. Giridharan, P. Stout, H.Q. Yang, M. Athavale, P. Dionne, A. Przekwas, *J. Model. Simul. Microsyst.* **2001**, *2*, 43–50.
4 M. Geier, A. Greiner, J.G. Korvink, in: *Proceedings of POLYTRONIC 2003, Montreux*; **2003**, p. 315.
5 T. Mautner, in: *Proceedings of Nanotech 2003*; **2003**, *1*, p. 242.
6 D.A. Wolf-Gladrow, *Lattice-Gas Cellular Automata and Lattice Boltzmann Models*; Lecture Notes in Mathematics 1725. Berlin: Springer, **2000**.
7 H. Chen, S. Chen, W.H. Matthaeus, *Phys. Rev. A* **1992**, *45*, R5339.
8 C.P. Lowe, S. Succi, in: *Bridging Time Scales: Molecular Simulations for the Next Decade*, P. Nielaba, M. Mareschal, G. Ciccotti (eds.); Lecture Notes in Physics 605. Berlin: Springer, **2002**, Chapter 9.
9 J.M. Buick, C.A. Greated, W.J. Easson, *Phys. Fluids* **1997**, *9*, 2585.
10 J.A. Cosgrove, J.M. Buick, S.J. Tonge, C.G. Munro, C.A. Greated, D.M. Campbell, *J. Phys. A* **2003**, *36*, 2609–2620.
11 G.-W. Peng, H.-W. Xi, C. Duncan, S.-H. Chou, *Phys. Rev. E* **1999**, *59*, 4675.
12 C. Manwart, U. Aaltosalmi, A. Koponen, R. Hilfer, J. Timonen, *Phys. Rev. E* **2002**, *66*, 016702.
13 R. Zhang, H. Chen, *Phys. Rev. E* **2003**, *67*, 66711.
14 I. Ginzburg, K. Steiner, *J. Comput. Phys.* **2003**, *185*, 61–99.
15 D.C. Rapaport, *The Art of Molecular Dynamics Simulation*; Cambridge: Cambridge University Press, **1995**.
16 J.M. Haile, *Molecular Dynamics Simulation. Elementary Methods*; New York: Wiley-Interscience, **1997**.
17 P.J. Hoogerbrugge, J.M.V.A. Koelman, *Europhys. Lett.* **1992**, *19*, 155–160.
18 P. Español, *Phys. Rev. E* **1995**, *52*, 1734.
19 P. Español, *Phys. Rev. E* **1996**, *53*, 1572.
20 C.A. Marsh, J.M. Yeomans, *Europhys. Lett.* **1997**, *37*, 511–516.
21 C.A. Marsh, G. Backx, M.H. Ernst, *Phys. Rev. E* **1997**, *56*, 1676.
22 P. Español, M. Serrano, I. Zuniga, *Int. J. Mod. Phys. C* **1997**, *8*, 899–908.
23 A.J. Masters, P.B. Warren, *Europhys. Lett.* **1999**, *48*, 1–7.
24 E.G. Flekkoy, P.V. Coveney, *Phys. Rev. Lett.* **1999**, *83*, 1775.
25 E.G. Flekkoy, P.V. Coveney, G. De Fabritiis, *Phys. Rev. E* **2000**, *62*, 2140.
26 U. Frisch, B. Hasslacher, Y. Pommeau, *Phys. Rev. Lett.* **1986**, *56*, 1505.
27 R.D. Groot, P.B. Warren, *J. Chem. Phys.* **1997**, *107*, 4423.
28 W. Dzwinel, D.A. Yuen, K. Boryczko, *J. Mol. Model.* **2002**, *8*, 33–43.
29 W. Dzwinel, D.A. Yuen, *Int. J. Mod. Phys. C* **2000**, *11*, 1–25.
30 P.V. Coveney, K.E. Novik, *Phys. Rev. E* **1996**, *54*, 5134.
31 E.S. Boek, P.V. Coveney, H.N.W. Lekkerkerker, P. van der Schoot, *Phys. Rev. E* **1997**, *55*, 3124.
32 I. Pagonabarraga, M.H.J. Hagen, D. Frenkel, *Europhys. Lett.* **1998**, *42*, 377–382.
33 C.P. Lowe, *Europhys. Lett.* **1999**, *47*, 145–151.

34 W.K. van Otter, J.H.R. Clarke, *Europhys. Lett.* **2001**, *53*, 426–431.
35 J. Bonet Avalos, A.D. Mackie, *Europhys. Lett.* **1997**, *40*, 141–146.
36 S.M. Willemsen, H.C.J. Hoefsloot, D.C. Visser, P.J. Hamersma, P.D. Iedema, *J. Comput. Phys.* **2000**, *162*, 385–394.
37 P.B. Warren, *Phys. Rev. E* 2003, 68, No. 066702.
38 J. Yepez, *Lattice-Gas Dynamics, Viscous Fluids*, USAF Technical Report PL-TR-96-2122 Vol 1, 1995
39 B.M. Boghosian, J. Yepez, F.J. Alexander, N.H. Morgolus, *Phys. Rev. E* **1997**, *55*, 4137–4147.
40 H. Chen, C. Teixeira, K. Molving, *Int. J. Mod. Phys. C* **1997**, *8*, 675.
41 S. Wolfram, *J. Stat. Phys.* **1986**, *45*, 471-526.
42 C.M. Teixeira, *Continuum Limit of Lattice Gas Fluid Dynamics*; PhD thesis, Massachusets Institute of Technology, **1992**.
43 S. Succi, *The Lattice Boltzmann Equation for Fluid Dynamics and Beyond*; New York: Oxford University Press, 2001.
44 H. Chen, J. Hoch, C. Teixeira, *Computer simulation of physical processes*; US Patent 6089744, **2000**.
45 P. Español, P. Warren, *Europhys. Lett.* **1995**, *30*, 191–196.
46 P. Español, *Europhys. Lett.* **1997**, *40*, 631–636.
47 C.A. Marsh, G. Backx, M.H. Ernst, *Phys. Rev. E* **1997**, *56*, 1676–1691.
48 M. Serrano, P. Español, *Phys. Rev. E* **2001**, *64*, 1–18.
49 P. Español, M. Revenga, *Phys. Rev. E* **2003**, *67*, 1–12.
50 M. Grmela, H.C. Öttinger, *Phys. Rev. E* **1997**, *56*, 6620–6632.
51 H.C. Öttinger, M. Grmela, *Phys. Rev. E* **1997**, 56, 6633–6655.
52 J.J. Monaghan, *Annu. Rev. Astron. Astrophys.* **1992**, *30*, 543.
53 J.M.V.A. Koelman, P.J. Hoogerbrugge, *Europhys. Lett.* **1993**, *21*, 363.
54 Y. Kong, C.W. Manke, W.G. Madden, A.G. Schlijper, *J. Chem. Phys.* **1997**, *107*, 592–602.
55 N.A. Spenley, *Europhys. Lett.* **2000**, *49*, 534–540.
56 K.E. Novik, P.V. Coveney, *Int. J. Mod. Phys. C* **1997**, *8*, 909.
57 M. Ripoll, P. Español, M.H. Ernst, *Int. J. Mod. Phys. C* **1998**, *9*, 1329–1338.
58 P. Español, in: *Micromechanics and Nanoscale Effects: MEMS, Multi-Scale Materials and Micro-Flows*, V.M. Harik, L.S. Luo (eds.); Dordrecht: Kluwer, **2004**, 213–236.
59 T.U. Benzler, *Pulverspritzgießen in der Mikrotechnik*; PhD thesis, Albert Ludwig University of Freiburg, **2001**.
60 M. Revenga, I. Zuñiga, P. Español, I. Pagonabarraga, *Int. J. Mod. Phys. C* **1998**, *9*, 1319–1330.
61 C. Trozzi, G. Ciccotti, *Phys. Rev. A* **1984**, *29*, 916–925.
62 D. Yao, B. Kim, *J. Micromech. Microeng.* **2002**, *12*, 604–610.
63 L.D. Landau, E.M. Lifshitz, *Fluid Mechanics*; Oxford: Pergamon Press, **1959**.

9
Nanofluidics – Structures and Devices

J. Lichtenberg and H. Baltes, PEL – ETH Zürich, Switzerland

Abstract
Nanofluidics research is focused on miniaturized fluidic structures with at least one dimension being confined to the sub-micrometer range. These devices offer a host of new possibilities and perspectives for both fluid handling and chemical and biochemical analysis systems, as they interact with the analyte molecule or the solvent matrix at or close to the molecular level. This chapter reviews both classical and unconventional fabrication methods used in nanofluidics. Underlying physical principles and application examples are discussed for liquid handling in nanochannels and biomolecule analysis.

Keywords
nanofluidics; nanofluidic structures; nanofluidic devices; nanochannels; fabrication; biological macromolecules.

Advanced Micro and Nanosystems. Vol. 1
Edited by H. Baltes, O. Brand, G.K. Fedder, C. Hierold, J. Korvink, O. Tabata

ISBN: 3-527-30746-X

9.1 Introduction

Over the past one and a half decades, microfluidics research has established a groundbreaking technology for analysis and synthesis in all fields of chemistry, ranging from drug screening to medical diagnostics and environmental chemistry [1–3]. While the pioneering publications in the field focused on shrinking and adapting conventional analytical techniques to the planar microchip format [4–6], recent research has introduced an increasing number of new approaches to analytical problems, which rely on unique properties of microscale fluidics. Diffusion-based filtering in laminar flows, pioneered in 1997 by Yager's group [7], is an early example of this breed. The principle relies on the fact that the majority of microfluidic devices operate under low Reynolds number conditions and mass transport between adjacent stream lines is driven by diffusion only. Clearly, the concept would be difficult to implement on a larger scale because of flow turbulences and the fact that diffusion time increases by the power of 2 with distance. However, if comparing microchannels – the key component of microfluidics – and conventional glass capillaries today, one finds that both are similar in size with bore diameters in the range from several micrometers to about 200 μm. For most microfluidic devices, miniaturization is predominantly achieved by planar integration of microchannel networks with zero dead volume interconnects and not necessarily by shrinking channel dimensions.

This has changed in recent years with the introduction of nanofluidic devices and their enabling fabrication technologies. Compared with the domain of solid-state nanotechnology, the fluidics community uses the adjunct *nano* in a less re-

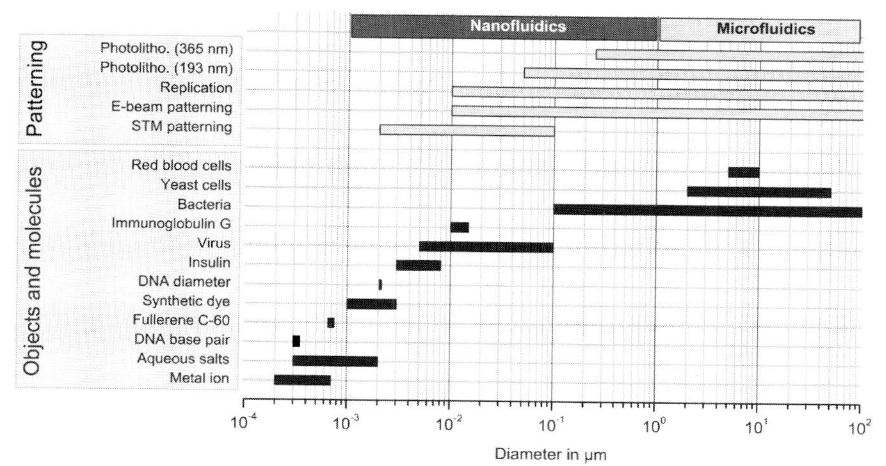

Fig. 9.1 Comparison chart for sizes of objects and molecules of biochemical and biological interest, also indicating the range covered by microfluidic and nanofluidic systems and the resolution achievable with different lithography techniques

strictive way and includes fluidic structures with sub-micrometer feature size in the field of nanofluidics. We therefore define the latter as the science of miniaturized fluidic structures whose system functionality is predominated by features confined to the sub-micrometer range in at least one dimension. Fig. 9.1 provides a size comparison for objects and molecules of biochemical and biological interest indicating the range covered by microfluidic and nanofluidic systems.

The transition from micro- to nanofluidics allows for the first time fluidic structures to interact with analyte molecules or the solvent matrix at or close to the molecular level. In some cases, this interaction is clearly visible (for example, for unzipping of double-stranded DNA molecules; see Section 9.4.2), but it holds essentially true for most nanofluidic devices. The ability to manipulate, sort and sieve single molecules and specifically to tailor molecular transport between adjacent structures provides researchers with a host of new opportunities to advance chemical analysis [8, 9]. It should also be noted that a number of approximations usually taken for granted for modeling flow and molecule physics are no longer valid at this level and a detailed analysis on the molecular scale is required.

In this chapter, we will introduce the reader to this young, interdisciplinary field of research in two steps. The first part of the chapter focuses on the fabrication methods at hand for creating nanofluidic features in a variety of substrates. The second part discusses applications of nanofluidics and the underlying physical and chemical principles. These include pressure-driven, electrokinetic and diffusion transport through submicrometer structures and the manipulation of biochemical molecules.

9.2 Fabrication Techniques for Nanofluidic Structures

The fabrication of microstructures for mechanical and fluidic applications relies essentially on methods developed by the planar microelectronics industry. These are generally known as thin-layer techniques and comprise deposition, photolithography and selective etching. This core set has been extended over two decades of microelectromechanical systems (MEMS) research by complementary processes such as anisotropic wet and dry etching for bulk micromachining, thick photoresist technology including LIGA and the epoxy-based SU-8, polymer-based molding methods and many more.

The following review of fabrication techniques is organized into five sections, starting with an introduction to high-resolution lithography, followed by the two classical techniques, bulk and surface micromachining, photopolymer-based fabrication and finally the fast growing area of replication methods. Given the close relation between many micro- and nanofabrication techniques, references to micrometer machining are included in the following discussion where applicable. As a full introduction to micromachining is beyond the focus of this chapter, the reader is referred to excellent text books on the fabrication technologies in general [10, 11] and on those used in microfluidics in particular [2, 12].

9.2.1 Lithography for Nanofluidic Structures

Structuring of a thin film on a planar substrate with an arbitrary pattern is typically achieved by photolithographic imaging through a partly transparent mask on to a photosensitive polymer layer. As a prerequisite, the resolution, R, of the lithographic equipment has to meet or exceed the minimum feature size of the pattern to be generated. The resolution is defined as the minimum distance between two distinguishable features and can be calculated for projection lithography from the wavelength, λ, used for exposure as

$$R = k\frac{\lambda}{NA} \qquad (1)$$

where k is an empirical parameter accounting for process variables (typically between 0.4 and 1) and NA the numerical aperture of the optical system. As a consequence, stricter requirements on the patterning resolution call for light sources of short wavelengths. Although 365 nm ultraviolet (UV) radiation generated by discharge lamps is still common in most research laboratories, ArF excimer lasers at 193 nm are already common in modern lithography equipment for industrial production. For the future, extreme UV lithography down to 13 nm using reflective optical systems is already in the development pipeline. Additionally, optical proximity correction and advanced phase shift masks push lithographic resolution down to below 65 nm for 193 nm equipment by taking advantage of precisely calculated diffraction and refraction effects.

However, as the cost of mask making is prohibitive for most research laboratories, projection lithography is rarely used for the generation of nanometer structures. As an alternative, direct writing techniques allow nanopattern generation on a photoresist-coated substrate without the need for an optical mask. Although the sequential writing process is slow, the fast turnaround times are advantageous for research purposes. Direct-writing particle beam lithography benefits from the high energy of the charged particles, which are used to pattern the resist layer instead of photons used in conventional lithography. Electron and ion beam pattern generators – widely used for writing chromium masks for UV lithography – are capable of focusing the incident electron beam to spots as small as 10 nm. These instruments are excellent lithography tools and find wide application in nanofluidic systems fabrication.

For particular applications, a number of additional lithography techniques are available, including nanolithography using a scanning tunneling microscope for localized surface modification [13] or laser interference lithography for the generation of periodic fringe patterns [14], an introduction of which is beyond the scope of this chapter.

However, as the discussion in the following sections will show, not all nanometer-sized structures require costly high-resolution lithography equipment. A number of, mostly 1D-confined, nanostructures can be fabricated using surprisingly traditional microtechnology techniques. This holds especially true for structures that are confined within the substrate plane, because controlled deposition and selective removal of thin material layers can be achieved down to the low nanometer range [15].

9.2.2
Bulk Micromachining

9.2.2.1 General Aspects

Despite a growing number of new techniques for rapid prototyping and mass fabrication of fluidic devices, bulk micromachining (BMM) of microchannels remains the most common fabrication approach. To describe the method briefly, the substrate is first protected by a suitable material layer, which is subsequently patterned photolithographically. Then, channels are chemically etched into the substrate where the protection layer has been opened. Finally, the protection layer is removed and the etched grooves are turned into closed channels by bonding a suitable cover plate to the structured substrate (Fig. 9.2). Owing to their excellent chemical stability, glass [5, 6] and quartz (fused silica) [16] have been main workhorses for microfluidics since its early days. Also, single-crystal silicon has found wide use, making good its shortcomings in terms of chemical inertness [17, 18] by the large number of dedicated fabrication methods available.

Fig. 9.3 a shows a typical cross-section of a channel formed by wet chemical etching in hydrofluoric acid (HF) in a glass substrate. Owing to the amorphous structure of the bulk material and to the isotropic nature of the etching chemistry, the channels have a smooth, semi-circular cross-section. As a consequence, the

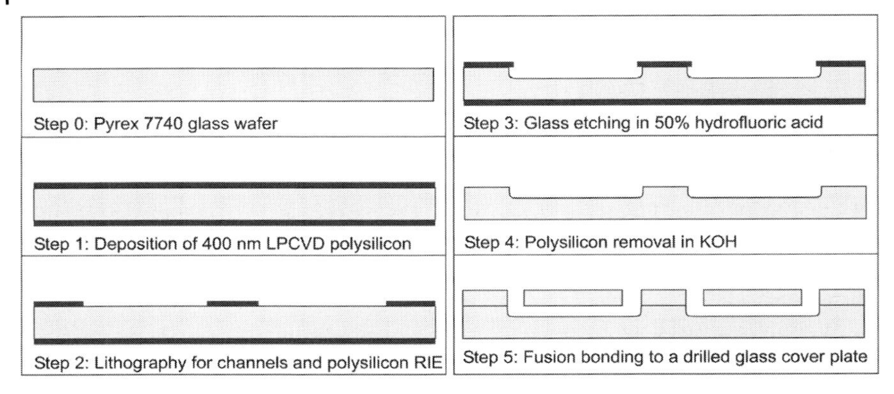

Fig. 9.2 A typical micromachining process for channel fabrication in glass substrates by wet etching in hydrofluoric acid (HF). Polysilicon is used as etching mask here; other masks such as metal layers (chromium–gold) or photoresist (for dilute HF solutions only) have been reported in the literature

width of the channel at its top is always at least twice its depth, which limits achievable aspect ratios (depth-to-width ratio).

Anisotropic etching techniques, both wet and dry, circumvent this limitation. For some monocrystalline materials such as silicon, selected wet chemical etchants show drastically increased etch rates for specific crystal planes whereas others are removed at a much lower rate [11]. Alkaline etchants such as potassium hydroxide (KOH), sodium hydroxide (NaOH) and tetramethylammonium hydroxide (TMAH) form distinct trapezoidal or triangular channel cross-sections in (100) silicon (Fig. 9.3 b) and rectangular ones in the (110) type. Note that different channel depths can be fabricated simultaneously as the etching comes to a nearly complete stop once the (111) planes meet up at the channel bottom.

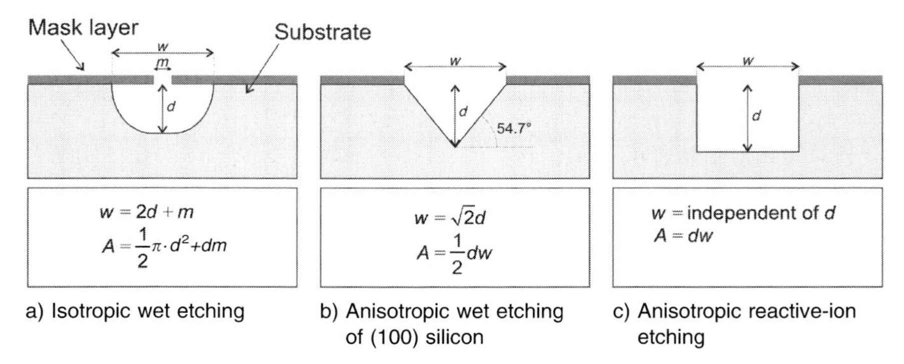

Fig. 9.3 Three typical bulk-micromachining etch processes: (a) isotropic etching of glass (in HF) or silicon (in HNO_3–HF–COOH mixtures); (b) anisotropic etching of (100) silicon in KOH or other alkaline etchants; (c) anisotropic dry etching of both crystalline and amorphous materials; w = channel width on top, A = cross-section area

Whereas wet etching yields anisotropic etching profiles only with single-crystal substrates, dry etching techniques are capable of providing anisotropy also with amorphous materials (Fig. 9.3 c) [19]. In contrast to purely chemical etching methods, a reactive-ion etching (RIE) process receives additional activation energy by spatially directed ion bombardment and thus achieves a certain degree of anisotropy even where chemical orientation dependence does not exist.

The protection layer or etch mask has to be chosen with regard to the etchant chemistry used. For HF etching of glass and quartz, polysilicon and chromium–gold bilayers are typically used, whereas Si_3N_4 and SiO_2 are ideal for silicon etching in alkaline solutions. Photoresist may be suitable for short, shallow wet chemical etching processes, but generally does not withstand longer etchant exposure. For RIE etching, however, photoresist is widely used as a mask layer. Only especially deep etching processes might require a more inert mask, such as electroplated metal [19].

For the sake of completeness, a number of other BMM channel fabrication techniques are noteworthy, including buried, closed channels [20, 21] and electropolishing [22, 23].

9.2.2.2 Nanochannel Fabrication

Conventional BMM processes can be used directly for the fabrication of 1D-confined nanochannels, i.e. sub-micrometer deep grooves with lateral dimensions in the micrometer and millimeter range (Fig. 9.4 a). In order to ensure good reproducibility of the channel depth, the areas to be etched have to be thoroughly cleaned before etching to avoid localized slow-down of the process leading to depth inhomogeneity. If the final channel depth is to be controlled by timing, the etch rate of the process used should be low (1–50 nm/min) to reduce the influ-

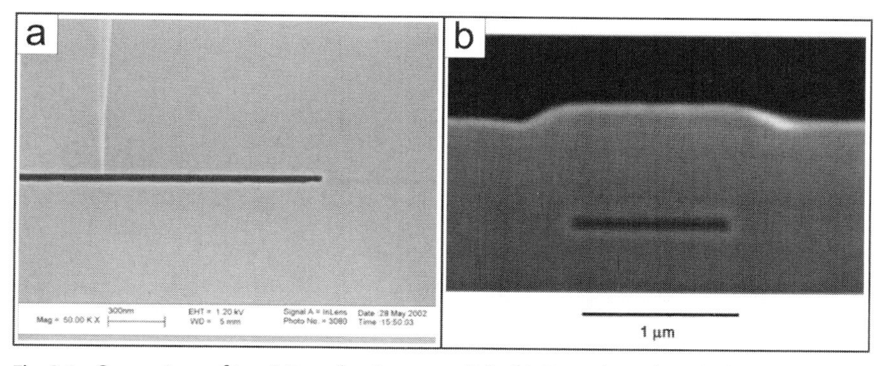

Fig. 9.4 Comparison of two 1D-confined nanochannels made by BMM and SMM. (a) Bulk-micromachined channel etched into a silicon substrate and covered with a second silicon wafer. Reproduced with permission from [24]. (b) Nanochannel made by removal of a sacrificial phosphosilicate glass layer covered by an SiO_2–Si_3N_4–SiO_2 sandwich. Reproduced with permission from [34]

ence of wafer immersion in the etchant. In the case of silicon substrates, Haneveld et al. [24] used a commercial developer solution for positive photoresists containing 2.5 wt% TMAH as etchant. At room temperature, the resulting etch rate was 3.7 nm/min for (110) silicon. As nanochannel fabrication requires only shallow etching, the native oxide layer of the silicon substrate (typically 2 nm) was sufficient to protect the substrate from dissolution. It was structured using photolithography followed by oxide patterning in 1% HF.

Similarly to most microchannels, for nanochannels made by BMM bonding to a cover plate is required to close the grooves after etching. This bonding process has to be well controlled in order to avoid bending or sagging of the cover plate, which might obstruct the inner space of the channels. As a consequence, applied pressure during bonding has to be limited in addition to the planar extent of the channel geometry in relation to its depth.

For 2D-confined nanochannels, e-beam patterning of a PMMA resist layer in conjunction with RIE has been used to machine 70 nm wide and 135 nm deep channels into a silicon wafer [25]. A different approach, adapted from [26], involves direct, localized depassivation of the hydrogen-terminated silicon surface by the e-beam followed by KOH etching. Although this direct-writing technique did not require photoresist coating and development, long exposure times were necessary to achieve the desired surface modification.

More efficient is the direct fabrication of 2D nanochannels in silicon and glass by focused ion beam milling (FIB), which circumvents the lithography step altogether. FIB systems ablate the substrate locally by physically extracting surface atoms upon bombardment with a focused beam of high-energy ions, e.g. gallium. If the beam is slowly scanned over the surface using computer-controlled deflection systems, grooves can be machined in any desired pattern. Using an FIB system with a spot size of 10 nm at 30 keV acceleration voltage, channels with cross-sectional dimensions from 500×500 down to 50×50 nm^2 were fabricated in glass wafers and evaluated using fluorescent Rhodamine B solution [27].

Porous silicon formation can also be used to form 2D-confined nanochannels, with the particularity that pores form perpendicular channels through low n-doped silicon wafers under an applied electric field [28–30], whereas all other processes described so far only allow the fabrication of planar channels. Despite the name, macropores can have sub-micrometer diameters [31] with lengths of several tens of micrometers, which makes them interesting for a number of applications such as drug release by controlled diffusion. Fabrication of defined pores requires that pore size and position are defined by patterning of an insulating layer on the wafer before the etching is carried out in an HF mixture [32]. A combination of electron–hole pair generation by illumination and carrier drift in an electric field provides and concentrates holes at the openings of the insulating layer. These holes are necessary for silicon dissolution in HF and therefore confine the etching locally. Owing to the applied electric field, the hole concentration can be focused at the bottom of the growing etch pit, which leads to a strong etching anisotropy and pore formation (Fig. 9.5).

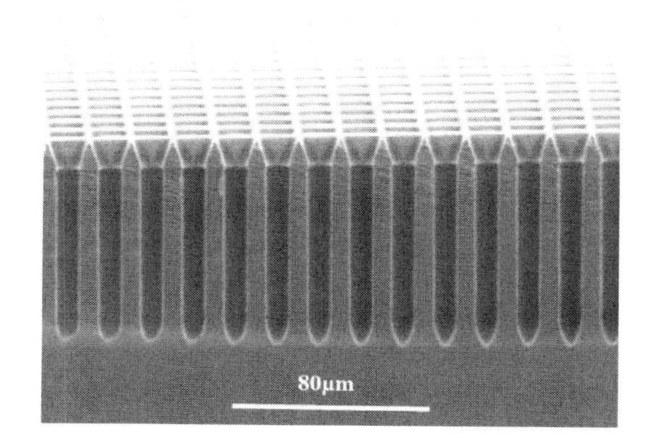

Fig. 9.5 Vertical channels formed by directed and aligned macropores etched electrochemically under illumination into n-type silicon in HF solution. Reproduced with permission from [32]

9.2.3 Surface Micromachining

9.2.3.1 General Aspects

In surface micromachining (SMM), structures are formed by a combination of deposition and selective removal of additional material layers without direct modification of the substrate. A typical SMM method is the sacrificial layer process, where the void inner portion of the channel is created by structuring a sacrificial thin film, covering it with a different material and selectively removing the sacrificial template again (Fig. 9.6 a). This results in channels with a depth equal to the thickness of the sacrificial layer, which are conveniently sealed by the cover layer and therefore do not require an additional bonding step for the cover plate (however, openings for fluidic access to the channels need still to be made by a suitable etching process).

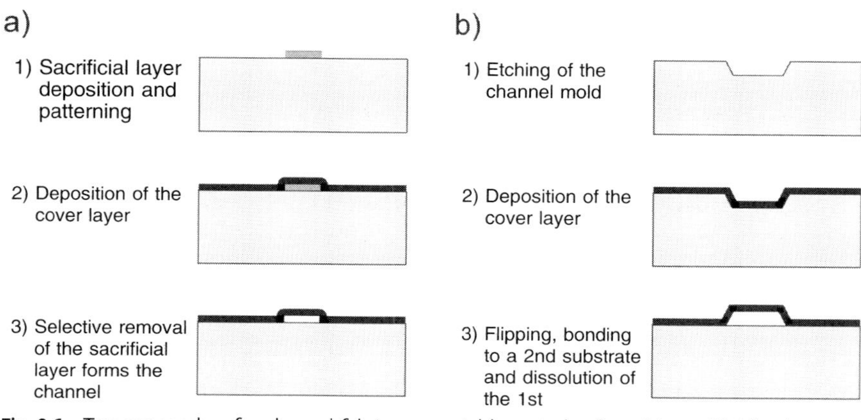

Fig. 9.6 Two approaches for channel fabrication using surface micromachining: (a) sacrificial layer technology (b) sacrificial substrate technology

The chemical selectivity of the available etchants limits the choice of sacrificial and cover layer materials; whereas the sacrificial layer should dissolve quickly, the cover layer should withstand the etching process unharmed. Typical sacrificial layers are phosphosilicate glass (PSG), a doped silicon oxide dissolving quickly in HF-based etchants [33] and polysilicon, which can be selectively etched in TMAH [34] or KOH [35]. For its stability in most etchants, vapor-phase deposited silicon nitride is often used as a cover layer.

A different sacrificial fabrication process for microchannels involving dissolution of the entire substrate after channel formation is illustrated schematically in Fig. 9.6 b [36].

9.2.3.2 **Nanochannel Fabrication**

The fabrication of 1D-confined nanochannels with depths between 20 and 100 nm can be achieved using amorphous silicon (a-silicon) [34] and polysilicon [37, 38] as sacrificial layers (Fig. 9.4 b). A thin silicon film of the desired channel height is vapor-deposited and patterned on a passivated silicon wafer and covered with a dielectric cover layer. After opening inlet and outlet ports at the channel end, the sacrificial layer can be removed in TMAH–water solutions. Replenishment of the etchant and removal of etching products are diffusion-limited, which causes long etching times (up to 80 h depending on the geometry [34]) taking into account a channel length of several millimeters, a width in the micrometer range and a thickness <100 nm. As a consequence, an Si etchant with high selectivity towards the dielectric cover layer of the channel is required (depending on temperature and concentration, TMAH has a selectivity of the order of 10 000:1 to SiO_2 and Si_3N_4). Still, sacrificially etched nanochannels may have a height profile slightly tapered along their main axis because the cover layer close to the ends of the channel is exposed to the etchant for a longer time than the middle section [37]. Nanostructured column/void PECVD silicon films have been proposed as sacrificial material for nanochannel fabrication, as their sponge-like structure facilitates dissolution [39].

After etching, the channels have to be rinsed and dried carefully to avoid etchant residues by replacement of the etchant with water, followed by 2-propanol, which is finally evaporated under rotation [37]. Fluid replacement is achieved by diffusion as pressure-driven flushing of solutions is difficult in practice.

Nanofluidic structures confined in two dimensions can be fabricated using sacrificial layers, if means to pattern the latter with sub-micrometer precision are available. The Craighead group has presented a variety of these structures, based on polysilicon as sacrificial material and vapor-deposited SiO_2 as cover layer [8]. E-beam lithography has been used to generate 400 nm deep channels containing a dense array of solid posts of 100 nm width. In order to solve the issue of time-consuming sacrificial layer removal by wet etching, Li et al. used the polymer poly(butylnorbornene) (PNB) as a sacrificial layer for the fabrication of 2D nanochannels in the 100 nm range [40]. The polymer can be structured by photolithography followed by RIE or by direct embossing (see Section 9.2.5.2). PNB has the interesting property that it withstands temperatures of up to 220 °C, allowing subsequent plasma-enhanced CVD processes to deposit an inorganic cover layer for the channels. Then, after opening access holes to the channels, the sacrificial layer can be removed by thermal decomposition at 440 °C in a nitrogen atmosphere without wet chemistry.

Whereas the techniques illustrated above rely on high-resolution lithography equipment, Tas et al. have demonstrated the fabrication of 2D-confined nanochannels by conventional methods [41]. The first method uses a sacrificial polysilicon nanowire to define the channel opening, which is deposited on the sidewall of a topography step by a combined CVD and backetch process. After covering the nanowire with silicon nitride, the sacrificial polysilicon is removed in KOH (Fig. 9.7 a). A second approach takes advantage of the adherence of the cover layer after the sacrificial layer has been removed. The thin cover is permanently dragged on to the substrate by capillarity during the drying step after etching, resulting in an asymmetric channel cross-section as illustrated in Fig. 9.7 b. The impact of the cross-sectional asymmetry on fluid flow in these nanochannels remains to be assessed.

The fabrication of vertical nanochannels with nanometer confinement in the horizontal direction requires more processing steps than its horizontal counterpart. Again, well-controlled layer formation techniques such as thermal oxidation or chemical vapor deposition are used to define the confined channel dimension. Fig. 9.8 shows a fabrication procedure developed by Lee et al. [42] where etching of vertical steps into a layer of amorphous silicon selects the position of the channels to be made, while the step height sets the channel depth. The channel width is defined in the following step by growing a precisely controlled thermal oxide layer under dry conditions on the amorphous silicon structure. While a 40 min oxidation step at 1000 °C results in an oxide thickness of 50 nm, oxide thicknesses

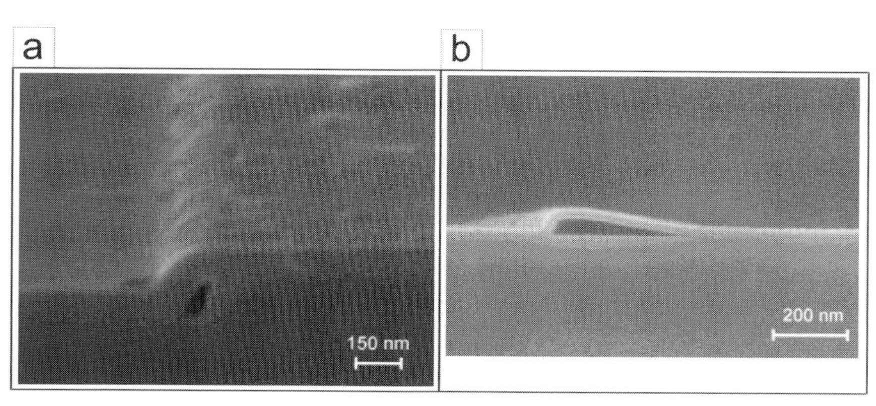

Fig. 9.7 2D-confined nanochannels fabricated using conventional thin-film and photolithography techniques. (a) Sacrificial layer etching of a polysilicon nanowire deposited on the vertical sidewall of a topography step. (b) Cover layer attached to the substrate by capillary forces during drying. Reproduced with permission from [41]

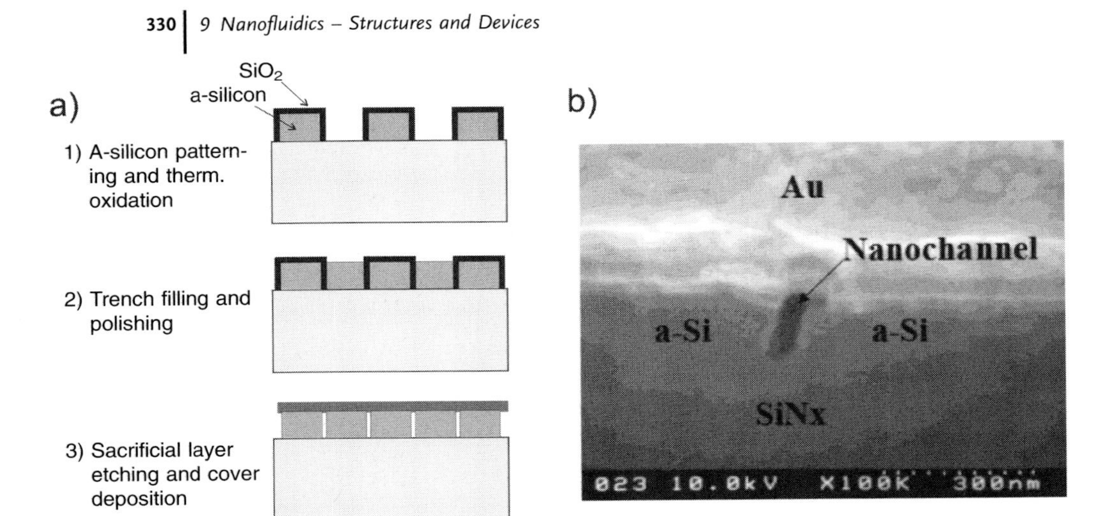

Fig. 9.8 Fabrication of vertical nanochannels. (a) Schematic fabrication procedure based on sacrificial layer technology and trench etching into amorphous silicon. Adapted from [42]. (b) SEM photograph of a 50 nm wide channel. Reproduced with permission from [42]

as small as 5 nm can be fabricated reliably. The remaining gaps are subsequently filled by a second amorphous silicon deposition and the whole substrate is planarized by chemical–mechanical polishing. Finally, the sacrificial oxide is removed by wet etching and the resulting vertical channels can be sealed by an additional silicon oxide or gold deposition.

9.2.4
Fabrication Based on Thin-layer Polymers

Photocurable polymers allow fast and efficient fabrication of structural patterns on a substrate without additional etching steps. However, as photoresists used in standard lithography are generally designed to be easily removed after pattern transfer to the underlying layer, their use for microfluidic applications is limited owing to the lack of chemical stability. However, a number of photocurable polymers are available that can be processed in a very similar fashion (spin coating, exposure, development) and that form comparatively inert structures upon cross-linking. The best known materials are the epoxy-based SU-8 photoresist [43] and photosensitive polyimide [44]. Fig. 9.9 shows as an example two SU-8 fabrication approaches for microchannels using either bonding to a cover plate or sacrificial layer removal with unexposed SU-8 assuming the role of the sacrificial material [45].

While these examples involve spin coating of the liquid polymer precursors, gas-phase deposition of polymer layers has also found application in microfluidics. The polymer parylene is inert to a wide range of solutions and can be deposited in the thickness range between 1 and 30 μm using pyrolysis of the monomer at 680 °C followed by conformal condensation and cross-linking on the substrate

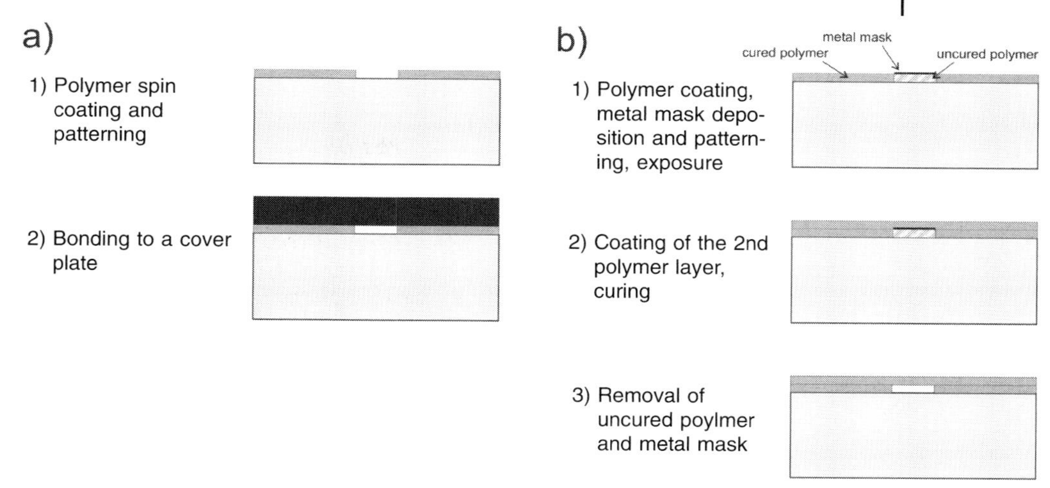

Fig. 9.9 Channel fabrication using photocurable polymers. (a) Microchannel fabrication based on lithography and channel sealing by a cover plate. (b) Sacrificial layer technique using non-cross-linked SU-8 as sacrificial layer: the first layer is exposed through a patterned thin-film metal mask, then cured, covered by a second resist layer, flood-exposed and cured again. Finally, the non-cross-linked SU-8 and the metal mask can be removed in developer and etching solutions, respectively. Adapted from [45]

at room temperature. Microchannels in parylene can be fabricated using photoresist as sacrificial layer sandwiched between two vapor-deposited polymer layers [46]. Ilic et al. fabricated 400–850 nm wide channels by depositing parylene in narrow trenches etched into a silicon wafer. As more polymer is deposited at the top of the trench than at the bottom, the material pinches off and seals the trench, forming a closed parylene tube [47].

9.2.5 Replication Technologies

Replication is advantageous for rapid prototyping and for cost-efficient mass production of microfluidic devices in polymers using a micromachined master structure [48–51]. Common to the different replication techniques used in micro- and nanofabrication is the rigid mold or master which contains a topography inverse of that desired for the replica. To replicate the mold topography into a mechanically soft substrate, both surfaces are brought into physical contact for a given time to allow for pattern transfer and are separated again.

All replication techniques allow inherently parallel pattern generation on a large number of substrates using a single mold. Many identical replicas can be fabricated from a single mold before it has to be replaced owing to deterioration. The importance that this technology has gained in recent years is underlined by the fact that commercial nanoimprinting and -embossing systems are available for routinely patterning in the sub-100 nm range.

9.2.5.1 Casting

Casting or molding of elastic polymer structures, one of the most widely used replication techniques in microfluidics, is very straightforward: the liquid polymer precursor is poured on to the master and subsequently cross-linked (see Fig. 9.10). It has been shown that highly complex microfluidic devices containing active elements such as valves and pumps can be fabricated in this fast and simple process [52]. Because of its good optical properties and self-sealing capability, the commercially available poly(dimethylsiloxane) (PDMS) is the elastomer of choice for research and industry [53–55]. As molds, etched silicon substrates, lathed metal sheets and thick photoresist structures are used. After curing and mold release, PDMS bonds spontaneously and reversibly to flat surfaces such as glass, silicon, some plastics or a second PDMS sheet. If permanent, mechanically stable sealing is required, a short oxygen plasma treatment can be used prior to bonding to activate the PDMS surface, resulting in a stronger bond.

Replica fabrication by casting of elastomers is capable of reproducing features down to tens of nanometers [56]. To assure defect-free separation of the master and the cured elastomer, bottom and sidewall roughness of the mold features has to be tightly controlled and the use of a surface treatment procedure (for instance, immersion in a chlorosilane solution) for improved mold release is recommended.

To machine 2D-confined PDMS nanochannels, Hug et al. [25] used e-beam lithography and PMMA resist for pattern generation, which was subsequently transferred into the silicon substrate by anisotropic wet etching in KOH. After casting and bonding of the replica to a glass cover, 100 nm deep and 300 nm wide channels in PDMS could be realized.

Owing to the increased surface-to-volume ratio, the surface properties of PDMS become more important with reduced channel dimensions. It has been reported [25] that spontaneous filling of PDMS nanochannels with aqueous solutions requires prior surface hydrophilization by oxygen-plasma treatment [57]. Also, it was observed that channels are easily obstructed at high width-to-depth ratios as the

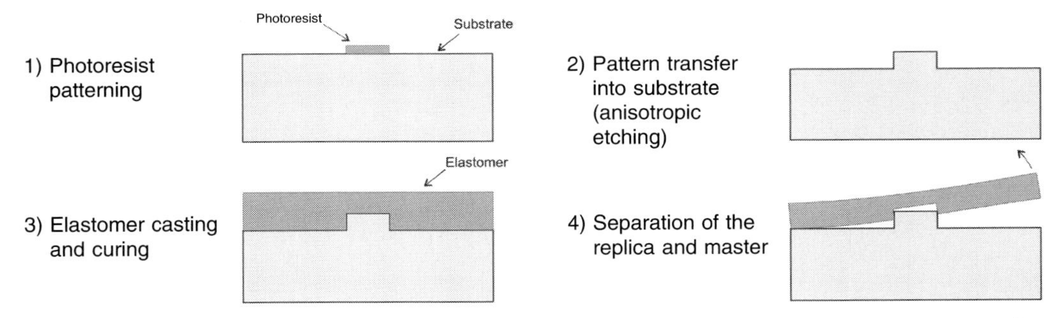

Fig. 9.10 A typical elastomer casting process using a structured substrate (e.g. a silicon wafer with reactive-ion etched features) as master. To facilitate separation between replica and master after curing, pretreatment of the master surface with a chlorosilane is recommended between steps 2 and 3

elastic channel walls sag or even collapse once they touch the opposite wall. This problem is also known from nanoprinting, where deformations of high-aspect ratio PDMS structures occur during stamping [51], and it makes the fabrication of 1D-confined nanochannels in PDMS difficult to achieve.

9.2.5.2 Embossing

Embossing allows pattern transfer into rigid, thermoplastic substrates or thin films, which are heated above their glass transition temperature, T_g (depending on the polymer used, between 50 and 150 °C) [48]. In a vacuum chamber, the master is pressed into the soft polymer (forces between 0.5 to 2 kN/cm^2) and held in place while the assembly cools below T_g, before both are separated again.

Nanoembossing was first reported to structure resist films (also known as nanoimprint lithography), which were used to fabricate metal structures in a lift-off process with 25 nm features [58]. After the embossing step, residual resist remains in the grooves of the patterned layer and an anisotropic etching step is required to transfer the pattern through the entire resist thickness. As an extension of this concept, Cao et al. recently reported the fabrication of 10 nm wide nanochannels by embossing and RIE, followed by nonuniform deposition of a few hundred nanometers of SiO_2 to cover and close the channels without bonding [59]. Patterning of a sacrificial polymer layer for nanochannel fabrication by embossing has been described [40]. Also, the issue of line edge roughness in the master due to scattering effects during lithography has been addressed by adapted etching procedures and thermal photoresist reflow [60].

In addition to its application in nanoimprint lithography, embossing can also be directly applied to form channel structures in thermoplastic substrates without further etching steps [61]. Using a customized hot embossing process for polymer pellets, Studer et al. fabricated nanopillar arrays down to 150 nm feature size [62].

9.2.5.3 Injection molding

Even more oriented to mass fabrication is injection molding, the dominant technique for the commercial fabrication of plastic parts of arbitrary shape and size (an example is the mass fabrication of movie DVDs with features 400 nm wide and 100 nm deep). Different from the previous method, the molten polymer (200–350 °C) is injected under pressure (600–1000 bar) into the molding chamber containing the mold insert, where it cools below its solidification temperature. Then, the two-part chamber opens and releases the replica, which can have feature sizes as small as 25 nm [63]. The application of injection molding to microfluidic device fabrication was first described by McCormick et al. for PMMA [64].

9.2.5.4 Printing

Printing of micro- and nanometer features, also referred to as soft lithography, has recently attracted increased attention for high-resolution patterning applications. Radically different from the conventional lithography approach, patterns are not generated by irradiation using light or particle beams, but by direct, physical transfer of a structured material layer on to the substrate [49, 51, 65].

Stamps are fabricated by PDMS casting on nanostructured masters similar to the procedures described earlier. The soft, elastomeric nature of PDMS is advantageous for the stamp, as the latter adapts itself to the surface roughness and topography of the substrate during printing. This conformal contact at the microscale has to be assured for successful patterning with low pinhole densities.

However, the softness of the stamp might compromise the accuracy of the pattern transfer owing to deformation under pressure. This issue can be solved for high-precision applications by using a stiffer elastomer for the patterned layer of the stamp [66–68]. Modified PDMS formulations, also called hard PDMS or h-PDMS, have been developed to respond to these requirements while maintaining the advantageous molding and printing properties of the material [69]. The thin, stiff stamp is supported by a thicker, soft elastomer support to allow practical handling of the tool.

The stamping process itself consist of three distinct phases: first, the stamp surface is covered with the ink solution, then it is brought in contact with the substrate to transfer the pattern, followed by separation of stamp and substrate. Upon contact, the functional species of the ink solution needs to adsorb to the substrate to assure successful pattern transfer. Especially for small structure widths, molecular diffusion of the ink molecules on the substrate surface is not negligible and can cause structure expansion by several hundred nanometers depending on the molecular weight of the ink [70].

A widely used ink–substrate system is based on alkanethiols, which form a well-ordered surface assembled monolayer (SAM) on gold substrates. The strong thiol bond ensures that the ink is transferred properly from the stamp to the gold substrate upon contact. Thiol-SAMs were shown to prevent molecular access to the gold surface and can therefore be used as etch mask for gold in various etching mixtures. A gold layer nanostructured in this way can serve as a functional structure, for instance as a nanowire or electrode, or it can be used as secondary etch mask for further etching of the underlying substrate [70, 71].

9.3
Fluid Flow and Molecular Transport in Nanochannels

Diffusion, pressure-driven flow and electrokinetic flow are the main processes that govern molecular transport in miniaturized fluidic devices. All are directly dependent on dimensional parameters, such as distance or surface-to-volume ratio. As a consequence, flow properties in nanostructures are essentially different from those with features of tens of micrometers as typically used in microfluidics.

9.3.1
Pressure-driven Flow in Nanochannels

Most macroscopic flow systems rely on one or several pressure sources for liquid transport, which allow precise control of flow-rates and flow paths in conjunction with valves. If a pressure difference ΔP is applied to both ends of a channel of length L, a flow Q is induced through this channel:

$$Q = \frac{\Delta P}{R_f L} \tag{2}$$

The parameter R_f is the hydraulic flow resistance of the channel geometry per unit length. It has been determined analytically for a number of cross-section geometries starting with the case of a circular tube with radius r (Hagen-Poiseuille):

$$R_f = \frac{8\eta}{\pi r^4} \tag{3}$$

The viscosity η is a temperature-dependent material property of the liquid in the channel, a typical value for dilute aqueous solutions at room temperature being 0.01 P. For triangular channels, such as they are formed by anisotropic wet etching of silicon, R_f has been determined to be (dimensional parameters as in Fig. 9.3 b) [72]

$$R_f = \frac{69.6\eta}{w^4} \tag{4}$$

A typical case for nanochannels is a rectangular cross-section (as depicted in Fig. 9.3 c). Under the condition that the depth d is much smaller than the width w, R_f can be written as

$$R_f = \frac{12\eta}{wd^3} \tag{5}$$

A more general treatment of Poiseuille flow in various duct geometries can be found, for instance, in [73].

Equations (3)–(5) underline the strong fourth-order dependence of the flow resistance on the cross-sectional dimensions, which makes pressure-driven flow less suitable as a driving mechanism for nanoscale fluidic systems. Fig. 9.11 compares flow-rates for the three channel geometries with dimensions ranging from 50 μm down to 50 nm. The resulting flow-rates span 12 orders of magnitude. However, for most applications, the time needed to exchange the liquid volume within a nanochannel is more important than the actual flow-rate. Although the flow-rates scale with r^4, the exchange time depends on the flow velocity of the liquid, which scales down only with r^2 (in the circular case) and is therefore less affected by reduced dimensions.

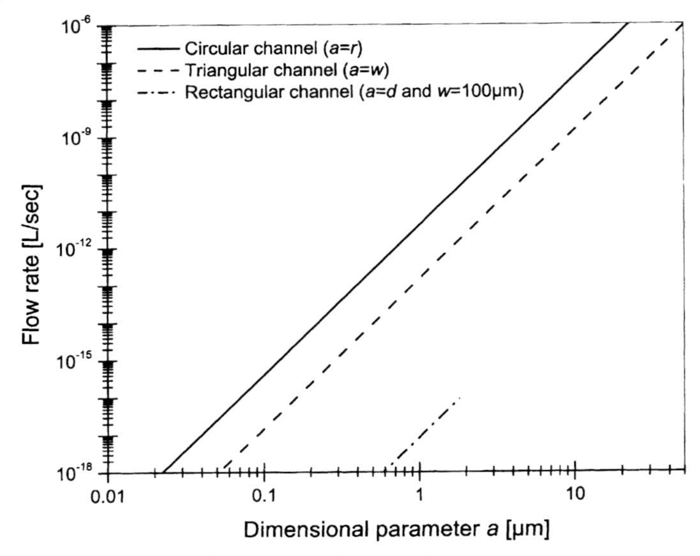

Fig. 9.11 Calculated flow-rate as a function of channel dimension for the three typical cases circular, triangular and rectangular cross-sections, based on Equations (2)–(5). As the pressure difference, a value of 100 kPa, is assumed, for a channel length of 1 mm. In the case of the rectangular channel, the width is assumed to be constant at 10 µm and the graph is only plotted for $a << w$

Filling of nanochannels with hydrophilic walls, such as glass or SiO_2, is typically accomplished by capillary action without external pressure. However, PDMS does not show spontaneous filling owing to the hydrophobicity of the surface [25]. Wetting properties of nanochannels are typically assessed using a microchannel array as test structure, which is filled with fluorescent dye solution [74]. The movement of the boundary between filled and empty channel can be monitored under a fluorescence microscope.

9.3.2 Electrokinetic Fluid Handling in Nanochannels

Electrokinetic fluid handling is a widely used technique in microfluidics for electrical manipulation of minute amounts of liquid reagents or particle suspensions [75, 76]. As the fluid flow follows an applied electrical field, a specific flow path can be chosen in a channel network without the need for micromechanical valves and pumps, which greatly facilitates device fabrication and operation. For nanofluidic applications, the technique has the additional advantage that flow is induced on the surface instead of in the bulk of the solution. As a result, electrokinetic pumping can be applied to nanochannels without major modification, while pressure-driven flow is greatly affected by downscaling of the channel geometry.

9.3.2.1 Electrokinetic Transport Phenomena

Although the field of electrokinetics comprises several effects related to charged solid–liquid interfaces, the two most important phenomena allowing molecular transport on the microscale are electrophoresis and electroosmosis [77]. Both describe the interaction between a statically charged solid surface and an electrolyte solution in an electric field.

Electrophoresis refers to the transport generated by the Coulomb force acting on a mobile, charged object (e.g. a polymer particle or an ion) suspended in a quasi-stationary solution in an electric field. Under an applied field, the object accelerates to a velocity where Coulomb force and viscous drag are equal:

$$qE = 6\pi \nu_{\text{ep}} r \eta \tag{6}$$

where q=object charge, E=electric field strength, ν_{ep}=electrophoretic velocity, r=radius of the object and η=buffer viscosity.

Equation (6), known as the Stokes-Einstein relation, is valid for particles, but can also be used as an approximation for an ion with its surrounding hydration shell [78].

A typical application of electrophoresis is the separation of molecules with different charge-to-size ratios in the field of chemical analysis. The electrophoretic mobility of an ion, μ_{ep}, relates electric field and resulting migration velocity, ν_{ep}, and can be derived from Equation (6):

$$\mu_{\text{ep}} = \frac{q}{6\pi r \eta} \tag{7}$$

and

$$\nu_{\text{ep}} = \mu_{\text{ep}} E \tag{8}$$

To give an order of magnitude, negatively charged amino acids have electrophoretic mobilities in the range -2.5×10^{-4} to -4×10^{-4} cm^2/V s [79], while smaller inorganic ions have mobilities ranging from $\pm4\times10^{-4}$ to $\pm8\times10^{-4}$ cm^2/V s depending on their polarity [80].

In *electroosmosis*, on the other hand, the picture is inverted and the charged surface remains fixed while the solution is mobile and can flow freely (Fig. 9.12). Electroosmosis refers to the bulk transport of a liquid volume enclosed by a charged, stationary surface in an electric field (electroosmotic flow, EOF). Most materials used for the fabrication of micro- and nanochannels exhibit a permanent surface charge, which upon filling of the channel leads to the formation of a layer of mobile electrolyte ions with opposite polarity close to the surface (also called electrolyte double layer, EDL). This layer can be forced to move by applying an electric field along the channel axis. As a result of viscous drag, the whole volume in the channels moves together with the thin surface layer, leading to a bulk

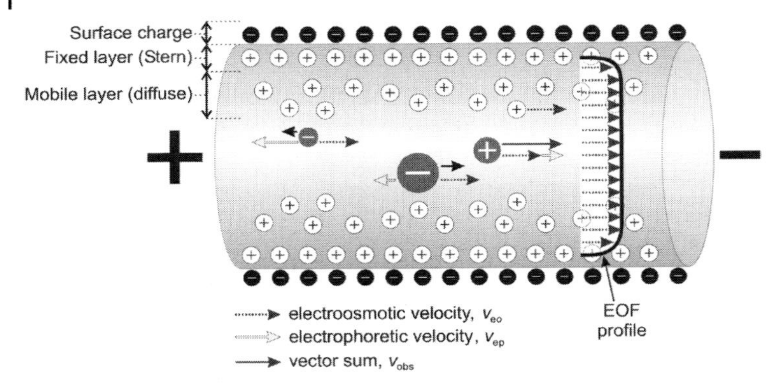

Fig. 9.12 Electroosmotic flow (EOF) in a microchannel or capillary. The observed velocity of the different ions is the vector sum of the EOF velocity and the ion-specific electrophoretic velocity. EOF develops a plug-like, rectangular flow profile

pumping mechanism. Similarly to μ_{ep} for ions or particles, an electroosmotic mobility can be attributed to the bulk solution:

$$\mu_{eo} = \frac{\varepsilon\zeta}{\eta} \tag{9}$$

where ε=dielectric constant of the solution and ζ=zeta potential of the surface–solution interface (see Section 9.3.2.2), and again

$$v_{eo} = \mu_{eo} E \tag{10}$$

For Pyrex glass, the electroosmotic mobility varies between 3.5×10^{-4} and 5×10^{-4} $cm^2/V\,s$ (with low values at low pH), while fused silica spans a range from 0.5×10^{-4} to 4.5×10^{-4} $cm^2/V\,s$ [81]. Polymers such as Teflon have generally a low zeta potential, leading to a μ_{eo} between 1×10^{-4} and 2×10^{-4} $cm^2/V\,s$. As a consequence, the velocity of EOF in the technically interesting field range of 10–1000 V/cm is between 10 µm/s and 5 mm/s.

The zeta potential describes the mobile portion of the EDL as described in the next section. Equation (9) is derived from the Helmholtz-Smoluchowski equation and is, strictly, valid only for cases where the extent of the double layer is small compared with the channel diameter.

In practice, electrophoresis and electroosmosis occur simultaneously and the observed velocity of an ion i can be written using the sum of both mobilities, μ_i:

$$v_i = E\left(\underbrace{\mu_{ep,i} + \mu_{eo}}_{\mu_i}\right) \tag{11}$$

Therefore, depending on the sign of μ_i, the ion can migrate in the direction of the EOF (co-electroosmotic transport) or against it (counter-electroosmotic transport).

9.3.2.2 **The Electrolyte Double Layer**

The structure of the EDL and its dependence on material and electrolyte parameters are very complex. A number of models exist to describe potential and charge distribution within the EDL, including the Helmholtz model, the Gouy-Chapman model and the Guy-Chapman-Stern model. Within the context of this chapter, we can give only a brief introduction to the physics of the EDL, and readers are referred to excellent text books on the topic for further study, e.g. [77].

The model which coined the term EDL was first put forward in the 1870s by Helmholtz. In this model, the surface holds a charge density due to an excess or deficiency of electrons and therefore causes a redistribution of ions close to the surface in order to ensure electroneutrality. Helmholtz assumed that the solution is composed only of electrolyte ions and that no electron transfer processes occur at the surface. As a result of the surface potential, excess ions form a charged layer on the surface, which is only as wide as an ion with its solvation shell. Hence the double layer is formed by the fixed surface charge on one side and the opposing ion charge in the solution on the other.

Because the Helmholtz model did not account for phenomena such as mixing, solvent interaction and molecular transport, it was refined over the years. First, Gouy and Chapman proposed in 1910 that excess ions are nonuniformly distributed in the vicinity of the surface, with a maximum concentration at the surface decreasing nonlinearly until they reach bulk concentration. The principle of this diffuse charge layer on the surface allowed, for instance, the dependence of the EDL structure on the electrolyte composition to be described. However, the model predicts an unrealistic surface concentration as it assumes that ions are infinitely small and can come infinitely close to the surface. It was therefore modified by Stern in 1923, introducing the concept of a fixed excess charge layer in close vicinity to the surface combined with a diffuse layer extending into the bulk solution.

Fig. 9.13 shows these two distinct zones, the thin Stern layer close to the wall, which is formed by immobile charges, and the diffuse or Gouy-Chapman layer, whose charges are mobile under an applied electric field. The interface between both layers at a distance δ from the wall separates moving and nonmoving liquid zones and is therefore called *plane of shear.* The potential at the plane of shear is the *zeta potential,* ζ, which is a function of electrolyte composition and wall material.

The Debye length λ_D is a measure of the width of the EDL:

$$\lambda_D = \sqrt{\frac{\varepsilon RT}{F^2 \sum_i c_i z_i^2}} \tag{12}$$

where ε=dielectric constant, R=gas constant, T=temperature, F=Faraday constant, (96 495 C/mol), c_i=concentration of ion i and z_i=valence of ion i.

In a physical interpretation, the parameter $1/\lambda_D$ (also known as Debye parameter, κ) represents the 1/e decay of the potential from the wall. Experimentally, λ_D can be modified by adjusting the ionic strength of the electrolyte solution, which

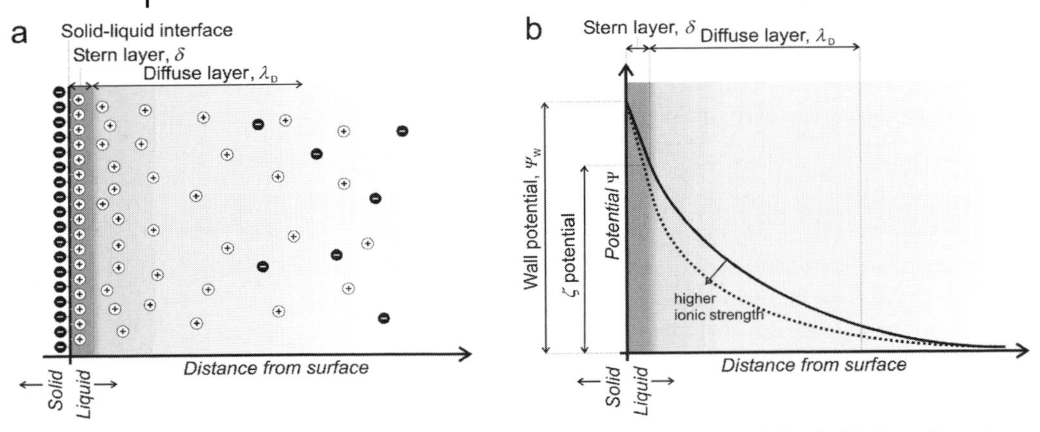

Fig. 9.13 Schematic representation of the Gouy-Chapman-Stern double layer model. (a) A fixed surface charge on the left attracts a positive counter-charge in the electrolyte to balance electroneutrality. The thin Stern layer of thickness δ is firmly attached to the surface and remains immobile, whereas cations in the loosely bound diffuse layer can be accelerated in an electric field. (b) The potential distribution through the double layer depends on the wall potential and on the ionic strength of the electrolyte. It is nearly linear in the Stern layer and decays exponentially in the diffuse zone. The potential at the plane of shear between Stern and diffuse layer is called zeta potential, ζ, and the distance at which the potential falls to $1/e$ is the Debye length, λ_D

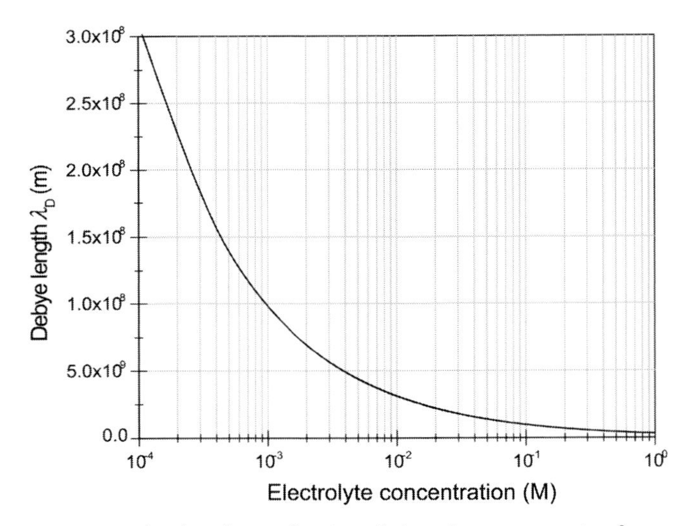

Fig. 9.14 Debye length as a function of electrolyte concentration for a monovalent aqueous solution at 25 °C

affects the sum term in Equation (12). Fig. 9.14 illustrates the dependence of λ_D on the concentration of a monovalent electrolyte solution.

It should also be noted that Equation (12) relies on the Debye-Hückel approximation for thin layers and is therefore only valid if the EDL is much smaller than the channel dimensions. Although this holds true in microfluidics with typical channel diameters ranging between 1 and 100 μm, nanochannels require a closer look at the double layer to predict EOF at these dimensions.

For most materials typically used in micro- and nanofluidic devices, such as glass [79] and polymers [57], the surface charge is negative for a wide range of pH values owing to deprotonation on the surface. As a consequence, the net charge of the EDL is positive under typical buffer conditions and electroosmotic flow is directed towards the cathode. An important property of EOF is the nearly rectangular flow velocity profile, which is caused by the fact that the driving force for the liquid movement is generated on the surface. This is in strong contrast to pressure-driven Poiseuille flow, which has a parabolic velocity profile with flow stagnation along the walls due to viscous shear. In practice, the rectangular profile has certain advantages, such as reduced mixing between adjacent liquid zones in separation techniques. In nanofluidics, it allows pumping through narrow channels or porous structures, which would require strong external pumps and connectors if pressure-driven flow were to be used.

9.3.2.3 Electroosmotic Pumping in Nanochannels

With regard to nanofluidics, EOF has scaling properties far superior to pressure-driven flow. While the flow velocity, v_p, at constant pressure decreases with the square of the channel radius r (i.e. $v_p \sim r^2$), the electroosmotic flow velocity, v_{eo}, is independent of r as long as the Debye-Hückel approximation is valid. As a result, liquid transport by pressure is difficult to achieve in extremely small channels or channels packed with a porous stationary phase.

However, Equations (9) and (10) do not yield correct values if the EDL thickness λ_D becomes a considerable fraction of the channel radius r. For cases where $r/\lambda_D \approx 1$, that is, for channels < 100 nm deep, the EDL spans across the whole channel cross-section and severely reduces EOF (illustrated schematically in Fig. 9.15). Mathematical models for electrokinetic flow and ion transport in nanochannels with $r/\lambda_D \approx 1$ have been developed for ultrafine capillary channels of rectangular cross-section [82], for symmetric and asymmetric nanochannels [83], for multivalent ions [84] and for silica nanotubes [85].

To verify the validity of these mathematical models for nanochannels, Ramsey's group has investigated electroosmotic flow in glass channels with depths between 83 nm and 10.4 μm using neutral, fluorescent markers to determine the EO flow velocity. The Debye length was modulated between 0.7 and 60 nm by using sodium tetraborate buffer at different concentrations according to Equation (12). While the electroosmotic mobility, μ_{eo}, follows a linear relationship with λ_D for larger channels, the impact of the double layer overlap is clearly visible in Fig. 9.16 for channels shallower than 100 nm.

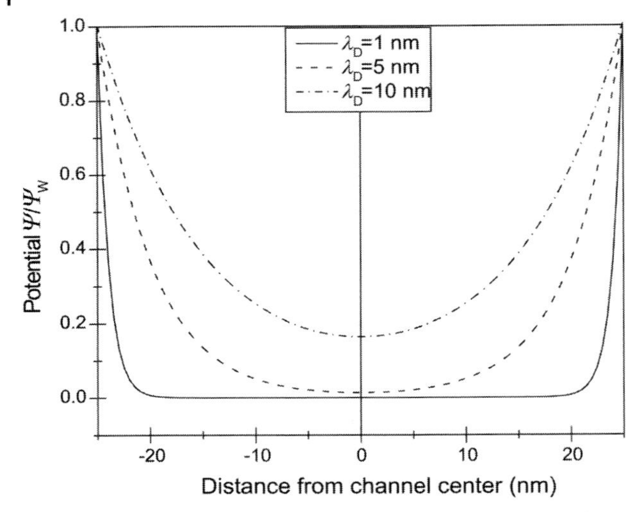

Fig. 9.15 Qualitative potential distribution normalized to Ψ_W from wall to wall in a 50 nm deep channel. For wider EDLs, the diffuse layers from both sides overlap

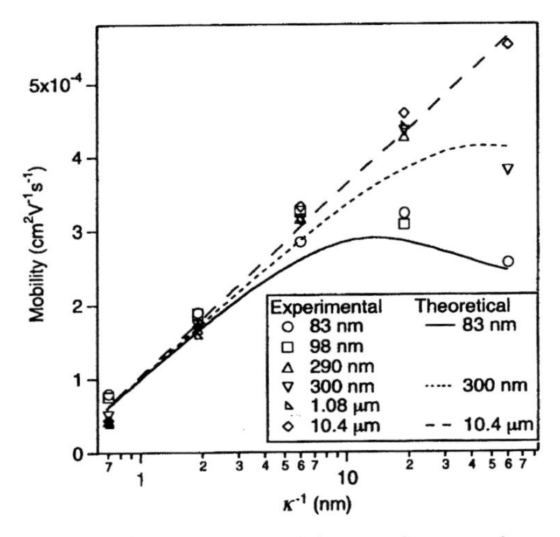

Fig. 9.16 Electroosmotic mobility as a function of double layer thickness for different channel depths. Reproduced with permission from [86]

This effect can be elegantly used to generate hydraulic pressure for microchip pumping applications in a simple channel configuration. A micrometer-size main channel serving as inlet and outlet is connected in a T-configuration with an auxiliary nanochannel. Once a driving voltage is applied between inlet and the nanochannel, an imbalance in electroosmotic flows occurs as the EO mobilities are different in the two portions of the system. The resulting difference in flow-rates builds up a hydraulic pressure which releases to the outlet channel in order to fulfil the mass-balance requirement. This electroosmotically induced, pressure-driven flow has the advantage of being electrical field-free, which allows, for instance, suspensions of objects such as biological cells, which are sensitive to electric potential gradients, to be pumped.

With the application of field-free EOF pumping on glass microchips in mind, the connecting nanochannel can also be fabricated by inducing a dielectric breakdown through the substrate between two closely adjacent channels [86, 87]. At electrical field strengths between 0.5 and 2 MV/cm, the bond interface between substrate and cover wafer cannot sustain the electric field and one or more narrow fractures appear at the closest point between the two channels. Depending on the time the voltage is applied after breakdown and on the electric field strength, these fractures can be in the nanometer to low micrometer range. Indirect EOF pumping and field-free sample injection into microfluidic electrophoresis devices have been demonstrated.

The dependence of electroosmotic flow on the r/λ_D ratio allows one to tailor molecular transport such that only molecules with a specific electrophoretic mobility are driven through the nanochannel, whereas others are held back [88]. Kuo et al. demonstrated gated fluidic transport through a PCTE membrane containing ion-track etched pores with diameters between 15 and 200 nm [89, 90]. The membrane is sandwiched between two microchannels and acts as a gate for flow and molecular transport between the two channels (Fig. 9.17a). In order for a mole-

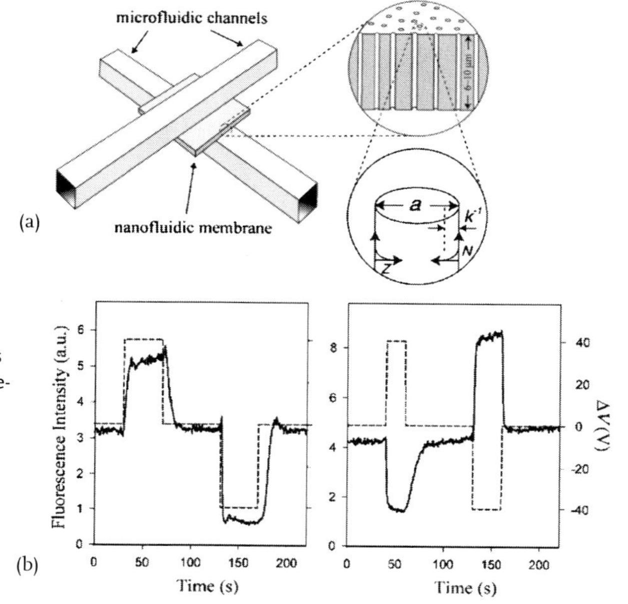

Fig. 9.17 (a) A PCTE membrane sandwiched between two microchannels serves as a nanocapillary array for gateable fluid control. (b) Reversion of the molecular transport direction of fluorescein in a pH 8 phosphate buffer for two different pore sizes (left, 15 nm; right, 200 nm). Reproduced with permission from [90]

cule to pass the gate under counter-electroosmotic flow conditions, its electrophoretic mobility, μ_{ep}, has to be higher than the EOF in the opposite direction. As the EOF depends on the pore size, the latter has an impact on the flow-rate for a specific analyte and can even reverse its flow direction (Fig. 9.17 b).

9.3.2.4 High-pressure Electroosmotic Pumps

For a variety of applications, miniaturized liquid pumps are required to generate minute flow-rates at high pressure ranging from a few hundred kPa to 10 MPa (e.g. for high-performance liquid chromatography). Whereas conventional, mechanical micropumps are only rarely capable of delivering output pressures of more than 50–100 kPa, EOF pumping can do so if two conditions are met. First, the internal surface of the pump available to generate electroosmotic flow has to be maximized as EOF is a surface phenomenon; second, the resistance to pressure-driven flow has to be increased to avoid reverse flow under high back-pressure conditions. Both conditions can be met by packing a microfluidic channel with a stationary phase that has pores in the nanometer range. Paul et al. showed that pressures as high as 55 MPa could be generated [91]. It is assumed that pumps based on the same principle will also be implemented in the future, taking advantage of defined nanochannel structures instead of porous media.

9.3.3
Nanoporous Membranes as Controlled Diffusion Barriers

Stable, well-controlled, nanoporous membranes are an enabling technology for the fabrication of implantable biosensors, biomolecular filtration and environmental analysis systems [92]. The purpose of these membranes is a strict separation between the analytes of interest and interfering sample matrix constituents, such as particulate organic and inorganic matter or unwanted molecules interfering with the analysis process. Nanoporous membranes, also referred to as molecular sieves, achieve this separation by reducing or preventing mass transport through their pore system for molecules above a certain cut-off size or weight. An ideal membrane has a narrow pore-size distribution, a large pore-to-surface ratio to facilitate sample through flow and is both biocompatible and nonfouling.

Membranes regularly used are based on polymer sheets perforated by stretching or ion-track etching. Owing to their polymeric nature and their tortuous porosity, these membranes often suffer cracking and clogging due to mineral deposits or protein absorption [93]. Also, the pore size distribution is comparatively wide for cast membranes (~30%), whereas ion-track etched membranes exhibit tighter specifications at the expense of low pore densities [94]. Nanostructures fabricated by deposition, patterning and selective removal of sacrificial thin films provide an interesting alternative to conventional membrane materials. As outlined in Section 9.2.3.2, sacrificial layer technology is advantageous for the fabrication of 1D-confined nanostructures as no nanolithography is required to achieve the desired pore sizes.

In an early example of the fabrication of microchip filters, Kittilsland et al. proposed a self-aligning process for creating tortuous flow paths through a silicon membrane fabricated by anisotropic wet etching in conjunction with a selective etch-stop technique on p^+-doped silicon. The cut-off size (50 nm) was defined by a sacrificial thermal SiO_2 layer [95].

Tortuous flow paths can be avoided by using different sacrificial layer systems, which provide straight vertical nanopores through a silicon membrane [94, 96]. Similarly to approaches used in nanochannel fabrication [42], vertical steps are etched into a silicon substrate, followed by sacrificial oxide growth and polysilicon refill of the remaining trenches. Before selective etching of the sacrificial oxide, the wafer is anisotropically etched from the back side to create a thin silicon membrane. As oxide growth is extremely well controlled (about ±1 Å for a 50 Å thick layer), pore sizes down to 10 nm could be fabricated with <5% variation over a 4 in wafer.

These nanoporous membranes can be used as biomolecular filters to protect glucose sensors from interfering proteins that might adsorb to the sensing surface. Using a silicon membrane with 25.4 nm wide pores, it is possible to retain the protein albumin [molecular weight (MW) 66 000] whereas glucose (MW 180) diffuses freely through the membrane [94]. Conversely, engineered nanopore membranes can also act as controlled-diffusion sources for the administration of medical substances [97].

9.3.4 Fluid Dynamics in Nanotubes

Carbon nanotubes (CNT), discovered in 1991 by Iijima, are one-dimensional carbon fullerenes, i.e. convex cages of atoms with only hexagonal or pentagonal faces with a cylindrical shape [98]. While the first nanotubes discovered consisted of several concentric graphite cylinders (multi-walled CNT), the fabrication of single-wall CNTs was developed shortly after [99]. Typically, their inner diameter is <10 nm whereas their length can be hundreds of micrometers. Apart from remarkable tensile strength, nanotubes exhibit varying electrical properties (depending on the way in which the graphite structure spirals around the tube and other factors) and can be insulating, semiconducting or conducting. Carbon nanotubes are fabricated in a carbon arc similarly to fullerenes.

It has been demonstrated that carbon nanotubes can be filled by various liquids, including liquid salts [100] and metals. Multi-wall carbon tubes have been used to study wetting, evaporation, condensation and expansion phenomena by transmission electron microscopy (TEM) [101]. In this case, hydrothermal nanotubes were filled with water in addition to dissolved and free gases, which were encapsulated during a modified fabrication procedure [102].

Fan et al. recently described the fabrication of silica nanotubes with internal diameters ranging from 5 to 100 nm and lengths of several micrometers [103]. A theoretical model describing ion transport in silica nanotubes has also been described [85].

Whether nanofluidic networks based on CNTs can be fabricated and deployed still remains to be proven. Although considerable progress towards localized and directed growth and deposition of CNTs has been made, interconnections between single CNTs and the interface to the macroscopic world remain issues to be solved.

9.4 Study of Biomolecules in Nanofluidic Devices

One of the driving forces behind micro- and nanofluidics is the study of biomolecules for applications including drug development and screening and medical diagnostics. Some of these molecules, for instance DNA fragments, have a considerable weight and their size is similar to the structural dimensions of the devices discussed in this chapter. As an example, under physiological buffer conditions, the 48 kbp bacteriophage λ-DNA has a radius of gyration of 0.5 μm when coiled up into a sphere, a persistence length of 50 nm and a total contour length of 16 μm when stretched out [104]. Nanofluidic structures are small enough to force a single DNA molecule to uncoil when it passes an obstacle in the flow path, thus affecting its migration velocity. It should be noted that these interactions between chain-like molecules and a solid phase have been exploited for many years in gel electrophoresis, but nanofluidics allows the study of the dynamics of these processes in detail and under reproducible conditions.

9.4.1 Sizing of Biological Macromolecules

While electric field-induced separation analysis is straightforward for many analytes, most macromolecules of biochemical interest have a constant charge-to-size ratio. As each base element of the macromolecule has the same charge, the high total charge of the larger molecule is compensated by its additional size. Therefore, electrophoretic separation is generally carried out in sieving matrices, which interact physically with the macromolecules and slow them while they migrate through the separation channel. This dynamic interaction between analyte and sieving matrix leads to a size-dependent separation. Gels such as polyacrylamide and agarose are used for electrophoretic separation of DNA molecules, both in conventional capillaries and on microchips [105]. Separation is not only required for the size analysis of DNA fragments, but also finds wide application as a purification step to isolate a certain fragment from an unknown mixture.

As an alternative to randomly arranged gel matrices, nanometer-sized sieving structures, also called artificial gels, can be created by direct nanostructuring of the substrate. These engineered structures are long-term stable compared with gels, which are typically prepared freshly for each analysis. Also, DC electrophoretic separations in these devices might result in faster separation of large DNA than the pulsed-field techniques used for gels [106].

An early example by Turner et al. is composed of a dense array of 100 nm wide posts with 100 nm wide gaps fabricated in a 400 nm deep, 800 μm long and 500 μm wide separation channel (see Fig. 9.18) [107]. The fabrication procedure is based on e-beam lithography for obstacle definition and sacrificial layer etching to yield covered nanochannels as described in Section 9.2.3.2. A mixture of two fluorescently stained DNA molecules, 43 and 7.2 kbp long, was separated by electrophoresis at voltages between 2 and 15 V. The electrophoretic mobilities for both molecules turn out to be electric field dependent, with a maximum ratio of 1.8 at 5 V.

In a follow-up study, the same group used pulsed electric fields to analyze the migration dynamics of DNA molecules in the post array [108, 109]. Video observations showed that smaller molecules were fully inserted into the artificial gel region during a pulse duration, while parts of the longer molecules still remain outside. Consequently, the inserted part of the larger molecules is pulled out of the post array by relaxation of the molecule during the time span between two pulses, thus preventing center-of-mass movement (Fig. 9.19). Short molecules, however, do not recoil between two pulses and therefore migrate in steps to the anode. The separation of an unknown mixture of different DNA fragments can be achieved by slowly increasing the pulse time while monitoring the number of molecules that cross the boundary between the open channel section and the post array. Shorter molecules migrate fully into the entropically unfavorable region at shorter pulse times than longer molecules. Using a similar structure fabricated in quartz, Kaji et al. recently achieved electrophoretic separation of DNA fragments ranging from 1 to 38 kbp in only 180 s [110].

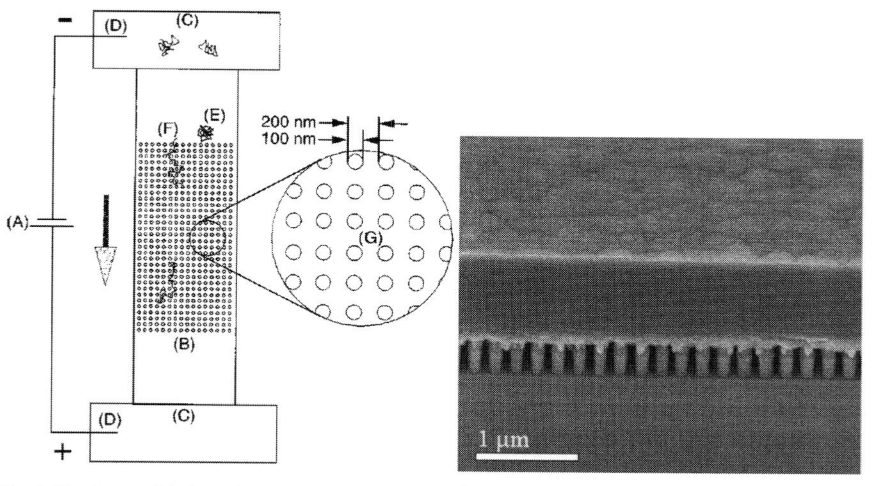

Fig. 9.18 A nanofabricated sieving structure for the separation of long DNA molecules. Left, top view with inlet and outlet regions and the separation sieve; right, close-up of the column array with 100 nm wide and 400 nm high posts. Reproduced with permission from [107]

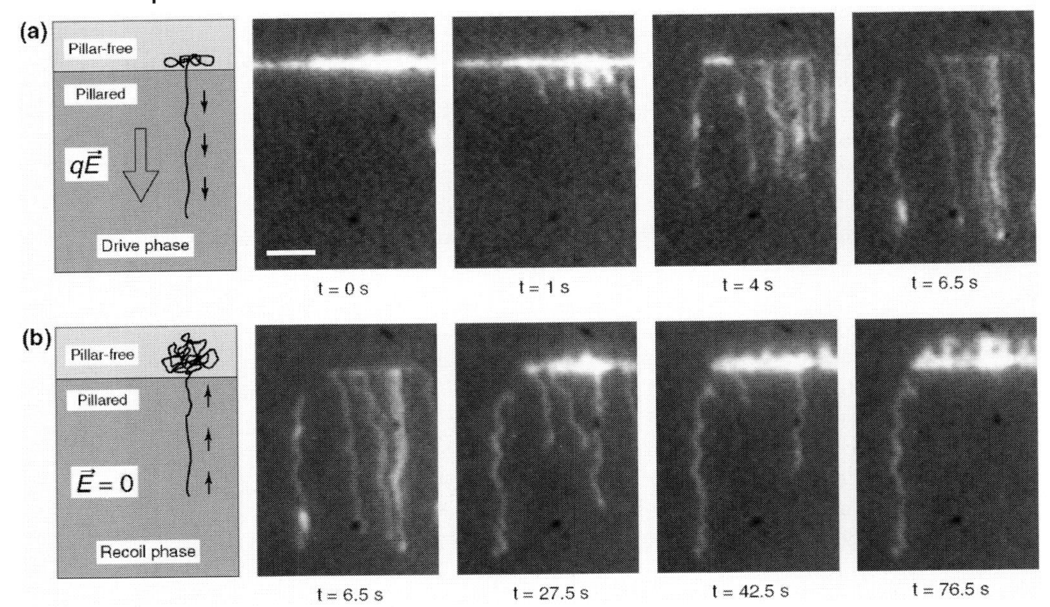

Fig. 9.19 (a) DNA molecules are driven into the dense pillar region, which occupies the bottom 80% of each frame. Six or seven molecules are shown, one (on the left) having been completely lodged in the unfavorable region. The scale bar represents 5 μm. (b) Entropic recoil at various stages ending when all the molecules have recoiled except the leftmost. Without the interaction between the two regions, this molecule experiences no center-of-mass motion and only slight contraction. Reproduced with permission from [109]

In a different device geometry with constrictions perpendicular to the substrate plane instead of posts, Han and Craighead achieved fast (~30 min) separation of large (5–200 kbp) DNA molecules by DC electrophoresis [111]. Entropic traps in the separation channel were formed by an alternation of deep (1.8 μm) and shallow (75–100 nm) regions. Surprisingly, smaller DNA fragments remain in the entropic traps for longer times than larger molecules and therefore have longer migration times (Fig. 9.20). This is in contradiction with conventional gel electrophoresis, where shorter molecules elute first. It is assumed that the part of the molecule in contact with the boundary of the thin region plays a crucial role in the escape of a DNA molecule from an entropic trap. Whenever Brownian motion introduces a sufficient number of DNA subunits into the thin region, which consequently has a high field strength, the escape of the whole molecule is initiated. Longer DNA molecules have a larger surface area in contact with the boundary and therefore have a higher probability to escape per unit time [112].

Although the methods mentioned above require separation, a direct length measurement on a single molecular level is also possible in nanofluidic channel networks. Introduction and stretching of fluorescently stained λ-DNA in nanochan-

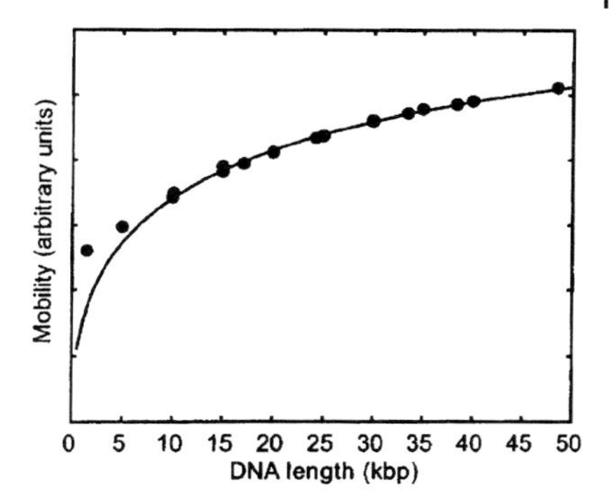

Fig. 9.20 Mobility of DNA in an entropic trapping channel plotted versus molecule length. The fit used (solid line) works well for larger DNA sizes, but fails for fragments shorter than 10 kbp. For these molecules, it is assumed that the radius of gyration approaches the depth of the shallow regions and trapping is reduced. Reproduced with permission from [111]

nels of 45–100 nm was successfully achieved by electrophoresis [40, 59]. Image analysis of video frames taken by a high-sensitivity CCD camera in an epi-fluorescent setup reveals the length of single molecules. Foquet et al. studied sizing of labeled DNA molecules in 1D nanochannels using fluorescence correlation spectroscopy at flow velocities of up to 5 mm/s [38]. By correlating the number of photon bursts with the DNA fragment length, unknown mixtures of DNA fragments can be analyzed in terms of size and molecular concentration in one run.

9.4.2 Migration of Biological Macromolecules Through Nanopores

While the examples presented in the previous section rely on fluorescence to detect DNA molecules, nanopores also allow a sensing scheme with a direct electrical readout for each molecule passing through the pore. If the pore is small enough and if the analyte molecule has different electrical conduction properties to the surrounding buffer, the electric current through the pore changes whenever the pore is blocked by a molecule passing through.

For successful operation, the pore membrane thickness has to be smaller than the molecule persistence length, 50 nm for double-stranded DNA, to prevent the molecule from passing through the pore in a folded configuration. Also, the pore diameter needs to be larger than the cross-section of the molecule ($\sim$2 nm). The challenges in fabricating such an instrument are tremendous, but considerable advances have been made recently by research groups at Harvard. Two approaches for nanopore formation were studied in parallel: natural ion channels inserted into lipid bilayers [113] and solid-state pores made by nanofabrication techniques [114]. Similar experimental setups were used in both cases, consisting of two polymer reservoirs filled with a buffer solution, which are separated by the membrane containing the nanopore. Analyte molecules are added to the cathode reservoir and are subsequently pulled through the nanopore by an electric field applied between the two reservoirs

Kasianowicz et al. deployed biochemical molecules as functional elements for pore fabrication. The pore is formed within minutes by self-assembly of an α-hemolysin channel obtained from *Staphylococcus aureus* bacteria, which is inserted into a 5 nm thick phospholipid bilayer separating the two reservoirs [113]. The resulting channel diameter is in the region of 1.5 nm, which lets single-stranded DNA pass but retains double-stranded DNA. It is experimentally possible to unzip double-stranded DNA physically by pulling only one strand through the pore while the other remains on the opposite side of the membrane [115]. Owing to the translocation mechanism inside the α-hemolysin channel, translocation times depend not only on the size of a fragment, but also on its composition. Polydeoxycytosines passed the pore three times faster than polydeoxyadenines of the same length [116].

Looking for a mechanically and chemically more robust solution, the formation of solid-state silicon nitride nanopores with diameters down to 3 nm was studied using ion-beam sculpting [117]. The process starts with focused-ion beam etching of an initial 100 nm pore into a 500 nm thick, low-stress Si_3N_4 membrane supported by a silicon wafer. This pore is subsequently closed by a lateral atomic flow of silicon nitride matter induced by FIB bombardment, which forms a thin membrane a few nanometers thick growing from the edge of the pore to its center. In order to stop the ion-beam sculpting process reliably before the pore closes completely, the ion current through the remaining pore is constantly monitored and fed back to the beam source. Fig. 9.21 a shows a transmission electromicrograph of the resulting pore [114].

Fig. 9.21 b shows the evolution of the electric current flowing through the nanopore at an applied bias of 120 mV. The current drops by ΔI_b for a period t_d due to a 1 μm long, 3 kbp double-stranded DNA fragment passing through the pore and reducing the electrical conductivity temporarily. Surprisingly however, it was observed that despite the small diameter of the pore, a minority of DNA fragments passed the pore in a folded 'hairpin' configuration. These events can be discriminated owing to their shorter blocking times and larger current drops.

Recently, Storm et al. introduced a method for the fabrication of SiO_2 pores with single-nanometer precision, which will allow handling of single-stranded DNA [118]. An electron beam is used to fluidize the silicon oxide of a larger (20 nm diameter) pore, causing a reduction in the diameter due to surface tension. The process can be monitored in real time under a transmission electron microscope, which is also used to provide the electron beam.

Extending the concept to single-molecule DNA sequencing could have a tremendous impact, as sequencing today is a laborious procedure involving multiple translation of the genetic information. If the pore membrane could be made thin enough to host only a few base-pairs, discrimination of the four base-types A, G, C and T based on their electrical conductivity could be possible. Alternatively, the integration of nanometer-scale electrodes into the pore for tunneling measurements perpendicular to the DNA chain is proposed. However, even if the technological issues can be solved, complex biological molecules such as DNA still need to be tamed to pass through the detector in a straight configuration.

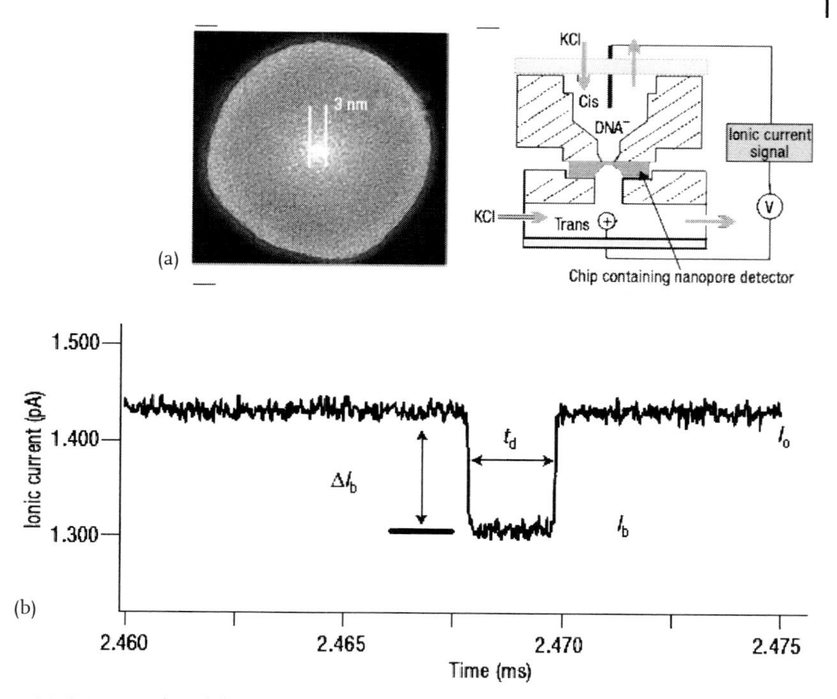

Fig. 9.21 (a) A 3 nm wide solid-state nanopore in a 5–10 nm thick silicon nitride membrane clamped as a separation between two liquid reservoirs can be used to detect DNA fragments passing the pore by monitoring the ionic current through the constriction. (b) Current blockade due to a 1 μm long, 3 kbp double-stranded DNA fragment passing through the pore. Reproduced with permission from [114]

9.5 Conclusions

The exciting aspect of fluidics in the nanometer range is the interaction between structural elements, solution chemistry and physicochemical analyte properties on a molecular length scale. In principle, the technology available today allows the inspection and manipulation of larger molecules one by one. This is the result of a common effort in the research community to combine and refine techniques from microelectronics, microtechnology and nanotechnology as successful tools for nanofluidic device fabrication. However, there is still a long way to go to understand fully the implications and the potential of this continuing reduction in feature size for physical and chemical processes.

Also, researchers should not forget that not all applications benefit from miniaturization. For an imaginary nanometer-scale capillary electrophoresis system, for example, the question arises of whether a sample volume of 1 aL (10^{-18} L or 100 nm^3) is still representative for the analysis. If the sample contains an analyte at a concentration of 10 μM, only six analyte molecules are present in the sample volume on average.

Despite these issues, nanofluidics provides us with a vast variety of opportunities in nanoscale liquid handling and single-molecule analysis. There are good indications that nanofluidics will establish itself as a strong field of research at the boundary between physics, engineering, chemistry and biology.

9.6 References

1 G.J.M. Bruin, *Electrophoresis* **2000**, *21*, 3931–3951.

2 E. Verpoorte, N.F. de Rooij, *Proc. IEEE* **2003**, *91*, 930–953.

3 D.J. Beebe, G.A. Mensing, G.M. Walker, *Annu. Rev. Biomed. Eng.* **2002**, *4*, 261–286.

4 A. Manz, D.J. Harrison, E.M.J. Verpoorte, J.C. Fettinger, A. Paulus, H. Ludi, H.M. Widmer, *J. Chromatogr.* **1992**, *593*, 253–258.

5 D.J. Harrison, K. Fluri, K. Seiler, Z.H. Fan, C.S. Effenhauser, A. Manz, *Science* **1993**, *261*, 895–897.

6 S.C. Jacobson, R. Hergenröder, L.B. Koutny, J.M. Ramsey, *Anal. Chem.* **1994**, *66*, 1114–1118.

7 J.P. Brody, P. Yager, *Sens. Actuators A* **1997**, *58*, 13–18.

8 H.G. Craighead, *J. Vac. Sci. Technol. A* **2003**, *21*, S216–S221.

9 G.M. Whitesides, *Nat. Biotechnol.* **2003**, *21*, 1161–1165.

10 W. Menz, J. Mohr, O. Paul, *Microsystem Technology*, Weinheim: Wiley-VCH, **2001**.

11 M.J. Madou, *Fundamentals of Microfabrication the Science of Miniaturization*, 2nd edn.; Boca Raton, FL: CRC Press, **2002**.

12 N.-T. Nguyen, S.T. Wereley, *Fundamentals and Applications of Microfluidics*; Boston: Artech House, **2002**.

13 H. Iwasaki, T. Yoshinobu, K. Sudoh, *Nanotechnology* **2003**, *14*, R55–R62.

14 S. Kuiper, H. van Wolferen, C. van Rijn, W. Nijdam, M. Krijnen, M. Elwenspoek, *J. Micromech. Microeng.* **2001**, *11*, 33–37.

15 E.Y. Wu, E.J. Nowak, A. Vayshenker, W.L. Lai, D.L. Harmon, *IBM J. Res. Dev.* **2002**, *46*, 287–298.

16 S.C. Jacobson, A.W. Moore, J.M. Ramsey, *Anal. Chem.* **1995**, *67*, 2059–2063.

17 D.J. Harrison, P.G. Glavina, A. Manz, *Sens. Actuators B* **1993**, *10*, 107–116.

18 A. Daridon, V. Fascio, J. Lichtenberg, R. Wutrich, H. Langen, E. Verpoorte, N.F. de Rooij, *Fresenius' J. Anal. Chem.* **2001**, *371*, 261–269.

19 L. Ceriotti, K. Weible, N.F. de Rooij, E. Verpoorte, *Microelectron. Eng.* **2003**, *67–68*, 865–871.

20 J.K. Chen, K.D. Wise, *IEEE Trans. Biomed. Eng.* **1997**, *44*, 770–774.

21 M.J. de Boer, R.W. Tjerkstra, J.W. Berenschot, H.V. Jansen, C.J. Burger, J.G.E. Gardeniers, M. Elwenspoek, A. van den Berg, *J. Microelectromech. Syst.* **2000**, *9*, 94–103.

22 R.W. Tjerkstra, J.G.E. Gardeniers, J.J. Kelly, A. van den Berg, *J. Microelectromech. Syst.* **2000**, *9*, 495–501.

23 J. Lichtenberg, G. Lammel, M. Ouveley, P. Renaud, E. Verpoorte, N.F. de Rooij, in: *Proceedings of the Eurosensors XIV, Copenhagen*; **2000**, pp. 259–261.

24 J. Haneveld, H. Jansen, E. Berenschot, N. Tas, M. Elwenspoek, *J. Micromech. Microeng.* **2003**, *13*, S62–S66.

25 T.S. Hug, D. Parrat, P.-A. Künzi, U. Staufer, E. Verpoorte, N.F. de Rooij, in: *Proceedings of the Micro Total Analysis Systems Conference, Squaw Valley, CA*; Transducers Research Foundation, Cleveland, OH, **2003**, pp. 29–32.

26 R. Qiao, N.R. Aluru, *J. Chem. Phys.* **2003**, *118*, 4692–4701.

27 J.P. Alarie, A.B. Hmelo, S.C. Jacobson, A.P. Baddorf, L. Feldman, J.M. Ramsey, in: *Proceedings of the Micro Total Analysis Systems Conference, Squaw Valley,*

CA; OH: Transducers Research Foundation, **2003**, pp. 9–12.

28 V. Lehmann, *J. Electrochem. Soc.* **1993**, *140*, 2836–2843.

29 H. Foll, M. Christophersen, J. Carstensen, G. Hasse, *Mater. Sci. Eng. R-Rep.* **2002**, *39*, 93–141.

30 V. Lehmann, *Electrochemistry of Silicon: Instrumentation, Science, Materials and Applications*; Weinheim: Wiley-VCH, **2002**.

31 V. Lehmann, H. Foll, *J. Electrochem. Soc.* **1990**, *137*, 653–659.

32 V.V. Starkov, *Phys. Status Solidi A* **2003**, *197*, 22–26.

33 H. Berney, A. Kemna, M. Hill, E. Hynes, M. O'Neill, W. Lane, *Sens. Actuators A* **1999**, *76*, 356–364.

34 M.B. Stern, M.W. Geis, J.E. Curtin, *J. Vac. Sc.i Technol. B* **1997**, *15*, 2887–2891.

35 J.W. Berenschot, N.R. Tas, T.S.J. Lammerink, M. Elwenspoek, A. van den Berg, *J. Micromech. Microeng.* **2002**, *12*, 621–624.

36 R.B.M. Schasfoort, S. Schlautmann, L. Hendrikse, A. van den Berg, *Science* **1999**, *286*, 942–945.

37 N.R. Tas, P. Mela, T. Kramer, J.W. Berenschot, A. van den Berg, *Nano Lett.* **2003**, *3(11), 1537–1540.*

38 M. Foquet, J. Korlach, W. Zipfel, W.W. Webb, H.G. Craighead, *Anal. Chem.* **2002**, *74*, 1415–1422.

39 W.J. Nam, S. Bae, K. Kalkan, S.J. Fonash, *J. Vac. Sci. Technol. A* **2001**, *19*, 1229–1233.

40 W.L. Li, J.O. Tegenfeldt, L. Chen, R.H. Austin, S.Y. Chou, P.A. Kohl, J. Krotine, J.C. Sturm, *Nanotechnology* **2003**, *14*, 578–583.

41 N.R. Tas, J.W. Berenschot, P. Mela, H.V. Jansen, M. Elwenspoek, A. van den Berg, *Nano Lett.* **2002**, *2*, 1031–1032.

42 C. Lee, E.-H. Yang, N.V. Myung, T. George, *Nano Lett.* **2003**, *3*, 1339–1340.

43 H. Lorenz, M. Despont, N. Fahrni, J. Brugger, P. Vettiger, P. Renaud, *Sens. Actuators A* **1998**, *64*, 33–39.

44 S. Metz, R. Holzer, P. Renaud, *Lab Chip* **2001**, *1*, 29–34.

45 B.E.J. Alderman, C.M. Mann, D.P. Steenson, J.M. Chamberlain, *J. Micromech. Microeng.* **2001**, *11*, 703–705.

46 P. Selvaganapathy, Y.S.L. Ki, P. Renaud, C.H. Mastrangelo, *J. Microelectromech. Syst.* **2002**, *11*, 448–453.

47 B. Ilic, D. Czaplewski, M. Zalalutdinov, B. Schmidt, H.G. Craighead, *J. Vac. Sci. Techno. B* **2002**, *20*, 2459–2465.

48 H. Becker, C. Gartner, *Electrophoresis* **2000**, *21*, 12–26.

49 G.M. Whitesides, E. Ostuni, S. Takayama, X.Y. Jiang, D.E. Ingber, *Annu. Rev. Biomed. Eng.* **2001**, *3*, 335–373.

50 X.M. Zhao, Y.N. Xia, G.M. Whitesides, *J. Mater. Chem.* **1997**, *7*, 1069–1074.

51 B. Michel, A. Bernard, A. Bietsch, E. Delamarche, M. Geissler, D. Juncker, H. Kind, J.P. Renault, H. Rothuizen, H. Schmid, P. Schmidt-Winkel, R. Stutz, H. Wolf, *IBM J. Res. Dev.* **2001**, *45*, 697–719.

52 M.A. Unger, H.-P. Chou, T. Thorsen, A. Scherer, S.R. Quake, *Science* **2000**, *288*, 113–116.

53 C.S. Effenhauser, G.J.M. Bruin, A. Paulus, M. Ehrat, *Anal. Chem.* **1997**, *69*, 3451–3457.

54 D.C. Duffy, J.C. McDonald, O.J.A. Schueller, G.M. Whitesides, *Anal. Chem.* **1998**, *70*, 4974–4984.

55 J.C. McDonald, D.C. Duffy, J.R. Anderson, D.T. Chiu, H.K. Wu, O.J.A. Schueller, G.M. Whitesides, *Electrophoresis* **2000**, *21*, 27–40.

56 T.W. Odom, V.R. Thalladi, J.C. Love, G.M. Whitesides, *J. Am. Chem. Soc.* **2002**, *124*, 12112–12113.

57 G. Ocvirk, M. Munroe, T. Tang, R. Oleschuk, K. Westra, D.J. Harrison, *Electrophoresis* **2000**, *21*, 107–115.

58 S.Y. Chou, P.R. Krauss, P.J. Renstrom, *Science* **1996**, *272*, 85–87.

59 H. Cao, Z.N. Yu, J. Wang, J.O. Tegenfeldt, R.H. Austin, E. Chen, W. Wu, S.Y. Chou, *Appl. Phys. Lett.* **2002**, *81*, 174–176.

60 Z.N. Yu, L. Chen, W. Wu, H.X. Ge, S.Y. Chou, *J. Vac. Sci. Technol. B* **2003**, *21*, 2089–2092.

61 H. Becker, L.E. Locascio, *Talanta* **2002**, *56*, 267–287.

62 V. Studer, A. Pepin, Y. Chen, *Appl. Phys. Lett.* **2002**, *80*, 3614–3616.

63 H. Schift, C. David, M. Gabriel, J. Gobrecht, L.J. Heyderman, W. Kaiser, S. Koppel, L. Scandella, *Microelectron. Eng.* **2000**, *53*, 171–174.

64 R.M. McCormick, R.J. Nelson, M.G. Alonso-Amigo, D.J. Benvegnu, H.H. Hooper, *Anal. Chem.* **1997**, *69*, 2626–2630.

65 Y.N. Xia, G.M. Whitesides, *Angew. Chem., Int. Ed. Engl.* **1998**, *37*, 551–575.

66 M. Tormen, T. Borzenko, B. Steffen, G. Schmidt, L.W. Molenkamp, *Appl. Phys. Lett.* **2002**, *81*, 2094–2096.

67 M. Tormen, T. Borzenko, B. Steffen, G. Schmidt, L.W. Molenkamp, *Microelectron. Eng.* **2002**, *61–62*, 469–473.

68 T.W. Odom, J.C. Love, D.B. Wolfe, K.E. Paul, G.M. Whitesides, *Langmuir* **2002**, *18*, 5314–5320.

69 H. Schmid, B. Michel, *Macromolecules* **2000**, *33*, 3042–3049.

70 E. Delamarche, H. Schmid, A. Bietsch, N.B. Larsen, H. Rothuizen, B. Michel, H.A. Biebuyck, *J. Phys. Chem. B* **1998**, *102*, 3324.

71 Y. Xia, X.-M. Zhao, G.M. Whitesides, *Microelectron. Eng.* **1996**, *32*, 255.

72 G.T.A. Kovacs, *Micromachined Transducers Sourcebook*; New York: McGraw-Hill, **1998**.

73 R.K. Shah, A.L. London, *Laminar Flow Forced Convection in Ducts. A Source Book for Compact Heat Exchanger Analytical Data*; New York: Academic Press, **1978**.

74 A. Hibara, T. Saito, H.B. Kim, M. Tokeshi, T. Ooi, M. Nakao, T. Kitamori, *Anal. Chem.* **2002**, *74*, 6170–6176.

75 A. Manz, C.S. Effenhauser, N. Burggraf, D.J. Harrison, K. Seiler, K. Fluri, *J. Micromech. Microeng.* **1994**, *4*, 257–265.

76 K. Seiler, Z.H. H. Fan, K. Fluri, D.J. Harrison, *Anal. Chem.* **1994**, *66*, 3485–3491.

77 D.J. Shaw, *Introduction to Colloid and Surface Chemistry*, 4th edn.; Oxford: Butterworth Heinemann, **1992**.

78 P.T. Kissinger, W.R. Heineman, *Laboratory Techniques in Electroanalytical Chemistry*, 2nd edn.; New York: Marcel Dekker, **1996**.

79 C.S. Effenhauser, A. Manz, H.M. Widmer, *Anal. Chem.* **1993**, *65*, 2637–2642.

80 V. Pacáková, P. Coufal, K. Stulík, B. Gas, *Electrophoresis* **2003**, *24*, 1883–1891.

81 D.N. Heiger, *High Performance Capillary Electrophoresis – An Introduction*; Palo Alto, CA: Hewlett-Packard, **1997**.

82 D. Burgreen, F.R. Nakache, *J. Phys. Chem.* **1964**, *68*, 1084–1091.

83 A.T. Conlisk, J. McFerran, Z. Zheng, D. Hansford, *Anal. Chem.* **2002**, *74*, 2139–2150.

84 Z. Zheng, D.J. Hansford, A.T. Conlisk, *Electrophoresis* **2003**, *24*, 3006–3017.

85 H. Daiguji, P. Yang, A. Majumdar, *Nano Lett.* **2004**, *4(1)*, 137–142.

86 J.M. Ramsey, J.P. Alarie, S.C. Jacobson, N.J. Peterson, in: *Proceedings of the Micro Total Analysis Systems Conference, Nara, Japan*; Dordrecht: Kluwer, **2002**, pp. 314–316.

87 R.M. Guijt, J. Lichtenberg, N.F. de Rooij, E. Verpoorte, E. Baltussen, G.W.K. van Dedem, *J. Assoc. Lab. Autom.* **2002**, *7*, 62–64.

88 P.J. Kemery, J.K. Steehler, P.W. Bohn, *Langmuir* **1998**, *14*, 2884–2889.

89 T.C. Kuo, D.M. Cannon, M.A. Shannon, P.W. Bohn, J.V. Sweedler, *Sens. Actuators A* **2003**, *102*, 223–233.

90 T.-C. Kuo, J. Donald M. Cannon, Y. Chen, J.J. Tulock, M.A. Shannon, J.V. Sweedler, P.W. Bohn, *Anal. Chem.* **2003**, *75*, 1861–1867.

91 P.H. Paul, D.W. Arnold, D.W. Neyer, K.B. Smith, in: *Proceedings of the Micro Total Analysis Systems Conference*: Dordrecht: Kluwer, **2000**, pp. 583–590.

92 J. Lichtenberg, N.F. de Rooij, E. Verpoorte, *Talanta* **2002**, *56*, 233–266.

93 F. Moussy, D.J. Harrison, R.V. Rajotte, *Int.J. Artif. Organs* **1994**, *17*, 88–94.

94 T.A. Desai, D.J. Hansford, L. Leoni, M. Essenpreis, M. Ferrari, *Biosens. Bioelectron.* **2000**, *15*, 453–462.

95 G. Kittilsland, G. Stemme, B. Norden, *Sens. Actuators A* **1990**, *23*, 904–907.

96 T.A. Desai, D.J. Hansford, L. Kulinsky, A.H. Nashat, G. Rasi, J. Tu, Y. Wang, M. Zhang, M. Ferrari, *Biomed. Microdev.* **1999**, *2*, 11–40.

97 F.J. Martin, C. Grove, *Biomed. Microdev.* **2001**, *3*, 97–108.
98 S. Iijima, *MRS Bull.* **1994**, *19*, 43–49.
99 S. Iijima, T. Ichihashi, *Nature* **1993**, *363*, 603–605.
100 D. Ugarte, A. Chatelain, W.A. de Heer, *Science* **1996**, *274*, 1897–1899.
101 Y. Gogotsi, J.A. Libera, A. Guvenc-Yazicioglu, C.M. Megaridis, *Appl. Phys. Lett.* **2001**, *79*, 1021–1023.
102 Y. Gogotsi, J.A. Libera, M. Yoshimura, *J. Mater. Res.* **2000**, *15*, 2591–2594.
103 R. Fan, D. Li, A. Majumdar, P. Yang, *J. Am. Chem. Soc.* **2003**, *125*, 5254–5255.
104 R. Verma, J.C. Crocker, T.C. Lubensky, A.G. Yodh, *Phys. Rev. Lett.* **1998**, *81*, 4004–4007.
105 J.P. Landers, *Handbook of Capillary Electrophoresis*; Boca Raton, FL: CRC Press, **1994**.
106 Y. Kim, M. Morris, *Electrophoresis* **1996**, *17*, 152–160.
107 S.W. Turner, A.M. Perez, A. Lopez, H.G. Craighead, *J. Vac. Sci. Technol. B* **1998**, *16*, 3835–3840.
108 M. Cabodi, S.W.P. Turner, H.G. Craighead, *Anal. Chem.* **2002**, *74*, 5169–5174.
109 S.W.P. Turner, M. Cabodi, H.G. Craighead, *Phys. Rev. Lett.* **2002**, *88*, 128103.
110 N. Kaji, Y. Takamura, Y. Horiike, H. Nakanishi, T. Nishimoto, Y. Baba, in *Proceedings of the Micro Total Analysis Systems Conference, Squaw Valley, CA*; OH: Transducers Research Foundation, **2003**, pp. 1315–1318.
111 J.Y. Han, H.G. Craighead, *Anal. Chem.* **2002**, *74*, 394–401.
112 J. Han, H.G. Craighead, *Science* **2000**, *288*, 1026–1029.
113 J.J. Kasianowicz, E. Brandin, D. Branton, D.W. Deamer, *Proc. Natl. Acad. Sci. USA* **1996**, *93*, 13770–13773.
114 J. Li, M. Gershow, D. Stein, E. Brandin, J.A. Golovchenko, *Nat. Mater.* **2003**, *2*, 611–615.
115 A.F. Sauer-Budge, J.A. nyamwanda, D.K. Lubensky, D. Branton, *Phys. Rev. Lett.* **2003**, *90*, 238101.
116 A. Meller, D. Branton, *Electrophoresis* **2002**, *23*, 2583–2591.
117 J. Li, D. Stein, C. McMullan, D. Branton, M.J. Aziz, J.A. Golovchenko, *Nature* **2001**, *412*, 166–169.
118 A.J. Storm, J.H. Chen, X.S. Ling, H.W. Zandbergen, C. Dekker, *Nat. Mater.* **2003**, *2*, 537–540.

10
Carbon Nanotubes and Sensors: a Review

J. R. Stetter, Sensor Research Group and Center for Electrochemical Science and Engineering, Illinois Institute of Technology, Chicago, IL 60616, USA
G.J. Maclay, Quantum Fields LLC, Richland Center, WI 53581, USA

Abstract

Carbon Nanotubes, CNTs, are currently the topic of an ever-increasing number of publications. Of the new device applications, sensors may be one of the earliest commercial successes although it is obvious that many devices with a variety of applications will emerge. Sensors, in general, do not demand an exact replication of electronic properties but rather a stable differential property with and without the measurand. This work describes the CNT and its origin with an emphasis on CNT use in the development of devices like sensors. Direct incorporation into structures, addition of reactive and passivation layers to tailor reactivity and properties, and CNT use as a template of nano-dimensions is discussed. Early results suffer from the lack of control experiments which cloud interpretations and hinder our understanding of the mechanisms of CNT device response. However, the promise of single molecule detection, ultra low power and size, and a new sensor platform with a potentially broad applicability in gases and liquids will encourage rapid progress in the use of CNT nanotechnology.

Keywords
nanotechnology; nanotubes; carbon; sensors, chemical; biological; microelectromechanical; systems, sensor arrays; CNT, properties of; applications of.

Advanced Micro and Nanosystems. Vol. 1
Edited by H. Baltes, O. Brand, G.K. Fedder, C. Hierold, J. Korvink, O. Tabata

ISBN: 3-527-30746-X

10.1 Introduction

Nanoscience and nanotechnology have created a great deal of excitement, activity, promise and expectation in the past few years. The immense academic interest has resulted in many publications and patents. Commercial interest has stimulated activities to produce practical nanomaterials and nanodevices, including sensors. This review discusses nanotechnology and sensors with an emphasis on carbon nanotubes (CNTs) and chemical and biochemical sensors made from carbon nanotubes.

In general, the literature on nanostructures, nanotubes, CNTs and sensor applications is growing exponentially. The literature on nanotubes for all applications is exploding and, like most new technologies, much of the growth has been first in the scientific and engineering novelties and later in more applied studies and ultimately realistic applications. There have been a number of reviews of the electrical and mechanical properties of CNTs and their preparation [1–6].

A recent report states that the global market for nanotubes in 2002 reached ~ $12 million. This market is predicted to grow exponentially over the next 3 years, reaching perhaps $700 million in 2005 [7]. Current commercial applications for CNTs include conductive polymers, advanced composites, fibers and displays. Industries already utilizing these applications include automotive, aerospace, household appliances, sporting goods, telecommunications equipment and medical products. Some of the interesting products include tennis racquets, fishing rods, soil-resistant fabric for pants, sun-block lotions and creams and heat shields for the space shuttle. CNTs are also seen as a potential foundation for a new generation of high-performance integrated electronic devices that will allow the electronics industry to keep Moore's law alive for another two decades. Because of their stability, high current capacity and low emission threshold, CNTs are expected to have application as field emission tips for flat panel displays, lamps, X-ray sources and microwave generators. Because of their large surface area, chemical stability, high electrical conductivity and high strength, CNTs are expected to be used in electrochemical devices including batteries, supercapacitors, fuel cells and hydrogen storage cells [8]. For some recent emphasis on nanotechnology, we need only note that the US National Nanotechnology Initiative (US NNI) [9] claims to be the first fundamental science-driven initiative announced by a US President and resulted in $470 million in funding in the USA in 2000 and will be $847 million in the President's 2004 budget. Other countries have similar initiatives and the US National Science Foundation (NSF) has predicted a $1 trillion market in 2015 for nanotechnology! In the past few years, worldwide industry has become a strong supporter of nanotechnology and we are beginning to see its potential for broader societal impacts.

However, the old approaches to catalysts and polymers were 'nanotechnologies' in the sense that the chemistry of tiny particles and large molecules has typical dimensions in nanometers, just as the first useful polymers were biopolymers. So is 'nanotechnology' new or old? Is it special or the same stuff with a new and catchy name? The answer is that both viewpoints contain some truth. In one sense, we look at the mature nanotechnologies and see them as improvements that were made years ago, processes that resulted in great adsorbents or catalysts, but we could not measure why. Today we see the new structures and materials under the scanning electron microscope (SEM) and examine the carbon nanotube knowing where every atom is placed. In fact, we always stand on the shoulders of those scientists and engineers who went before us. We have only recently been able to 'see' clearly the nanoworld, enabled by the improvements in SEM and transmission electron microscope (TEM) instrumentation and the invention of the atomic force microscope (AFM) and related techniques. New materials and methods of fabrication such as fullerenes and even the nanotubes themselves preceded the term 'nanotechnology', but working with them at atomistic dimensions and utilizing their quantum properties is indeed new. Tab. 10.1 provides a possible clarification of the 'new' and 'old' nanomaterials.

Tab. 10.1 Comparison of some new and old 'nanomaterials'

Old	*New*
Carbon black	Carbon nanotubes
Fused silica	Quantum dots
Aerogels, polymers	Molecular wires
TiO_2 and powders	Thin films
Molecular sieves	SWCNT sensors
Nanocrystalline SnO_2 sensor	Unprecedented properties
Improvements	Disruptive technology

So what is next? Is there a lesson or useful observation here? Perhaps two. First, a product such as the SnO_2 nanocrystalline sensor did not wait for the 'invention' of nanotechnology or the 'declaration' of its arrival! Successful products are always 'outside-in' driven, and this means that they are developed from technology but are pulled through by the market. We do not need to see what unique structures our process produces to use it, but we do learn to control a process and a structure so it becomes useful. The same will be true for nanotechnology. Experiments and research will lead to more technology and a broader base from which to pull products to the market, but the unique new sensor products will be pulled through by outside influences and demands even if the base technology is not well understood and it may take many more years to understand and unravel its many mysteries completely.

The second observation is that perhaps a useful distinction between new and old nanotechnology comes in the form of differentiating between incremental and revolutionary improvements. Revolutionary developments cause a paradigm shift in thinking and ultimately in societal behavior. The current use of nanomaterials allows us to explore the limiting behavior of macroscopic systems as the characteristic dimensions are reduced into the nanometer regime, where typical dimensions are 1–100 nm. As the dimensions are reduced, the usual bulk behavior of materials disappears and new phenomena appear, often providing advantages for attaining engineering goals. The recent focus on and funding for 'nanotechnology' has led to unique materials, new processes and novel structures that enable us to reach into uncharted territory in both fundamental and applied research areas. The truly new tools and research are examining 'nano' particles and structures from the ground up, atom by atom, and not as the result of a somewhat blind process as may have been true in the 'old' nanotechnology.

The new nanostructures are of substantial interest academically and technologically. Recent CNT publications have discussed the preparation, characterization, functionalization, manipulation and applications in fibers, fabrics and electronic and optical devices including sensors. Since sensing depends on a differential response, i.e. the difference in the response with and without analyte, we predict that applications in chemical and biochemical sensing will be realized sooner than applications such as transistors where exact control of the absolute properties of each CNT may be demanded. In the near future, the CNT will touch all manner of chemical and biochemical sensing.

We present here a brief introduction to the CNT, the structures themselves, their properties and devices with special attention paid to the characteristics that will yield practical applications. Then we discuss some of the fundamental electronic structures made from CNTs and the application of the structures to make devices, sensors and particularly chemical and biochemical sensors.

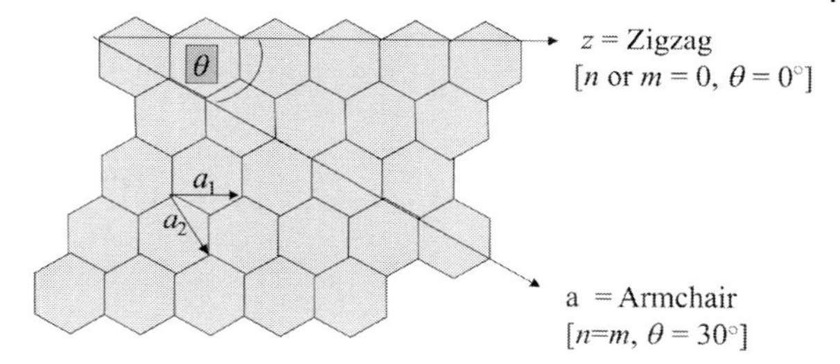

Fig. 10.1 Formation of a single-walled CNT structure. Roll the graphene sheet on an axis *a* or *z* to form a tube. CNT properties depend on chiral vector (na_1+ma_2) and chiral angle θ. Tube length is perpendicular to the roll axis. When rolled on the axis where $\theta=30°$, an armchair structure results

10.2 Structure and Properties of CNTs

10.2.1 Structure

CNTs are rolled sheets of graphene, a cylindrical fullerene. The formation of a single-walled carbon nanotube (SWCNT) is illustrated in Fig. 10.1. The armchair and the zigzag CNTs result from rolling the graphene on an axis with a chiral angle of 30° or 0°, respectively. The CNTs grown from metal catalysts are normally open at the ends because the catalyst particle is located at the end. CNTs can be closed at the ends but this requires the inclusion of five-membered rings in which each carbon atom is connected to one five- and two six-membered rings, the buckeyball structure, as shown in Fig. 10.2. The CNT structure is unique because each CNT is a single molecule with an exactly known conformation of each atom in the structure. The rolled graphene sheet can be described by the chiral

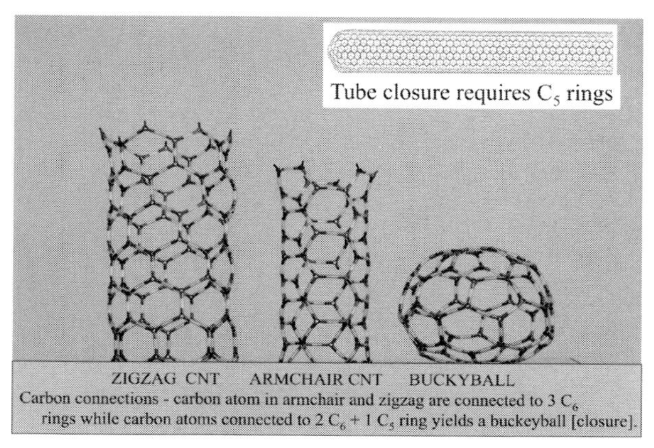

Fig. 10.2 Fullerene CNT structures

vector **C** that connects two crystallographically equivalent sites and is expressed as the sum of two unit vectors, $\boldsymbol{a}_1$ and $\boldsymbol{a}_2$, such that $\mathbf{C} = m\mathbf{a}_1 + n\mathbf{a}_2$. The values of n and m determine the diameter, D, and the chiral angle, θ, of the tube [10a]:

$$D = (3^{1/2}/\pi) d_{CC} (n^2 + m^2 + nm)^{1/2} \tag{1}$$

$$\theta = \arctan\{(n - m)/[3^{1/2}(n + m)]\} \tag{2}$$

The observed electrical and spectroscopic properties depend upon (n,m) or equivalently on the nanotube diameter and chiral angle. Nanotubes typically have diameters from 1 to 100 nm and lengths up to hundreds of micrometers have been reported [10b].

The CNT structure can be altered dramatically by defects and the physical change that is observed when the six-membered ring pattern is interrupted is illustrated in Fig. 10.3. A regular insertion of five-membered rings such that each carbon atom is a member of one five- and two six-membered rings result in closure. The buckeyball shape, as mentioned above, and CNT endcaps formed in this way. Insertion of a seven-membered ring defect results in divergence of the structure and these have not been observed experimentally but can indeed be envisioned. The inclusion of a C_7 + C_5 defect has also been postulated. Other defects of carbon are possible and there are a host of impurities in the current structures and nearby surroundings. Nanotubes have also been made from other materials such as BN and zinc oxide, leading to speculation about nanotubes, dopants and other defects that are important in sensor design and performance.

Since the structure determines the properties to a large extent, efforts have been made to synthesize all types of nanotubes and some examples are given in Fig. 10.4, taken from various websites as indicated. Illustrated are tubes with single walls (SWCNTs), multiwalled carbon nanotubes (MWCNTs) that are comprised of concentrically nested SWCNTs and CNTs with endcaps and fillings (nanotestubes). Addi-

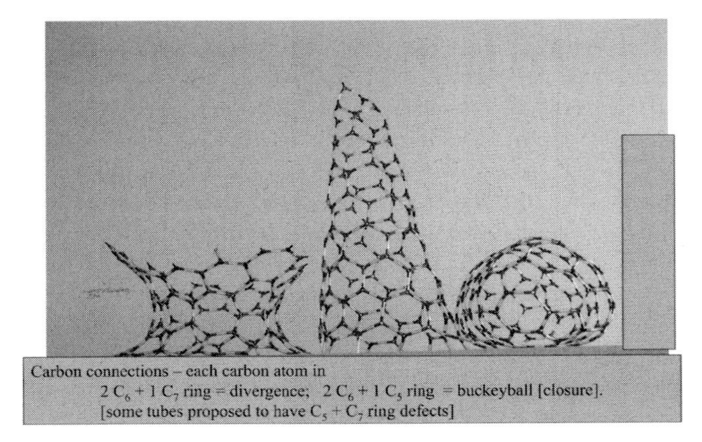

Fig. 10.3 CNT structures and defects

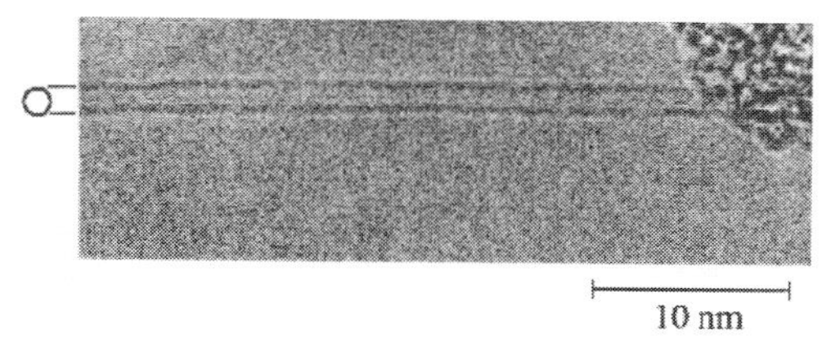

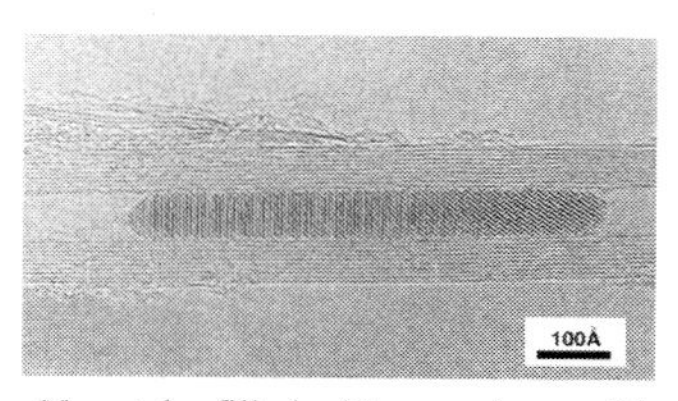

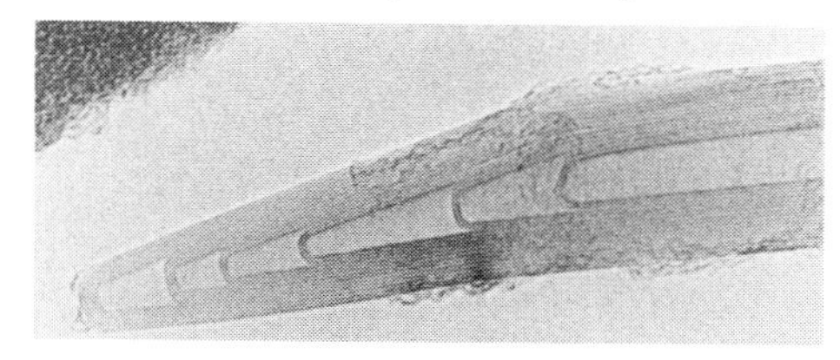

Fig. 10.4 Examples of different types of nanotubes

tional structures, including nanoropes, consisting of numerous intertwined SWCNTs, nanohorns, nanofibers and torroids and cones, have also been observed. The use of different conditions for the nanotube synthesis leads to different tube structures and different physical properties. The diameters of the SWCNTs are typically from 0.4 to ~5 nm, whereas the MWCNT have diameters from ~1.5 to 100 nm or more [8].

MWCNTs have received less attention than SWCNTs since they are more complex and each carbon shell may have different electronic properties and chirality and there are interactions between the shells. One of the ongoing challenges in the fabrication of CNTs is the difficulty of growing them in the desired location with well-defined properties, including diameters, chirality, purity and defects.

10.2.2 Properties

Because of their new and interesting properties, CNTs have become a topic of increasing interest. The properties can be isolated into two camps: (1) those that are theoretically calculated which illustrate the possibilities for CNTs as materials and (2) those that have been measured or in some way experimentally verified. Currently there is a gap between the two camps, probably due in large measure to the large variety of CNT structures and to inadequate controls in the fabrication procedures. Although we believe it very important to be aware of both camps, in our discussions of sensors we will tend to emphasize the experimentally verified properties as chemical and biochemical sensors are truly experimental systems with a straightforward purpose. The theoretical predictions are very important since they provide a guide for experimental results in areas in which the experiments are difficult.

Of all the structures of molecules, CNTs are one of the few that are known precisely at the atomic level, where each atom is located and held in place. This is in contrast to the free structures of polymers and proteins for non-rigid systems. The CNT is analogous to a crystal in its repetitive regularity, although the CNT may have fewer defects than most three-dimensional crystals. The fact that the CNT has interesting properties that might be exploited in electrodes in electrochemical reactions, catalyst supports or functionalized templates for molecular wires that join the chemical and electronic worlds, are an interesting if not compelling reason to explore uses for CNTs.

The molecular structure of the CNT leads to the prediction of some unique physical, electrical and chemical properties (see Tab. 10.2). The SWCNT is a one-dimensional conductor wherein all of the electrons are confined to move in one atomic layer. The direction of electrical conduction is normally measured along the tube and perpendicular to the tube-rolling axis. The conduction electron wavelength around the circumference of a nanotube is quantized due to periodic boundary conditions and only a discrete number of wavelengths can fit around the tube. Along the tube, the electrons are not confined. The structure has a unique aspect ratio, being only ~ 1 nm in diameter but micrometers in length. Further, all of the atoms in the SWCNT structure are surface atoms, making it a

Tab. 10.2 Properties of Single-wall CNTs [3, 8, 16]

Mechanical	*Strength, toughness, flexibility, surface/volume*	
Composites have a CNT Young's modulus 1 TPa, 5 times that of steel, and tensile strength 45 GPa, 20 times that of steel, a density of 1.4 g/cm^3 (Al: 2.7 g/cm^3); and a strength/weight ratio 500 times greater than Al, steel and Ti and an order of magnitude greater than graphite/epoxy CNTs have reported linear elasticity of up to 5–10%. Concentric MWNTs can expand like a telescope with non-Hooke's law spring forces and rotate The largest possible surface-to-volume ratio		
Chemical	*Bonding, reactivity*	High chemical stability
Chemical and biological reactivity can be obtained by functionalization; CNTs possess stability in solvent, acids and bases		
Thermal	*Insulators, conductors*	High temperature stability
Higher stability than graphite and amorphous carbons. Theory predicts thermal conduction is 6000 W/m K (Cu is 400) to 3 kW/mK, which is greater than that of diamond (2 kW/mK)		
Electrical	*Conductivity*	High electronic conductivity
Suitable for microelectronics, can be semiconducting or metallic CNTs with high current-carrying capacity stable at $J \approx 10^9$ A/cm^2 (1000 times greater than Cu); suitable for field mission tips. Can oscillate tips electrostatically		
Optical	*Absorption, reflectivity*	High bandwidth
Smallest of fibers and filters or waveguides appear possible; light affects conductivity, field emission tip generates x-rays, IR detection/emission possible		

one-dimensional conductor and a single molecule wire, i.e. a molecular wire of the same size as or smaller than polymers such as DNA. The CNT is much smaller than a biological cell and closer in size to the building blocks of the cell.

Chemically, CNTs are very stable, with covalent carbon bonds and with properties such as stability that are generally thought of as between those of graphite, which has sp^2 atoms in a 2D plane, and diamond, which has sp^3 atoms in a 3D space. The CNT, and also the fullerenes, are in a third stable (or metastable) form of a carbon lattice structure. The larger the diameter of the CNT, the more the carbon bonds are planar and sp^2 in character. To the extent that the sp^2 bond vectors are bent into a third dimension, the bond character is more like sp^3. Hence it is not surprising that CNTs exhibit electrical and mechanical characteristics between those of graphite and diamond [11]. Infact we could predict that it is possible to use CNTs as the seed for the growth of diamond. Unlike in silicon, there are no free bonds in CNTs, except possibly at the ends of the nanotube, leading to a chemically stable surface which can be functionalized for chemical or biological sensing, e.g. by plasma activation [12]. The ends can be terminated with carboxyl groups or futher functionalized [13, 14]. The 'all-surface-atom' structure gives the highest possible ratio of surface to volume for catalyst, adsorbent or sensor applications. The CNTs are thermally stable up to $\sim 300\,^{\circ}\mathrm{C}$ with a thermal conductivity greater than that of diamond. CNTs are tough and flexible, having a measured Young's modulus of 0.3–1.5 TPa (cf. steel, 0.2 TPa). The variability is thought to be due to the different fabrication procedures [8, 15] and typical values of properties are given in Tab. 10.2.

A slight change in the winding of the hexagons along the axis of the tube can change the tube from acting as a metal to a large-gap semiconductor. Experimentally, about two-thirds of the SWCNT tubes produced are semiconducting and one-third are metallic. This result fits with theoretical calculations which conclude that the behavior of a particular CNT with indices (n, m) would be metallic provided $n-m=3i$, where i is zero or an integer and semiconducting otherwise, with a bandgap E_g that goes inversely with the diameter D, $E_g=(4\hbar v_F/3D)$ where v_F is the Fermi velocity. Bandgaps vary from about 10 meV to 1 eV. Density of states measurements and spectroscopic measurements on CNTs have verified the features predicted in the theoretical band structure calculations [4]. The armchair tube with its six-membered ring units in a line (see Fig. 10.2) is a metallic conductor whereas the zigag CNTs are semiconducting. It appears that MWCNTs tend to behave as metallic conductors [8] or at least to behave in accordance with the expected properties of the outer shell if side-bonded and operated at low bias [5]. On the other hand, the SWCNTs can be either metallic or semiconducting, as indicated above. The electrical conductivity can be very high [17, 18]. The conduction is predicted to be ballistic under certain conditions, resulting in a very low resistance and high current-carrying capacity [19, 20].

Although improvements in the fabrication of CNTs appear to be yielding reduced defect densities and higher conductivity, the defects remain present. In fact, the presence of defects is predicted from theory and can even be used to modulate the conductivity of metallic CNTs [21]. Measured room temperature resistances of the best recent SWCNT FET structures tend to be about 10 kΩ or more, which should be compared with the corresponding theoretical quantum mechanical resistance for a 1D metallic SWCNT of 6.5 kΩ. The additional resistance is attributed to the metal–CNT contact resistance, nonidealities, impurities and finite temperature effects. The contact resistance can be greater, particularly for semiconducting CNTs, and can dominate the resistance of the fabricated device. For the semiconducting CNTs, the CNT–metal contacts are actually Schottky barriers and their properties can have a major influence on the device behavior [5, 22–24]. Improved procedures for making contacts are needed and are still under development.

CNTs respond to light, not only with a change in conductivity but also by changing shape [25]. Exposing filament bundles of SWCNTs to visible light causes them to bend elastically. The light also caused current to be generated and the filaments to move between two electrodes. Calculations of the optical properties based on an sp^3 tight binding approximation have been made [26]. It appears that the CNTs offer the promise to be both controlled emitters and absorbers of electromagnetic radiation, allowing optical device and sensor possibilities.

10.3 History, Preparation and Cleaning

10.3.1 History

In 1991, Iijima at NEC is credited with the discovery of CNTs with very large length-to-diameter ratios in carbons made in a carbon arc discharge [27]. However, as with many great discoveries, there were even earlier reports of carbon filaments of nanometer dimensions from the 1970s [6, 10a] and of course the discovery of fullerenes [28] and even an article published in 1978 that apparently described carbon nanotubes as carbon fiber layers on the arc electrodes themselves [29]. Over the ensuing years, many types of tubes and preparative methods for nanotubes and related structures have been published, and more appear each month.

10.3.2 Methods of Preparation

A variety of methods exist to prepare CNTs with different levels of purity, sizes and shapes, multi-walled and single-walled geometries. Generally, it has been reported that multiwall tubes do not require a catalyst to grow and single-walled tubes are grown using a catalyst, which keeps them uncapped at the ends. Today the catalytic processes result in a wide variety of chemical purities. Specific impurities are a function of the process and include amorphous carbon, other fullerene structures, organic materials or inorganic materials such as catalyst support and metal.

The current view of the CNT growth process comes from observations. If we disperse a nanoparticulate metal catalyst such as Fe on the surface, then subject it to 900 °C and a source of carbon in a reducing atmosphere, the nanotubes grow like grass up from the surface with the tiny metallic nanoparticle on the top of the tube. When the metal nanoparticle is dissolved with HCl, an open-ended tube results. The explanation is that the nanoparticle is in a molten state and the source of carbon saturates the metal with carbon, which then begins to precipitate out. Since the source of carbon is at the top of the particle, the precipitation occurs at the particle bottom by the substrate in the form of the nanotube. The nanoparticle is thereby lifted up or pushed along the surface as the tube is knitted. It has been observed that the nanoparticle metal catalysts are on the ends of the tubes away from the surface and that the nanotube diameter is a function of the nanoparticle diameter [30]. The earliest arc discharge methods for CNT preparation made many different structures simultaneously such as mats, ropes and tubes [27]. Later, laser ablation methods gave some measure of improvement [31] but the catalytic pyrolysis of hydrocarbons [32] and other CVD methods [33] made vast improvements in the control and uniformity of the CNTs produced. A typical process requires a catalyst of nanosized metal particles dispersed on the dielectric surface, a carbon source, temperatures of about 850–1100 °C, pressure control of each gas, a reducing atmosphere and reaction times of 5–60 min. With such a process, preferential growth of single-walled, long carbon nanotubes is observed. Low-pressure methods and rf-PECVD CNT growth recipes have been reported [34].

Pulsed laser and arc discharge techniques [35], CVD on doped Si-wafer substrates [33] and templated approaches have been used to make the CNTs grow preferentially in certain areas [36]. Other ideas such as a gel-casting foam method for CNT growth [37], a method on SiO_2 [38], a variation of electric arc discharge [39] and a method using modification with ion beams to obtain CNT alignment on the surface [40] have been reported. Recent modifications to processes incorporate novel YAlFe perovskite catalysts [41] and an HiPCO disproportionation process [42].

Highly ordered arrays of carbon nanotubes, with densities as high as 10^{10} cm^{-2}, have been made using anodized aluminum nanopore templates [43] (see Fig. 10.5). Sen-

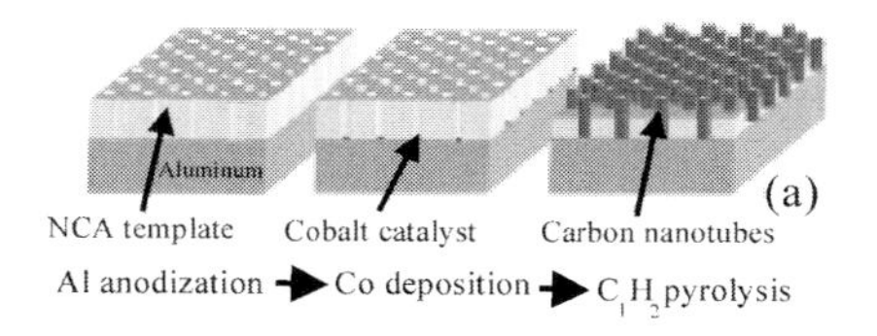

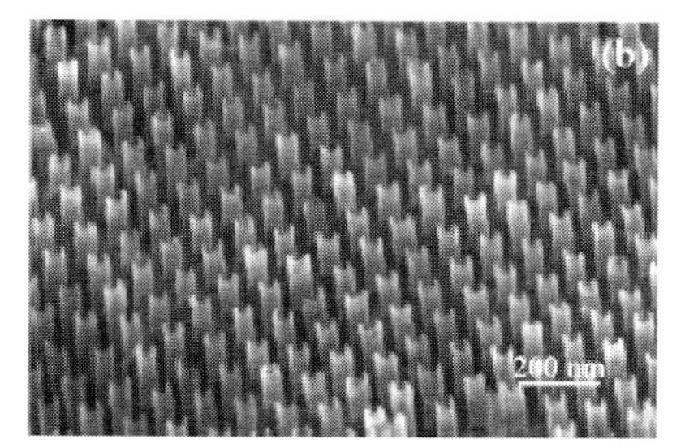

Fig. 10.5 (a) Schematic of fabrication process and (b) SEM image of the resulting hexagonally ordered array of CNTs on an anodized alumina template with nanopores [43]

sing electrodes for glucose oxidase redox activity have been made from these functionalized electrodes [44].

Of all the methods of preparation, it is becoming routine to produce carbon nanotubes with a specific diameter within a narrow range and in the location of interest defined on the micrometer or hundreds of nanometer size scale. There are now several companies that sell, in addition to the CNTs themselves, CNT fabrication equipment [45] complete with recipes to produce CNTs. These unique nanostructures can be incorporated into device structures. However, the challenge remains to obtain the CNT performance desired, and this depends upon many factors in addition to the CNT being present.

10.3.3 Nanotube Purification and Cleaning

The CNT purity and the substrate surrounding the CNTs are topics not often discussed in publications to date. The lack of control experimentation is disturbing and often clouds the interpretation of results. How do we know we are measuring the property of the nanotube or its impurities or the surroundings? We do not know, and CNT synthesis is today a chemically impure process and also one that leads to significant structural defects. This subject is important especially to the understanding and construction of practical sensors.

Cleaning processes are beginning to be developed for CNTs. In general, cleaning processes can utilize high temperatures (a few hundred degrees Celsius) in air to remove amorphous carbon since the CNT is stable to several hundred degrees. Cleaning can also use organic or mineral acid solvents with or without chromatographic separation to remove organics and inorganic/ionic material. However, the yield, level of purity, cost and uniformity of CNTs vary widely with today's techniques. Costs of commercially available CNTs depend strongly on the purity of the product.

Only a few publications have started to report the purity of the CNTs produced. Purification by acid treatment, filtration, water wash and vacuum drying has been reported to reduce the metal content to 3.5%. This procedure also reduces the porosity of the CNTs. High-temperature vacuum annealing [47] and HCl washing [48] have been used to obtain similar reductions in metal content and the latter procedure resulted in a 65% increase in the N_2 adsorption BET surface area to 861 m^2/g. It is clear that the washing changes the surface area available for interaction with gases and the nature of the available interface to substrate and fluids. The total surface area for a single-walled carbon nanotube can be calculated assuming that all surface atoms are available, giving about 3000 m^2/g [49]. Since no reported areas are yet this large, we can assume that there are some parts of the surface that are as yet blocked. The highest surface area that carbons can achieve is $\sim$1000–1500 m^2/g and the highest experimentally reported CNT area is 1587$\sim m^2/g$ [50], which was achieved by washing the HiPco prepared CNTs in solvent followed by acid washing and wet oxidation to lower the metal content to $<1\%$ while removing amorphous carbon.

10.4 Characterization of Carbon Nanotubes

The electrical and mechanical properties depend on the diameter and the chiral angle of the CNT. Techniques have been developed to measure these properties using STM (scanning tunneling microscopy), STS (scanning tunneling spectroscopy), AFM (atomic force microscopy), SEM (scanning electron microscopy) and TEM (transmission electron microscopy). The first reported STM measurements on multi-wall CNTs show an atomically resolved pattern affected by the two outermost layers of the tubes [51, 52]. In 1996, Lin et al. accurately determined the chiral angle of a multi-walled nanotube [53]. The first STM measurements on SWCNT date from 1994 [54] and 4 years later the groups of Lieber and Dekker independently succeeded in recording atomically resolved images of SWCNT [55, 56]. Images of SWCNT produced by the arc discharge method have been reported [57] and images of intramolecular nanotube junctions and the corresponding topological defects have also been measured [58]. Even with excellent spatial resolution (better than 0.5 Å), extracting reliable information from experimental images is not an easy task because of distortions of the nanotube when current is flowing [3]. In order to obtain the best images from STM, both low positive and negative bias images are needed [59].

STS measurements are performed by keeping the STM tip stationary above the nanotube, switching off the feedback mechanism and recording the current I as a function of the voltage V applied to the sample. The differential conductance, dI/dV, of the measured current–voltage curve gives the density of states involved in the tunneling conduction and measurements compare well with theoretical calculations, at least for biases below about 0.75 V [3]. From the density of states obtained from STM, the bandgap for a semiconducting CNT can be obtained, which can be used to calculate the diameter. Thus the STS electrical measurement determines the diameter of semiconducting CNT. For metallic nanotubes, the width of the metallic plateau in the density of states, defined as the energy between the first van Hove singularities located on each side of the Fermi energy, is inversely proportional to the tube diameter, as with the semiconducting CNT. When STS is used with STM, it it possible to perform a determination of the nanotube structure [60, 61]. It is difficult to obtain good measurements of diameter without the STS measurements because of tip-shape convolution effects in STM [62]. STM and AFM have been combined to image SWCNTs [63].

Some of the vibrational modes in nanotubes can be excited with Raman spectroscopy. At the right frequency, phonons can be excited in SWCNT of specific diameter [64]. The position of the breathing modes shifts with the diameter, hence the resonant frequency can be used to determine the CNT diameter [65]. Resonant Raman scattering has become a powerful tool to determine the distribution of CNT diameters in bulk samples [66, 67]. Raman spectroscopy has also been used to detect strain in SWCNTs [68]. Electrostatic force microscopy (EFM) has provided noncontact methods for measurement of conductance of a CNT and DNA molecules [69], the electric potential in CNT circuits [70, 71] and the ferroelectricity of CNTs [72].

10.5 Electronic Devices Incorporating CNTs (Not Sensors)

Before a review of sensors, it is useful to discuss the structure and properties of devices that have incorporated CNTs. These devices and the structures therein can be used to implement sensor designs.

10.5.1 Carbon Nanotube Field-effect Transistors (CNTFETs)

The progress in terms of nanotube applications is extraordinarily rapid, as evidenced by the appearance of publications of devices for FETs, nanologic and nanomemory circuits, electron emitters, electromechanical actuators and chemical and biochemical sensors. The present status of the efforts to make CNTFET devices for use in electronic circuits has been reviewed [73]. Other electronic and electromechanical structures, including mechanical oscillators, have also been reviewed [74].

One of the first uses of a single CNT was as a conductive channel in a FET-type structure that operates at room temperature [75]. Several years later, a room temperature FET was made that can detect the transfer of a single electron to the gate [76]. Kruger et al. [77] exposed the gate of a CNT FET to a drop of electrolyte, which caused large shifts in the Fermi level and large resistance changes, with the conclusion that nanotubes are possibly the most sensitive FETs for environmental applications. These first devices had CNTs on the surface of silicon dioxide, with each end of the CNT lying on top of a metallization pad. The gate dielectric was the silicon dioxide below the CNT, so the gate connection was made to the substrate. FET structures with both SWCNT and MWCNT were made and the MWCNT devices showed no variation in conductivity with the gate voltage. Electron transport was diffusive rather than ballistic [78]. Although having the CNT exposed to the ambient may be a good sensor design, for a transistor this exposure leads to unstable behavior. The situation is analogous to that of FET chemical sensors in which the open gate provides both the sensitivity and the instability.

Recently, top-gated CNTFETs have been made by the deposition of an oxide and metallization on top of the CNT [79]. The threshold voltage for the top-gated FETs is about –0.5 V, compared with –12 V for the older bottom-gated devices. The top-gated devices behave electrically much more like typical FET structures, with a drive current that is about three to four times higher and a transconductance that is about four times higher than that of state-of-the-art 15–50 nm long channel silicon or SOI MOSFETs. The newer top-gated structures also have a 200-fold increase in the transconductance and significant improvements in the source and drain contacts to the CNT. In the first devices the CNTs were just laid on the metal pads and the contact resistances were >100 kΩ. After annealing of Ti contacts deposited on top of the CNT, the contact resistance is below ~ 30 kΩ. The ability to make good contact is improving rapidly and may already have been achieved by the time this article is published!

The CNTFET devices as fabricated show behavior that is similar to that of a p-type silicon MOSFET device. It has been shown that this is due to the reversible

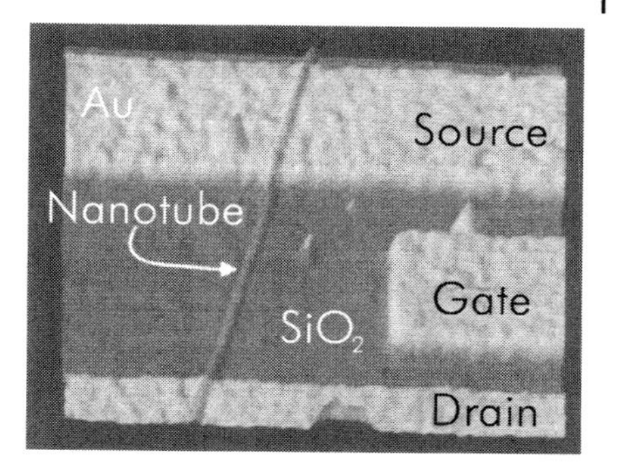

Fig. 10.6 FET with carbon nanotube. From P. Avouris website [5, 6]

exposure to oxygen during the fabrication [80]. It is possible to transform a p-type CNTFET to an n-type CNT by annealing in vacuum and return to the p-type after exposure to oxygen. To detemine if this behavior is due to a doping phenomena, CNTFETs were doped with potassium, an electron donor, to make the devices n-type [81]. Although n-type devices were obtained, the CNTFET behavior with respect to threshold voltage and subthreshold current–voltage characteristics was not the same as that observed with oxygen exposure. The present interpretation is that the transport and switching behavior of CNTFETs is determined by Schottky barriers present at the source and drain contacts where the metal and semiconducting carbon meet. Oxygen affects these contacts in a reversible manner, opening the door for the design of a chemically sensitive CNTFET. Support for this interpretation is provided by theoretical modeling and the temperature dependence of the current. This oxygen sensitivity was actually used to make a CMOS-like inverter using an n-type CNTFET (vacuum annealed and covered with PMMA to protect it from ambient oxygen) and a p-type CNTFET (vacuum annealed and not covered with PMMA) [82] (see Fig. 10.6). We expect that much of the sensing behavior attributed to CNTs will later be found to be substrate and contact effects.

The fabrication of FET structures from carbon nanotubes is a challenging task, recently reviewed [83]. One approach is to try to grow the CNTs where one wants them to build a device and the other approach is to grow the CNT is the best way and then assemble the CNT into devices. The newest tools for the latter approach include multiple degrees of freedom nanorobotic manipulators with the capability to position CNTs with nanometer resolution.

10.5.2 Field Emission Devices

The unique properties of CNTs include carrying high currents, high chemical stability, low emission threshold, high thermal conductivity and the ability to fabricate arrays of CNTs using templates all make CNTs attractive for use as field emission devices in displays and related applications [84–86]. A theoretical approach has also been published [87], together with the use of CNTs as luminescent elements [88]. CNTs have been used as cathodic field emission tips in the generation of X-rays [89] and also in the generation of microwave power [90].

10.5.3
Miscellaneous Devices and Applications of CNTs

In general, the Hamiltonian contains three parts: there are terms for the atomic contribution, the electronic contribution and an electron–phonon interaction contribution. For the CNTs, the electron–phonon interaction term is particularly large, suggesting the presence of strong electromechanical interactions. This manifests, for example, in the observed optically induced deformation [91]. Strong electromechanical interaction was shown in a experiment in which two sheets of billions of nanotube bundles were laminated into a double-layer film configuration with two-sided Scotch tape [92]. When a DC potential was applied in aqueous NaCl, the double layer deflected. The direction of deflection depended on the polarity of the voltage. The behavior is attributed to differential expansion due to electrochemical double-layer charging. Cantilevered MWCNTs were deflected using electrostatic potentials and could be excited into resonance [93]. A theoretical analysis of the data indicated that the induced charge was all at the tip of the CNT. Ordered arrays of parallel CNTs have been used in the detection of IR radiation [94]. Carbon nanotubes have been used as very robust tips in atomic force microscopes, with and without chemical functionalization [95]. The high surface area and chemical stability of CNTs have been exploited to store hydrogen [96 a]. Adsorption of CO_2 has also been studied [96b] and found to involve physisorption on the sidewalls of SWCNTs.

10.6
CNT Sensors for Chemical and Biochemical Applications

10.6.1
Sensor Properties, Designs and Structures

There are certain CNT properties that are particularly relevant to sensors, e.g. the surface area, size and shape, electrical conductivity, chemical reactivity and optical properties. However, clearly a chemical sensor will require a well-characterized and stable CNT device structure in order to:

1. obtain a constant and predictable CNT surface and signal;
2. have a surface of high purity for reliable functionalization;
3. understand the effect of variables, e.g. temperature, pressure, relative humidity (RH) and chemical contamination, on sensor response; and
4. design controlled surfaces for optimum response.

While fabrication and growth of the CNT are significant and necessary accomplishments, they are not sufficient for the successful design and construction of a sensor. Sensors based on CNTs will require work on substrates, contacts, functionalization chemistries at the interfaces with and without encapsulants, methods of measurement, annealing, stabilizing and packaging to house the sensing behavior. The

CNT properties, purities and contacts are still variable. In many ways the situation is reminiscent of the Schottky barrier devices of the late 1960s, where lack of purity and high-performance vacuum systems was a major problem affecting the repeatable performance of a device whose properties depended on reproducibility and cleanliness of an interfacial layer a few atoms thick. Today there are still considerable hurdles to the making of CNT sensors and even larger hurdles to their ultimate commercialization. Having said this, the near future for devices incorporating nanotubes is most promising for a variety of reasons including the promise of single-molecule sensitivity and the molecular engineering of responses with high selectivity and extremely fast response time in the tiniest of packages with power consumption orders of magnitude lower than current sensing devices.

Even though the CNT structures have been available for less than a decade, all manner of sensors have been created and tested in one form or another. There is sensor technology in the use of the CNTs as direct molecular probes, as platforms for surface chemical layers and as electrodes in analytical measurements. At this point it is best to examine several ways in which the CNTs are used in sensors to make measurements.

There are a few successful batch fabrication or microfabricated structure designs that have been used for chemical and biochemical sensing and variations are illustrated in Fig. 10.7. These sensor output depends on the conductivity of the device, which may include a CNT with a coating, for example. In practice, the CNTs are grown on the surface in a CVD process at 900 °C. The structure may be fabricated in either of two ways: (1) deposit the CNTs first and then perform metallization or (2) grow the nanotubes on surfaces with electrodes sufficiently refractory to withstand the CNT growth procedure. The CNTs can be individual or multiple SWCNTs or networks that are contacted by electrodes of various geometry. There is a difference

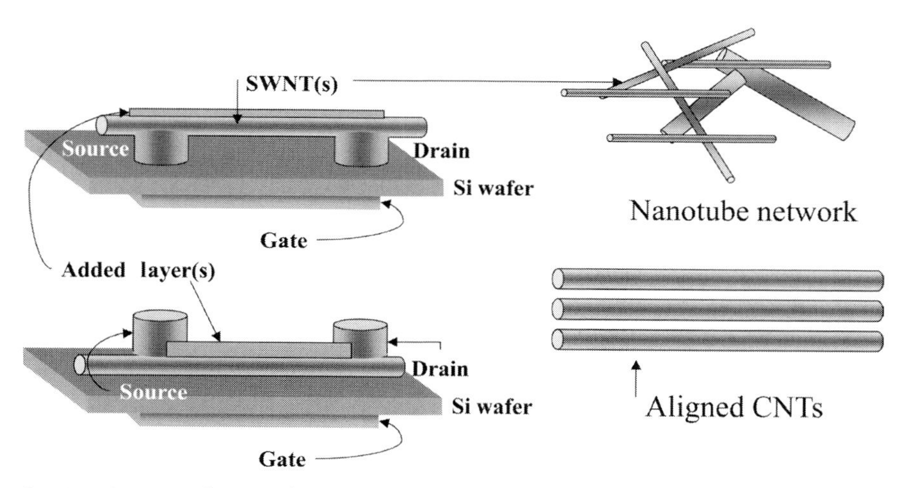

Fig. 10.7 Structures for CNT chemical sensors. Grow CNTs, then perform metallization (bottom) or place electrodes on the surface and then deposit CNTs (top). CNTs can be single or multiple tubes or CNT networks

between multiple tubes, aligned or not aligned, each spanning the electrode gap and the network nanotube structure of either uniform or variable CNTs that span the gap between contacts. In the case of networks, the multiplicity of CNT–CNT contacts across the electrode gap leads to different conduction character. All of these device structures allow one to install a bottom gate and operate the sensor as a FET as well as a simple chemiresistor. Other versions of the sensor structure with top gates or side gates have also been envisioned and fabricated. Another process to make CNT sensors is to use a single electrode or a multiple electrode structure, e.g. in a lock and key configuration, and deposit CNTs from a suspension onto the surface of the electrodes [96c]. Networks can be easily made in this manner and the added advantage is that the CNTs can be presorted and washed for greater physical and chemical purity (Fig. 10.8). Techniques for manipulation and alignment of CNTs on surfaces and electrodes are an important area of study in sensor development.

10.6.2
Added Layers: Functionalization, Decoration, Adornment

The sensor is a multi-layered structure having a substrate, CNT and chemical sensitizing and encapsulating layers. At each layer there is an interface of variable complexity that is intimately connected to the observed sensor response. As indicated in Fig. 10.8, CNTs can be functionalized with added layers to change their reactivity and thereby achieve different performance characteristics as sensors. Functionalization is a word used to describe adding or bonding something to the CNT in order to impart one or more important characteristics, i.e. a characteristic that improves the sensor function toward its intended application. From the point of view of the sensor, reasons for functionalization include doping the CNT with states of a particular reactivity and/or coating the CNT to block unwanted reactivity. Coatings can reduce RH dependence or passivation layers can improve stability over time or conditions of use. From the point of view of the semiconductor CNT, functionalization includes adding intrinsic or extrinsic dopants. Intrinsic would be those functionalizations included in the fabrication of the CNT itself, such as impurities or different atoms or missing atoms in the semiconductor or carbon arrangements that alter the electronic states such as the five-membered rings involved in tube end closure or the pair of five- and seven-membered ring defects.

Functionalization can also be described by the chemistry of attachment of the layers. Tab. 10.3 illustrates a categorization of CNT sensor functionalization. The general categories include metals and nonmetals and liquids and solids. Metals such as Pd can make the CNT sensitive to molecular hydrogen and polymeric organic films can make the CNT sensitive to electron donor or acceptor molecules. In general, any dielectric layer placed next to an SWCNT can alter its conductivity in some manner. Indeed, then any molecule that alters the dielectric behavior of the ad-layer that is immediately next to the CNT wall has a good chance of altering the observed CNT device electrical characteristics. Although many reports of CNT sensor response are given, rarely is the part of the structure responsible for the change given.

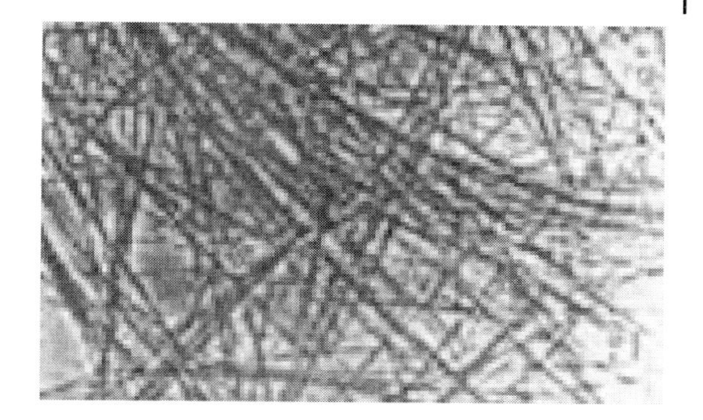

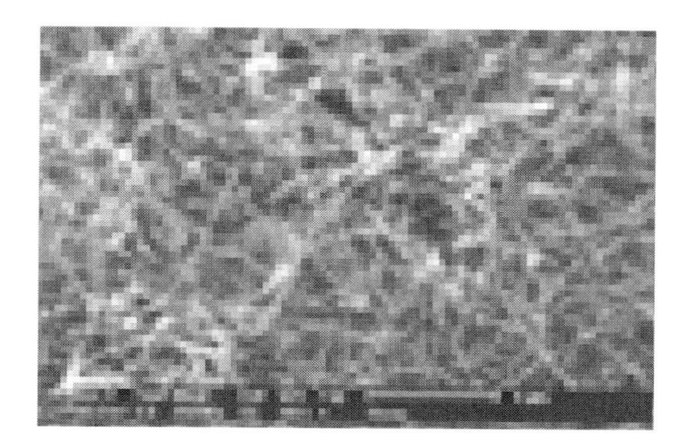

Fig. 10.8 Examples of (top) treated–purified multi-walled CNTs and (bottom) treated–purified SWCNTs. From www.sesres.com

Specific chemical functionalization of CNTs [97] can lead to reversible behavior of the coatings [98] toward analytes or to permanently reacted layers of metal oxides [99]. Boron nitride nanotubes have also been functionalized [100] in an attempt to tailor the reactivity of the resulting CNT device. Noncovalent methods for decoration of CNT layers for biosensors [101] have been surveyed and large sensor arrays have been studied [102]. The array work leads to future versions of electronic noses and tongues for imaging in chemical and biochemical space [103]. Chemical bonding to CNTs results in specific atoms and molecules [104] decorating the outerwall surface.

Methods for coatings must be suitable for the material, e.g. e-beam or thermal evaporation can be used for many materials including insulators such as SiO_2 [105] or metals. An interesting method of spontaneous metal deposition [106] has been reported that appears to be a simple corrosion reaction of the surface Fe-metal impurities on the insulator and the more conductive CNTs that results in localized electrodeposition of the metal. Intrinsic doping with oxygen has been studied from a theoretical perspective [107 a] and a survey of functionalization for SWCNT focused on phthalocyanine [107 b] has been reported.

Nanotube–polymer systems have recently received much attention for solubilizing and for coating CNTs [108–111]. Polymers can bind noncovalently to CNTs

Tab. 10.3 Functionalization of CNTs

Intrinsic doping
IMPURITY
Atom missing, 5–7 ring pairs, 5 ring closure, or
Any atom or atom pair in the wall that dopes the
Semiconductor
Extrinsic doping
BOND ENERGY
Weakest:
Van der Waals
π-Bonding, dipole–dipole interactions
Hydrogen bonding
Covalent bonding
Strongest:
Ionic bonding
MATERIALS
Metals, semimetals, semiconductors – Pd, Pt, alloys, ...
Metal oxides – SnO_2, ZnO_2, ...
Organic layers – polymers: PEI, DNA, Teflon, Nafion, ...
Composite layers or liquid layers – Nafion/CNTs, electrolytes, ...

[112] and yet dramatically alter the characteristics of nanotube FET devices [113]. PEI coatings were found to shift FET device characteristics from p- to n-type semiconducting CNTs and this was attributed to the electron-donating ability of amine groups in the polymer [114]. However, covalently bound polymers have also been studied [115] and in addition to polymers coated on CNTs, the CNTs have been put on to the polymers [116].

10.6.3 Examples of CNT Sensors

The review could be divided into gas-, liquid- and solid-phase sensors or into physical, chemical and biosensors. However, all manner of sensors have been attempted and so we will review examples that illustrate structure, materials and methods employed for the various sensing and analytical applications.

10.6.3.1 Gases

After the fabrication of the first FET devices in 1998 [75], it was quickly discovered that these bottom-gated structures were extremely sensitive to gases such as nitrogen oxides and ammonia [117]. The exposure history to oxygen [118] was very important and oxygen chemisorption was used to explain the p-type behavior of these FET devices [119, 120]. These FETs were used in large sensor arrays [121] and in biosensors [122–124]. Short vertically grown nanotubes or 'nanoturf' sensors have been made on a surface that contained a Pt interdigitated electrode

(IDE) and used for sensing NO_2 [125] with cross-sensitivity to ammonia, ethanol and humidity. The hystersis in the FET current–voltage curve behavior, consistent with mobile ion effects [114], has been used for RH sensing [126] when the FET is covered with a very thin Nafion film. A recent version of the FET uses a network of CNTs as the source to drain channel and is applied to trace detection of nerve agents [127] and in this design the CNTs were functionalized with a selective polymer. CNTs were physically deposited from solution on Au IDEs and used in a simple chemiresistor method to measure gases and vapors [128]. The CNTs were purified prior to use and sensitivity to benzene was reported that is not typically reported for in situ grown CNTs. This CNT sensor in the chemiresistor format, where R/R_0 is the sensor signal, also exhibited a linear response with analyte concentration not seen from the CNT FET platform. Clearly the FET and chemiresistor approaches are using different CNT properties for the measurement.

The gas that is sensed by the CNT device depends not only upon the selection of CNTs and the measurement method but also on the substrate and functionalizations that are used. One of the earliest gases detected was hydrogen and the CNTFET was activated with a Pd metal layer [129]; the Pd allows the formation of atomic hydrogen that can alter the FET characteristics. The presence of molecular hydrogen does not produce a signal. Low-level sub-ppm concentrations of NO_2 have been detected [130] with SWCNTs and also ammonia [131]; other environmental contaminants [132] and gas-sensing applications [133] have used multi-walled CNTs [132].

10.6.3.2 **Liquids**

Electrocatalytic oxidation of the hormone NO [134] and the anodic oxidation of hydrazine [135] have been reported. The detection of liquids can be performed at CNTs since they are basically fullerenes and stable in electrolytes over a wide range of pH values. However, the electrodes are often fabricated as CNT powder electrodes, as has been applied to the detection of cysteine [136], or as composites, as has been applied to the detection of glucose [137]. One of the most important types of sensing in liquids is for biological purposes and so it is only logical that many of the examples of sensing in liquids are for biological materials. Other biodirected protein analyses with modified CNT electrodes include xanthine and uric acid [138] and technology for enzyme modification of CNT electrodes has been reported [139]. A review of electrochemical sensing has appeared [140] and the direct detection of DNA hybridization has been reported [141]. Successful biosensing has been performed with Teflon composite electrodes [111]. The CNT electrodes in liquids have been used to detect phenomena such as Li^+ ion intercalation [142] and soluble nitrites [143]. One can even find carbon modified by CNTs [144] which comprises a carbon on carbon electrode and there have been reports of novel CNTs and other nanomaterials for biosensing [145]. Finally, the use of the nanotube structure for a template [146] offers exciting possibilities to the device experimenter.

10.6.3.3 **Physical Sensors**

There is, of course, the whole gamut of physical sensors for temperature, pressure [147, 148], stress, strain [149, 150], flow [151–153], position and viscosity and other variables that has not been discussed. Electronic sensing and field emission sensors [154] have likewise not been covered here. There is great excitement about potential robotic applications and one can envision the potential for the combination of chemical, biochemical and physical sensing to produce artificial sensing systems and human-like robots. However, these sensors and systems are beyond the scope of this review, which has focused on the CNTs themselves, devices and chemical and biochemical sensors.

10.7 Conclusion

We have discussed a variety of sensors utilizing CNTs and specifically SWCNTs. Most of the sensors that have been discussed in the literature are based on changes in electrical properties, although other sensing mechanisms are possible, such as the shift in resonance of a CNT mechanical structure or variation in optical properties. Scientists and engineers are just exploring many of the properties of CNTs and how to use them to nanoengineer specific device properties. It is without doubt that new sensing mechanisms and novel variations of old approaches will be explored in the search to find out what sensor methods really fit the CNT technology. This exploration will take some time. Two decades ago we began the euphoric exploration of silicon-based microfabricated and micromachined sensors which were hailed as the panacea for the then current sensor problems. In the last 20 years, some silicon sensors have had great success in the market place, such as accelerometers, pressure sensors and blood gas analyzers, but most of the microfabricated silicon sensors described in the literature did not have the competitive advantages and performance required for successful commercialization. A similar shakeout is expected with CNT sensors. It is our hope that with a more fundamental understanding of CNTs and the relationship between performance and molecular design that not just a few great applications will be found by trial and error, but that numerous great applications will be consciously identified and developed successfully.

Sensors are 'layered' devices and each layer influences the overall function. For example, layers consist of the substrate, the CNT itself or the platform device, which may be configured as a FET or chemiresistor, Schottky barrier, etc. that allow the changes in the layers to be detected. The layers that functionalize the CNT control the reactivity and the species, such as the analyte or moisture, that reach the CNT. The various layers are chosen specifically to obtain the required sensor parameters of sensitivity, selectivity, response time and stability. Understanding and controlling the complex interactions that will make a practical CNT sensor remain a topic of intense investigation and one of lively debate. Of course, there is also the measurement circuit, which, it is hoped, will be able to correct all of the shortcomings of the sensor and produce a measurement system that performs as needed. And if the circuit fails to correct the sensing behavior, there are always smart algorithms. However, no circuit or software can make up for a poor sensor and so it is always best to obtain the best sensor performance.

It is exciting that it is possible to utilize these new and unique nanomaterials in practical applications such as sensors. It remains a proper question for the science community whether the recent torrent of descriptive, anecdotal and theoretical publications about nanotubes amounts to understanding. It is odd somehow that so much has been learned about this system but so little is understood about its basic function and how to engineer this technology into sensors. The use of the CNT in any truly practical system has been very limited to date and its use as a template has not yet been explored. The sensor technology community is aware that control of the electrical properties, the contacts and the placement on substrates are some of the key factors in using these structures in sensor designs. The study of the unique properties that result at the interfaces of these materials with substrates and added layers is truly an additional exciting area that will have a profound impact on the unfolding of this technology into sensors and allied applications.

10.8 References

1 H. Dai, *Acc. Chem. Res.* **2002**, *35*, 1035–1044.

2 P.M. Ajayan, *Chem. Rev.* **1999**, *99*, 1787–1799.

3 J. Bernholc, C. Brenner, M. Nardelli, V. Meunier, C. Roland, *Annu. Rev. Mater. Res.* **2002**, *32*, 347–375.

4 C. Dekker, *Phys. Today* **1999**, May, 22–28.

5 P. Avouris, J. Appenzeller, R. Martel, S.J. Wind, *Proc. IEEE* **2003**, *91*, 1772–1784 (see *www.research.ibm.com/nanoscience/*).

6 M. Dresselhouse, P. Avouris (eds.), *Carbon Nanotubes*; Berlin: Springer, **2001**.

7 *Carbon Nanotubes – Worldwide Status and Outlook: Applications, Applied Industries, Production, R&D and Commercial Implications*; Fuji-Keizai USA, **2002**; Printed from *www.researchandmarkets.com/reports/1130*.

8 T. Fukuda, F. Arai, L. Dong, *Proc. IEEE* **2003**, *91*, 1803–1818.

9 The US National Nanotechnology Initiative; a Visionary R&D Program and Infrastructure that Complements Industry with Education, Long-range Research and Cross-cutting Centers of Excellence; a report by Neal Lane, Science Advisor to William Clinton. January **2000** (see *www.ostp.gov/NSTC*).

10 (a) M. Dresselhouse, G. Dresselhouse, P. Eklund, *Science of Fullerenes and Carbon Nanotubes*; San Diego, CA: Academic Press, **1996**; (b) S. Huang, C. Cai, J. Liu, *J. Am. Chem. Soc.* **2003**, *125*, 5636–5637.

11 J. Xu, *Proc. IEEE* **2003**, *91*, 1819.

12 Q. Chen, L. Dai, M. Gao, S. Huang, A. Mau, *J. Phys. Chem. B* **2001**, *105*, 618.

13 S.S. Wong, E. Joselevich, A.T. Woolley, C.L. Cheung, C.M. Lieber, *Nature* **1998**, *394*, 52–55.

14 A. Guiseppi-Elie, C. Lei, R. Baughman, *Nanotechnology* **2002**, *13*, 559.

15 E. Wong, P. Sheehan, C. Lieber, *Science* **1997**, *277*, 1971.

16 *Nanotech Briefs* October **2003**, *1(1)*, 8; see *www.nanotechbriefs.com*.

17 A. Yao, C. Lane, C. Dekker, *Phys. Rev. Lett.* **2000**, *61*, 2941.

18 P. Collins, M. Hersam, M. Arnold, R. Martel, P. Avouris, *Phys. Rev. Lett.* **2001**, *86*, 3128.

19 C. White, T. Todorov, *Nature* **1998**, *393*, 240.

20 J. Guo, M. Lundstrom, S. Datta, *Appl. Phys. Lett.* **2002**, *80*, 3192.

21 M. Bockrath, W. Liang, D. Bozovic, J. Hafner, C. Lieber, H. Park, *Science* **2001**, *291*, 283.
22 R. Seidel, M. Liebau, G. Düsberg, F. Kreupl, E. Unger, A. Graham, W. Hönlein, *Nano Lett.* **2003** (April).
23 H. Soh, C. Quate, A. Morputgo, C. Marcus, J Kong, H. Dai, *Appl. Phys. Lett.* **1999**, *75*, 627.
24 M. Anantram, S. Datta, Y. Xue, *Phys. Rev. B* **2000**, *61*, 14219.
25 Y. Zhang, S. Iijima, *Phys. Rev. Lett.* **1999**, *82*, 3472.
26 F. Shyu, *Phys. Rev. B* **2002**, *67*, No. 045405, 1.
27 S. Iijima, *Nature* **1991**, *354*, 56.
28 H.W. Kroto, J.R. Heath, S.C. O'Brien, R.F. Curl, R.E. Smalley, *Nature* **1985**, *318*, 162.
29 P. Wiles, J. Abrahamson, *J. Carbon* **1978**, *16*, 341. Y. Li, W. Kim, Y. Zhang, M. Rolandi, D. Wang, H. Dai, *J. Phys. Chem. B* **2001**, *105*, 11424–11431.
31 M. José-Yacamán, M. Miki-Yoshida, L. Rendón, *Appl. Phys. Lett.* **1993**, *62*, 202.
32 H. Cheng, F. Li, G. Su, H. Pan, L. He, X. Sun, M. Dresselhaus, *Appl. Phys. Lett.* **1998**, *72*, 3282.
33 J. Kong, H.T. Soh, A.M. Cassell, C.F. Quate, H. Dai, *Nature* **1998**, *395*, 878.
34 E. Gamaly, T. Ebbesen, *Phys. Rev. B* **1995**, *52*, 2083.
35 T. Guo, et al., *Chem. Phys. Lett.* **1995**, *243*, 49.
36 G. Che, B.B. Larghmi, E.R. Fisher, C.R. Martin, *Nature* **1998**, *393*, 346.
37 Rul, et al., *J. Eur. Ceram. Soc.* **2003**, *23*, 1233–1241.
38 Hafner, Leiber, et al., *J. Phys. Chem. B* **2001**, *105*, 743.
39 P. Ajayan, T. Ebbesen, *Rep. Prog. Phys.* **1997**, *60*, 1025.
40 Gohel, et al., *Chem. Phys. Lett.* **2003**, *371*, 131.
41 T. Guo, et al., *Scr. Mater.* **2003**, *48*, 1185.
42 Nikolaev, et al., *Chem. Phys. Lett.* **1999**, *313*, 91.
43 J. Li, C. Papadopoulos, J. Xu, M. Moskovits, *Appl. Phys. Lett.* **1999**, *75*, 367.
44 J. Xu, *Proc. IEEE* **2002**, *91*, 1819.
45 See websearch and *www.atomate.com or www.firstnano.com for equipment and www.sesres.com* as examples of sources for nanotubes.
46 M. Bronikowski, P. Willis, D. Colbert, K. Smith, R. Smalley, *J. Vac. Sci. Technol. A* **2001**, *19*, 1800.
47 M. Yudasaka, et al., *Nano Lett.* **2001**, *1*, 487.
48 C.M. Yang, et al., *Nano Lett.* **2002**, *2*, 385.
49 Y.F. Yin, et al., *Langmuir* **1999**, *15*, 8714.
50 M. Cinke, et al., *Chem. Phys. Lett.* **2002**, *365*, 69–74.
51 M. Gallager, D. Chen, B. Jacobson, D. Sarid, L. Lamb, F.A. Tinker, J. Jiao, D.R. Huffman, S. Seraphim, D. Zhou, *Surf. Sci. Lett.* **1993**, *281*, L335.
52 Z. Zhang, C.M. Lieber, *Appl. Phys. Lett.* **1993**, *62*, 2792; M. Ge, K. Sattler, *Science* **1993**, *260*, 515.
53 N. Lin, J. King, S. Yang, N. Cue, *Carbon* **1996**, *34*, 1295.
54 M.H. Ge, K. Sattler, *Appl. Phys. Lett.* **1994**, *65*, 2284.
55 J.W.G. Wildoer, L. Venema, A. Rinzler, R. Smalley, C. Dekker, *Nature* **1998**, *391*, 59.
56 T. Odom, J. Huang, P. Kim, C. Lieber, *Nature* **1998**, *391*, 62.
57 Y. Maruyama, T. Takase, M. Yoshida, K. Kogure, K. Suzuke, *Fullerene Sci. Technol.* **1999**, *7*, 211.
58 M. Ouyang, J.-L. Huang, C. Cheung, C. Lieber, *Science* **2001**, *291*, 97.
59 W. Clauss, D. Bergeron, M. Freitag, C. Kane, E. Mele, A. Johnson, *Europhys. Lett.* **1999**, *47*, 601.
60 P. Kim, T. Odom, J. Huang, C. Lieber, *Carbon* **2000**, *38*, 1741.
61 I. Wirth, S. Eisebitt, G. Kann, W. Eberhardt, *Phys. Rev. B* **2000**, *61*, 5719.
62 G. Mark, L. Biro, J. Gyulai, *Phys. Rev. B* **1998**, *58*, 12645.
63 W. Clauss, M. Freitag, D.J. Bergeron, et al., *Carbon* **2000**, *38*, 1735–1739.
64 A. Rao, E. Richter, S. Bandow, B. Chase, P. Eklund, K. Williams, S. Fang, K. Subbaswamy, M. Menon, A. Thess, R.E. Smalley, G. Dresselhaus, M. Dresselhaus, *Science* **1997**, *275*, 187.
65 R. Saito, T. Takeya, T. Kimura, G. Dresselhaus, M. Dresselhaus, *Phys. Rev. B* **1999**, *59*, 2388.
66 X. Liu, *Phys. Rev. B* **2002**, *66*, No. 045411, 1–8
67 Bandow, et al., *Phys. Rev. Lett.* **1998**, *80*, 3779.
68 M.D. Frogley, Q. Zhao, H.D. Wagner, *Phys. Rev. B* **2002**, *65*, 113413 (1–4).

69 M. Bockrath, N. Markovic, A. Shepard, M. Tinkham, L. Gurevich, L.P. Kouwenhoven, M.W. Wu, L.L. Sohn, *Nano Lett.* (Communication), **2002**, *2(3)*, 187–190.

70 P.J. de Pablo, C. Gómez-Navarro, A. Gil, J. Colchero, M.T. Martinez, A.M. Benito, W.K. Maser, J. Gómez-Herrero, A.M. Baró, *Appl. Phys. Lett.* **2001**, *79*, 2979.

71 A. Bachtold, M. Fuhrer, S. Plyasunov, M. Forero, E. Anderson, A. Zettl, P. McEuen, *Phys. Rev. Lett.* **2000**, *84*, 6082.

72 C. Ahn, T. Tybell, L. Antognazza, K. Char, R. Hammond, M. Beasley, Ø. Fischer, J.-M. Triscone, *Science* **1997**, *276*, 1100.

73 P. Avouris, J. Appenzeller, R. Martel, S.J. Wind, *Proc. IEEE.* **2003**, *91*, 1772.

74 J. Xu, *Proc. IEEE* **2003**, *91*, 819.

75 S. Tans, A. Verschueren, C. Dekker, *Nature* **1998**, *393*, 49.

76 H. Postma, T. Teepen, Z. Yao, M. Grifoni, C. Dekker, *Science* **2001**, *293*, 76.

77 M. Kruger, M. Buitelaar, T. Nussbaumer, C. Schoenenberger, *Appl. Phys. Lett.* **2001**, *78*, 1291.

78 R. Martel, T. Schmidt, H. Shea, T. Hertel, P. Avouris, *Appl. Phys. Lett.* **1998**, *73*, 2447.

79 S. Wind, J. Appenzeller, R. Martel, V. Dercyke, P. Avouris, *Appl. Phys. Lett.* **2002**, *80*, 3187.

80 P. Collins, K. Bradley, M. Ishigami, A. Zettl, *Science* **2000**, *287*, 1801.

81 P. Avouris, J. Appenzeller, R. Martel, S.J. Wind, *Proc. IEEE* **2003**, *91*, 1772–1784.

82 V. Derycke, R. Martel, J. Appenzeller, P. Avouris, *Nano Lett.* **2002**, *2*, 929.

83 T. Fudaka, F. Arai, L. Dong, *Proc. IEEE* **2003**, *91*, 1803–1818.

84 W.A. de Heer, A. Châtelain, D. Ugarte, *Science* **1995**, *270*, 1179–1180.

85 D.-S. Chung, S.H. Park, H.W. Lee, J.H. Choi, S.N. Cha, J.W. Kim, J.E. Jang, K.W. Min, S.H. Cho, M.J. Yoon, J.S. Lee, C.K. Lee, J.H. Yoo, J.-M. Kim, J.E. Jung, Y.W. Jin, Y.J. Park, J.B. You, *Appl. Phys. Lett.* **2002**, *80*, 4045.

86 A. Maiti, *CMES-Comp. Model. Eng.* **2002**, *3*, 589–599.

87 A. Maiti, J. Andzelm, N. Tanpipat, *Abstr. Pap. Am. Chem. Soc.* **2001**, *222*, 204; A. Maiti, J. Andzelm, N. Tanpipat, *Abstr. Pap. Am. Chem. Soc.* **2001**, *222*, 39.

88 J.M. Bonard, *VIDE* **2001**, *56*, 251.

89 G. Yue, *Appl. Phys. Lett.* **2002**, *81*, 355.

90 P. Siegel, T. Lee, J. Xu, *The Nanoklystron: a New Concept for THz Power Generation;* NASA Jet Propulsion Lab, New Technology Report NPO 21014, **2000**.

91 Y. Zhang, S. Iijima, *Phys. Rev. Lett.* **1999**, 82, 3472.

92 R. Baughman, C. Cui, A. Zakhidov, Z. Iqbal, J. Barisci, G. Spinks, G. Wallace, A. Mazzoldi, D. Rossi, A. Rinzler, O. Jaschinski, S. Roth, M. Kertesz, *Science* **1999**, *284*, 1340.

93 P. Poncharal, Z. Wang, D. Ugarte, W. deHeer, *Science* **1999**, *283*, 1513.

94 J. Xu, *Infrared Phys. Technol.* **2001**, *42*, 485.

95 H. Dai, J. Hafner, A. Rinzler, D. Colbert, R. Smalley, *Nature* **1996**, *384*, 147; C. Lieber, et al., *Nature* **1998**, *394*, 52.

96 (a) Q.K. Wang, C.C. Zhu, W.H. Liu, et al., *Int. J. Hydrogen Energ.* **2002**, *27*, 497–500; (b) M. Cinke, J. Li, C.W. Bauschlicher, A. Ricca, M. Meyyappan, *Chem. Phys. Lett.* **2003**, *376*, 761–766; (c) J. Li, Y. Lu, Q. Ye, M. Cinke, J. Han, M. Meyyappan, *Nano Lett.* **2003**, *3*, 929–933.

97 S.B. Sinnott, *J. Nanosci. Nanotechnol.* **2002**, *2*, 113–123.

98 J.B. Cui, M. Burghard, K. Kern, *Nano Lett.* **2003**, *3*, 613–615.

99 W.Q. Han, A. Zettl, *Nano Lett.* **2003**, *3*, 681–683.

100 W.Q. Han, A. Zettl, *J. Am. Chem. Soc.* **2003**, *125*, 2062–2063.

101 R.J. Chen, S. Bangsaruntip, K.A. Drouvalakis, *Proc. Natl. Acad. Sci. USA* **2003**, *100*, 4984–4989.

102 Q.F. Pengfei, O. Vermesh, M. Grecu, *Nano Lett.* **2003**, *3*, 347–351.

103 J.R. Stetter, W.R. Penrose (eds.) *Artificial Chemical Sensing: Olfaction and the Electronic Nose (ISOEN 2001)*; Pennington, NJ: Electrochemical Society, **2001**, *2001-15*, 1–227.

104 S.B. Fagan, A.J.R. da Silva, R. Mota, R.J. Baierle, A. Fazzio, *Phys. Rev. B* **2003**, *67*, 033405, and *Phys. Rev. B* **2003**, *67*, 205414.

105 Q. Fu, C.G. Lu, J. Liu, *Nano Lett.* **2002**, *2*, 329–332.

106 H.C. Choi, et al., *J. Am. Chem. Soc.* **2002**, *124*, 9058–9059.

107 (a) D.J. Mann, M.D. Halls, *J. Chem. Phys.* **2002**, *116*, 9014–9020; (b) G. del la Torre, W. Blau, T. Torres, *Nanotechnology* **2003**, *14*, 765–771.

108 M. J. O'Connell, et al., *Chem. Phys. Lett.* **2001**, *342*, 265–271.

109 A. Star, et al., *Angew. Chem., Int. Ed. Engl.* **2001**, *40*, 1721–1725.

110 J. Wang, et al., *J. Am. Chem. Soc.* **2003**, *125*, 2408–2409.

111 J. Wang, M. Musameh, *Anal. Chem.* **2003**, *75*, 2075–2079.

112 A. Star, et al., *Macromolecules* **2003**, *36*, 353–360.

113 M. Shim, et al., *J. Am. Chem. Soc.* **2001**, *123*, 11512–11513.

114 K. Bradley, J. Cumings, A. Star, J.-C. P. Gabriel, G. Gruner, *Nano Lett.* (Communication) **2003**, *3 (5)*, 639–641; W. Kim, A. Javey, O. Vermesh, Q. Wang, Y. Li, H. Dai, *Nano Lett.* (Communication) **2003**, *3(2)*, 193–198.

115 F. J. Gomez, et al., *Chem. Commun.* **2003**, 190–191.

116 H. T. Ng, et al., *Langmuir* **2002**, *18*, 1–5.

117 J. Kong, N. Franklin, C. Zhou, M. Chapline, S. Peng, K. Cho, H. Dai, *Science* **2000**, *287*, 622–625.

118 P.G. Collins, K. Bradley, M. Ishigami, A. Zettl, *Science* **2000**, *287*, 1801–1804.

119 S.-H. Jhi, S.G. Louie, M.L. Cohen, *Phys. Rev. Lett.* **2000**, *85*, 1710–1713.

120 H. Ulbricht, G. Moos, T. Hertel, *Phys. Rev.* **2002**, *66*, 075404.

121 P. Qi, O. Vermesh, M. Grecu, A. Javey, Q. Wang, H. Dai, *Nano Lett.* **2003**, *3*, 347–351; H. Dai, et al., *Pure Appl. Chem.* **2002**, *74*, 1753.

122 R. J. Chen, S. Bangsaruntip, K. A. Grouvalakis, N. Wong, S. Kam, M. Shim, Y. Li, W. Kim, P. J. Utz, H. Dai, *Proc. Natl. Acad. Sci. USA* **2003**, *100*, 4984–4989.

123 A. Star, *Nano Lett.* **2003**, *3*, 459–463.

124 K. Besteman, T.-O. Lee, F. G. M. Wiertz, H. A. Heering, C. Dekker, *Nano Lett.* **2003**, *3*, 727–730.

125 C. Cantalini, L. Valentini, I. Armentano, L. Lozzi, T. M. Kenny, S. Santucci *Sens. Actuators B* **2003**, *95*, 195–202.

126 A. Star, T.-R. Han, V. Joshi, J. R. Stetter, *Electroanalysis* **2004**, *16 (1–2)*, 108–112.

127 J. P. Novak, E. S. Snow, E. J. Houser, D. Park, J. L. Stepnowski, R. A. McGill, *Appl. Phys. Lett.* **2003**, *83*, 4026.

128 J. Li, Y. Lu, M. Cinke, J. Han, M. Meyyappan, *Nano Lett.* **2003**, *3*, 929.

129 J. Kong, M. G. Chapline, H. J. Dai, *Adv. Mater.* **2001**, *13*, 1384–1386.

130 L. Valentini, I. Armentano, J. M. Kenny, C. Cantalini, L. Lozzi, S. Santucci, *Appl. Phys. Lett.* **2003**, *82*, 961–963.

131 S. Chopra, A. Pham, J. Gaillard, A. Parker, A. M. Rao, *Appl. Phys. Lett.* **2002**, *80*, 4632–4634.

132 W. Purwanto, *MRS Bull.* **2001**, *26*, 157.

133 O. K. Varghese, P. O. Kichambre, D. Gong, *Sens. Actuators B* **2001**, *81*, 32–41.

134 F. H. Wu, G. C. Zhao, X. W. Wei, *Electrochem. Commun.* **2002**, *4*, 690–694.

135 Y. D. Zhao, W. D. Zhang, H. Chen, *Talanta* **2002**, *58*, 529–534.

136 Y. D. Zhao, W. D. Zhang, H. Chen, *Sensors and Actuators B-Chem.* **2003**, *92(3)*, 279–285.

137 M. Gao, L. Dai, G. G. Wallace, *Synth. Met.* **2003**, *137*, 1393–1394.

138 Y. Y. Sun, J. J. Fet, K. B. Wu, *Anal. Bioanal. Chem.* **2003**, *375*, 544–549.

139 K. Yamamoto, G. Shi, T. S. Zhou, *Analyst* **2003**, *128*, 249–254.

140 Q. Zhao, Z. H. Gan, Q. K. Zhuang, *Electroanalysis* **2002**, *14*, 1609–1613.

141 H. Cai, X. N. Cao, Y. Jiang, *Anal. Bioanal. Chem.* **2003**, *375*, 287–293.

142 J. Zhao, Q. Y. Gao, C. Gu, *Chem. Phys. Lett.* **2002**, *358*, 77–82.

143 P. F. Liu, J. H. Hu, *Sens. Actuators B* **2002**, *84*, 194–199.

144 R. S. Chen, W. M. Huang, H. Tong, Z. L. Wang, J. K. Cheng, *Anal. Chem.* **2003**, *75(22)*, 6341–6345.

145 S. Sotiropoulou, V. Gavalas, V. Vamvakati, *Biosens. Bioelectron.* **2003**, *18*, 211–215.

146 H. Matsui, R. MacCuspie, *Nano Lett.* **2001**, *1*, 671–675.

147 J. R. Wood, M. D. Frogley, E. R. Meurs, *J. Phys. Chem. B* **1999**, *103*, 10388–10392.

148 J. R. Wood, H. D. Wagner, *Appl. Phys. Lett.* **2000**, *76*, 2883–2885.

149 Q. Zhao, M. D. Frogley, H. D. Wagner, *Compos. Sci. Technol.* **2002**, *62*, 147–150.

150 Q. Zhao, M. D. Frogley, H. D. Wagner, *Polym. Adv. Technol.* **2002**, *13*, 759–764.

151 N. Sen, *Curr. Sci. India* **2003**, *84*, 269–270.

152 S. Goosh, A. Sood, N. Kumar, *Science* **2003**, *299*, 1042–1044.

153 A. Maiti, A. Svizhenko, M. P. Anantram, *Phys. Rev. Lett.* **2002**, *88*, 126805.

154 A. Maiti, *CMES-Comp. Model. Eng.* **2002**, *3*, 589–599.

11
CMOS-based DNA Sensor Arrays

R. Thewes, F. Hofmann, A. Frey, M. Schienle, C. Paulus, P. Schindler-Bauer, B. Holzapfl, and R. Brederlow, Infineon Technologies, Corporate Research, Munich, Germany

Abstract

The invention of the transistor some 50 years ago, the subsequent realization of integrated circuits and today's availability of complete systems-on-a-chip have had an essential impact on today's way of life. The next revolution of similar impact may arise from developments in the area of biotechnology and life sciences. In this field, tools are required which can – in some cases only – be provided using techniques established by the semiconductor manufacturing world. Ongoing developments which intend to merge the know-how and potential of both disciplines, bio- and semiconductor technology, envision a way towards intelligent biosensor and actuator chips. In this chapter, CMOS-based DNA sensor chips with fully electronic readout are considered. The chapter is organized as follows. First, an introduction on basic DNA microarray techniques is given. Applications are briefly discussed and also the related technical requirements in terms of number of test sites, sensitivity, specificity and dynamic range. In Section 11.2, we address the motivation to develop electronic detection systems and the economic and technical boundary conditions for CMOS-based sensor arrays. A concrete example of a CMOS-based DNA sensor array is considered in detail in Section 11.3. There, after an introduction concerning the detection principle, CMOS process-related issues, architecture and circuit design issues, the interdependence of these areas in this nonstandard CMOS application and system aspects are discussed. A brief outlook towards label-free detection methods is given in Section 11.4 and finally, in Section 11.5, the chapter is summarized.

Keywords

DNA; sensor arrays; CMOS; DNA microarrays; DNA chips; redox-cycling detection.

Advanced Micro and Nanosystems. Vol. 1
Edited by H. Baltes, O. Brand, G. K. Fedder, C. Hierold, J. Korvink, O. Tabata

ISBN: 3-527-30746-X

11.1 Introduction

11.1.1 DNA Microarrays

The development of DNA microarray sensor chips in recent years has opened the way to high parallelism and high throughput in many biotechnology applications [1–9]. The most widely known fields are genome research and drug development; exploration of these tools in the field of medical diagnosis is under development. The purpose of DNA microarrays is to enable the parallel investigation of a given analyte concerning the presence of specific DNA sequences. Depending on the particular application, requirements range from relatively simple 'presence or absence' evaluations to quantitative analyses with a high dynamic range. In the following, the basic setup, the basic operation principle, the state-of-the-art optical readout method and application-driven specifications and requirements are briefly reviewed.

11.1.2
Basic Operation and Detection Principles

A DNA microarray is a slide or a 'chip' typically made of glass or a polymer material [1–9]. Silicon is also used as carrier material, since the huge know-how on silicon manufacturing allows silicon-based chips to be obtained with specific advantageous properties [10–12]. Within an area of the order of square millimeters to square centimeters, single-stranded DNA receptor molecules are immobilized at predefined positions on such chips. These so-called probe molecules consist of different sequences of typically 20–40 bases. As depicted schematically in Fig. 11.1, these probe molecules can be deposited using microspotters [13, 14]. Today, such spotters are able to handle volumes in the sub-nanoliter range.

In Fig. 11.2a and b, two different sites within an array are considered after the immobilization phase. For simplicity, only five bases are drawn in this schematic illustration. As shown in Fig. 11.2c and d, in the next step the whole chip is flooded with an analyte containing the ligand or target molecules. Note that these molecules can be up to two orders of magnitude longer than the probe molecules. In case of complementary sequences of probe and target molecules, this match leads to hybridization (Fig. 11.2c). If probe and target molecules mismatch as shown in Fig. 11.2d, this chemical binding process does not occur. Finally, after a washing step, double-stranded DNA is obtained at the match positions (Fig. 11.2e) and single-stranded DNA (i.e. only the probe molecules as at the beginning of the whole procedure) remain at the mismatch sites.

Since the receptor molecules are known, the information whether double- or single-stranded DNA is found on the different test sites reveals the composition of the analyte. Hence the remaining demand is to make sites with double-stranded DNA visible. In the widely used state-of-the-art optics-based readout technique, the target molecules are labeled with fluorescence molecules before the analyte is applied to the chip. After hybridization and a subsequent washing step, the whole chip is illuminated or scanned with monochromatic light with a wave-

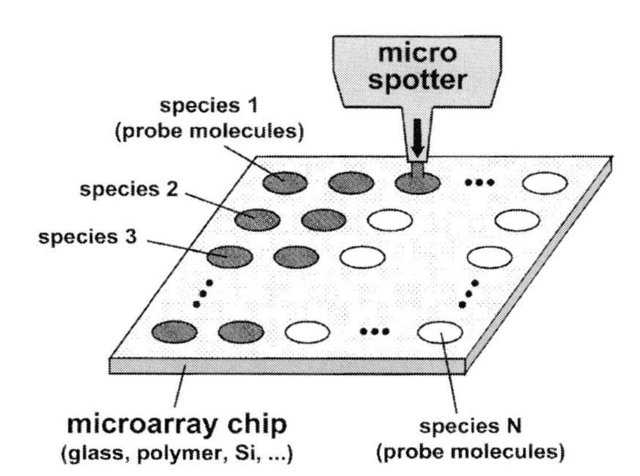

Fig. 11.1 Schematic plot showing a DNA microarray during the functionalization process. A microspotter deposits single-stranded DNA probe molecules (known DNA sequences) on the surface of the chip at predefined positions

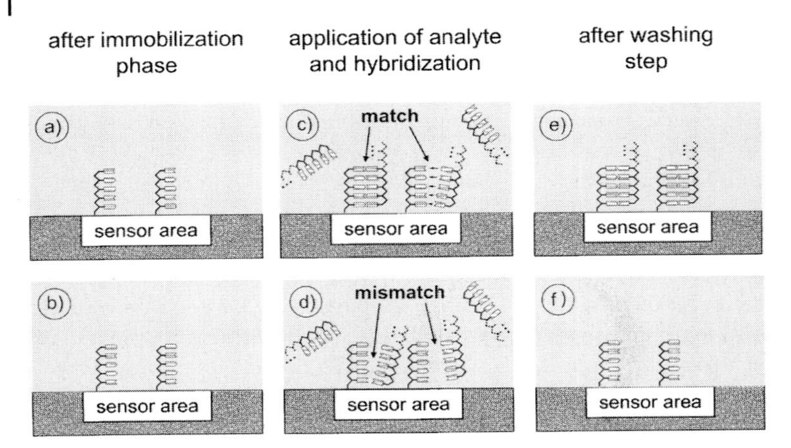

Fig. 11.2 (a), (b) Schematic consideration of two test sites after the immobilization process. For simplicity, probe molecules with five bases only are shown here. (c), (d) Hybridization phase. An analyte containing target molecules to be detected is applied to the whole chip. Hybridization occurs in the case of matching DNA strands (c). In the case of mismatching molecules (d) chemical binding does not occur. (e), (f) Situation after washing step

length matched to the absorption profile of the marker molecules (Fig. 11.3). A camera system with a blocking filter for the excitation wavelength takes an image of the array chip. Fluorescence light emitted at a considered position reveals successful hybridization and double-stranded DNA at this position.

We should mention that the spotting technique discussed here is only one of the methods to functionalize a microarray chip. In this case, the probe molecules

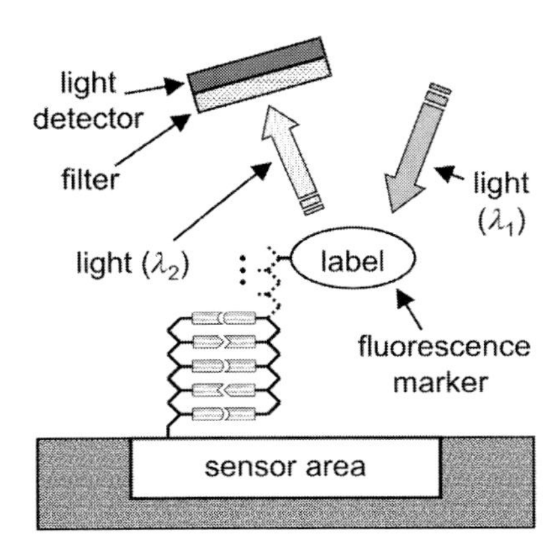

Fig. 11.3 State-of-the-art optical DNA microarray readout method (schematic plot). The target molecule is labeled with a fluorescence marker. The chip is illuminated with light of a defined wavelength λ_1. A camera system scans the whole chip or takes an image. The marker molecule which is only present at matching positions emits fluorescence light with a wavelength $\lambda_2 > \lambda_1$ which is detected by the camera system. Since the intensity of this light source is orders of magnitude lower compared with the stimulating light source, a filter with very good suppression characteristics at λ_1 is used to shield the camera system

are synthesized off-chip. This technique is adequate for low- and medium-density arrays with pitches of order 100 μm.

Using a lithography-based mask technique similar to that known from the semiconductor manufacturing world, in situ synthesis of the receptor molecules can be performed [15, 16]. There, the probe molecules are synthesized base-by-base on-chip. Ligation of a base at the strands under construction is triggered or blocked by the presence or absence of light at the respective sites. The required mask count is approximately equal to the number of different bases (cytosine, guanine, adenine, thymine) × length of the probe molecules. This optical technique, used by Affymetrix, is specifically advantageous if a large volume of high-density chips with a high or very high number of test sites (e.g. $\geq$100 000) is to be manufactured.

Febit has developed a system [17] where the light-triggered chemistry behind the in situ synthesis of the probes is similar to that of Affymetrix. The main difference compared with the Affymetrix approach is that no masks are required but illumination patterns are dynamically generated using a digital projector-based system.

The Nanogen principle allows one to move off-chip synthesized receptor molecules to their on-chip target position using a dielectrophoresis technique [18–20]. For this purpose, some logic CMOS circuitry is provided on the same chip to apply the required voltages at the sensor sites. Readout is based on optical detection.

Recently, Combimatrix has suggested a system for on-chip in situ synthesis applying electrical potentials to the test sites during the functionalization process [21, 22]. The required potentials are controlled by CMOS logic gates. Readout is done using an optical technique.

11.1.3 Applications and System Requirements

The goal of this section is roughly to define specifications and requirements for the different fields of DNA microarray application which must be considered during the configuration of a chip-based system. Two main application areas, gene expression profiling and genome-related investigations, are identified.

In case of gene expression chips [23–25], the change of cell metabolism functions is monitored after application of stress or drugs. The response of cell tissue is monitored using cDNA as the information-carrying species. The expression profiles achieved, for example, reveal whether cancer-suppressor genes are up- or down-regulated after a chemical compound is applied. Such applications require arrays with at least several hundred or even up to several thousand sites. Since the amount of regulation – or expression – is the main parameter of interest, quantitative analyses are necessary. The dynamic range should exceed 2.5 decades in most applications.

On the other hand, in the area of medical diagnosis, single nucleotide polymorphism (SNP) detection is of high interest [26–28]. SNPs, for example, determine the reaction of a patient's metabolism after application of a certain drug. In

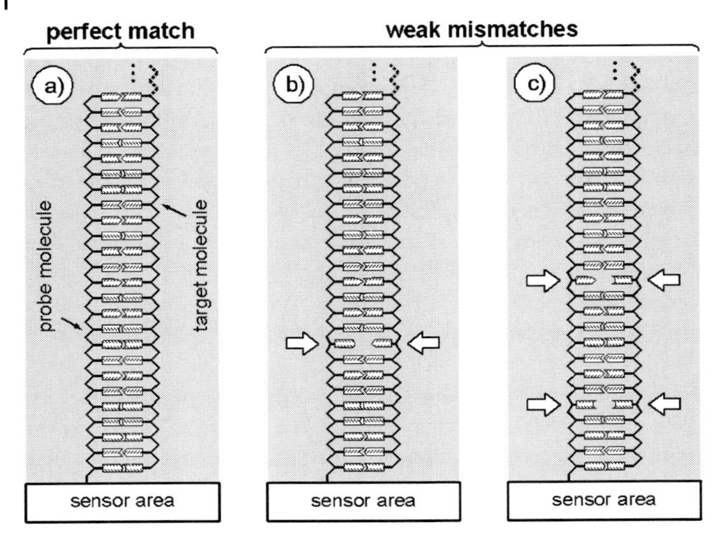

Fig. 11.4 Schematic plot depicting an example of the different situations to be distinguished in an SNP detection experiment. (a) Perfect match; (b) weak mismatch (one of 25 bases); (c) weak mismatch (two of 25 bases)

the future, SNP characterization may provide the doctor with the information on which drug and which dose of that drug should be applied. This scenario of individual personalized medication may significantly help to lower patients' suffering, to lower adverse effects and to reduce costs. The cancer drug Herceptin is a first example where an SNP test has to be performed before application [29, 30].

Assuming probe molecules with 25 bases as an example, SNP detection platforms must be able to distinguish between a perfect match (where each of the 25 bases finds the complementary base at the correct position) from so-called weak mismatches (i.e. from hybridization events, where only 24 or 23 bases of probe and target molecule match) (see Fig. 11.4). The number of available target molecules is usually relatively high in such tests, since they are provided by a biochemical amplification process, the polymerase chain reaction (PCR) [31–35]. Hence the main requirement for such sensor arrays is high specificity, whereas high sensitivity or large dynamic range are of less importance. The number of sites required is typically between about 50 and a few hundred.

Finally, it should be emphasized that the use of DNA microarrays alone does not help to develop a new drug or to heal a patient. A DNA chip is only a tool which helps to provide us efficiently and at reasonable cost with a huge amount of valuable data. The sampled information, however, must be interpreted using the methods from the field of bioinformatics. Exploration of the full potential of microarrays therefore also depends on further progress in the already ongoing parallel development in both areas.

11.2
Electronic DNA Chips

11.2.1
Motivation

So far, DNA array chips based on *optical readout techniques* have been discussed. DNA sensor array chips with *fully electronic readout* promise several advantages over the optical type, since they allow easier handling by the user and avoid expensive optical setups comprising CCD cameras, lenses, etc. These properties have the potential to allow access to new fields of application and to new markets (e.g. diagnosis in hospitals and doctors' offices, point-of-care and outdoor applications, food control). On the other hand, today's status of development is less advanced compared with the optical platforms.

11.2.2
CMOS-based DNA Chips

11.2.2.1 Economical Boundary Conditions

There are numerous suggestions in the literature for electronic DNA sensors focusing on the realization of a few test sites per chip only. These approaches utilize a suitable biocompatible substrate material carrying the transducer elements. The electrical terminals of the electronic sensor(s) on-chip are connected to an off-chip read-out apparatus which provides and measures all necessary electronic signals to operate the chip. Due to the fact that such chips do not contain any active electronic device they will be referred to as 'passive chips' in the following.

This is the most cost-effective concept as long as only few sites per investigation are required so that the amount of interconnects to the external read-out apparatus is not too high. Although a CMOS chip would help to further enhance the sensor signal(s) on-chip, there is no mandatory need for this measure in most applications. The costs of a CMOS chip are usually considered to be too high in these low-density applications to compete with passive, low-cost, test stripe-like electronic systems.

With increasing number of test sites per chip, an interconnect problem arises for passive chip-based approaches. The large number of interconnects needed to contact a medium-density passive chip lowers the available area per contact pad on-chip. As a consequence, reliability and yield decrease. Since the total area of such chips is limited, the available area and thus the signal strength per sensor also decrease. The lowered interconnect reliability at decreasing signal strengths finally leads to a complete loss of signal integrity. In this case 'active chips' (with active on-chip circuitry) are required, to amplify and process the weak sensor signals on-chip, i.e. in the direct neighborhood where the sensor signals are generated. Moreover, adequate on-chip circuitry allows operation of large microarray chips with a small number of contact pads independent of the numbers of test sites per chip.

Tab. 11.1 Economic aspects and technical boundary conditions of passive and active DNA sensor arrays

	Passive electronic DNA chips (only electronic transducer elements + interconnect lines on chip)	*Active electronic DNA chips (electronic transducer elements +active on-chip circuitry for signal pre-processing, electronic multiplexing, ...)*
Density	Low	Medium–high
Test sites per chip	~10	≥100
Costs per chip	Low	Increased processing costs: – CMOS processing costs – Process to provide transducer elements must be compatible with CMOS process
Cost per data point	Approximately constant	Decrease with: – Increasing number of test sites per chip – Increasing number of required data points per investigation
Electrical performance Electronic signal integrity	Medium – Limited robustness – Loss of signal integrity at high test site count per chip	High – By far increased robustness – Independent of number of test sites per chip

Note that in this case higher costs per chip are not only acceptable owing to occurrence of a technical brick wall, but that another metric must be considered here: Whereas the *costs per chip* are increasing, the *costs per data point* decrease. Consequently, in the case of applications where a number of data points of the order of 100 or higher is required, active CMOS-based sensor array chips represent the technically and economically better choice compared with their passive counterparts.

A summary of these arguments is given in Tab. 11.1.

11.2.2.2 Technical Boundary Conditions

The realization of active electronic biosensor arrays requires the integration of biocompatible interface, sensor and transducer materials into standard CMOS environments [36–40]. The chips have to be equipped with a passivation, which allows them to be operated in contact with wet media. Moreover, the material used to realize the transducer must be provided. Unfortunately, the materials available in standard CMOS lines are often not suitable to fulfill all biological and chemical requirements. In the case of sensor arrays based on electrochemical principles, the transducer material is usually a noble metal, in many cases gold. Gold, however, and other noble metals are not part of standard CMOS production lines. Integration of these materials in a CMOS production line is crucial, since they may lead to contamination problems which have a significant impact on the performance of the CMOS devices and on the yield.

A reasonable concept to deal with this challenge is to fabricate CMOS wafers using the full standard CMOS process without any extra steps up to standard passivation. Standard or slightly modified Si_3N_4 passivations from CMOS production lines often meet the biological requirements, i.e. they are suitable to shelter the chip from the wet media which are to be analyzed. The specific transducer materials are then provided in a post-CMOS process implemented outside the CMOS production line.

In addition to the already discussed wafer- and chip-related manufacturing aspects, innovative packaging solutions are needed which differ from standard concepts used in the semiconductor world. The total sensor area, i.e. the area with the fluid interface, must be isolated from the region with pads used to contact the chip electrically. Moreover, the package must provide an electronic and a fluidic interface to the readout apparatus.

A further condition for robust and reliable operation of chips with a large number of positions is to provide array-compatible high-precision analog circuitry. The challenges usually arise from relatively low signals to be pre-processed within a given area per sensor site at sufficient dynamic range. Control logic and digital circuitry are also needed on-chip, but the total area consumption and design challenges of the pure digital parts are relatively low.

11.3 CMOS-based DNA Sensor Array Using a Redox-cycling Detection Technique

In this section, concrete development steps of a fully electronic CMOS-based DNA microarray platform are described. Starting with the detection principle used, experimental data concerning the extended CMOS process, circuit design issues, DNA experiments and system integration aspects are discussed.

11.3.1 Detection Principle

The electrochemical sensor principle used is shown schematically in Fig. 11.5. It is based on an electrochemical redox-cycling technique [41–45]. A single sensor (Fig. 11.5, left) consists of interdigitated gold electrodes (generator and collector electrode). Probe molecules are spotted and immobilized on the surface of the gold electrodes and chemical bonding is achieved by, e.g. thiol coupling. The target molecules in the analyte which is applied to the chip are tagged with an enzyme label (alkaline phosphatase). After the hybridization and washing phases, a suitable chemical substrate (*p*-aminophenyl phosphate) is applied to the chip. The enzyme label, available at the sites where hybridization occurred, cleaves the phosphate group and the electrochemically active *p*-aminophenol is generated (Fig. 11.5, right).

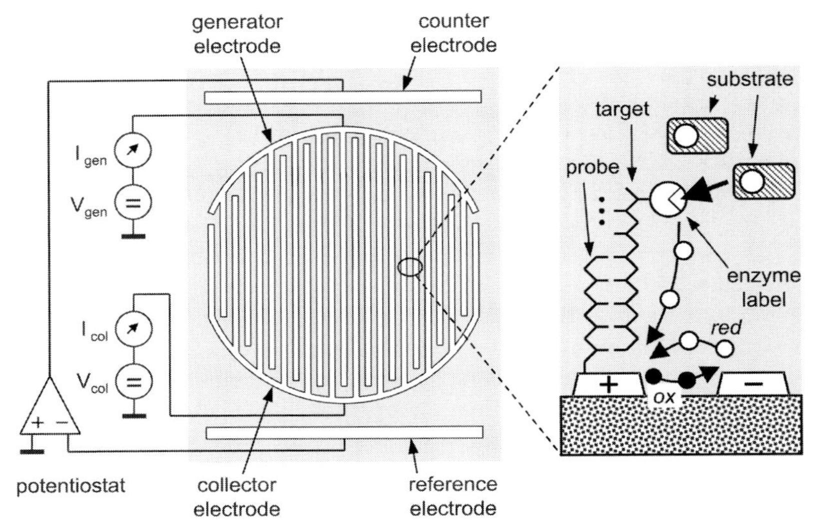

Fig. 11.5 Redox-cycling sensor principle and sensor layout. Left: single sensor consisting of interdigitated gold electrodes and potentiostat circuit with counter and reference electrodes. Right: blow-up of a sensor cross-section showing two neighboring electrodes after successful hybridization. For simplicity, probe and target molecule are only shown on one of the electrodes. After a washing step, *p*-aminophenol phosphate is applied and electrochemically active compounds are created by the enzyme label molecule (alkaline phosphatase) bound to the target DNA strands. Applying simultaneously an oxidation and a reduction potential to the sensor electrodes, a redox process is started at the electrodes and a current flow occurs at both electrodes

Applying simultaneously an oxidation and a reduction potential to the sensor electrodes (V_{gen} and V_{col} in Fig. 11.5, e.g. +300 and −100 mV with respect to the reference potential), *p*-aminophenol is oxidized to quinonimine at one electrode and quinonimine is reduced to *p*-aminophenol at the other. The activity of these electrochemically redox-active compounds translates into an electron current at the gold electrodes and the electronic measurement devices or circuits connected to the electrodes (symbolized by I_{gen} and I_{col} in Fig. 11.5).

This idealized schematic description of the redox-cycling process suggests a balanced charge situation at the two working electrodes. However, in reality not all particles oxidized at the generator reach the collector electrode. A measure for this phenomenon is the collection efficiency [44, 45]:

$$ce = I_{col}/I_{gen} \tag{1}$$

Measured data reveal values down to 80–90%. A potentiostat circuit, whose output and input are connected to a counter and to a reference electrode, respectively, provides the difference currents to the electrolyte. In particular, it forms a regulation loop which holds the potential of the electrolyte at a constant value. The complete four-electrode system is shown in Fig. 11.5 (left).

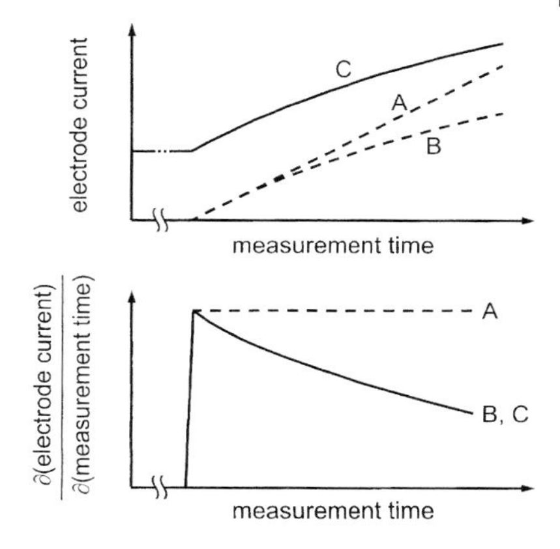

Fig. 11.6 Schematic plots demonstrating the impact of different contributions to the sensor signal. (a) Electrode current; (b) derivative of the electrode current with respect to the measurement time. Curves A, ideal case; curves B, with saturation effects due to increasing concentrations of *p*-aminophenol and quinonimine at the considered sensor site; curves C, with saturation effects and further electrochemical artifacts

The current flow at the sensor electrodes is a function of the contribution initially generated by the enzyme label and of the redox-cycling related contribution at the sensor electrodes. These two mechanisms lead to a simple differential equation whose solution predicts an electrode current which increases in proportion to the measurement time (Fig. 11.6, top, curve A). With increasing concentration of *p*-aminophenol, saturation of the measurement currents occur due to diffusion effects in the analyte (Fig. 11.6, top, curve B). Owing to electrochemical artifacts, an offset current may also contribute to the detection current (Fig. 11.6, top, curve C). For this reason, as suggested in [41], often the derivatives of the sensor current with respect to the measurement time, $\partial I_{col}/\partial t_{meas}$ and $\partial I_{gen}/\partial t_{meas}$, are evaluated instead of the absolute values (Fig. 11.6, bottom, curve C).

11.3.2 Extended CMOS Process

In Fig. 11.7, the process used to provide the gold electrodes is schematically depicted [39]. We start on the basis of a 6 inch *n*-well CMOS process specifically optimized for analog applications (high-ohmic polysilicon resistors, poly–poly-capacitors). The minimum gate length is 0.5 μm, the oxide thickness is 15 nm and the supply voltage is 5 V.

The steps depicted in Fig. 11.7 show the process flow beginning at the point where processing of the standard CMOS flow is completed. The CMOS flow ends with encapsulation of the second aluminum layer by an oxide (SiO_2), application of a chemical–mechanical planarization (CMP) step and nitride (Si_3N_4) passivation (Fig. 11.7 a). In the next step, via holes are etched down to the aluminum layer (Fig. 11.7 b). A Ti/TiN barrier layer is deposited and the holes are filled with tungsten by a CVD process (Fig. 11.7 c). The tungsten is etched back stopping at the barrier layer. The Ti/TiN layer at the surface of the wafer is also etched with an RIE process (Fig. 11.7 d). Finally, the gold electrodes are fabricated in a lift-off process after evaporation of a Ti/Pt/Au stack with layer thicknesses of 50/50/300–500 nm each (Fig. 11.7 e). Gold metallization is also used for the output pads.

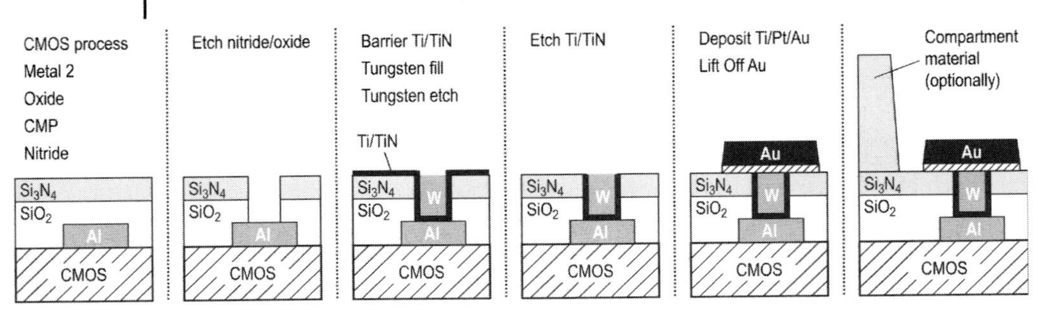

Fig. 11.7 Schematic description of the extra process flow after standard CMOS to provide the Au sensor electrodes and optionally a sensor compartment

Optionally, a compartment can be provided surrounding the sensor area of each test site (Fig. 11.7 f). In our case, it is made of polybenzoxazole (PBO) and processed as a resist, i.e. exposed, developed and baked. The purpose of this hydrophobic compartment is to avoid contamination of neighboring sensors during the spotting process. In particular, the need for processing the compartment depends on the specific spotting technique used [13, 14].

The SEM photograph in Fig. 11.8 shows the second aluminum layer from the CMOS process, via contact and sensor electrode. Fig. 11.9 shows a tilted SEM cross-section photograph with sensor finger electrodes and CMOS elements after the complete process run. (The nitride layer on top of the sensor electrodes is only used for preparation of the SEM photograph.) A top view of a sensor with interdigitated gold electrodes embedded within a PBO ring is depicted in Fig. 11.10. Width and spacing of the gold electrodes is 1 μm in all cases.

Fig. 11.11 a shows a fully processed demonstrator chip with 128 (16×8) positions and Fig. 11.11 b shows a blow-up of the sensor area of a fully processed demonstrator chip with 32 (8×4) positions.

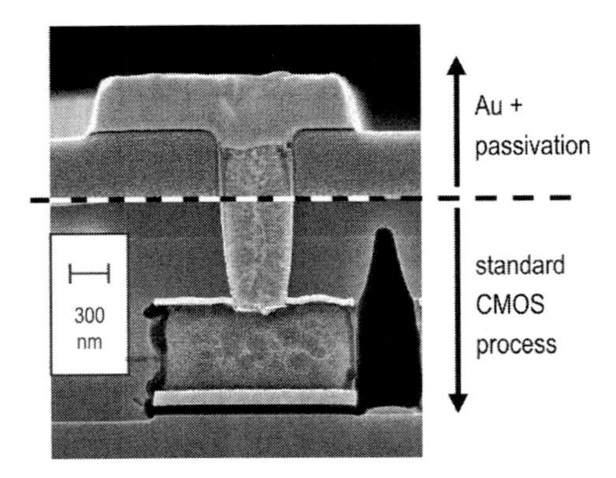

Fig. 11.8 SEM photograph showing the aluminum 2 layer from the CMOS process, via contacts and Au sensor electrode

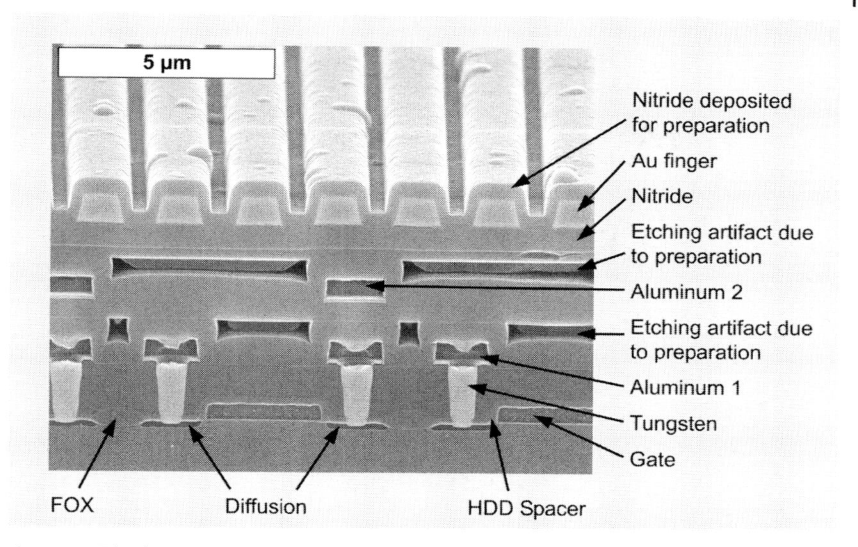

Fig. 11.9 Tilted SEM cross-section photograph with sensor electrodes and CMOS elements after the complete process run. Note that the nitride layer on top of the sensor electrodes is used only for preparation purposes. The width and spacing of the sensor electrodes are 1 µm

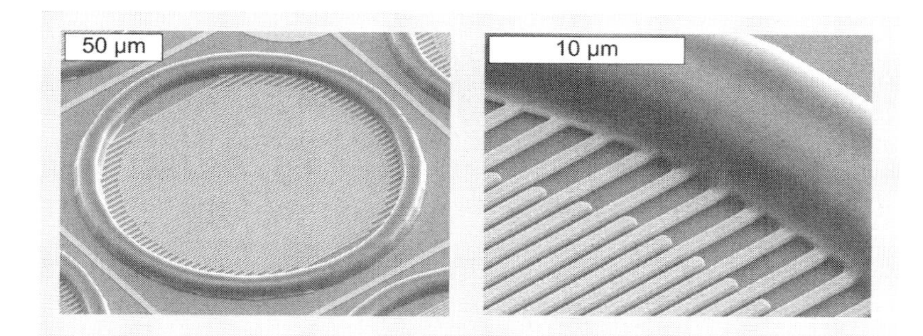

Fig. 11.10 Top view of a sensor with interdigitated gold electrodes embedded within a PBO ring. The width and spacing of the sensor electrodes are 1 µm

11.3.3
Demonstrator Chip: Architecture and Circuit Design Issues

The architecture of the demonstrator chip depicted in Fig. 11.11a is shown in Fig. 11.12. Sensor site selection is done by column and row decoders and a multiplexer. A common potentiostat circuit is used for all sensor sites in parallel, measuring the potential of the electrolyte at its input (reference electrode) and driving a counter electrode at its output to keep the potential of the electrolyte at a prede-

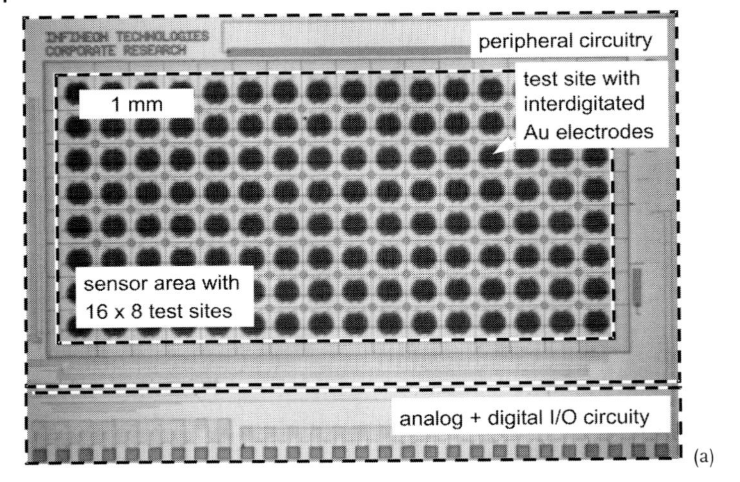

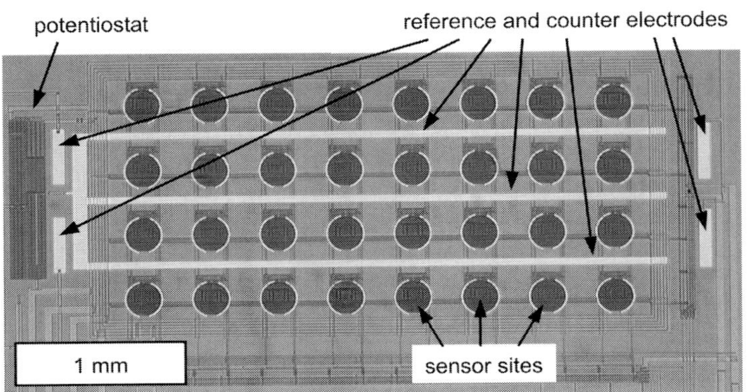

Fig. 11.11 Chip photographs of fully processed demonstrator chips. (a) Configuration with 16×8 positions, sensor pitch = 300 μm, single sensor diameter = 200 μm; (b) configuration with 8×4 positions, sensor pitch = 400 μm, single sensor diameter = 200 μm

fined voltage. Considering stability, this circuit is designed to be robust against the widely varying electrical properties of the electrolyte. Further detailed discussions on CMOS potentiostat circuits can be found elsewhere [46–49].

The potentials of the sensor electrodes of each sensor are controlled by the circuits located underneath the sensor sites (Fig. 11.13). In order to cover a wide band of possible applications, the circuit is designed to be operated with sensor currents from 1 pA to 100 nA. It consists of two regulation loops to control the potentials of both electrodes, whose currents are both recorded. They are amplified by a factor of 100 using two cascaded current mirrors in series within each branch.

The transfer characteristics of each pixel can optionally be calibrated. This is achieved by forcing reference currents into both branches of the circuit through tran-

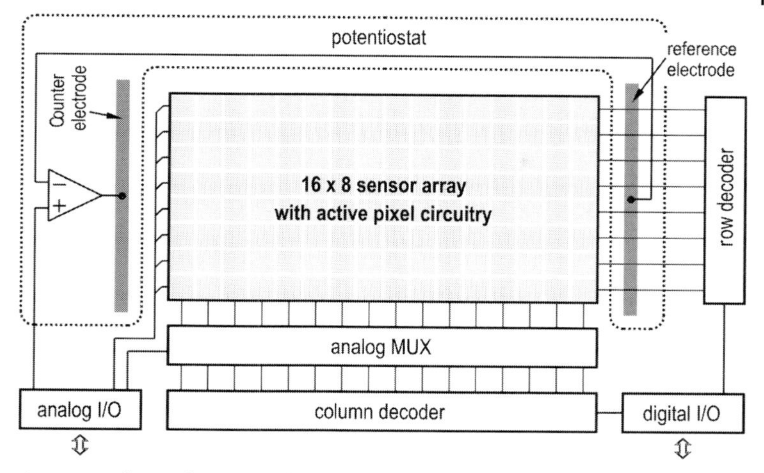

Fig. 11.12 Chip architecture of a demonstrator chip with 16×8 sensor sites

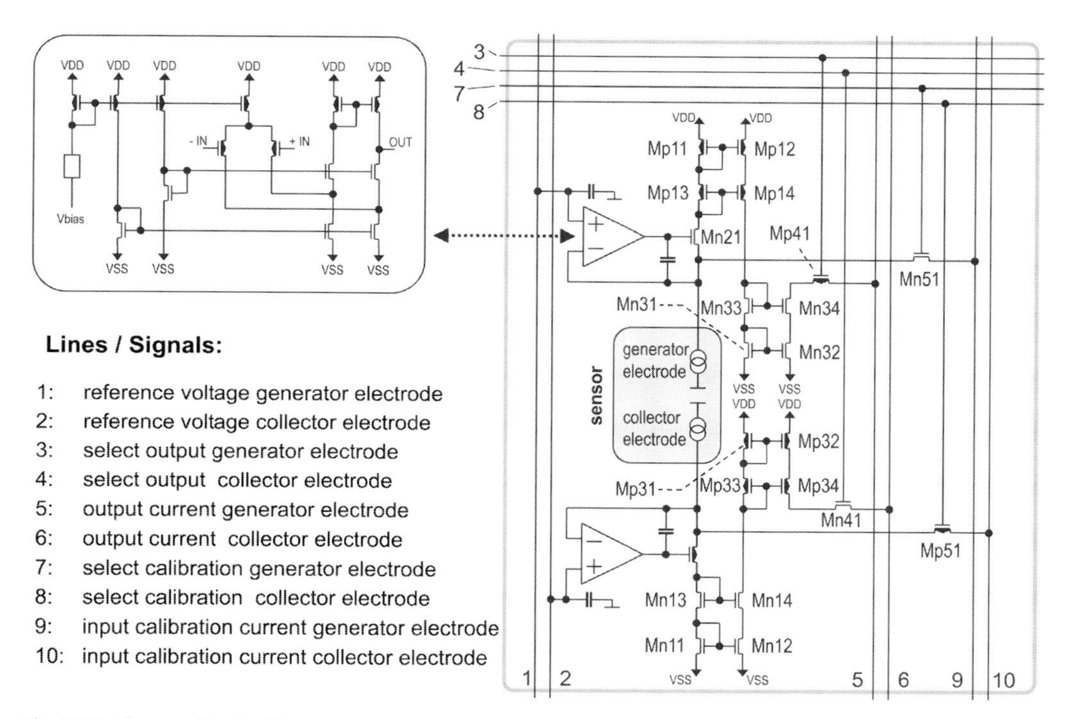

Fig. 11.13 Sensor site circuit

sistors Mn51 and Mp51 at the same nodes, where the sensor currents are applied [38, 39, 50] and measuring the circuit response (output current) for different operating points.

The variation of the transfer characteristics is determined by the variation of the current gain of the current mirrors (Mn11–Mn14, Mp11–Mp14, Mn31–Mn34, Mp31–Mp34). The variation of the current gain originates mainly from the variation of the transistor threshold voltage, V_t, which can be modeled as [51, 52]

$$\sigma(V_t) = \frac{A(V_t)}{\sqrt{W \times L}} \tag{2}$$

where $\sigma(V_t)$ is the transistor area (width $W \times$length L)-dependent standard deviation of the threshold voltage and $A(V_t)$ is the process-dependent threshold voltage related matching constant. For the considered process, $A(V_t) = 14$ mV μm holds for both n- and p-MOS transistors [53].

Owing to the very high dynamic range of five decades in current required here, the transistors of the current mirrors are operated in the subthreshold regime in most cases. In the subthreshold region, the transistor drain current, I_D, can be modeled as

$$I_D = I_0 \times \frac{W}{L} \times 10^{(V_G - V_t)/S} \tag{3}$$

where I_0 is a process and device type (n- or p-MOS)-dependent constant, V_G the transistor's gate voltage and S the transistor's subthreshold slope (a constant weakly dependent on the process and the device type).

Since the devices Mn13 and Mn14, Mp13 and Mp14, Mn33 and Mn34, and Mp33 and Mp34 only ensure that the devices Mn11 and Mn12, Mp11 and Mp12, Mn31 and Mn32, and Mp31 and Mp32 are operated at similar drain voltages, respectively, the main contribution to the mismatch of the current gain is provided by the latter group of devices.

Considering the collector electrode related part of the circuit as an example and using Equation (3) we can calculate the gain of the two current mirrors Mn11–Mn14 and Mp31–Mp34 in series:

$$\frac{I_{out}}{I_{cal}} = \frac{W_{Mn12}}{W_{Mn11}} \times \frac{W_{Mp32}}{W_{Mp31}} \times 10^{[(\Delta V_{t,Mn12,Mn11}/S_n)+(\Delta V_{t,Mp32,Mp31}/S_p)]} \tag{4}$$

The relative error of the gain of the two current mirrors is given by

$$\frac{\Delta(I_{out}/I_{cal})}{I_{out}/I_{cal}} = 10^{[(\Delta V_{t,Mn12,Mn11}/S_n)+(\Delta V_{t,Mp32,Mp31}/S_p)]} \tag{5}$$

where S_n and S_p are the n- and p-MOS transistor's subthreshold slopes, $\Delta V_{t,Mn12,Mn11}$ and $\Delta V_{t,Mp32,Mp31}$ are the threshold voltage differences of devices Mn11 and Mn12 and Mp31 and Mp32, respectively. Length L is constant for all devices and parameter variations of L can be neglected.

An important output of Equations (4) and (5) is that the gain mismatch is independent of the transistors' gate voltages. As a consequence, the gain error is constant independent of the sensor current/the circuit's output current as long as the transistors in the current mirrors are operated in the subthreshold region.

Assuming that all current mirror transistors are operated under strong inversion conditions, which is given for sensor currents in the upper region,

$$I_D = k \times \frac{W}{L} \times (V_G - V_t)^2 \tag{6}$$

holds for the transistors' drain currents, with k being a process and device type (n- or p-MOS)-dependent constant. On the basis of this relation, the gain of the two current mirrors is approximately given by

$$\frac{I_{out}}{I_{cal}} = \frac{W_{Mn12}}{W_{Mn11}} \times \frac{W_{Mp32}}{W_{Mp31}} \times \left[1 - \frac{2 \times \Delta V_{t,Mn12,Mn11}}{V_{G,n} - V_{t,n}} - \frac{2 \times \Delta V_{t,Mp32,Mp32}}{V_{G,p} - V_{t,p}}\right] \tag{7}$$

where $V_{t,n}$ and $V_{t,p}$ are the mean values of n- and p-MOS transistor threshold voltages, respectively. Since the denominator in the last two terms in brackets in Equation (7) is proportional to the square root of the transistor current (following Equation (6)), we can describe the relative error of the gain of the two current mirrors as

$$\frac{\Delta(I_{out}/I_{cal})}{I_{out}/I_{cal}} = \frac{a}{\sqrt{I_{cal}}} \tag{8}$$

with a being a process- and design-dependent constant.

On the basis of Equations (5) and (8), the gain error can be easily modeled for the whole operating range: In Fig. 11.14, the gain error is plotted schematically as a function of $1/\sqrt{I_{cal}}$. The behavior can be approximated by two straight lines: one in the inversion region increasing in proportion to $1/\sqrt{I_{cal}}$ and the other one in the subthreshold region with zero slope. In the weak inversion regime, i.e. where the transistors' operation points change from strong inversion to the subthreshold region, a transition of the two lines into each other occurs.

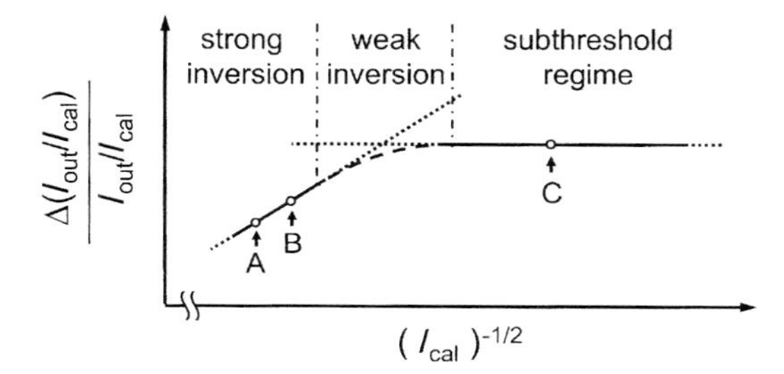

Fig. 11.14 Schematic plot of the gain error as a function of (calibration current)$^{-1/2}$ induced by parameter variations of the current mirror transistors of the circuit depicted in Fig. 11.13

These relations allow easy calibration of the circuit and of the whole array. Using two values for the calibration current the behavior in the inversion regime can be determined. These currents, labeled A and B in Fig. 11.14, must be chosen in the upper region of the range of specified sensor currents. In the subthreshold regime, one calibration data point (labeled C in Fig. 11.14) is sufficient. The value of this current may, for example, lie in the middle or in the lower half of the specified current window. Storing the measured output currents for these three calibration points for each test site, the whole array can be calibrated.

11.3.4 Process Optimization

Using the calibration option described above, the fabricated chips can also be electrically tested [38, 39, 50]. The chips are then operated without a fluid or biological material applied, but in the same electrical configuration as in the calibration mode. Here we use the option of pure electrical operation to evaluate the CMOS process properties.

Figs. 11.15 and 11.16 show the measured transfer characteristics and the gain as a function of the input current of all test sites from a 16×8 chip as shown in Figs. 11.11 a and 12. Processing is performed as described in Section 11.3.2; PBO compartments are not processed here, characterization is done at wafer level. The plots reveal that for input currents below 10 pA a deviation of the gain is obtained compared with higher input currents. This effect coincides with a bending of the transfer characteristics in Fig. 11.15 in the region below 10 pA.

This artifact results from a degradation of the transistor properties compared with the pure CMOS process. Using charge-pumping characterizations [54, 55],

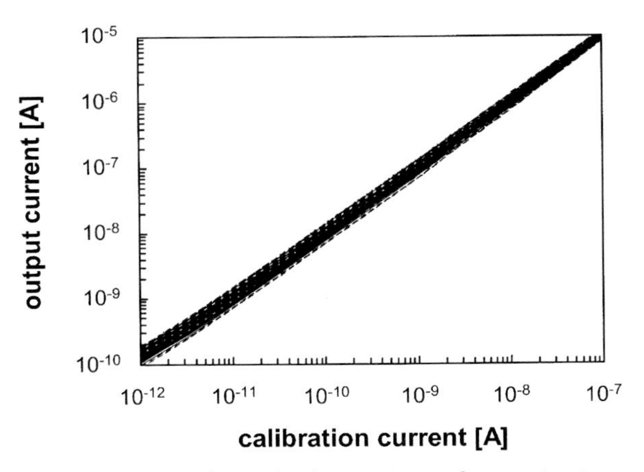

Fig. 11.15 Measured transfer characteristics of a test site circuit as function of the input (calibration) current of all test sites from a 16×8 chip as shown in Fig. 11.11 a. No annealing after Au process

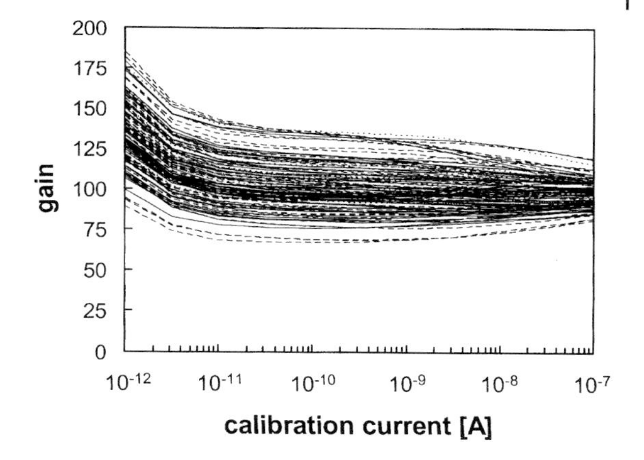

Fig. 11.16 Measured gain of a test site circuit as function of the input (calibration) current of all test sites from a 16×8 chip as shown in Fig. 11.11 a. No annealing after Au process

the interface state density N_{it} of the transistors is measured. For the pure CMOS process without extra process steps, a value of $N_{it}=10^{10}\,cm^{-2}$ is found for both p- and n-MOS transistors [39]. This is a reasonable value; values of the same order are found for most standard commercial CMOS processes. After gold processing, a more than 20-fold increase is measured ($N_{it}>2\times10^{11}\,cm^{-2}$). This excessive value translates into an increased, i.e. deteriorated, subthreshold slope S of the transistors, worsened off-state characteristics and thus to increased leakage currents. These increased leakage currents lead to the bending of the transfer characteristics in Figs. 11.15 and to the increase of the ratio of output and input current in Fig. 11.16. Consequently, this effect must not be interpreted as a true increase in the gain.

In standard CMOS processes, a forming gas (N_2, H_2) anneal is applied to reduce damage at the MOS transistor interface. In order to avoid the effects described above, such an annealing step is also applied here after the gold process module. N_2/H_2 annealing is performed at different temperatures for 30 min. Characterization of the interface state density reveals an excellent value of $N_{it}<2\times10^{9}\,cm^{-2}$ after annealing at 400 °C; 350 °C annealing leads to slightly higher but still good values. As shown in Fig. 11.17, the decreased number of interface states turns into proper transfer characteristics.

We should also discuss the standard variations of transfer characteristics and gain at this point. The statistical variations in Figs. 11.15–11.17 are approximately independent of the input current between 1 pA and 1 nA; for higher currents a slight decrease is found. Comparing these measured data with analytically calculated values on the basis of Equations (2)–(8) using transistor matching parameters extracted for the pure CMOS process [53], very good agreement is found. For example, the standard variation of the measured gain for operating points in the subthreshold region amounts to ~20%; the calculated value is 18%. We therefore conclude that the gold process does not influence this important analog device property.

Up to now, we have considered CMOS process front-end parameters in this section. However, the characteristics of the gold electrodes with and without annealing

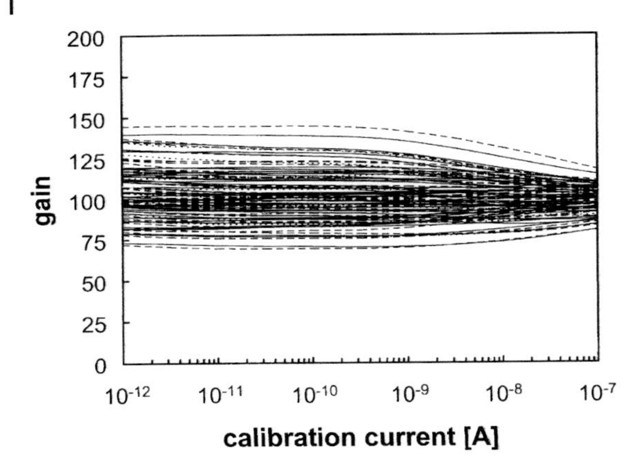

Fig. 11.17 Same plot as in Fig. 11.16, but with an N_2/H_2 annealing step after Au process at 350 °C for 30 min

Tab. 11.2 Resistances of gold and aluminum 2 lines and of the related via connections without and with annealing steps at different temperatures after the Au process

Anneal (N_2/H_2)	*Square resistance Au lines (mΩ/square)*	*Resistance via holes (0.8×0.8 μm) (mΩ)*	*Square resistance Al 2 lines (mΩ/square)*
Without	48	370	79
350 °C, 30 min	51	360	76
400 °C, 30 min	61	340	74

must also be investigated. Measured resistance data for gold and aluminum 2 lines and of the related via connections are given in Tab. 11.2. The data without an annealing step and with annealing at 350 °C are similar for all parameters. With annealing at 400 °C, a 20% increase in the gold resistance occurs. The SEM photographs in Fig. 11.18 reveal that this increase coincides with a rearrangement of grains within the gold layer. On the other hand, there is no change to the as-deposited Au stack up to 350 °C annealing. Consequently, annealing at 350 °C is chosen as a process window where both device and electrode properties are optimized.

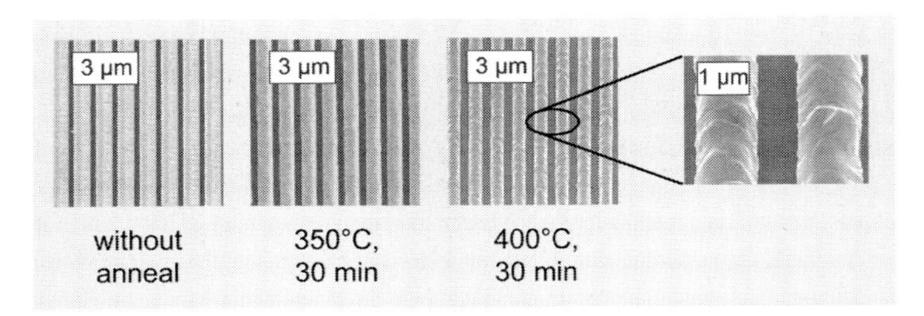

Fig. 11.18 SEM photographs showing the Au sensor electrodes without and with annealing steps at different temperatures after the Au process

11.3.5
Measured Results

In this section, results with biology applied to the chips are reported. We start with a simple oligo experiment. In this experiment, on one part of the sensor sites single-stranded DNA oligo molecules are immobilized which are exactly complementary to DNA oligo sequences in the analyte. On the other part of sensor sites, non-specific random sequences of probe DNA are immobilized. The experiment is performed at wafer level. The fluidic contact is provided via a flow cell positioned above the chip under test.

Two positions with matching strands and two positions with random sequences are considered. Since the absolute values of collector and generator current look very similar, only the collector current is shown in Fig. 11.19. As can be seen, match and mismatch clearly lead to different amounts of currents, but the offset currents (at $t_{\mathrm{meas}} \approx 38$ s, where the dashed line is plotted) for the matching positions differ by a factor of 2 (see discussion in Section 11.3.1). Using the derivative $\partial I_{\mathrm{col}}/\partial t_{\mathrm{meas}}$ (Fig. 11.20), as discussed in [41], the data for the two matches and for the two mismatches are in very good agreement.

Further measurements with PCR products performed by project partners have demonstrated proper operation down to 10^{-13} mol.

11.3.6
System Design Aspects

11.3.6.1 Extended System Architecture

The results and achievements discussed above describe a demonstrator chip. This configuration is adequate to perform experiments from proof-of-principles towards

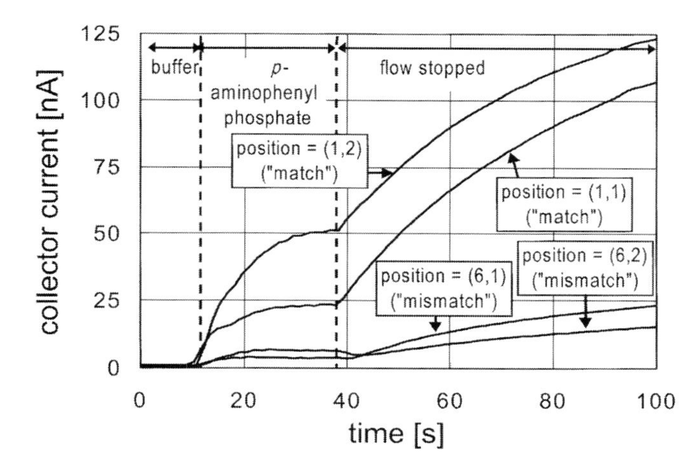

Fig. 11.19 Measured sensor currents for two positions with matching strands and two positions with mismatching random sequences. Data to be interpreted arise at $t_{\mathrm{meas}} \approx 38$ s (dashed line)

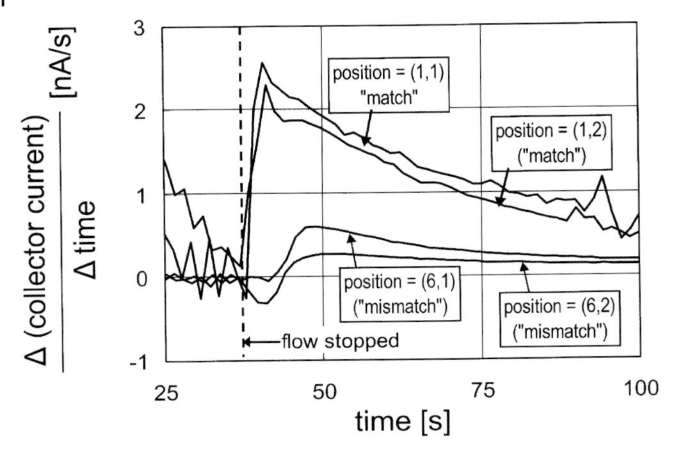

Fig. 11.20 Derivatives $\partial I_{col}/\partial t_{meas}$ of the sensor currents in Fig. 11.19

benchmarking investigations with competitive approaches. Furthermore, the architecture shown in Fig. 11.12 allows one to study many further important properties of this CMOS-based approach and to collect a huge amount of information which is required to design a user-friendly prototype version.

The next generation of these chips will consist of the kernel described here and further on-chip peripheral circuits to increase functionality and to provide robust and easy applicability. For this purpose, analog signal processing will be completely kept on-chip thanks to the addition of reference circuits and analog-to-digital and digital-to-analog converters. The introduction of further control logic circuitry allows the chip to be operated with a pure digital interface. Moreover, the number of external interconnects will be reduced by this approach (e.g. down to six or eight) independent of the number of test sites per chip.

11.3.6.2 **Alternative Circuit Design Approaches**

Alternative circuit design and architecture approaches may be required if larger arrays are needed (e.g. with 1000 test sites or more). In this case, the sequential analog readout of each test site (e.g. using the circuit in Fig. 11.13) results in the fact that the data are sampled in different time intervals. Owing to the very low minimum currents to be monitored, the readout time interval per test site cannot fall below a given minimum value.

A concept which uses the same detection principle but circumvents this problem independent of the array size is described in the following [54]. CMOS process and sensor design are the same as above. The concept is based on the realization of a specifically adapted analog-to-digital converter within each sensor site. The realized sensor circuit consists of two complementary parts for the generator and collector electrodes. The principle is shown in Fig. 11.21. The voltage of the sensor electrode is controlled by a regulation loop via an operational amplifier and

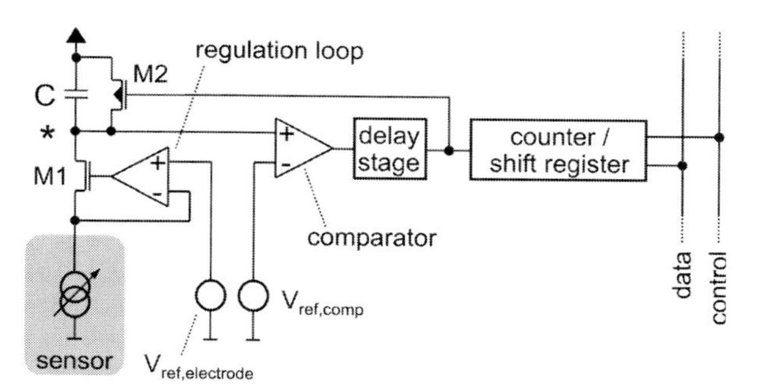

Fig. 11.21 Sensor circuit principle using analog-to-digital conversion based on a sawtooth generator concept within each sensor site

transistor M1. For A/D conversion, a sawtooth generator concept is used, where an integrating capacitor C is charged by the sensor current. When the switching level of the comparator is reached, a reset pulse is generated and the capacitor is discharged by transistor M2 again.

Equation (9) describes the frequency behavior as a function of the sensor current including parasitic and device mismatch effects:

$$\frac{1}{f} = \frac{(V_{\text{ref,comp}} + V_{\text{offset,comp}}) \times C}{I_{\text{electrode}} + I_{\text{leak}}} + t_{\text{delay}} \tag{9}$$

where f is the oscillation frequency, $V_{\text{ref,comp}}$ the comparator switching level, $V_{\text{offset,comp}}$ the comparator input offset voltage, C the total capacitance (i.e. including parasitic capacitances), $I_{\text{electrode}}$ the sensor electrode current, I_{leak} the leakage current (e.g. due to transistor junction leakage) at the circuit node connected to the comparator input (labeled with an asterisk in Fig. 11.21) and t_{delay} the comparator delay time. Equation (10) gives an approximation for the measured frequency:

$$f \approx I_{\text{electrode}}/(V_{\text{ref,comp}} \times C) \tag{10}$$

The chosen sawtooth amplitude $V_{\text{ref,comp}}$ is 1 V here and $C \approx 140$ fF, so that frequencies between 7 Hz and 700 kHz are obtained for a sensor current range from 10^{-12} to 10^{-7} A. The number of reset pulses is counted with an in-sensor-site counter (24 stages in [54]). For readout, the counter circuit is converted into a shift register by a control signal and the data are provided to the output.

For test purposes, an array with 16×8 test sites is fabricated. Two different comparator circuit designs are used for the generator and for the collector branch. Both comparators are based on a Miller-type operational amplifier. They differ in the reset pulse-shaping circuit.

These test arrays also provide a test/calibration current input similar to the case of the arrays discussed above. This option is used to characterize the electrical behavior of the chips. Evaluation is done for a specified range of currents between 1 pA and 100 nA. Fig. 11.22 shows the result of an experimental evaluation of the homogeneity (i.e. the relative accuracy) of the array. The 3σ standard variations of the measured frequencies are plotted for the different comparator circuits and both electrodes. As can be seen, the 3σ values are <2% in the current range from 10^{-10} to 2×10^{-8} A. In the range from 10^{-11} to 10^{-7} A, the 3σ variations are <6% for all circuits. At 10^{-12} A, 3σ values between 12 and 20% are obtained depending on the circuit type considered.

This behavior can easily be understood using the Gaussian equation for error propagation:

$$\sigma^2[F(p_1, p_2, p_3, \ldots)] = \sum_i \left[\left(\frac{\partial F}{\partial p_i} \right)^2 \times \sigma^2(p_i) \right] \quad (11)$$

with F being a function of parameters p_1, p_2, p_3, ... Applying this relation to Equation (9), the relative standard deviation of the measured frequencies is obtained:

$$\frac{\sigma(f)}{f} \approx \sqrt{\frac{\sigma^2(I_{\text{leak}})}{I^2_{\text{electrode}}} + \frac{\sigma^2(V_{\text{offset,comp}})}{V^2_{\text{ref,comp}}} + \frac{\sigma^2(C)}{C^2} + \sigma^2(t_{\text{delay}}) \times \left(\frac{I_{\text{electrode}}}{V_{\text{ref,comp}} \times C} \right)^2} \quad (12)$$

In the mid-frequency/mid-current range, the variations are dominated by the current- and frequency-independent terms in Equation (12), i.e. by mismatch of the integrating capacitor and by the comparator offset voltage. For low currents/low frequencies, leakage current mismatch leads to a contribution proportional to $1/I_{\text{electrode}}$. In the high-current/high-frequency region, the delay mismatch dominates because the contribution of this parameter increases in proportion to $I_{\text{electrode}}$.

Overall, the data from Fig. 11.22 represent tolerable levels for most applications without calibration at a high dynamic range. If higher accuracy is required, calibration of the individual sensor site circuits as described above can be performed to compensate for these deviations.

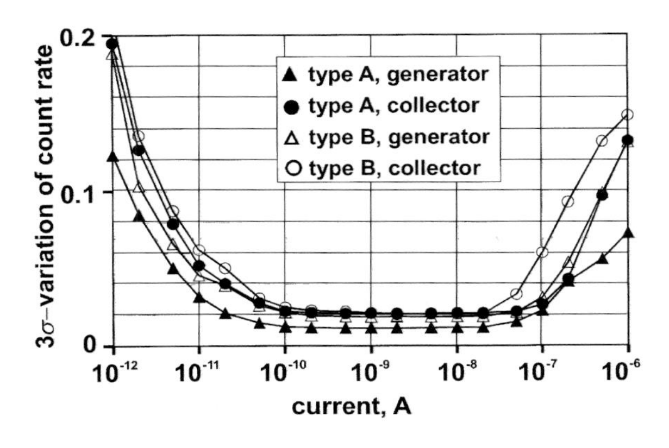

Fig. 11.22 Standard variations of the measured frequencies as a function of the input (calibration) current for the two different comparator circuits (types A and B) and both electrodes. The plot is shown for input currents up to 1 μA (specified range: 1 pA–100 nA) to demonstrate the frequency dependence in the high current region more clearly

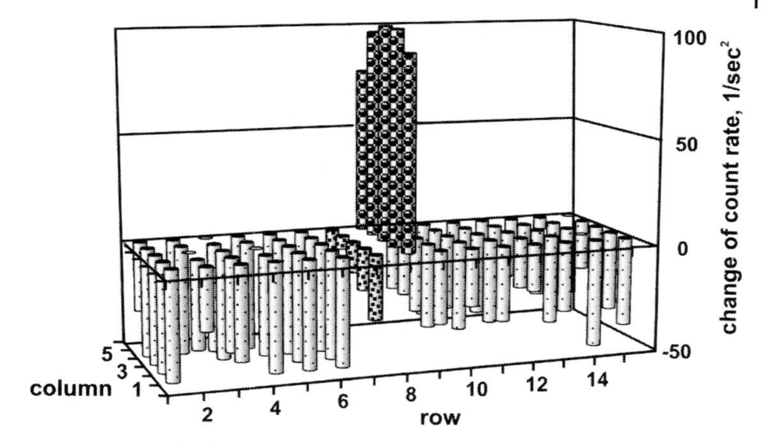

Fig. 11.23 Result of a DNA experiment using the circuit concept from Fig. 11.21. Row 8, matching probes; row 7, mismatching probes; all other positions not functionalized

In order to demonstrate biological measurements, two sensor rows are spotted with two different DNA probe sequences. An analyte containing target molecules matching one of the probes is applied. A flow cell is used in this experiment which covers only 5 columns and 16 rows of the sensor field, so that only 80 positions are shown in Fig. 11.23. There, a clear increase in count rate is observed at the matching positions (row 8), whereas no response is seen at the mismatching positions (row 7) as compared with the uncovered positions. The gradient in the column direction is assumed to originate from microfluidic nonidealities in the flow cell.

11.4 Label-free Detection Methods

The readout methods discussed so far are based on labeling the target molecules. In this section, we briefly address approaches which allow electronic label-free detection. The goal of such approaches is the simplification of the biochemical procedure needed to operate a DNA sensor since the labeling step applied to the target molecules is avoided here.

It is not the purpose of this section to discuss comprehensively all known suggestions concerning label-free sensor principles including all possible advantages and disadvantages. Here, only some of the most promising and advanced approaches in this area are briefly sketched.

The basic principle of impedance based approaches is depicted schematically in Fig. 11.24. There, the electrical impedance between a sensor electrode and the electrolyte (or between neighboring electrodes) is measured [55–57]. Hybridization events lead to a change in the electrical properties of the electrode–electrolyte in-

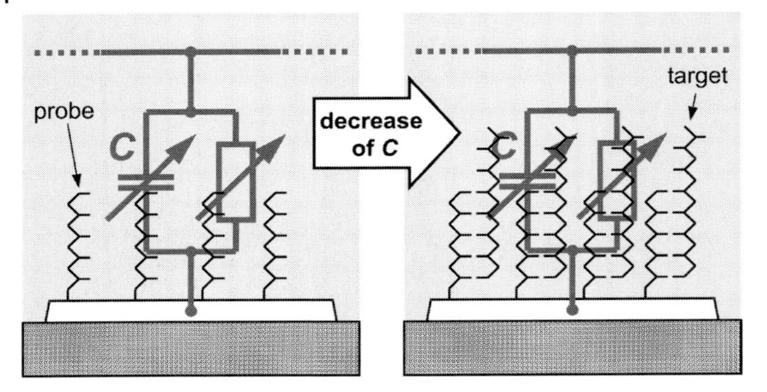

Fig. 11.24 Schematic plot demonstrating the basic operation principle of impedance-based DNA sensors. Left, before/without hybridization; right, after/with hybridization

terface. Using the first-order equivalent circuit also shown in the figure, hybridization results in a decrease in the capacitive part of the impedance.

The impedance and the impedance change of such sensors are not only determined by a capacitive effect, but also a resistive contribution is found. Both contributions depend on the quality of the layer of probe molecules. If the density of this layer is not sufficient, the resistive element may even shunt the capacitive contribution so that evaluation of this parameter is aggravated. For this reason, phase-sensitive amplifiers (lock-in amplifiers) are usually used to measure amplitude and phase of the sensor signal [57], or resistive and capacitive contributions are distinguished using other discriminating measurement principles [58].

The realization of mass-sensitive approaches [60–63] is usually based on the change of the resonance frequency of an electromechanical sensor due to the increase in the mass attached to the sensor surface in the case of binding events at the sensor surface. A further contribution to the change of the oscillation frequency originates from viscosity changes at the sensor surface after binding events.

An example is depicted in Fig. 11.25, where the topology of a film bulk acoustic wave resonator (FBAR) is shown [60]. A piezoelectric layer is sandwiched between two metal electrodes. The top electrode is coated with a biochemical coupling layer. The resonance frequency of the resonator is determined by the thickness of the piezoelectric layer and the mass of the electrodes. To prevent leakage of acoustic energy into the substrate, an acoustic mirror (similar to an optical Bragg reflector) is formed by several layers with alternating low and high acoustic impedances.

The resonance frequencies of the FBAR oscillator circuit in water and after forming a BSA layer on the sensor surface are shown in Fig. 11.26. A shift of 850 kHz compared with the pure water case is observed. Since frequencies are measurable with very high accuracy, this approach promises highly sensitive sensors. Moreover, the principle is robust against imperfect properties such as pinholes of the probe molecule layer.

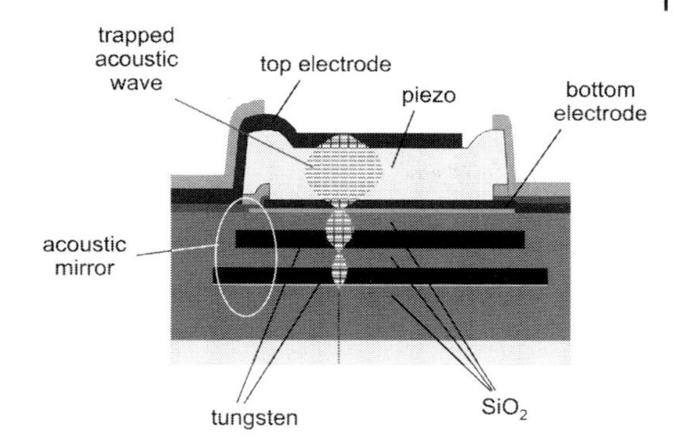

Fig. 11.25 Schematic cross-section of an FBAR sensor [60]. Below the piezoelectric layer an acoustic Bragg mirror is deposited using several layers of materials with different acoustic velocities. This mirror reflects vibrational energy back to the piezoelectric layer and therefore strongly enhances the quality factor of the sensor

A very similar sensor principle uses cantilevers instead of film bulk acoustic wave resonators [62, 63]. In particular, it has been demonstrated [63] how active CMOS circuitry is used to enhance significantly the quality factor (*Q*) of oscillation-based systems.

11.5 Conclusions

The development of CMOS-based DNA sensor arrays with fully electronic readout promises to provide tools with several advantages over DNA microarrays based on optical readout techniques, but today's state of development is lower for the electronic devices. Recent achievements, however, envision application scenarios using CMOS sensor arrays for the near future, such as individual medication, point-of-care DNA diagnosis and high dynamic range gene expression at attractive

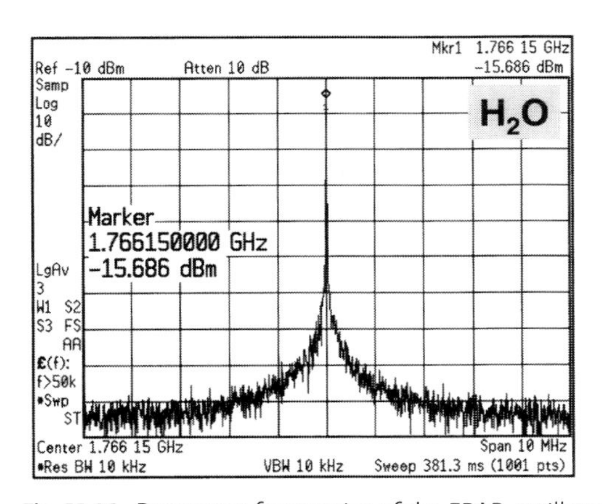

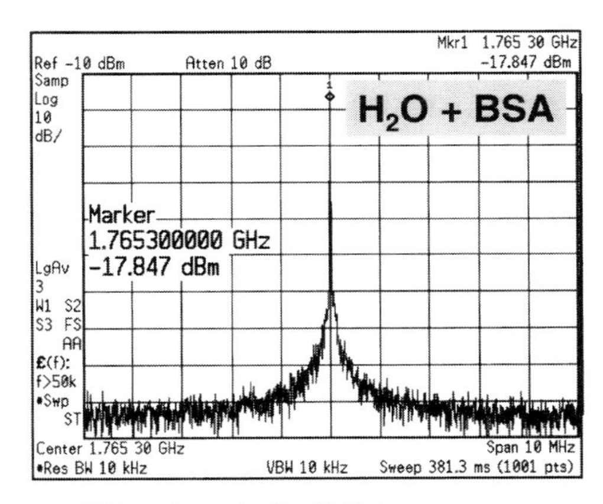

Fig. 11.26 Resonance frequencies of the FBAR oscillator circuit [60] as shown in Fig. 11.25 in water and after forming a BSA layer on the sensor surface

costs. In this chapter, the potential of a fully electronic CMOS-based DNA sensor array based on an electrochemical redox-cycling method has been shown. The integration of specific transducer materials for building the sensor and sensor-related circuit design issues have been discussed in detail. The future challenge is to develop a complete, robust and user-friendly system consisting of an advanced chip, a related package, the readout apparatus including software and biochemical assays specifically developed for the different applications. The next-but-one step may be to switch to label-free detection-based sensors to reduce further the complexity of biotechnological pre-processing.

11.6 Acknowledgment

The authors acknowledge fruitful cooperation and discussions with Hans-Christian Hanke, Thomas Haneder, Michaela Fritz, Alfred Martin, Volker Lehmann, Stephan Dertinger, Werner Simbürger, Hans-Jörg Timme, Robert Aigner, Stephan Marksteiner, Lüder Elbrecht (Infineon Technologies, Munich, Germany), Rainer Hintsche, Eric Nebling, Jörg Albers (Fraunhofer Institute for Silicon Technology, Itzehoe, Germany), Walter Gumbrecht, Gerald Eckstein, Manfred Stanzel (Siemens, Munich/Erlangen, Germany), Jörg Hassmann, Dirk Kuhlmeier, Jürgen Krause, Ugur Ülker, Jürgen Schülein (November, Erlangen, Germany), Wolfgang Goemann (Eppendorf Instrumente, Hamburg, Germany) and Stefan Zauner and Arpad L. Scholtz (Technical University of Vienna, Austria).

Part of the work presented in Section 11.3 was funded by the Bundesministerium für Bildung und Forschung, Germany. The support of Jürgen Herzog (Deutsches Zentrum für Luft- und Raumfahrt e.V., Projektträger im DLR) is gratefully acknowledged.

11.7 References

1 E.M. Southern, *Anal. Biochem.* **1974**, *62*, 317–318.
2 The Chipping Forecast, *Nat. Genet.* **1999**, *21* (1 Suppl.), January.
3 *http://www.nature.com/ng/chips_interstitial.html.*
4 M. Schena (ed.), *DNA Microarrays: a Practical Approach*; Oxford: Oxford University Press, **2000**.
5 M. Schena (ed.), *Microarray Biochip Technology*, Natick, MA: Eaton Publishing, **2000**.
6 D. Meldrum, *Genome Res.* **2000**, *10*, 1288–1303.
7 F. Bier, J.P. Fürste, in: *Frontiers in Biosensorics I*, F. Scheller, F. Schubert, J. Fedrowitz (eds.); Basel: Birkhäuser, **1997**.
8 T. Vo-Dinh, B. Cullum, *Fresenius' J. Anal. Chem.* **2000**, *366*, 540–551.
9 P. Hegde, R. Qi, K. Abernathy, C. Gay, S. Dharap, R. Gaspard, J.E. Hughes, E. Snesrud, N. Lee, J. Quackenbush, *Biotechniques* **2000**, *29*, 548–556.
10 B.J. Cheek, A.B. Steel, M.P. Torres, Y.Y. Yu, H. Yang, *Anal. Chem.* **2001**, *73*, 5777–5783.

11 A. Steel, M. Torres, J. Hartwell, Y.-Y. Yu, N. Ting, G. Hoke, H. Yang, in: *Microarray Biochip Technology*, M. Schena (ed.); Natick, MA: Eaton Publishing, **2000**, pp. 87–117.
12 *http://www.infineon.com/bioscience.*
13 E. Zubritsky, *Anal. Chem.* **2000**, *72*, 761A–767A.
14 V.G. Cheung, M. Morley, F. Aguilar, A. Massimi, R. Kucherlapati, G. Childs, *Nat. Genet.* **1999**, *21* (1 Suppl.), 15–19.
15 S.P.A. Fodor, R.P. Rava, X.C. Huang, A.C. Pease, C.P. Holmes, C. L. Adams, *Nature* **1993**, *364*, 555–556.
16 *http://www.affymetrix.com.*
17 *http://www.febit.com.*
18 M.J. Heller, A. Holmsen, R.G. Sosnowski, J. O'Connell, in: *DNA Microarrays: a Practical Approach*, M. Schena (ed.); Oxford: Oxford University Press, **2000**, pp. 167–185.
19 M.J. Heller, *IEEE Eng. Med. Biol. Mag.* **1996**, *15*, 100–104.
20 *http://www.nanogen.com.*
21 K. Dill, D. Montgomery, W. Wang, J. Tsai, *Anal. Chim. Acta* **2001**, *444*, 69–78.
22 *http://www.combimatrix.com.*
23 D.J. Duggan, M. Bittner, Y. Chen, P. Meltzer, J.M. Trent, *Nat. Genet.* **1999**, *21* (1 Suppl.), 10–14.
24 D.D. Bowtell, *Nat. Genet.* **1999**, *21*(1 Suppl.), 25–32.
25 D.J. Lockhart, E.A. Winzeler, *Nature* **2000**, *405*, 827–836.
26 S.V. Tillib, A.D. Mirzabekov, *Curr. Opin. Biotechnol.* **2001**, *12*, 53–88.
27 R. Favis, J.P. Day, N.P. Gerry, C. Phelan, S. Narod, F. Barany, *Nat. Biotechnol.* **2000**, *18*, 561–564.
28 J.D. Chiche, A. Cariou, J.P. Mira, *Crit. Care* **2002**, *6*, 212–215.
29 *http://www.herceptin.com.*
30 N.E. Hynes, D.F. Stern, *Biochim. Biophys. Acta* **1994**, *1198*, 165–184.
31 K. Mullis, F. Falcomer, S. Scharf, R. Snikl, G. Horn, H. Erlich, *Cold Spring Harbor Symp. Quant. Biol.* **1986**, *51* (Pt. 1), 263–273.
32 M.A. Innis, D.H. Gelfand, J.J. Sninsky (eds.), *PCR Applications – Protocols for Functional Genomics*; San Diego, CA: Academic Press, **1999**.
33 M. Huber, A. Mundlein, E. Dornstauder, C. Schneeberger, C.B. Tempfer, M.W. Mueller, W.M. Schmidt, *Anal. Biochem.* **2002**, *303*, 25–33.
34 M.A. Shoffner, J. Cheng, G.E. Hvichia, L.J. Kricka, P. Wilding, *Nucleic Acids Res.* **1996**, *24*, 375–379.
35 J. Cheng, M.A. Shoffner, G.E. Hvichia, L.J. Kricka, P. Wilding, *Nucleic Acids Res.* **1996**, *24*, 380–385.
36 M. Tartagni, A. Fuchs, N. Manaresi, L. Altomare, G. Medoro, R. Guerrieri, R. Thewes, in: *Sensors Update*, H. Baltes, G.K. Fedder, J.G. Korvink (eds.); Weinheim: Wiley-VCH, **2004**, *13*, 155–200.
37 P. Cailat, M. Belleville, F. Clerc, C. Massit, in: *Tech. Dig. International Solid-State Circuits Conference (ISSCC)*; **1998**, pp. 272–273.
38 R. Thewes, F. Hofmann, A. Frey, B. Holzapfl, M. Schienle, C. Paulus, P. Schindler, G. Eckstein, C. Kassel, M. Stanzel, R. Hintsche, E. Nebling, J. Albers, J. Hassman, J. Schülein, W. Goemann, W. Gumbrecht, in: *Tech. Dig. International Solid-State Circuits Conference (ISSCC)*; **2002**, pp. 350–351 and 472–473.
39 F. Hofmann, A. Frey, B. Holzapfl, M. Schienle, C. Paulus, P. Schindler-Bauer, D. Kuhlmeier, J. Krause, R. Hintsche, E. Nebling, J. Albers, W. Gumbrecht, K. Plehnert, G. Eckstein, R. Thewes, in: *Tech. Dig. International Electron Device Meeting (IEDM)*; **2002**, pp. 488–491.
40 M. Xu, J. Li, Z. Lu, C. Feng, Z. Zhang, P.K. Ko, M. Chan, in: *Tech. Dig. International Solid-State Circuits Conference (ISSCC)*; **2003**, pp. 198–199.
41 R. Hintsche, M. Paeschke, A. Uhlig, R. Seitz, in: *Frontiers in Biosensorics I*, F. Scheller, F. Schubert, J. Fedrowitz (eds.); Basel: Birkhäuser, **1997**.
42 M. Paeschke, F. Dietrich, A. Uhlig, R. Hintsche, *Electroanalysis* **1996**, *8*, 891–898.
43 M. Paeschke, U. Wollenberger, C. Köhler, T. Lisec, U. Schnakenberg, R. Hintsche, *Anal. Chim. Acta* **1995**, *305*, 126–136.

44 A.J. Bard, J.A. Crayston, G.P. Kittlesen, T.V. Shea, M.S. Wrighton, *Anal. Chem.* **1986**, *58*, 2321–2331.

45 A.J. Bard, L.R. Faulkner, *Electrochemical Methods*; New York: Wiley, **2001**.

46 A. Frey, M. Jenkner, M. Schienle, C. Paulus, B. Holzapfl, P. Schindler-Bauer, F. Hofmann, D. Kuhlmeier, J. Krause, J. Albers, W. Gumbrecht, D. Schmitt-Landsiedel, R. Thewes, in: *Proc. International Symposium on Circuits and Systems (ISCAS)*; **2003**, pp. V9–V12.

47 E. Lauwers, J. Suls, G. Van der Plas, E. Peeters, W. Gumbrecht, D. Maes, F. Van Steenkiste, G. Gielen, W. Sansen, in: *Tech. Dig. International Solid-State Circuits Conference (ISSCC)*; **2001**, pp. 244–245.

48 R. Turner, D. Harrison, H. Baltes, *IEEE J. Solid-State Circuits* **1987**, *SC-22*, 473–478.

49 R. Kakerrow, H. Kappert, E. Spiegel, Y. Manoli, in: *Proc. Transducers 95, Eurosensors IX*; **1995**, Vol. 1, pp. 142–145.

50 A. Frey, F. Hofmann, R. Peters, B. Holzapfl, M. Schienle, C. Paulus, P. Schindler-Bauer, D. Kuhlmeier, J. Krause, G. Eckstein, R. Thewes, *Microelectron. Reliab.* **2002**, *42*, 1801–1806.

51 M. Pelgrom, A. Duinmaijer, A. Welbers, *IEEE J. Solid-State Circuits* **1989**, *24*, 1433–1440.

52 K. Laksmikumar, R. Hadaway, M. Copeland, *IEEE J. Solid-State Circuits* **1985**, *20*, 657–665.

53 C. Paulus, R. Brederlow, U. Kleine, R. Thewes, in: *Proc. 8th IEEE International Conference on Electronics, Circuits and Systems (ICECS)*; **2001**, pp. 107–112.

54 G. Groeseneken, H.E. Maes, N. Beltran, R.F. de Keersmaecker, *IEEE Trans. Electron Devices* **1984**, *31*, 42–53.

55 P. Heremans, J. Witters, G. Groeseneken, H.E. Maes, *IEEE Trans. Electron Devices* **1989**, *36*, 1318–1335.

56 M. Schienle, A. Frey, F. Hofmann, B. Holzapfl, C. Paulus, P. Schindler-Bauer, R. Thewes, A fully electronic DNA sensor with 128 positions and in-pixel A/D conversion, in: *Tech. Dig. International Solid-State Circuits Conference (ISSCC)*; **2004**, 220–221.

57 V. Mirsky, M. Riepl, O. Wolfbeis, *Biosens. Bioelectron.* **1997**, *12*, 977–989.

58 C. Guiducci, C. Stagni, G. Zuccheri, A. Bogliolo, L. Benini, B. Samori, B. Ricco, in: *Proc. European Solid-State Device Research Conference (ESSDERC)*; **2002**, pp. 479–482.

59 F.K. Perkins, S.J. Fertig, K.A. Brown, D. McCarty, L.M. Tender, M.C. Peckerar, in: *Tech. Dig. International Electron Device Meeting (IEDM)*; **2000**, pp. 407–410.

60 R. Gabl, M. Schreiter, E. Green, H.-D. Feucht, H. Zeininger, J. Runck, W. Reichl, R. Primig, D. Pitzer, G. Eckstein, W. Wersing, in: *Proc. IEEE Sensors*; **2003**, pp. 413–414.

61 R. Brederlow, S. Zauner, A.L. Scholtz, K. Aufinger, W. Simbürger, C. Paulus, A. Martin, M. Fritz, H.-J. Timme, H. Heiss, S. Marksteiner, L. Elbrecht, R. Aigner, R. Thewes, Biochemical sensors based on bulk acoustic wave resonators, in: *Tech. Dig. International Electron Device Meeting (IEDM)*; **2003**, 992–994.

62 R. Raiteri, M. Grattaroly, R. Berger, *Mater. Today* **2002**, January, 22–29.

63 Y. Li, C. Vancura, C. Hagleitner, J. Lichtenberg, O. Brand, H. Baltes, in: *Proc. IEEE Sensors*; **2003**, pp. 244–245.

11.8 List of Symbols

$A(V_t)$	[mV μm]	transistor threshold voltage-related matching constant
a	$[A^{1/2}]$	process- and circuit design-dependent constant used to describe the relative error of the transfer characteristics of a sensor test site circuit
C	[F]	capacitance
ce	[1]	collection efficiency
$\Delta V_{t,Mn12,Mn11}$	[V]	difference of transistor threshold voltages of n-MOS transistors Tn11 and Tn12
$\Delta V_{t,Mp32,Mp31}$	[V]	difference of transistor threshold voltages of p-MOS transistors Tp32 and Tp31
$F(p_1, p_2, p_3, \ldots)$		function of parameters $p_1, p_2, p_3, \ldots$
f	$[s^{-1}]$	oscillation frequency
I_0	[A]	process- and device type (n- or p-MOS)-dependent constant used to model the transistors subthreshold current
I_{cal}	[A]	calibration current used to calibrate the sensor site related gain stages
I_{col}	[A]	collector electrode current
I_D	[A]	transistor drain current
$I_{electrode}$	[A]	sensor electrode current
I_{gen}	[A]	generator electrode current
I_{leak}	[A]	leakage current
I_{out}	[A]	output current of sensor site-related gain stage
k	$[A/V^2]$	process- and device type (n- or p-MOS)-dependent transistor constant
L	[μm]	transistor length
λ_1, λ_2	[nm]	light wavelengths
N_{it}	$[cm^{-2}]$	gate oxide interface state density
S	[mV/decade]	transistor subthreshold slope
S_n	[mV/decade]	n-MOS transistor subthreshold slope
S_p	[mV/decade]	p-MOS transistor subthreshold slope
σ		standard deviation
t_{delay}	[s]	delay time
t_{meas}	[s]	measurement time
V_{col}	[V]	collector electrode voltage
V_{gen}	[V]	generator electrode voltage
V_G	[V]	transistor gate voltage
$V_{G,n}$	[V]	n-MOS transistor gate voltage
$V_{G,p}$	[V]	p-MOS transistor gate voltage
$V_{ref,comp}$	[V]	comparator switching level/sawtooth signal amplitude

$V_{offset,comp}$	[V]	comparator input offset voltage
V_t	[V]	transistor threshold voltage (mean value)
$V_{t,n}$	[V]	n-MOS transistor threshold voltage (mean value)
$V_{t,p}$	[V]	p-MOS transistor threshold voltage (mean value)
W	[μm]	transistor width
W_{Mn11}	[μm]	width of n-MOS transistor Mn11
W_{Mn12}	[μm]	width of n-MOS transistor Mn12
W_{Mp31}	[μm]	width of p-MOS transistor Mn31
W_{Mp32}	[μm]	width of p-MOS transistor Mn32
Q	[1]	quality factor (in oscillating systems)

Index